Consciousness Inside and Out: Phenomenology, Neuroscience, and the Nature of Experience

Studies in Brain and Mind

VOLUME 6

For further volumes:
http://www.springer.com/series/6540

Richard Brown
Editor

Consciousness Inside and Out: Phenomenology, Neuroscience, and the Nature of Experience

Springer

Editor
Richard Brown
Philosophy Program
LaGuardia Community College, CUNY
Long Island City, NY, USA

ISBN 978-94-007-6000-4 ISBN 978-94-007-6001-1 (eBook)
DOI 10.1007/978-94-007-6001-1
Springer Dordrecht Heidelberg New York London

Library of Congress Control Number: 2013947097

Printed on acid-free paper

Springer is part of Springer Science+Business Media (www.springer.com)

Contents

Chapter 1
Introduction

Richard Brown

Most papers in this volume come from the 3rd Online Consciousness Conference, which was held February 18–March 4 2011. While the original papers, presentation materials, and discussion, both from this and previous conferences, remain online at http://consciousnessonline.com, most papers have been extensively revised in light of the discussion at the conference. In addition, commentators provided new commentaries and in most cases the author provides a new response. What emerges from this are conversations that are highly integrated. This makes the contents of this volume more of a product of the online consciousness conference than a snapshot of what happened.

As I write this I am in the midst of the 5th conference which runs February 15–March 1st 2013. It is hard for me to believe that this conference has been as successful as it has been, especially considering that it has been done, for the most part, without any money. It is my hope that this inspires others to try online conferences, as I was myself inspired by the original Online Philosophy Conference that came before me. When I learned that just two people had put together those conferences I figured that one person should be able to do it as well. Luckily for me, this specious reasoning worked out! I do not want to see online conferences replace traditional face-to-face conferences but I do hope that the record of conference publications from Consciousness Online and the Online Consciousness Conferences serves as a model for how open, rigorous discussion can serve to move debates forward and produce high-level resources for those working on understanding consciousness.

R. Brown (✉)
Philosophy Program, LaGuardia Community College, CUNY, Thomson Ave. 31-10, 11101 Long Island City, NY, USA
e-mail: onemorebrown@gmail.com

R. Brown (ed.), *Consciousness Inside and Out: Phenomenology, Neuroscience, and the Nature of Experience*, Studies in Brain and Mind 6, DOI 10.1007/978-94-007-6001-1_1,

The book is organized into ten parts each of which contains chapters consisting of a target paper, commentaries and, in most cases, an author response. The papers come from a conference and so range over many different areas in the philosophy of mind and neuroscience. Given this there are many ways that they could be grouped.

Ruth Millikan presents an epistemological problem for phenomenology. Over the course of her career Millikan has defended a broadly Sellarsian account of the nature of our concepts, filtered through the lens of evolutionary theory. If one is convinced, or even sympathetic to, a theory of this kind, then one faces the following puzzle. How can we have accurate concepts of our own phenomenology? Millikan argues that we cannot, rather what we have is a flawed lay theory. In true heterophenomenological spirit, there may merely seem to be phenomenology. Gualtiero Piccinini and Corey Maley respond by arguing that one can endorse Millikan's program without being agnostic on whether or not there are sensory qualities if one accepts their 'self-measurement' view. On this view scientists treat subjects as measuring instruments and take their reports in the way they would the read-outs of a self-measuring instrument. Thus even though it may be the case that subjects are unable to form the right kind of concepts about their own experience, that is no bar to the scientific study of phenomenology.

Paul Churchland argues that arguments against physicalism based on a priori reasoning fail by their own standards. He first points out that many different theorists have started from the armchair and come to very different conclusions. This in and of itself should suggest that a priori reasoning is not great at letting us know how the actual world really is. Churchland is happy to admit that, for all he knows, some form of dualism may be true. But he is betting that science will show that it isn't and that once we get clear on the arguments for dualism they will loose their air of being rationally compelling. He begins by discussing Nagel's kind of argument based on the subjective/objective distinction. He argues that there are two different kinds of knowledge here, but not two distinct properties. He then goes on to argue that both the dualist and the physicalist are committed to the existence of apparently simple qualitative properties. The question for Churchland is whether the fact that we **seem** to encounter simple qualitative properties in our experience is right. How do we know that when we are experiencing pure phenomenal red, say, that it doesn't merely seem to us that we are in contact with a simple unanalyzable property instead of it being the case that we really are in contact with one. That is, we can know a priori that there must be some limit to how far we can decompose the elements of our experience, whether or not that limit is merely due to our epistemic situation we cannot tell, since whatever we don't know about, we don't know about!

He then argues that, given we see the apparent qualitative simples as neutral ground, dualism is to be thought of as an explanatory theory of our phenomenal experience, but when we evaluate it on that ground it looses out big-time to the emerging neuroscientific explanations. Thus, when we compare the two theoretical accounts side-by-side the physicalist has an explanatory advantage. One well known problem is how one could come to know about one's consciousness if it is not physical and can have no causal impact on the physical world. Dualists at this point usually appeal to knowledge by acquaintance, and we will come back to that when we discuss Philip Goff's paper, so I will put that issue aside for now. The second problem Churchland sees is that unless one is a substance dualist it is unclear who is actually doing the apprehending in these cases. Who is the conscious subject that is directly acquainted with consciousness if not just the brain or some non-physical substance?

Torin Alter responds by pointing out that a large number of Churchland's criticisms do not threaten the knowledge argument or the conceivability argument (he leaves the bat out of it). The chief complaint of Churchland's paper is that property dualism cannot give an explanation that is at least as good as the physicalist explanation. But the kinds of things that Churchland cites are the kinds of things that the property dualist expects to find. That is, they expect there to be law-like regularities that connect physical and functional facts up with the phenomenological facts. One way to read Churchland, however, is as endorsing the claim that by postulating identities between, say pain and certain neural functioning, we then get to explain how pain, the qualitative feel of it, causes us to do various things. Read in this way Churchland is not merely claiming we can explain these kinds of structural properties, but that they allow us to explain how the mind causes behavior, which he claims is at the core of our common sense conception of consciousness. It is because property dualism cannot explain that, whereas the physicalist can, that Churchland claims that there is explanatory power in the physicalist's theory that is lacking in the dualist's theory.

Alter then goes on to discuss Churchland's distinction between the two kinds of knowledge in his debunking of the knowledge argument. Alter denies that this response works. The knowledge argument depends on two claims. The first is that Mary could not deduce what it is like to experience red just from the (completed) neuroscientific facts. The second is what Alter calls non-necessitation, which is the idea that there are truths which are not necessitated by our fundamental physical theory as traditionally conceived. Roughly speaking the idea of the knowledge argument is to move from non-deducibility to non-necessitation and from that to the falsity of physicalism. This argument may be controversial but it does not seem to commit the fallacy that Churchland points out. That is, at no point in the argument does it assume that scientific knowledge must somehow constitute the thing it is knowledge of. Alter goes on to say that the knowledge argument does rely on something related to this principle, which he calls the Propositional Knowledge Claim, which is just the idea that Mary's knowledge can be expressed in such a way that it can be evaluated as true or false. Churchland could reformulate his argument in terms of the Propositional Knowledge claim but he does not. Also, as Alter notes, one would need to give an argument that Mary does not learn something that can be evaluated for truth or falsity and Churchland does not give a convincing argument for this.

Instead Churchland responds by objecting to the formulation of the argument in terms of deducibility. It is a mistake, he claims, to demand that from the physicalist, since the identities must be postulated in order to allow the deduction to take place. So, it is no objection to the account that Churchland wants to defend that someone who was ignorant of the bridge laws would be unable to make these kinds of deductions; this happens all of the time according to him and is exactly why the identities are postulated in the first place. Secondly, Churchland rejects the formulation of the argument in terms of necessitation. Rather, he prefers to stick to the formulation where the question is whether the fact that she learns something new (which Churchland admits) has ontological consequences. To make his point he notes that Mary would be just as surprised when she learned what it was like for her to have a certain brain sate, but that is certainly physical! Churchland also rejects the notion of reduction that is at work in Alter's formulation of the argument. All in all Churchland seems to endorse what is known as Type-Q materialism, which denies the modal apparatus needed to make the anti-physicalist arguments work.

Philip Goff argues that plausible commitments of the standard property dualist commits them to panpsychism. The argument roughly goes as follows. In order for the standard anti-physicalist arguments to work they are committed to what Goff calls transparency, which is the claim that introspection reveals the real nature of conscious experience. The reasoning is straightforward. If we are truly to draw metaphysical conclusions from epistemological considerations then it must be the case that we have epistemic access to the metaphysical nature of conscious experience. The property dualist, Goff continues, is also committed to the claim that consciousness is a sharp concept, which means that there are no fuzzy or halfway cases. You either consciously see red or you don't. Given these two commitments Goff considers a typical sorites case where we start with you consciously seeing red at one end and a pillar of salt at the other end. The property dualist must either say that consciousness is vague or that it suddenly disappears at some point. But neither option is appealing so the best conclusion is that the property dualist must conclude that the pillar of salt is conscious, which is panpsychism. One can see this as an argument against property dualism if one thinks that panpsychism is sufficiently beyond the pale.

William Robinson argues that one can be a property dualist and resist Goff's argument. One does this, roughly, by holding that it is changes in the neural substrate of the brain that seem to matter, as opposed to changes in fundamental particles. If one does this then one would expect a change in conscious experience if one has a change in the neural underpinning of that experience. So if one thinks of an experience of a sound it is plausible to think that this experience can fade out, and if we find a good correlation between that fading and the fading neural activity then we have found that consciousness can be vague. Jonathan Simon argues that Goff has not succeeded in showing that the standard arguments against physicalism are committed to phenomenal transparency. At most they seem to be committed to a form of what Goff calls translucency. That is, to the claim that phenomenal concepts reveal some but not all of the essential features of their objects. Secondly Simon goes on to argue that Goff is wrong in thinking that phenomenal transparency commits one to consciousness not being vague.

David Chalmers himself, at the online consciousness conference, has denied that the 2D argument against materialism depends on this kind of transparency. Suppose that our phenomenal concepts are translucent in Goff's sense, then there is an aspect of our conscious experience which is hidden from us. But now consider a modified zombie world, one where there is a 'that's all' clause so that it is a mere physical duplicate of the actual world. If that world is possible then we know that there is an aspect of consciousness that is not physical and that is enough to refute physicalism. The reason for this is that, though physicalism may be true for consciousness, it will not be true for whatever aspect of consciousness is missing at the zombie world, and as it happens that aspect is the one that we are acquainted with! But, as Goff points out, transparency is required in order to get the first premise of the zombie argument. Thus, one can be a Type Q physicalist, as Churchland seems to be, or one can argue that phenomenal concepts are radically opaque, as Millikan seems to. Or one could hold that our concepts are translucent and deny that zombies are conceivable.

Benj Helli argues for a version of direct realism. What Hellie wants to defend is the claim that when two subjects are in different rational positions they must have different phenomenal experiences. He argues that when one consciously sees red one accept a kind of sentence in which the phenomenal experience itself is a part. Thus there is no way to accept it without the sentence being true. He calls this kind of sentence 'situatedly analytic'. He contrasts four cases. In one case we are awake and perceiving verdically. In another case we are asleep and perceiving verdically. This case involves lucid dreaming. In a lucid dream one is experiencing red, say, and is conscious that one is dreaming. On the other side we have the bad cases. We have cases of dreaming and not knowing that we are dreaming and cases of hallucination while not knowing that we are hallucinating (or of being awake and thinking we are having a lucid dream). Hellie takes it to be the case that in the case of lucid dreaming we can tell that our experiences are not the same as they are when they are awake. This is, at least in part, how we know that we are not dreaming. He uses this to argue that in the bad cases the subject holds contradictory attitudes. One accepts a sentence of 'I see red,' which has red as one of its parts, but you also deny that you accept that. Or to put it another way you accept a sentence like 'I am seeing a red similacrum' which has the red-thingy as a part, but you also deny that you accept that since you think that you are really seeing red. Thus, on Hellie's view the person who is hallucinating is no longer able to be made sense of from the point of view of rational psychology. They get 'exculpation,' but they do so only from the second person point of view.

Jacob Berger argues for perceptual justification outside of consciousness. He contends that whether one is an externalist or not about phenomenal character we have good reason to think that we sometimes make judgments on the basis of unconscious perceptions. This evidence comes from experimental cases, like blindsight, as well as common sense cases. Berger then explores possible replies from Hellie. The first may be to attack the claim that judgments of blindsight patients are fully rational. Or it may be the case that Hellie thinks that the states in question are sub-personal and hence unable to count as part of one's rational psychology.

Heather Logue argues against the McDowellian inspired thesis that we cannot evaluate a person who is in one of the so called bad cases in terms of rational psychology. The person in Hellie's version of the bad cases may believe something that is contradictory but there are none the less beliefs that she would be justified in accepting. Hellie responds that we can reinterpret talk of justification in terms of which beliefs will be caused. Logue considers a mismatch case where one is actually veridically seeing a red tomato but believes that one is hallucinating. In this case Logue contends, it would irrational of you to believe that you were veridically seeing a red tomato, and so rational psychology does apply, even in mis-match cases. Given this one must either reject Hellie's claim that someone having incoherent beliefs excludes them from the norms of rationality or that the person in the mis-match cases is truly incoherent. Logue closes by exploring the idea of partial justification. It may be the case that someone in a mis-match case has partial justification for believing that there is a tomato present.

Jeff Speaks focuses on the relationship between a belief and a sensation. In particular he takes up the question of what it means for a representation to be self-referential in the way that Hellie needs. The problem is that it seems that the instantiation of any property will result in that property self-representing itself, but this can't be right. What is needed, then, is a full account of the kind of self-representation that Hellie has in mind. Moving on to the issue of perceptual justification Speaks poses a problem. The relationship between the self-representational sentence one accepts and one's belief must be the kind that allows one to be mistaken, as this is what happens in the mis-match cases. Yet, on the account that Hellie has developed it is hard to see how it is that we could be mistaken. Or to put it the other way around, we do not usually form the belief that we are dreaming when we are, yet on Hellie's account we should.

Kathleen Akins begins the discussion by challenging a distinction that seems unchallengeable. Her aim is to undermine the distinction between black and white vision on the one hand and color vision on the other hand. In particular she wants to show that it is a mistake to think of black and white vision as simply the same as color vision yet minus the color. Or that it is a mistake to think that adding color vision is simply adding colors on top of a black and white gray-scale image. Following Sellars, Aikins argues that this distinction is first learned from the way that we actually produces images (dating back to pre-historic cave paintings according to Aikins) and then applied to conscious visual experience. In the visual system we find a luminance system and a chromatic system. The analogy that Aikins wants to dispel is that the luminance system provides a black and white representation which is then colored in by the chromatic system. To make this argument she pays close attention to what are known as rod achromats, which are people who only have rods and so who only have the luminance system. The first step of her argument is to try to show that a rod achromat's vision will not be like our normal black and white vision. If this is right then our own experience of luminance may not be as we think that it is. In this way one can see Aikins as providing a specific argument for the kind of position advocated by Millikan. Aikins argues as follows. When we learn the details of the luminance system we find out that the visual system does not represent intensity of light. Since a black and white image just is one that represents light intensity at each point on the image, it follows that human luminance systems are not producing anything like a black and white image. To make the point more vivid Aikins appeals to a very creative art instillation called RGB by the artist Carnovsky. In this exhibit images are printed in three different colors of ink and then viewed under different lights. This makes some of the images invisible, while others stand out. Aikins argues that our experience in this kind of setting is more what the rod achromat experiences, and it is not a world in black and white. For Aikins the real difference between the luminance and chromatic systems is in the filters they apply in processing contrast information. Thus adding the chromatic system does more than merely add colors to a pre-existing black and white image. It allows a greater range of contrasts.

Peter Mandik poses what he calls Akins problem: can there be a visual experience that lacks both color phenomenology as well as black and white phenomenology? Akins' paper can be seen as arguing for a yes answer, but what does that mean? Mandik suggests that we can make sense of her claim as a version of conceptualism. The conceptualist takes the view that phenomenology consists in conceptual representations. If one has that view it is easy to see how there can be conscious visual experiences that have neither hue nor shade. Mandik cites 'seeing a rectangular mat' but we might also cite peripheral vision as well.

Adam Pautz argues that the science of taste, smell, sound, and pain suggest that phenomenal externalism is false. In particular he presents detailed psychophysical and neuroscientific evidence that there is in some sense a bad correlation between the structural relationships between experiences and physical properties of objects. While there is a good correlation between these properties and internal brain states. For instance in the case of taste Pautz points to evidence that suggests that taste experience correlate with the pattern and intensity of activation in ensembles of neurons and that they correlate badly with external properties. The situation is even worse for smell. When it comes to pain Pautz presents evidence that the properties we experience in pain do not correlate with the size or severity of the wound or with the intensity of activity of nociocepters. On the other hand we see a very good correlation between reported pain experiences and firing of neurons in pain areas. After going through many different sources of evidence from many different sensory modalities where there seems to be a conflict, he extends this to an argument making the conflict explicit. The first argument he calls the internal dependence argument and his goal is to construct a counter-example to tracking intentionalism. Pautz argues that the empirical results are not enough since the opponents can claim that one of these cases is an illusion or they might say that the two creatures are tracking different properties of the physical objects. To avoid these issues Pautz provides cases that are not actual but are based on actual examples and do not involve anything which is scientifically implausible. Each case starts with two creatures that optimally track the same property but which have different neural activations. In taste the two creatures are Yuck and Yum who both optimally track the same physical substance but have different neural activations. Given what we know about the science we would predict that they should have different experiences but the externalist has to say that they have identical experiences. For smell it is Sniff and Snort, for pain Mild and Severe, for sound Loud and Soft. This culminates in his official statement of the argument:

1. If tracking intentionalism is true, then in every possible coincidental variation case, the right verdict is Same Experiences.
2. But it is much more reasonable to suppose, in at least some coincidental variation cases the right verdict is Different Experiences; call this internal-dependence.
3. So tracking intentionalism is (probably) mistaken.

After presenting this Pautz turns to his second argument, which he calls 'the structure argument'. This is a more general argument which aims to cast doubt on any version of objectivisim about the sensory qualities. The basic idea behind this argument is that, given the bad external correlations, people will make systematically mistaken judgments about the nature of the external world. For instance, if they have a burning pain that is twice as intense as one had a moment ago one will conclude that there is something about the world that isn't there. In the final section of the paper Pautz extends his argument from tracking intentionalism to most forms of externalism about sensory qualities.

David Hilbert and Colin Klien respond by suggesting that Yuck and Yum track different aspects of the same property and so there is no problem, at least for their version of phenomenal externalism.

Jason Leddington argues for the claim that we hear non-sounds in hearing sounds, which is a version of the view advanced by Heidegger. On this view we directly hear the events in the hearing of the sound. This is contrasted with the view advanced by Berkeley, namely that we never actually hear the non-sounds directly. We hear the non-sounds indirectly. Leddington argues that phenomenological considerations mediate in favor of the Heideggerian view. His claim is that in auditory experience we experience the sounds as being bound to the events that make those sounds. Given the background assumption that the only two ways to hear non-sounds are the Heideggerian and Berkeleyian views (a claim that Leddington labels 'Sonicism') this constitutes an argument for the Heideggerian view. One powerful reason for thinking that we hear sounds as being fused with events that generate them is that it explains why sound sources are available for demonstrative reference. It is because I hear the tear in the bag as it is happening that I am able to think 'that bag is tearing!' Leddington argue that the Berkeleyian view has trouble explaining this without rejecting sonicism. This is because the Berkeleyian view cannot allow that I can directly refer to a non-sound via a sound. I can only indirectly refer to a non-sound. Another worry is that the Berkelyian view seems at odds with phenomenology of the locatedness of sounds. A further worry is that the Berkelyian view has sounds as appearing to be only contingently related to the events that produced them. But this is not the way that we experience sounds.

Casey O'Callaghan responds by arguing that he accepts Phenomenological Binding and also suspects that one could reject sonicism. O'Callaghan accepts a version of the phenomenological binding claim, so he does admit that there is some sense in which sounds are heard as being fused with their originators. But he denies that this is the same way in which colors are seen as fused with their objects. That is he wants argue that sounds are heard as distinct individuals that posses properties of loudness and pitch. On O'Callaghan's view sounds are heard as parts of the events that they compose.

Matthew Nudds responds in a similar way. He too views sounds as individuals that posses properties and so views them as being experienced as in some sense independent from their sources. But he also makes a distinction between the sounds themselves and our experiences of those sounds. He claims that our experiences of sounds represent them as having two kinds of properties. The first is that they are in some sense independent of their sources, and the other is that they are produced by their sources. The sense in which they are independent of their sources, on Nudds view, is that they do not appear in our experience to be properties of their sources in the way that the color of an object appears to us to be a property of that object. Thus, on Nudds view, one can endorse both of the claims that Leddington advances. Our experience of sounds does represent them as being produced by their sources but it also represents them as being independent of their sources in an important way. This explains, for Nudds, how it is we can non-veridically represent. In the good cases we represent the sound and the source, but there are cases where we correctly represent the sound (getting its pitch correct say) but mis-represent its source (we experience it as being produced by the dummy's mouth and not the ventriloquist's mouth).

Kevin Connolly takes up the question of our phenomenal experience, which seems to combine many sensory modalities. When we are at a concert, say, and we can see the musicians playing, we experience the music as originating from the movements of the musicians. Connolly's question is whether we need to appeal to specific multimodal contents or whether the usual ones will do. Connolly gives arguments against several different ways of trying to establish truly multimodal contents. He then suggests an alternative account of multimodal experience. On his view we can think of different modalities as families of quality spaces and then we can think of multimodal experience as our coming to associate properties in one quality space with the properties in the other quality spaces (e.g. sounds with lip movement).

Matthew Fulkerson explores the issues by distinguishing two senses in which one might be a conservative about the content of multimodal experiences. One way to make the claim is to hold that no sensory content is shared among the senses. Another way is to hold that the content of any given perceptual experience consists only in the sensible features found in the individual modalities.

Berit Brogaard presents evidence for a kind of visual seeming that is not based in the visual areas of the brain. Using synesthesia as a case study she presents cases where there is robust visual phenomenology but no change in the activity of the visual areas. She argues that this is evidence for a kind of visual seeming that is conceptual in nature. She also argues against the standard debunking of this kind of high-level conceptual experience, namely that the high-level conceptual content changes the first-level activity.

Ophelia Deroy in her commentary on Brogaard carefully considers ways in which we might tease apart these various notions of seeing. She then presents an alternative reading of the evidence presented by Brogaard. Instead of thinking that there is a kind of seeing that is neither perceptual nor imagistic Deroy suggests that there may be a kind of visual experience that is a blending of perceptual experience and imagistic experience.

Miguel Sebatsian argues that the most plausible neural implementation of higher-order thought theory is that it is reflected in activity of the dorsal lateral prefrontal cortex. In particular he appeals to the work of Hakwan Lau's lab to show that selectively interfering with this area produces blindsight-like performance in a visual discrimination task in normal subjects. We know independently that this area is relatively deactivated during REM sleep. Given that we think that REM sleep is when we have dreams and that dreams are conscious, then there seems to be some tension. If dreams are conscious and occur when the dorsal lateral prefrontal cortex is relatively inactive then it seems as though the higher-order thought theory is in trouble.

In response Josh Weisberg raises several worries. On the one hand one might doubt that dream are conscious. This seems bizarre, but is hard to rule out. More worrisome, though, is the claim that dreams are conscious, but less vividly so as waking conscious experience. If so then we would expect that the areas related to conscious experience would show some level of deactivation. In addition, Weisberg argues, there are other candidates for the neural realizer of higher-order thoughts. These include Caruthers' claim that they are connected to the Theory of Mind module (postulated to be in the medial prefrontal region), Damosio's theory that they are a kind of self-consciousness and are found in the anterior cingulate cortex, and Flohr's proposal that they are distributed neural assemblies involving NMDA-sensitive synapses.

Matt Ivonowich further presses this issue by arguing that the dlpfc is not a good candidate for the realization of higher-order thoughts.

Part I
First-Person Data and the Science of Consciousness

Chapter 2
An Epistemology for Phenomenology?

Ruth Garrett Millikan

2.1 Introduction

There is a tendency to assimilate so called "consciousness studies" to studies of the phenomenology of experience, and it seems to me that this is a shame. It is a shame, I think, because there is no such thing as a legitimate phenomenology of experience whereas there certainly is such a thing as consciousness. So long as people assimilate studies of consciousness to studies of phenomenal experience, they are side stepping the real issues – the ones for another lifetime.

What then are the problems I see with phenomenology? In outline, they are as follows.

First, if one holds a Sellarsian view of cognition, ideas are not given in perception. If you can describe or know in some way about your phenomenal experience, you must have ideas that apply to it, say, applicable empirical concepts. But on a Sellarsian view, the origins and certifications for such ideas are not Humean or Russellian. Concepts are not obtained merely by copying or by naming or abstracting from sensory data, by giving names to directly experienced properties. A theory of what concepts are – or, in classical idiom, preferred for reasons to be explained later, a theory about the nature and origin of ideas – is needed before one can begin to discuss phenomenology. Only with such a theory in hand can it be legitimate to ask how ideas pertaining to phenomenal experience might be obtained, and whether there is reason to think we have or could have any adequate ones.

Second, the theory of the nature and origin of ideas I would advocate implies that adequate empirically-based ideas can be developed and validated only through ongoing experience both over time and over a variety of perspectives. But the phenomena that phenomenology purports to investigate *cannot* be studied over time

R.G. Millikan (✉)
Department of Philosophy, University of Connecticut, Mansfield, CT, USA
e-mail: ruth.millikan@uconn.edu

R. Brown (ed.), *Consciousness Inside and Out: Phenomenology, Neuroscience, and the Nature of Experience*, Studies in Brain and Mind 6, DOI 10.1007/978-94-007-6001-1_2,

and over a variety of different perspectives. This makes phenomenology inherently wide open to the breeding and feeding of chimaeras.

Third, I think a coherent and empirically respectable theory can probably already be sketched to explain what really is going on when people *think* they are describing their phenomenal experience, a theory that explains away the chimaeras. I will describe such a candidate theory, and although I am not committed to arguing for any of its neurological details. If I should be right about empirical ideas more generally, then that some theory of this general kind is right about phenomenology becomes highly plausible.

The upshot of the whole would be, of course, that Dan Dennett is right – that the closest we can get to a legitimate phenomenology of experience is what he calls "heterophenomenology" (1991, 2003).

2.2 Introducing Unicepts

I'll start by jumping right in to explain the picture of empirically-based "ideas" that underlies my skepticism about phenomenology.[1]

Consider an extraordinary ability that you have, the ability to recognize, for example, your mother, or a sibling, your spouse, your best friend. You can do this by seeing that person across the room, 20 m up the street, perhaps at 1,000 m by his or her walk, certainly at 30 cm, from the front, from the back, from the left side or the right or most any other angle, half hidden behind another person or a chair or a table or a book, sitting, standing, lying down, yawning, stretching, running, eating, holding still or moving in any of various ways, in daylight, candlelight or moonlight, under a street lamp, through a fog, in a photograph, on TV, through binoculars, by hearing their voice from any of many distances or as it passes through a variety of media such as lightweight walls, under water, over the phone, over many kinds of masking sounds such as wind, or rain, or other people talking, and so forth.

Now generalize the ordinary notion of recognizing a person just a bit so that it encompasses your wider ability to keep track of when information is arriving at any of your various senses about this same person. You might recognize them, or signs of them that enabled you to gather information about them, by recognizing their signature or handwriting, their style of prose or humor or, perhaps, of musical interpretation or of some other activity, by the sound of the instrument they play coming from the next room or the hammering that accompanies their current home project, also by recognizing their name when someone talks about them or when it is written, by hand or in any of a 100 fonts, and so forth.[2] You could recognize that

---

[1]The next few paragraphs are adapted from "Accidents," *Proceedings of the American Philosophical Association* November 2012.

[2]In Millikan (2012a, b) I have argued that the sense of "information" involved these various cases is univocal.

the information arriving is about them through many hundreds of descriptions: the person who was or did this or that, about whom this or that is true. Or you might recognize whom the information is about using various kinds of inference, induction or abduction. If these latter ways of recognizing a person seem to you to divide off rather sharply from recognizing them "in the flesh," recall that recognizing a person by their looks or voice is also gathering in information about them *through signs*. The light that strikes your eyes, the vibrations that strike your ears, are merely signs of what you see or hear. It may also help to consider intermediate cases, such as seeing the mirror, hearing over the telephone, recognizing in a video or through a telescope.

You possess then a complex, extraordinarily versatile, skill – the ability to bring to one focus innumerable small bits of natural information arriving in the form of a hugely diverse set of proximal stimulations impinging on your various sensory surfaces, all of which happen to carry natural information about just one thing, the same person. This allows you to bring these scattered bits of information to bear one on another, via mediate inference and practical learning over time, and to use the results during later encounters when you again recognize this person or come across new information about them. And so, of course, with many of your other friends or with individual objects of your acquaintance. You are enabled to bring to a single focus information about the same thing that has been widely dispersed over time and space through diverse media and that has affected your senses in widely diverse ways.

Our remarkable abilities to reidentify – more generally, to "coidentify," since various methods of recognition may be employed simultaneously, supplementing and reinforcing one another – are not, of course, restricted to individual objects. We also have abilities to recognize various properties, say, shapes or colors or distances, under a wide variety of external conditions. Think of the variety of proximal visual stimulations – what actually hits the eye – to which a given shape may give rise when viewed from various angles, from different distances, under different lighting conditions, through various media such as mist or water, when colored different ways, when partially occluded. How shape constancy is achieved by the visual system, the capacity to recognize the same shape as the same under a wide range of proximal stimulation conditions, is a problem of enormous complexity on which psychologists of perception are still hard at work. And shape is coidentified by the haptic systems, feeling the shape of a small object your hand a variety of ways, with these fingers or those, when the object is turned this way or that way, perhaps by using two hands, by merely holding the object or by actively feeling or stroking it, by exploring with larger motions that involve your arms, body and perhaps legs, employing the touching surfaces of any of a wide variety of your body parts. This kind of perception of shape, which involves the coordination of information about the exact positions of one's body parts with information about what touches these parts, is of such a complex nature that, psychologists have hardly begun to study it. Similarly, the variety of ways which color constancy, texture constancy, size constancy, place constancy, distance constancy, sound constancy, phoneme constancy are achieved are enormously complicated matters. Recalling again that even the most direct perception is perception through signs, we can include also information received about various properties through the use of all kinds of measuring instruments and scopes, and through the use of different kinds of inference. All of these are ways of bringing back to one focus the scattered bits and pieces of information about the properties of a thing that have been dispersed over space and time through diverse media, finally to impinge on our outer sensory organs.

I have recently coined the term "unicept" for the mental/neural vehicle that holds this information in focus, taken along with the repertoire of input methods that the person harboring the unicept knows to employ.[3] "Uni" is for one, of course, and "cept" is from Latin *capera*, to take or to hold. One's unicept for an object, or property, or kind, or relation etc., takes in many proximal stimulations and holds them as one distal entity. A developed unicept reaches through a radical diversity of sensory impressions to find the same distal thing again. It may also have to sort through similar or identical sensory impressions that have diverse distal things behind them. It funnels information collected by many coidentification methods into storage such that it is marked to interact in inference and action guidance an appropriate way, a way that "takes" it all to concern a single thing. A unicept is a specific individual *faculty* developed for a very specific purpose, the purpose of collecting and integrating information about some particular thing.

Unicepts, I believe, are the fundamental units of cognition. They form the fundamental components of empirical beliefs. They are *not* "concepts," *at least not concepts of a kind recognized by any familiar tradition* – this for several important reasons. Unicepts are what we have instead of empirical concepts as traditionally understood.

First, unicepts are not things that people share. Each of us has our own private stock of unicepts. Many of your and my unicepts do, of course, succeed in gathering up information about exactly the same things in the world, but they do this, pretty unexceptionally, in somewhat different ways, often utilizing many overlapping input methods but also many that are distinct. (Hellen Keller's unicepts succeeded in gathering information about many of the same things yours do, but in ways most of which were very distinct from your ways.)

Second, many of our unicepts involve abilities to coidentify through prior recognition of words that, with context, carry information about these things, these words, in context, indicating to us what we are receiving information about. But the fact that you and I may have unicepts for the same thing, and that these unicepts may include our abilities to recognize that thing when manifested through the same word, does not strictly imply any further similarities between our two unicepts. (Helen Keller spoke English too.) There is no reason to suppose that extensional words need to correspond across people who use them competently to psychological similarities, to similar or even to overlapping input methods, or to similar or even

[3]The predecessors of unicepts in my writings were called "empirical concepts." The next paragraphs make clear why I have withdrawn that term favor of "unicepts."

to overlapping inferential patterns.[4] The meaning of an extensional term is often purely referential or extensional. (Here I depart from Sellars, of course, opening some pretty wide disagreements.)

Third, and most relevant for us, is that having a unicept is *a practical achievement*; it involves having a certain kind of ability or *capacity* to deal, *successfully*, with an aspect of the natural world. Prior to adequacy in beliefs is adequacy in unicepts relied on forming beliefs. A unicept is no good – perhaps we would want to say it is no unicept at all – unless what it pulls in information about is indeed one and only one thing. If it pulls together information about many things, using this as though it were about one thing, then, *if it can be called a real unicept at all,* it is an *empty* unicept (a "vacucept") or at best an *equivocal* unicept (an "equivocept").

2.3 The General Epistemological Question

The huge question that immediately arises is what evidence we ever have that a certain unicept is really a unicept, a genuine capacity to tag only information that really is about the same as information about the same, rather than being a vacucept or an equivocept. What evidence do I have that it is indeed the same person, day after day, that I think of and call "Don" (my husband) or the same property that I think of and call "red," or the same real kind that I differentiate, reidentify, think of and call "dog" or "cat."[5] These are not things that I know a priori. That should be apparent. It is not a matter of logic, say, but of natural law that distal objects and properties cause just the variety of proximal stimulations that they do, under these or those conditions. It is a highly empirical matter, for example, what visual stimulations hound dogs send back to me from a distance when running through dappled shade crossways in my visual field. It is a highly empirical matter what Don's voice does to my auditory nerves and how that changes through the medium of the telephone or through a wall. Clearly it has to be learned, somehow, which proximal stimulations go with which, which are caused by the same distal things. It has either to be learned by the individual or some of it has, perhaps, to have been learned by the species. But how?

Learning how to reidentify various perceptual objects, properties and relations under a variety of different conditions probably begins with the ability to track objects for short times with the eyes and head, also ears and hands, as these objects

---

[4]In Millikan (2010) I explain why this remark applies not only to proper names and names of empirical properties and relations but to most kind terms as well.

[5]When this question concerns reidentification of kinds, its relevance and importance is not obvious unless the right sort of realism about kinds has been introduced. I have argued for an ontology of "real kinds" that separates them sharply from classes and makes clear why there both are and must be many alternative ways to recognize the members of any real kind, making the question of correct reidentification central (Millikan 1984, 1998, 2005, 2009, and especially 2010).

rotate, become displaced in relation to oneself, and move through a variety of perceptual conditions such as different lighting conditions, occlusions. masking sounds and so forth. For it seems that the very first project, at least of the visual system, is to notice and keep track of various objects as we and they move about, not by noting and then reidentifying their properties as such, but by tracing continuities in path over short periods of time (Pylyshyn 2007). Reidentification of objects and kinds after breaks in tracking is probably accomplished in large part by attending to patterns of stimulus correlation. But the epistemological question we have raised is not directly addressed by these mechanisms, which might be viewed, strictly speaking, as methods of hypothesis formation rather than methods of confirmation. The epistemological problem concerns evidence that these methods of attempting to learn reidentification techniques result in reidentifications that are truly objective, distal objects, properties and kinds that really are the same again being correctly identified as such.

There are at least two different methods that seem to be used to address this basic epistemological issue. We might call these the "practical" method and the "theoretical" method. The practical method explains why it is possible for many non-human animals to acquire a modest collection of unicepts, indeed, how evolution through natural selection may even build some unicept skeletons into animals, perhaps also into humans. The theoretical method, on the other hand, is probably peculiar to humans, helping to explain why humans have concepts in numbers several orders of magnitude beyond those of any nonhuman animals.

The practical test is merely that one can learn, over time and repeated identifications, how, productively, to be guided by the identified object or kind during practical activity. Evidence for a dog that it can indeed recognize its master is that it is able to learn, over time, how to behave in rewarding ways in its master's presence; evidence that it is indeed able to distinguish squirrels from rabbits is that it has learned successfully how to fit the chase to the quarry, heading squirrels away from trees, heading rabbits away from hedgerows and so forth.

The theoretical method involves the capacity to make propositional judgments, to entertain thoughts having subject-predicate structure, the predicate being subject to negation, or that can at least be expressed this medium. It requires a sensitivity to contradiction, and a disposition to alter unicept input methods when contradictions begin to arise. Obvious examples come from the development of empirical science, discovering the objectivity of the temperature scale, for example, by successfully devising diverse kinds of instruments that agree in measuring it, as well as many ways of predicting it – identifying it ahead of time – by inference using theory. But a more universal and fundamental way of testing the adequacy of ones unicepts is the home method, the use and understanding of language, finding that one agrees with other people who have come to recognize the same facts but from different perspectives, perhaps using different unicept input means from those one commands oneself. Arguably it is exactly the use of this latter method that sets our cognitive capacities so far apart from other animals.

Very much more needs to be said about the use of propositional judgment – of thought and/or language that has subject-predicate structure and is sensitive to a negation transformation (e.g., Millikan 1984, 2000, 2004, Chs. 18–19). But for our purposes, the main lesson to be remembered is merely that in both the practical and the propositional judgment cases, unicept adequacy is something that is learned and tested over time and over a variety of perspectives. Adequate unicepts are earned. If there are any unicepts, or perhaps skeletons for them, or dispositions to pick them up on quick exposure that are supplied to us natively, they will surely have been earned through a history of natural selection, and can be presumed not to be idle but to have significant functions.

2.4 The Epistemological Question for Phenomenology

Uniceptual capacities are *representational* capacities. I am working here with a representational theory of mind. "Phenomenal experience" is something many philosophers have beliefs about. These beliefs purport to be representations in *thought* of real properties of another real thing called "experience". We need to understand then, *in a way that is consistent with our more general views on epistemology*, how a person can develop the necessary ideas/unicepts with which to think about and have knowledge of these properties and of this experience. I am posing the epistemological question for phenomenology as a question how the unicepts applied during the description of phenomenological experience acquire their credentials. What is the origin of these ideas? What evidence is there that they are unicepts, rather than vacucepts (caloric, pholgiston) or equivocepts ("heaviness," before mass and weight were distinguished)?

Important to keep in mind here is the Sellarsian warning that the fact that an idea is directly applied in observation judgments does not guarantee its nonemptiness. That caloric could be directly felt, for example, is no argument for its existence. An excellent and totally convincing argument to this effect that does not, incidentally, presuppose anything in Sellars, may be found in Churchland (1986, Ch. 2).

A second thing to notice is that it would be really weird to suppose that we have some special innate capacities to form the ideas of phenomenal properties and phenomenal experience, capacities to form adequate unicepts for these things on demand. What would be the evolutionary point of such an ability? What life- or society-preserving activities would our ancestors have been using these abilities and the resulting unicepts for? It seems clear that we must be using just our ordinary unicept forming capacities in the generation of our ideas that concern the phenomenology of experience, thus leaving it open, and appropriate, to ask whether and how these ideas are or have been validated.

An important epistemological principle in the case of ordinary empirical ideas, ordinary unicepts, is that the likelihood that one's unicept for a thing is nonempty and univocal goes up with the variety of ways one knows to reidentify that thing so as to confirm one's judgments. It goes up with the variety of perspectives from which one is able to identify that thing. And it goes up with the number of occasions on which one finds opportunity to test a unicept's input methods against one another.

How are we to gain such perspectives and opportunities in the case of unicepts for phenomenal properties and objects? How do we know we are thinking of anything real when we appear to ourselves to be thinking of such things?

That's the epistemological problem. I will not press it further. What I will do instead is to begin to construct a candidate theory, consistent with the description of unicepts outlined above, about what "phenomenological description" really is. This will require a little background, however. First I must introduce a proposal about the development of our ideas/unicepts for various ordinary perceptual properties, such as red and sour.

2.5 Our Ideas of Some Ordinary Perceptual Properties

Begin by considering for what our perceptual capacities were designed. Like the rest of us, our minds evolved. They were built up by tinkering, building newer capacities out of older ones, by using these older capacities in new ways. Newer mechanisms often control the activities of older ones more sensitively, or redeploy them for new purposes. Our own minds were built on top of animal minds, almost literally, the upper and more frontal parts of our brains having evolved last. We still have animal minds, though we have remodeled a bit and built on some fairly spacious additions.

The function of perception in the higher animal species prior to man appears to be quite exclusively guidance of immediate practical activity – navigation among objects in the immediate environment, initiation of action towards or away from objects, the manipulation of objects for practical purposes. That is, its fundamental use is in the perception of, as J.J. Gibson put it, *affordances* of various kinds, perception for action. That, likely, is the first function of perception for humans as well. It is interesting, however, that many of the most obvious perceptual properties, taken one by one, are of no immediate use at all in guiding action. The colors, the sounds, the tastes, and the smells of things, and the internal relations among these properties – roughly, the classical "secondary qualities" and their internal relations – are none of them of much help in guiding immediate practical activity. There is nothing that being red is good for as such, nor having emitted a certain sound or odor. There is nothing about the internal relations among wave lengths for colors, or the internal relations among physical sounds, that carries direct significance for guiding action. Contrast these properties and relations with the classical "primary" properties and relations, for example, with shape, size, and weight. The values of and relations among of these latter properties, taken in relation to the animal's own physical properties and capacities, *do* very much matter to an animal who would manipulate objects or navigate among them.

It has been thought, though the matter remains under dispute, that there is a division within the visual and perhaps also the auditory systems of higher animals (even hamsters) into a dorsal system, which achieves perception of the relations of objects to the animal's body as needed to guide approach, retreat, object-manipulation and so forth, and a ventral system, which allows an animal to identify objects and object kinds, so as to decide which actions are appropriate to which objects. Whether or not these two functions are actually divided into separate neural processing streams, it remains clear that they are of somewhat separate kinds, and that they require the registration of different though overlapping sets of properties. Given this, it seems reasonable to speculate that capacities to discriminate among colors, sounds, odors and so forth were originally developed for use merely in identifying objects and object kinds. For it was the identities and differences among *objects*, not among these secondary perceptual properties themselves, that were important for deciding what needed to be attended to in the environment. The original things recognized in completed perception for action would be contrarily affording things and stuffs, things that would need to be treated or responded to incompatibly. Notice that the existence of color metamers and their analogues, for example, for taste would not interfere with mere object identification purposes in any more significant way than does the fact that different objects and kinds can have the same reflectances. Natural selection yields mechanisms that suffice for their purposes, and the purposes here are not precise.

Just as the edge detectors, vertical line detectors, motion detectors and so forth in early visual cortex are not used in the direct guidance of action but only in guiding construction of more meaningful representations of objects and properties, the original use of color discrimination, taste discrimination and so forth must have been merely in implementing the reidentification of objects. Although they have no practical significance themselves, the reflectance properties of an object and the odors and sounds it emits, when put together with other bits of information, may be crucial for reidentifying the object or the kind of object being encountered. Obviously the properties of things are causally involved in any perceiver's abilities to differentiate among affording things, but this does not imply that they are represented in perception-for-doing *as* attributes of substances. Similarly, no one has supposed that the gradients and edges of early visual perception are represented as such in the final products of visual perception. That secondary properties are not the first things evident in perception is suggested, for example, by the fact that there are languages that have few or no words for colors and that children learn color words quite late. Similarly, we do not have words for sounds or odors but describe them by reference to what they are of – the smell of bacon, a rasping sound, a bell-like sound. When merely smoothly acting and not reporting or reflecting – when not using propositional tools – I suspect that we do not represent sounds, or sound qualities, but rather doors closing, people shouting, or perhaps a *something* over there (not a *sound* over there) that we hear but can't make out. We do not, in the first instance, smell odors, but rather pine trees or bacon cooking. We do of course feel and see shapes, but not as attributes of things but merely as guides to identifying them or handling them. We see how to move or to pick up a thing given its position and shape, how to walk on it if it is rough or slippery, and so forth.

In sum, there is no propositional structure in mere perception for action. Compatibly, negations do not occur there. Perception for action does not involve perception of colors, sounds, odors and tastes *as such*, but only perception of the objects and kinds they help to signify. I offer this suggestion not as a bit of phenomenology but as speculation on what the end products in perceptual neural representation actually amount to for animals and also for humans during absorbing action.

What might we say then about the underlying systems, noted above, that account for perceptual constancies, shape constancy, size constancy, color constancy, sound-at-source constancy and so forth? What seems reasonable is that during the process of development of our perceptual systems through evolution or learning, distilled out in the background, taking their various places upstream in addition to such things as gradient, edge and motion detectors, were more sophisticated detectors of various simple object properties, recognition of which could be recombined for use in helping to identify a great variety of different useful things. I am thinking here, for example, of the way NETtalk, in learning to turn written text into phonological sequences, managed to distill out underneath in its operations something like individual vowels and consonants (Sejnowski and Rosenberg 1988). We might think of these underlying property-constancy mechanisms as like proto unicepts, abilities to reidentify the same distal properties through multivarious proximal stimulation, but without involvement yet in information storage regarding these properties. They are originally involved at a level of information processing well below the level either of perception for action or propositional judgment.

We suppose then that *much later* these underlying proto-unicepts are *redeployed*, probably by humans only, in processes leading to perceptual propositional judgments about properties of objects. They are taken up in the formation of thoughts with subject-predicate structure thus becoming involved, for the first time, in the operation of true unicepts for properties and relations. These emerging unicepts, we further speculate, were (and still are in children) developed along with *linguistic skills* that allow communication about objects having as yet no names but that need to be identified to hearers. That is, we assume that they do not develop until there is a use for them, and that this use involves judgment and communication. Indeed, quite generally the development of unicepts for propositional judgment would seem to ride piggyback on the earlier development of practical unicepts, unicepts of the kind, say, that dogs employ when they recognize their masters or recognize a rabbit. That these unicepts would sometimes redeploy representations from earlier stages of neural processing that had supported perception-for-action seems natural. They may involve redeployment of chemical property detectors (taste, smell) or distal color and shape detectors (color and shape constancy) or sound-at-source detectors and so forth. In the case of taste and smell there are no constancy mechanisms. So in developing propositional unicepts of tastes and smells, more direct neural mechanisms prior to object detection would have been reused.

These mechanisms were redeployed in the attempt to develop ideas that could serve as predicate unicepts for propositional judgment about distal objects. The general purpose of such unicepts would be identification and reidentification of objective distal properties, as evidenced through stability in judgment. The identities of such things as the objective colors and shapes of things are highly confirmed this way, and not merely by one's individual reexamination over times and perspectives but, importantly and powerfully, through agreement in judgments with other people.

In the case of tastes and smells, however, agreement in judgments with other people was the only way that more than one perspective (other than temporal) could be obtained. Coordinately, I think that the apparent objectivity of tastes and smells has had a fairly slender hold even on the common mind. Tastes and smells are not so insistently thought as really "in" the objects tasted and smelled. When people are being careful, these properties are often thought of as objective but relational.

Then modern science arrived, sporting a variety of new ways to input many of our unicepts through theory (inferential ways of observing) and sophisticated apparatuses designed for the study of light, of sound, of chemical composition and so forth. It became apparent that many of our well-established, simple observation unicepts were, in fact, equivocal. There were some color metamers, hard to illustrate in nature, of no practical significance for reidentification of objects, but none the less real. In this way, our unicepts for colors were discovered to be a bit blurred, equivocal on certain edges – a bit like having, mixed in with our information about Aristotle, a tiny bit of information about a previously unknown brother of his. For taste, however, there emerged the analogue of dozens of metamers. And just what should be said about smell remains rather a mystery.

In the case of color, a particularly instructive case emerged. Relying on our color constancy mechanisms, unicepts for certain relations among colors had been developed and, apparently, highly confirmed through agreement in judgments. Objective colors had been thought to lie next to one another in similarity in such a way as to form a circle, or taking into account saturation and lightness, within a three dimensional space, with some being at opposite poles from others, and so forth. It turns out, however, that there are in fact no such uniform continuity relations or polar relations among the distal colors. Ignoring metamers (count them just as illusions), reidentification of the same color again is reidentification of a real thing, namely, of the same or a similar reflectance property. But the apparently observed relations among the colors are not real. That is, the reidentifications we make of same-color-again are pretty good. Mostly we get it right. But our thoughts that these distal properties have certain objective relations to one another are confused. The relations that we seem to be observing and reidentifying – red "opposite" green, blue "opposite" yellow, purple "closer to" blue than to orange – are not out there. Like caloric, they may indeed be "observed," but they are chimerical. Given the above reflections on the possible redeployment of early perceptual processes in the development of propositional unicepts, we could tell a general story about how this kind of thing might (indeed, roughly how it actually did) come about.

Suppose that though some accidental quirk in my computer's design, every other word that I typed came out red, the in-between words in blue. The relations of identity and difference in word color would be obvious to you, but you should not take them to indicate differences in the ideas I was expressing with the words. Similarly, relations of kind and degree of similarity between the neural vehicles of different representation do not necessarily, simply as such, *represent* these relations as holding between their corresponding representeds. They will represent these relations only if *used*, downstream, in a manner that requires it. They will represent these relations only if they are interpreted that way. Neural representations that are used merely as tags for simple reidentification of objects and kinds might be a lot like one another in some ways and different in others without these representing similarities or differences in content. Certain relations, say, among the neural representations of colors, among the neural representations of odors, and so forth – the dimensions and distances in this or that neural similarity space – though they might *in some cases* carry a certain amount of *natural* information about relations among the real distal properties represented, might carry no intentional information at all, no information that the brain had been designed to *use*. They would not then *represent* any relations among the things represented, just as the relations among "cat" and "bat" and "rat" and "sat" do not represent relations. But we can imagine that in later reuse of these vehicles, in the attempt to use them in the development of propositional unicepts, the relations among them might be erroneously interpreted as naturally indicating relations among their representeds. Agreement with other people on the occurrences of these relations would apparently seal the matter.

2.6 Phenomenological Description

That was a very lengthy introduction to what will now be a very short discussion of phenomenology. I have suggested a mechanism by which our unicepts even of so-called perceptual properties such as tastes and the relations among colors may have come to be equivocal, confusing together a diversity of distinct actual properties or inventing chimerical relations. Thus we can understand how what is apparently known by the most direct possible observation may be worse than false. It may be senseless. Let me now tell a story that, as I understand it, was once roughly J.J. Gibson's story on the status of "the visual field" (in which he did not believe).[6] The story makes out apparent facts about phenomenal experience as erroneously represented – as fictions.

Suppose that you are looking through a window at the scene outside, but a friend (perhaps a British empiricist) has convinced you that the scene you see is really inside, projected onto the flat real two dimensional surface you had mistakenly thought before was a transparent window pane. You and your friend each proceed, with great care, to try to describe the shapes and colors of the patterns on the window pane. Both of you find this exercise quite difficult, but considerable agreement between you emerges on bold features. (I imagine that people who are good painters find this kind of thing easier than I do.) That's the original exercise that was called "phenomenological description" for vision, description of "the visual field." It would seem to involve the redeployment of certain normally far-upstream

[6] "The visual field, I think, is simply the pictorial mode of visual perception, and it depends in the last analysis not on conditions of stimulation but on conditions of attitude. The visual field is a product of the chronic habit of civilized men of seeing the world as a picture." Gibson (1952), p. 148.

sensory detectors, further upstream even than the output of the perceptual constancy mechanisms – possibly the same that are employed in a painter's re-envisionment of a scene order to paint it? – the attempt to identify objects and properties in a hypothesized inner realm posited by philosophers convinced of a certain queer theory of knowledge. One symptom of what's strange about this, incidentally, is that the description is done with everyday words, not special ones developed for the purpose, as one might have thought necessary for describing some totally new kind of stuff or entities in some totally new ontological realm.

How one is supposed to produce phenomenological descriptions of *heard* scenes or *felt* scenes or *tasted* or *smelled* scenes is less clear. (Similarly, I imagine it would be very unclear just how phenomenological description for visual experience is supposed to be done if you were an adult who had had no experience with paintings.) What is a description of the phenomenology of smell, for example, besides just a naming according to what one would normally take the smells to be of? Perhaps it involves an application of one's ordinary unicepts for odors while pretending to oneself not to know anything about present conditions, such as what's really in front of one's nose or whether one has a cold? One uses terms that would describe what one *supposes one would suppose* one was smelling given no outside information, pretending to withhold, as well, any, ontological commitment (Husserl's epoche)? In the case of touch, perhaps one concentrates on what one would take the apparently touched item to be doing to oneself, pressing on one, pricking one, rather than what properties one would take the touched items themselves to have. We may tend to ask, "How do I feel when I touch it?" not "What properties can I feel it to have?" When you feel how rough or smooth the road is under your tires as you drive (compare Fulkerson 2012) and then turn to think instead about what is happening to your bottom, does the phenomenology change? Are you sure? How do you know?

However one does it, the descriptions one comes up with are likely to express representations, unicepts for the same sort associated with any other deeply mistaken scientific or lay theory. For nothing whatever helps to certify that the apparent unicepts one is using are not empty.

The alternative to these skeptical reflections, I believe, is to embrace Russell's 1912 sense data as the foundation for your epistemology.

References

Churchland, P. 1986. *Scientific realism and the plasticity of mind*. Cambridge: Cambridge University Press.

Dennett, D.C. 1991. *Consciousness explained*. Boston: Little, Brown & Company.

Dennett, D.C. 2003. Who's on first? Heterophenomenology explained. *Journal of Consciousness Studies, Special Issue: Trusting the Subject? (Part I)*, No. 9–10: 19–30.

Fulkerson, M. 2012. Touch without touching. http://www.philosophersimprint.org/012005/

Gibson, J.J. 1952. The visual field and the visual world. *Psychological Review* LIX: 148–151.

Millikan, R.G. 1984. *Language, thought and other biological categories*. Cambridge, MA: The MIT Press.

Millikan, R.G. 1998. A common structure for concepts of individuals, stuffs, and basic kinds: More mama, more milk and more mouse. *The Behavioral and Brain Sciences* 22(1): 55–65. Reprinted in E. Margolis and S. Laurence eds, *Concepts: Core Readings*, MIT Press 1999, pp. 525–547.

Millikan, R.G. 2000. *On clear and confused ideas*. Cambridge, MA: The MIT Press.

Millikan, R.G. 2004. *Varieties of meaning*. Cambridge, MA: MIT Press.

Millikan, R.G. 2005. Why (most) concepts are not categories. In *Handbook of categorization in cognitive science,* ed. Henri Cohen and Claire Lefebvre, 305–316. Elsevier. http://www.elsevier.com/wps/find/bookdescription.cws_home/705263/description

Millikan, R.G. 2009. Embedded rationality. In *Cambridge handbook of situated cognition*, ed. Murat Aydede and Philip Robbins, 171–181. Cambridge UK: Cambridge University Press.

Millikan, R.G. 2010. On knowing the meaning; With a coda on Swampman. *Mind* 119(473): 43–81.

Millikan, R.G. 2012. Accidents. *Proceedings and Addresses of the American Philosophical Association*, November 2012: 92–103.

Millikan, R.G. 2012a. Natural information, intentional signs and animal communication. In *Animal communication theory: Information and influence*, ed. Ulric Stegmann. Cambridge: Cambridge University Press.

Millikan, R.G. 2012b. Natural signs. In *Computability in Europe 2012, lecture notes in computer science*, ed. S. Barry Cooper, Anuj Dawar, and Benedikt Loewe, 497–507. New York: Springer. http://www.amazon.com/How-World-Computes-Conference-Computability/dp/3642308694/ref=sr_1_1?s=books&ie=UTF8&qid=1374243864&sr=1-1&keywords=3642308694

Pylyshyn, Z. 2007. *Things and places*. Cambridge, MA: The MIT Press.

Sejnowski, T.J., and C.R. Rosenberg. 1988. NETtalk: A parallel network that learns to read aloud. In *Neurocomputing,* ed. J.A. Anderson and E. Rosenfeld. Reprinted: Cambridge, MA: MIT Press.

Chapter 3
From Phenomenology to the Self-Measurement Methodology of First-Person Data

Gualtiero Piccinini and Corey J. Maley

Ruth Millikan argues that there is no "legitimate phenomenology of experience": that there is no method—not even a fallible or partially reliable one—for accurately describing our experiences in the first-person. The reason is that there is no method for checking that the ideas we think we have about experience are about anything at all. Like phlogiston, there may be no such things as the properties we take experience to have.

Millikan's problem with phenomenology is threefold. First, we need a substantive theory of ideas in order to explain how we can know about or describe our experience (unless we think that ideas are somehow "given" in perception, a view that—nowadays anyway—virtually nobody holds). Phenomenology cannot be well grounded without such a theory. But then, the substantive theory that Millikan believes can do this job employs an assumption about ideas that is incompatible with the assumptions of phenomenology. Roughly, phenomenology assumes that the units of experience can be validated at one time and from one perspective, while Millikan makes the plausible case that "adequate empirically-based ideas" must be validated over time and across perspectives. Finally, Millikan sketches a theory that explains "what is really going on when people *think* they are describing their phenomenal experience." The upshot of Millikan's theory is that the phenomenology of experience is explained away.

The central figure in Millikan's theory is the *unicept*, what she calls "the fundamental units of cognition." Although they are similar to concepts, unicepts

G. Piccinini (✉)
Associate Professor, Department of Philosophy, University of Missouri – St. Louis, 599 Lucas Hall, 1 University Blvd, St. Louis, MO 63121, USA
e-mail: piccininig@umsl.edu

C.J. Maley
ABD Graduate Student in the Department of Philosophy at Princeton University, and a Research Associate in the Department of Philosophy and the Center for Neurodynamics at the University of Missouri – St. Louis, 599 Lucas Hall, 1 University Blvd, St. Louis, MO 63121, USA
e-mail: maleyc@umsl.edu

R. Brown (ed.), *Consciousness Inside and Out: Phenomenology, Neuroscience, and the Nature of Experience*, Studies in Brain and Mind 6, DOI 10.1007/978-94-007-6001-1_3,

are not concepts, not even in the regimented sense used in psychology. Unicepts are not shared, and even if two people have a unicept about the same thing, there is no guarantee that they have access to the same information about that thing. So what *is* a unicept? Roughly, a unicept is that bit of our mental apparatus that allows us to perform amazing feats of object constancy. It is Adam's unicept of his mother that allows him to recognize her by her voice, her face, her gait, and so on, all in imperfect sensory conditions. Unlike having a concept of something, having a unicept of something implies that one has the ability to recognize and reidentify the thing that the unicept is about.

Millikan offers a speculative but compelling story about the evolution and development of unicepts for ordinary perceptual properties. We know that in the visual system, there are detectors for various properties: there are edge-detectors, color-detectors, motion-detectors, face-detectors, and so on. Some of these, like face-detectors, are further downstream and take input from upstream detectors. Millikan suggests that, in a similar way, there are detectors for simple object properties, which could then feed into a mechanism for combining these into "proto-unicepts." Once we add in the ability to deploy proto-unicepts in propositional judgment and communication about objects, we have unicepts of properties and relations: a real candidate for a basic unit of cognition.

Modern science has changed the kinds of inputs available to our unicepts, Millikan argues, showing some of our unicepts to be chimerical. Where we once thought that certain relations objectively held among colors, suggested by their apparent similarity, we now know that this is not the case. It would simply be a mistake—as Millikan puts it—to think that the similarity between unicepts for two colors indicates similarity between those colors, just as similarity between the words "cat" and "rat" does not indicate similarity between cats and rats. And as previously mentioned, we have, via the sciences, methods for determining whether our unicepts refer to anything at all (such as phlogiston), or to more than one thing (such as heaviness into weight and mass). In sum, with the methods of empirical science, we can determine whether our unicepts are genuine, referring to one thing and one thing only, and whether the relations that seem to hold among unicepts in our experience reflect objective relations among the things represented. As Millikan puts it: "the likelihood that one's unicept of a thing is nonempty and univocal goes up with the variety of ways one knows to reidentify that thing so as to confirm one's judgments. It goes up with the variety of perspectives from which one is able to identify that thing. And it goes up with the number of occasions on which one finds opportunity to test a unicept's input methods against one another."

The problem with phenomenology, according to Millikan, is that there is no way to validate unicepts that purport to refer to properties of our experience. In the case of ordinary unicepts referring to public objects and properties, we compare different judgments under different conditions, and judgments made by different people, in an attempt to insure that all such judgments converge on one and the same public objects and properties. If they do, their unicepts are validated. But in the case of experience there are no public objects or properties to be referred to, so there is no way to compare our judgments under different conditions, let alone judgments by different people, to insure that all such judgments are about one and the same thing. Thus, there is no way to validate unicepts that purport to be about experience, which is what phenomenology requires. Thus, phenomenology is impossible.

What remains possible, and what Millikan embraces, is *heterophenomenology*—a methodology articulated and defended by Dennett (1991, 2003, 2007). Millikan does not explain how heterophenomenology gets around the failure of phenomenology. We will now sketch how her reasoning might go.

Heterophenomenology holds that when subjects utter first-person reports, which purport to describe their experiences, the heterophenomenologist must remain neutral as to their truth value. Instead of interpreting first-person reports as reports about experience, the heterophenomenologist interprets first-person reports as descriptions of the subjects' *beliefs* about their experience. About such beliefs, subjects are deemed incorrigible.[1] The heterophenomenologist can then use the appropriately interpreted reports as evidence that, in combination with other scientific evidence, can ground a theory of consciousness. According to Dennett, heterophenomenology is the method scientists currently follow when they use first-person reports as sources of data.

From Millikan's standpoint, what seems important is that the heterophenomenologist avoids any direct inference from first-person reports to the properties of experience. Whatever subjects utter in purporting to describe their experience is reinterpreted by the heterophenomenologist as expressing *beliefs* about experience. As a consequence, the heterophenomenologist need not take any unicepts to refer to properties of their experience. Since the heterophenomenologist remains neutral about the truth value of the subject's utterances, she avoids the problem that affects phenomenology.

Like Millikan, we doubt the viability of phenomenology. But unlike Millikan, we will argue that there is a way to validate unicepts that describe our experience. As a result, while we agree with Dennett that first-person data are public data on a par with other scientific data, we maintain that heterophenomenology can be improved in such a way that first-person data can be interpreted in terms of (typically) conscious mental states.

We grant that there is a sense in which unicepts that purport to be about our experiences lack a public object or property that can guide comparisons between judgments. But the same holds for unicepts that are legitimately employed in the sciences to refer to objects and properties that are not directly observable—unicepts for electrons, neutrinos, black holes, and the like. Physicists manage to validate unicepts for electrons, neutrinos, and black holes, while invalidating—and eventually rejecting as empty—unicepts for phlogiston, epicycles, the ether, and the

[1] Unfortunately, Dennett is somewhat equivocal about the status of the subjects' beliefs about their experience. Sometimes he describes them as the causes of first-person reports, which presumably means they are real, while at other times he describes them as merely constituting a fictional heterophenomenological world narrated by the subject. This makes Dennett's claim that subjects are incorrigible equivocal between a substantive empirical claim about the causes of first-person reports, which we take to be false, and a claim that is true by definition about the fiction narrated by the subject, which is true but trivial (cf. Schwitzgebel 2007).

like. By the same token, it may be possible for subjects to validate their unicepts for at least some properties of experience while, perhaps, invalidating and thus eventually rejecting others.

How might that work? Consider a child learning to use mentalistic language, say, the word 'sad'. She can observe public manifestations of sadness in her and other people's facial expressions, posture, gestures, tone of voice, etc. She may be told that she looks sad or is acting sadly and asked why. By engaging in conversations about sadness, she may begin to notice that something about her experience correlates with her public manifestations of sadness, and she may begin to use the term 'sad' to refer to that aspect of her experience. By engaging in conversations about sadness, initially based on overt manifestations of sadness, she will be able to compare different judgments under different conditions, and judgments made by different people, eventually becoming reliable at insuring that such judgments converge on one and the same property of her experience. She can do the same thing for other mental state unicepts (cf. Piccinini 2003, Section 2).

The crucial difference between the process of unicept validation we just sketched and phenomenology is that phenomenology is generally assumed to be conducted by a single subject working on her own, whereas unicept validation requires the public coordination of judgments by different people based largely on publically observable manifestations of mental states. We may think of mentalistic unicept validation on the model of instrument calibration.

One of us has argued that, when we do the methodology of first-person data, rather than thinking of the subjects who generate first-person data (through either first-person reports or other first-person behaviors) as observers reporting on what they experience, it is better to think of these subjects as self-measuring instruments (Piccinini 2009). On this view, it is not the responsibility of the subjects in scientific studies to eliminate biases or to determine how reliable they are; rather, that is the responsibility of the investigators, just as it would be their responsibility with any other scientific instrument. Furthermore, it is the investigators' responsibility to make sure that their instruments produce data that can be checked by other investigators in a public way. Thus, first-person data need not be private (which would be antithetical to scientific investigation), but can be public in the same way that data acquired via any scientific instrument is public.

This self-measurement methodology of first-person data improves on heterophenomenology in a number of ways (for more detailed discussion, see Piccinini 2010):

First, we should not be tempted into thinking that the first-person reports central to phenomenology must concern consciousness. Many psychologists use first-person reports to study things other than consciousness, such as memory or morality; and even if many of these phenomena are conscious in the sense that subjects are reporting on conscious mental states, there is no *a priori* reason to rule out that first-person reports can be about unconscious phenomena, interpreted and described by trained psychologists. Further, one should not be tempted into thinking that first-person *reports* exhaust the sources of first-person data. Scientists routinely use subjects' button presses as sources of data about a variety of psychological phenomena, where the subjects may be human or non-human primates.

Another way in which we break from heterophenomenology is that we reject agnosticism about first-person reports. We think it best for scientists to take subjects' reports at face value, treating such reports as one would in an everyday context. People are not generally agnostic about whether what others tell them is true; rather, they use their best judgment to decide whether and to what extent to believe other people, or whether to reinterpret their claims in appropriate ways. Determining whether we should doubt or reinterpret someone's claims is a matter of using whatever other evidence we have at hand. This practice is not infallible: there are pathological liars, for example, who lie without any particular reason that we might discover. But if we were to always withhold judgment about whether people were truthful, we would never learn anything from their first-person behaviors, and this holds true in the scientific use of first-person reports as well.

We also reject the heterophenomenologist's suggestion that first-person reports can only tell us about subjects' beliefs. Rather, first-person reports can tell us about other kinds of mental states, which subjects may or may not have any beliefs about. It is problematic to ascribe a person who reports that she feels, say, ashamed with nothing more than a *belief* that she feels ashamed. She may have such a belief, but that belief may not have been the cause of her report: she may have formed the belief *after* expressing the report. The idea that all first-person reports are caused by beliefs and that all first-person reports express beliefs (and only beliefs) is unjustified: what causes first-person reports (and first-person data more generally) is an open, empirical question. Much like the previous point, it is better to take what first-person reports are about at face value. And again, there may be special reasons to doubt that a person who says she is ashamed is actually ashamed, but in general it is inadequate to infer only that such a person merely believes she is ashamed.

Having rejected the heterophenomenologist's interpretation of first-person data solely in terms of beliefs, we also reject the heterophenomenologist's insistence that the subject be deemed incorrigible about such beliefs—and hence about her "heterophenomenological world". First-person data often give us useful information about experience, and about mental states more generally, but there is nothing incorrigible about first-person data any more than there is something incorrigible about any other data. It is up to psychologists and neuroscientists to investigate which first-person data are reliable about which mental states under which circumstances.

Our final point responds to Dennett's claim that heterophenomenology licenses the same experiments as phenomenology. On the contrary, a sound methodology of first-person data makes a significant difference to scientific practices. Unlike traditional phenomenology, which relies on the introspecting subject to avoid biases and errors (a dubious expectation), we recommend that psychologists and neuroscientists who collect first-person data exert the utmost care and rigor in eliciting, processing, and interpreting their data. Of course, this is not a surprising recommendation: we should expect all scientists to exercise such care when it comes to their data and their instruments. But there are specific steps that can and should be taken in the case of first-person data. A sound methodology can help to uncover and highlight them.

If Millikan's account of unicepts is correct, then she seems right about phenomenology: individual observers are simply not in a position to validate their mentalistic unicepts and the apparent relations among them from inside their minds (nor could they ever be in such a position). But if we are right, subjects can still validate their mentalistic unicepts with the help of others and by observing public manifestations of their mental states, possibly under the guidance of psychologists and neuroscientists who use them as subjects. As a consequence, first-person data may be used in scientific studies of mental phenomena. When circumstances allow it, first-person data may be interpreted directly in terms of experience. Like other investigations into how scientific instruments work, Millikan's theory—if correct—might shed light on some fundamental limitations of first-person data. But that will depend on the phenomenon being investigated, plus whatever other information we can marshal about that phenomenon.

In conclusion, there is a difference between the phenomenology of experience, which—we agree with Millikan—is a flawed methodology, and the scientific study of phenomenal experience. If we are right, the latter is a legitimate part of science, to be conducted in accordance with the self-measurement methodology of first-person data.

References

Dennett, Daniel C. 1991. *Consciousness explained.* Boston: Little, Brown & Company.

Dennett, Daniel C. 2003. Who's on first? Heterophenomenology explained. *Journal of Consciousness Studies* 10(9–10): 19–30.

Dennett, Daniel C. 2007. Heterophenomenology reconsidered. *Phenomenology and the Cognitive Sciences* 6: 247–270.

Piccinini, G. 2003. Data from introspective reports: Upgrading from commonsense to science. *Journal of Consciousness Studies* 10(9–10): 141–156.

Piccinini, G. 2009. First-person data, publicity, and self-measurement. *Philosophers' Imprint* 9(9): 1–16.

Piccinini, G. 2010. How to improve on heterophenomenology: The self-measurement methodology of first-person data. *Journal of Consciousness Studies* 17(3–4): 84–106.

Schwitzgebel, Eric. 2007. No unchallengeable epistemic authority, of any sort, regarding our own conscious experience—contra dennett? *Phenomenology and the Cognitive Sciences* 6: 107–113.

Part II
Phenomenal Properties and Dualism

Chapter 4
Consciousness and the Introspection of 'Qualitative Simples'

Paul M. Churchland

4.1 Introduction

Philosophers have long been familiar with the contrast between predicates or concepts that denote or express "qualitative simples," as opposed to predicates or concepts that denote or express "structural, relational, causal, or functional" features. The tendency has been to think of these two classes of properties as being ontologically quite different from each other. Paradigm examples of the former would be features such as the redness of a tomato, the sweetness of sugar, the low pitch of a sound, and the warmth of a hearth. These particular examples, all features of things in the objective physical world, would be joined by a further population of presumed qualitative simples, features displayed in the conscious states of a human or other cognitive creature, features such as the qualitative character of your visual sensation of a tomato, of your gustatory sensation of a sugar cube, of your auditory sensation of a sound, and of your tactile sensation of a glowing hearth. Indeed, some may want to insist that the features displayed in this *private* cognitive domain are the *only* genuinely simple qualitative features, on grounds that their external brethren all turn out to admit of a structural, relational, causal, or functional analysis of some kind after all.

Whether or not this secondary claim is correct, we shall address anon. Let me continue this opening exploration of the contrast at issue by pointing out that predicates or concepts that denote *non*-simple properties are typically supposed to be *analyzable* or *definable* in terms of sundry other features, and in terms of the characteristic configuration of relations that severally unite those other features. Thus, the property of *being explosive* can be analyzed in terms of having the disposition to burst outwards suddenly, under suitable conditions of ignition. The

P.M. Churchland (✉)
Department of Philosophy, UCSD, 9500 Gilman Drive #0119, La Jolla, CA 92093, USA
e-mail: pchurchland@ucsd.edu

R. Brown (ed.), *Consciousness Inside and Out: Phenomenology, Neuroscience, and the Nature of Experience*, Studies in Brain and Mind 6, DOI 10.1007/978-94-007-6001-1_4,

property of *being in motion* can be analyzed in terms of continuously changing one's spatial position relative to some background frame of reference. The property of *being a unicorn* can be analyzed in terms having a horse-like bodily configuration plus large wings and a white coat. And so forth. The great majority of our concepts are said to fall into this latter ontological category. The qualitative simples, by contrast, form a comparatively tiny elite, distinguished by their *not* being subject to any such definition or to any such decompositional analysis. We acquire these simple concepts – we learn the meaning of these simple predicates – by *ostension*, it is said, rather than by composition from concepts or predicates that we already command. This special semantic status, it is said, is further reflected in the special *epistemological* status enjoyed by the judgments that record the occurrence of these qualitative simples. We do not recognize an instance of redness, for example, by way of recognizing the peculiar configuration of some more basic features that collectively constitute a case of redness. We simply recognize a case of redness directly or immediately. We need not be *infallible* in apprehending instances of such qualitative simples, but our apprehension of them, whether in perception or in introspection, is decidedly non-inferential and deeply inarticulable. We cannot say *how* we recognize such a simple feature. We just can.

Altogether, the apparent *ontological simplicity*, the *semantic autonomy*, and the *epistemological immediacy* of this smallish family of qualitative properties might well suggest that we are here looking at an ontologically special family of features, features that enjoy a unique status among the elements of reality. Certainly many philosophers have been inclined to claim a special ontological status for them, based on the several considerations just explored, and on various thought-experiments that are supposed to draw out their metaphysical consequences. In particular, a number of contemporary philosophers have argued that these qualitative simples, at least in their internal incarnation as features of our conscious experiences, are forever immune to the sorts of reductive/explanatory assimilations frequently displayed in the physical sciences. All parties may agree that water is H_2O, that light is electromagnetic waves, and that stars are thermonuclear furnaces. *These* 'intertheoretic reductions,' at least, are well-established parts of human knowledge, and they have successfully relocated water, light, and stars (and a great many other commonsense things) within a conception of physical reality that is broader and deeper than the everyday conceptual framework that was their original and more modest home. But a worthy minority of our profession deems it profoundly unlikely that the qualitative simples at issue above, the internal ones anyway, will ever find a similar fate. By their very nature, it is claimed, they are immune to intertheoretic reduction in terms of the properties embraced by the physical sciences.

It is not difficult, perhaps, to appreciate their position here. After all (and speaking fairly roughly), to reduce any given property to something recognized by the physical sciences is to successfully *reconstruct* the peculiar (structural, causal, relational, functional) profile displayed by the target property at issue, in terms of the conceptual resources of the particular physical theory that aspires to achieve that reduction. To use the examples already cited, the property of being water has a broad and characteristic profile of causal properties, as does the property of being light, and of being a star. And these complex causal profiles are precisely what modern chemistry, electromagnetic theory, and gravitational and nuclear physics, respectively, have successfully reconstructed in such illuminating detail. But on the face of it, at least, the target properties at issue in the preceding paragraphs *have no* such characteristic profile for the aspirant reducing theory to even try to reconstruct. They are, after all, qualitative *simples*; their unanalyzable qualitative character is what is essential to their identity; and so they present themselves as smooth-walled mystery to the reconstructive ambitions of at least the physical sciences. There is simply nothing there, apparently, for those sciences to get a reconstructive grip on.

Over the past 40 years, considerations such as these have motivated a family of closely related arguments to the effect that the qualitative dimension of our internal conscious experience is forever beyond the reductive/explanatory reach of the physical sciences of the brain. Thomas Nagel was perhaps the first to pose the challenge by pointing out – quite plausibly, to judge by the paper's reception – that no matter how much one might know about the physical structure, operations, and states of the brain of a bat, one would still not know "what it would be like" to have the experiences of a bat (Nagel 1974). That is, objective information of the former kind, no matter how complete, would not suffice to specify the subjective *qualitative* character(s) of the bat's experiences.

Frank Jackson provided a similar argument focussed on an imaginary neuroscientist named Mary who was entirely colorblind or otherwise color-deprived from birth. Mary might come to know, he argued, everything there is to know about the physical operations of the brains of color-normal people, but, being colorblind herself, she would still fail to know what it is like to see the color red (Jackson 1982). Only if her colorblindness were somehow reversed could she gain access to relevant qualitative character. At about the same time, Joe Levine published an essay that described the apparently unbridgeable "explanatory gap" between the resources of the physical sciences and the peculiar character(s) of our subjective mental "qualia" (Levine 1983).

And David Chalmers subsequently produced a comprehensive book that celebrated these earlier arguments and added an argument or two of his own to underscore their collective point (Chalmers 1996). He, too, points to the "absence of an analysis" of any of the qualitative simples at issue, an absence he sees as diagnostic of their ontologically special nature. And he has us imagine a race of 'zombies,' creatures whose physical makeup (and physical behavior) is identical to ours – that is, they share with us *all* of the same physical/functional/causal/relational properties – but whose subjective *qualitative* mental life is simply absent. The fact that this scenario is at least conceivable, he argues, shows that what is *essential* to *our* internal qualitative states is something beyond what mere physical reality can hope to provide. Altogether, we have here a gathering consensus that the qualitative dimension of our conscious experience is something that the physical sciences, such as modern neuroscience, will never explain.

This conclusion, let us own at the outset, may be true. Conceivably, some form of Property Dualism will turn out to be the correct theory of mind, just as these authors severally suggest. And yet, one may want to pause here, to express amazement that such a spectacularly important factual claim should be legitimately established by arguments that arise entirely from the armchair, arguments based on preemptive and wholly *a priori* 'analyses' of the crucial properties involved, via considerations that are available to anyone who merely shares our *current* conceptual framework for comprehending conscious experience. Would that our theoretical understanding of some of the Universe's deepest mysteries were always so easily achieved.

My hesitation here, as many readers will appreciate, is not new. But in my earlier writings on this topic my impulse has always been to focus on either (1) the conditions actually required for successful intertheoretic reduction (Churchland 1985), a matter of some complexity and ongoing dispute even now, or on (2) the genuine virtues of the emerging neuroscientific accounts of human sensory experience (Churchland 2005), another unfamiliar matter of considerable complexity, or on (3) the history of science, and the presumptive lessons that past scientific episodes provide for the issues confronting us in the present case (Churchland 1996). All three approaches place serious demands on the scientifically marginal reader, and they may have been rather more opaque, to many, than I allowed myself to believe at the time. I take back nothing said in any of those papers, but on the present occasion I wish to take a more general and more philosophical approach to the anti-reductionist arguments at issue, in hopes of deflating their appeal in a more accessible manner. Those arguments not only run into trouble with the philosophy of science, with emerging neuroscience, and with the history of science generally. They lack integrity even by the standards of purely analytic philosophy. Or so I shall argue.

4.2 On the Determination of Essences

This undertaking is more difficult, more time-consuming, and altogether more entertaining than one might at first suppose. Let us agree that properties *have* essential features – following tradition, and without too much prejudice, we might call them 'defining' features or 'necessary' features. And let us agree also that it is a major part of the human cognitive adventure to discover the categorical structure of reality, to discover, that is, what *properties* the universe displays, and to discover what invariant or timeless relations unite and divide them. Learning about the world, after all, is not just a matter of determining which particulars happen to instantiate which properties, properties drawn from some antecedently settled population of universals. We, and all other cognitive creatures for that matter, have to *learn* the relevant properties or universals if we are ever to make contentful judgments about which particulars instantiate them. That is, we must *learn* the world's *general* features. We must learn the similarities and differences that collectively configure those diverse features. And we must learn the many nomic or causal relations that often unite them in prototypical temporal sequences. The result of such a learning process is a *conceptual framework*, a background 'map' of the timeless structure of the universe. With such a map in place, one's sense organs can help one to locate, at some appropriate place within that structured map of categorical possibilities, whatever particulars or processes one happens to encounter. One then knows, assuming that the background map is accurate, what to expect of those local particulars as one's experience of them unfolds.

Constructing a conceptual framework that is even roughly adequate to the demands of one's practical experience is a major accomplishment, and it does not happen overnight. Humans spend years – indeed, decades (indeed, millenniums) – developing conceptual frameworks that are adequate to comprehending and navigating new domains of ever-increasing empirical complexity. Individuals at different stages of this long developmental process will display quite different conceptions of the universe's abstract structure, culturally or individually idiosyncratic conceptions that are different in the breadth, in the depth, and in the accuracy with which they portray that objective categorical structure. Improvement in any of these three representational dimensions constitutes *conceptual progress*. Such progress is a supremely important kind of empirical learning. Indeed, it is the single most important kind of all, because it originally provides, and continually changes, the very concepts we attempt to apply in any and all of our singular judgments. Accordingly, trying to determine the essences of things is not the obscure and occasional indulgence of cloistered philosophers. Rather, it is the basic aim of the learning process in all cognitive creatures, and it is the basic aim of the empirical sciences generally.[1]

The essential character of our *conscious mental states* has become the focus of much interest and theoretical speculation in recent centuries. In a critical response to the impressive but alarming development of the "mechanical philosophy" in the Seventeenth Century, Descartes outlined a form of Substance Dualism which claimed, for *res cogitans* ("thinking stuff"), an essence or ontological status forever distinct from that of *res extensa* ("spatially extended stuff"). Notably, Descartes' argument at the time *also* used a thought-experiment, concerning what he could and could not *imagine*, in order to divine the presumed essence in question. Curiously, his argument was just the reverse of Chalmer's zombie exercise: Descartes thought he could successfully imagine *himself* without any of his *physical* features, and concluded that those physical features were thus not a part of his essential nature.

But by the time we reached the first half of the twentieth century, physicalism was once again asserting itself, and very powerfully, into the domain of the mental. For example, in a reaction against Descartes, the Logical Behaviorism of Ryle and Wittgenstein had begun to sweep through most of the philosophical profession, insisting that the essence of any given kind of mental state was simply the unique profile of observable-physical-circumstances-as-input/observable-physical-behavior-as-output relations that possession of the relevant state implied. The 'ghost within the machine' was thereby exorcised as an unnecessary ontological extravagance.

---

[1]What this most fundamental learning process consists in, at the neurobiological level, is explored in some detail in Churchland (2012).

But with the ghost went all of our internal states as well, apparently, and their appropriately internal qualitative properties with them. Despite the background ideological pressure of the then-dominant Logical Positivism – with its emphasis on the epistemological and semantic primacy of *objective* observables – this blanket exorcism was persistently difficult to accept, even for us anti-dualists. Fortunately, Functionalism emerged in the 1960s, thanks to philosophers such as Putnam and Fodor, apparently to save the day. For these philosophers insisted that internal mental states were perfectly acceptable. They needed not to be exorcised from our ontology, but to be properly knit within it. They could be welcomed back into our ontology, it was argued, by acknowledging their role as *causal intermediaries* between sensory inputs and behavioral outputs, intermediaries with complex and mutually embracing causal profiles of their own, unique profiles that constituted the essential nature of each distinct kind of internal mental state. Those characteristic causal profiles might be realized in diverse physical substrates, cautioned the Functionalists, but those possibly diverse substrates were not what was important. What was essential to the shared essence of these internal mental states was not the metaphysically simple qualities they displayed, nor the underlying medium in which they were realized – it was their shared causal/functional profile. It hardly mattered whether those profiles were realized in neural brain-stuff, in Cartesian mind-stuff, or in electronic computer-stuff, although Functionalists typically claimed that the empirical evidence was converging massively in favor of the first alternative, massively *against* the second alternative, and that the third alternative was growing as a future technological possibility.

What is noteworthy here is that once again we find genuinely gifted philosophers taking a close look at the domain of the mental, as that domain is currently comprehended within our existing conceptual framework, and then announcing, on the basis of that arm-chair examination, that the *essential* features of the elements of that domain are . . . (place your favorite ontological prejudices here). No substantive experiments are cited to sustain the analysis. No empirical theories are proposed or evaluated. And yet the 'analysis' here proposed was advanced with considerable confidence and authority, *despite* the fact that it stands in stark opposition to the 'analysis' proposed by Descartes, who possessed the same conceptual framework for the relevant internal domain that we do, but whose take on what were and were not its essential features was diametrically opposed to that of the Functionalists.

That philosophers can disagree is not news. But this is ridiculous. A possible explanation for this situation is that both forms of analysis are actually correct, but they comprehend *distinct dimensions* of our inner lives – the subjective/conscious/qualitative dimension in the one case, and the objective/causal/functional dimension in the other. This is in fact the line taken by the contemporary tradition that runs from Nagel through Chalmers discussed earlier, and it has at least a *prima facie* appeal. Not just because it resolves an awkward dilemma, but because the ontological division it proposes is antecedently plausible, at least to some. This line also constitutes, note well, yet a *third* 'analysis' of the structure of essential features within the domain of the mental, one no less born of the armchair than were the first two. It just bifurcates the elements within that domain, in ways that neither Descartes and Putnam were apparently willing to embrace. If one's confidence in the armchair route to understanding the mental were already growing thin, how likely is it to recover in the face of yet another such analysis, one that simply pastes together two antecedent failures?

Still, and its armchair origins aside, this third analysis might be correct. For the sake of argument, let us suppose that it is. What would follow about the real nature of our mental states? Absolutely nothing. For we still have to address the question of whether or not our current conception of the domain of mental states is an *accurate* or *faithful* portrayal of the actual elements and the real nature of that domain. Even if we have finally gotten it straight what our current conceptual convictions and commitments *are*, it remains a separate question whether those convictions and commitments are *correct*. After all, we humans have repeatedly been forced, by developments in the natural sciences, to *reconceive* a variety of things that were and remain central to our dealings with the world. We used to think that the Earth was essentially motionless: indeed, it was thought to be the essential background bedrock or reference frame against which all genuine motions had to be reckoned in the first place. But it isn't. We used to think that Light was essentially that-which-made-things-visible. But the vast majority of kinds of light – i.e., all wavelengths outside the tiny 'optical window' – do no such thing, at least for humans. And even within that tiny window, making environmental information available to terrestrial creatures is an extremely peripheral feature of light, hardly its essence.

We used to think that the essence of Life was some kind of Soul or Vital Spirit. But it isn't. We used to think, without question, that Mass and Length were simple, one-place properties. But they both turned out to be more penetratingly and accurately reconstructed as *two*-place predicates, denoting a variable relation between a thing and a variety of reference-frames. And so on. Evidently, being a constituting element of one's current conceptual framework is hardly a guarantee of genuine essentiality, or even of bare truth, come to that.

At this point one might anticipate that I am going to argue that our current conceptual framework for mental states is defective in some major way. It might be, and I have so argued in the past, at least where the propositional attitudes are concerned (Churchland 1981). But that is not my purpose in this essay. The focus of the authors cited earlier is on the ontological status of mental states with a distinctive *qualitative* character, and that will be my focus also. My aim is to eviscerate their arguments that the qualitative characters of these states are forever beyond an explanatory reduction in terms of the physical dimensions of brain activity.

4.3 Subjective Knowledge Versus Objective Knowledge

I begin with Nagel's original argument, which leaned so hard on the distinction between subjective knowledge and objective knowledge. The basic idea was that the *objects* of these two kinds of knowledge, respectively, are completely disjoint and mutually exclusive, as are the two kinds of knowledge themselves. Accordingly, since the knowledge supplied by the physical sciences is always (and essentially?) objective, science can never give us subjective knowledge, nor, therefore, knowledge of its typical objects, namely, subjective qualitative characters.

Arguing for fundamental ontological distinctions on the basis of the idiosyncratic, historically relative, and changeable profiles of our supposed *knowledge* of those ontological categories is a dubious undertaking on its face, especially when, as in the present case, our knowledge of those categories is relatively paltry. But this is a general complaint, and I wish to register a highly specific objection to Nagel's prima facie compelling argument. *The typical objects of the two forms of knowledge at issue are not at all disjoint and mutually exclusive, even by the standards of current common sense.*

To begin, a great many objective and plainly physical facts and features are made directly available to oneself by introspection. The current configuration of one's body, for example, is something of which one is directly and continuously aware, such as being in a sprinter's starting crouch, or being seated, or having one's arms folded in front of one. The (wholly physical) state of contraction and tension of every muscle in the body is made continuously available to the brain by the body's *proprioceptive* system. (And a good thing, too, if the brain is to exercise continuous control over an ever-changing bodily configuration.) One knows one's own bodily configuration in a way that no one else can or ever will. Those others have such subjective access to *their own* bodily configurations; but not to yours.

Similarly, even with a constant bodily configuration, one knows directly if one is being gently rotated (as in a barber's chair) or gently rocked to the left and right (as on a large ferry boat), even if one's eyes are closed and one's tactile senses are disabled. The *vestibular* apparatus of the inner ear (the innervated semicircular canals) provides the brain with a super-sensitive monitor of any rotational changes in the head's position. One is aware of such changes in one's own case in a way that no one else is. Others may see you rotate in various ways, but they will never access your rotation as you access it. And yet, and as with static bodily configurations, such rotations are as objective and as physical as can be.

To continue, one knows that one's stomach is full, or that one's bladder is full, in ways that no one else can know it. Yet these are objective facts about physical things. As well, one knows when one's muscles are seriously overtired, as after protracted stressful work (they are then awash in lactic acid, a chemical byproduct of biological energy use) in a way that no one else knows their weary condition. And one knows, with arresting introspection, when that condition occasionally produces a 'cramp,' a spontaneous maximal contraction that freezes a given muscle into excruciating immobility. One knows that one's sinus cavities are swollen, as with the common cold, in a way that is difficult to articulate, but unambiguous even so. And no one else will know their swollen state as you do.

These examples can be multiplied indefinitely. They are all instances of what psychologists and neuroscientists call *interoception*. Evidently, one knows a great deal about what is going on inside oneself, and knows it in ways forever denied to creatures not identical with oneself, even though those introspectively accessible states and activities are ostentatiously physical. Introspective apprehension, accordingly, is clearly *not* confined to states of a non-physical nature, let alone to states that are 'qualitative simples'. It includes physical facts of substantial complexity, although that complexity is often only partially or only dimly apprehended, depending on the particular physical state involved. Indeed, the degree to which one spontaneously appreciates the complexities that may be involved in these internal states can *vary* as a function of how much one has *learned* about the relevant kinds of physical states.

For example, a young infant's proprioceptive apprehension of its own bodily configuration is presumably of much lower 'resolution' than that of an older child's, save perhaps where the mouth, lips, and tongue are concerned, elements of the infant's body that are already hard at work at feeding time. Similarly, the proprioceptive apprehension, by a pianist, guitarist, or harpist, of his complex hand-and-finger positions while playing his instrument is markedly superior to the same apprehension by one untrained in musical performance. A skilled typist shows the same advantage over a non-typist. And in the case of a professional ear-nose-and-throat doctor, her interoceptive appreciation of the details of her sinus infection (e.g., which of the several cavities is affected, and how) will be rather greater than in the case of a naïve youngster. In sum, there simply is no dividing-line that excludes physical facts from the domain of introspectively accessible facts. That domain includes a multitude of physical facts and physical things. And *how far* these undoubtedly physical facts intrude into that 'favored' epistemological domain *varies* both with time and with increasing knowledge.

We can illustrate, experimentally, this sort of epistemological intrusion without your even having to put aside the page you are now reading. When one reads black text on a white background, one's visual system is chronically fixed on some one or other horizontal line of *text*, rather than on the empty (white) horizontal spaces that separate those successive lines of dark graphical elements. Your eyes move left and right and up and down, to be sure, but while reading text, your visual field contains a roughly constant grid of white and dark horizontal lines. The white lines, being vertically fixed in your visual field, and being brighter than the lines of text, produce a form of fatigue or adaptation in the visual neurons that code what you are seeing, a fatigue that is confined to the neurons that chronically code for those white lines. The result is that, when you suddenly shift your vision to a surface of *uniform* brightness (such as the empty margin at the bottom of this page), the fatigued neurons all fail to respond normally to the relevant parts of the now-uniform surface. They represent those parts as being darker than they really are. The result, subjectively speaking, is an *after image* of light and dark horizontal lines, an image that is brightness-inverted relative to the original page of text. To see this vividly for yourself, simply fixate rigidly on some word in the middle of this line for 10 s or so (count slowly), and then relocate your gaze on the empty page-margin below. The after-image will be obvious, although it will begin to fade within a second or two as the relevant neurons begin to recover from their induced fatigue.

What you are then noticing, perhaps for the first time, is the fatigued or energy-depleted state of a specific subset of the neurons in your visual system, an entirely physical condition. You know your condition subjectively, that is, in a way that makes it evident to you, but to no one else. After all, it is your after-image. And you would probably never have known of your neuronally fatigued condition, but for the physical description I gave you, and the physical instructions that went with it. But in principle, this case is no different from the familiar case of knowing, introspectively, that your *muscles* are fatigued.

In fact, such visually evident fatigue-patterns are a constant intrusion into our visual experience, but the brain tends, quite rightly, to ignore them or look past them as regrettable noise. We are mostly unaware of them. But once they have been pointed out to you, you start to notice them almost everywhere, especially in cases where one's external visual experience involves sharp brightness-contrasts.

Subjective knowledge, then, is *not* confined to some ontologically special class of nonphysical sensations. It regularly concerns the condition of one's stomach, one's viscera, one's muscles, one's visual nervous system, one's sinuses, one's overall skeletal configuration, and one's bodily motions – physical conditions all.

Still, it will be said, subjective knowledge itself remains distinct from the various forms of objective knowledge, even if their typical objects frequently overlap. And those epistemological objects that are knowable *only* subjectively, if there are any, might yet form an ontologically special class of things. That is, if there is a domain of phenomena that *cannot* be known by any objective means, then perhaps the fortunes of nonphysical qualia might be worth betting on after all.

Sensations themselves and their qualitative characters (as opposed to the physical conditions that they frequently signal) are the preferred candidates for this supposed role. These things, it is often claimed, are known *only* subjectively, never objectively. But this claim is false on its face. I have systematic and ongoing knowledge of the sensational states of the people near and dear to me – of their pains, their hunger, their anxieties, the warmth they enjoy or the cold they endure, the tastes they encounter (and like or dislike), even (as we saw) the detailed quality of the visual after-images they may have. To be sure, I do not know these things in the way that they do, but I certainly have knowledge of these things. That same knowledge governs much of my daily behavior. I can even *explain*, in neurophysiological terms, some of the more interesting facts about at least some of their subjective lives. Short of a blanket skepticism about our knowledge of Other Minds, then, we seem once again denied any uniform essence that would mark off the domain of sensations as ontologically distinct in some way. One knows about physical conditions both objectively and subjectively. And one knows about phenomenal conditions both objectively and subjectively. So far, no dividing essence has emerged.

But we have not yet addressed the most salient and the most widely cited element of Nagel's overall argument, the element that motivated his paper's title. Once again, it involves a thought experiment, but an admittedly compelling one. He asks you to imagine that you have somehow come to know *everything* about the physical nature and the physical activities of a bat's brain. However, and despite your exhaustive command of the physical details of bat cognition, you still wouldn't know *what it is like* to *be* a bat-style cognizer. You still wouldn't have the *subjective* knowledge of the bat's cognitive and sensory life that the bat himself has. From this, he concludes that there must be something missing – something real and something important –

from the purely physical story that you have learned. Knowing the complete physical theory of bat-style cognition wouldn't *make you* a bat-style cognizer.

Indeed it wouldn't. What is required to make you a bat-style cognizer – to make you enjoy the special dimensions of a bat's subjective cognitive activity here at issue – is that the complete physical theory of bat-style cognition *be true of you.* (Which, of course, it isn't.) Whether or not you happen to *know* that theory is utterly irrelevant to whether or not you actually have bat-style cognitive states. The natural-born bat doesn't have any inkling of that theory either (even though it is true of him), and yet he clearly doesn't need it to enjoy the subjective dimensions of his own existence. And you may know the physical theory in exhaustive detail, as in Nagel's thought-experiment, but that wouldn't give you what the bat has. Simply knowing the theory doesn't make the theory *true* of you, not in this case, and not in any other case either.

To cite parallel examples, having complete knowledge of the physical nature of superconductivity doesn't make you a superconductor. (That would require that the theory of superconductivity be true of you.) Having complete knowledge of the physical nature of pregnancy doesn't make you pregnant. (That would require that the theory of pregnancy currently be true of you.) Having complete knowledge of the physical nature of diabetes doesn't make you a diabetic. (That would require that the theory of diabetes currently be true of you.) The fallacy involved in these parallel cases is immediately obvious. Why wasn't the fallacy in the case of bat-cognition similarly obvious? Because the bat-case concerned your gaining, or rather, *failing* to gain, a certain form of *knowledge*, in a circumstance where your scientific/physical/objective knowledge of bat-style cognition was supposedly complete. The failure here, accordingly, looked like a failure in the reach of that scientific/physical/objective knowledge, at least where subjective phenomena are concerned. But it isn't a failure of that kind at all. The proper *test* of that scientific/physical/objective theory of bat-style cognition is whether, when that theory happens to be genuinely *true of* some given creature, then the creature actually *has* the subjective experiences of a bat. And nothing in Nagel's paper suggests, even for a second, that a complete theory of bat neurophysiology would fail *this* test.

In sum, Nagel is implicitly demanding or expecting that mere possession of a certain body of *theoretical* knowledge should *constitute* (as opposed to describe or explain) a quite *distinct form* of knowledge: bat-style subjective cognition. But there is not the remotest reason to expect any such thing, and no ontological lessons to be drawn from the utter failure of the neurophysiological account to actually *provide one with* bat-style cognition, subjective *or* objective. As is illustrated in the parallel cases just listed, that expectation is wholly unreasonable in the first place, and its failure to be fulfilled is entirely without significance for the adequacy of the particular theory involved. Most especially, the 'failure,' in every case, is without any *ontological* significance for anything.

These same considerations undermine Jackson's closely similar argument concerning the effect (or the lack of it) of possessing complete physical knowledge (of the brain activities of color-normal people) on the visual experiences of color-blind

Mary. Mary, you will recall, was supposed to be neuroscientifically omniscient, but despite this distinction, she *still* didn't know what it was like to have the subjective visual experience of red. As it was often argued, "She knows all of the physical facts, but there is still something she does *not* know; so there must be some *non*physical facts."

But here again, Jackson is expecting, quite wrongly, that one form of knowledge should *constitute* a quite different form of knowledge. He is expecting that explicit/discursive/scientific knowledge should somehow *constitute* subjective knowledge of visual experiences. But that expectation is unreasonable on its face. We might as well expect that your exhaustive discursive knowledge of the micro-organization of a professional golfer's motor cortex would constitute actual practical 'knowledge,' on your part, of how to hit a golf ball 200 yards down the middle of the fairway, even if you have never swung a golf club before in your life.

This last analogy gives a specific voice to a classic objection to Jackson's original argument, namely, that it is formally invalid by reason of equivocating on the term "knows." Nemirow (1980) and Lewis (1983) pointed out that the first occurrence of the term "knows," in the brief argument quoted in the preceding paragraph, denotes explicit or discursive knowledge, whereas the second occurrence of the term concerns a suite of cognitive *abilities* (to recognize red visually, to imagine red, to remember red, etc.). Given this equivocation, the case for 'nonphysical facts' evaporates.

One may indeed wonder exactly how to understand or analyze the distinct nature of 'knowing' in the subjective case, and in the intervening years much space in the philosophy journals has been spent pursuing that question. But even in advance of a settled analysis, it is plain that we are here looking at two distinct kinds of knowledge. Having discursive scientific knowledge of anything requires having a *language*. Knowing what it is like to see red requires nothing of the sort. As well, some three decades later, it now seems that the most promising place to *find* a discursive account of what-it-is-to-know-the-colors-subjectively lies in the emerging *science* of how biological brains actually represent the domain of colors, and how the visual system activates specific representations therein. Such neuronal accounts already exist, and they do not derive their plausibility from the armchair. More on such accounts anon.

4.4 Back to the Sensations Themselves

If our sensations and their properties are the true claimants to some special ontological distinction, then perhaps we should focus on *them* in order to reveal that distinction, rather than on the not-so-distinctive profile of how we happen to know them. This is the line that Chalmers takes, and we need now to examine his rather different approach to our question.

Chalmers also focuses our attention on the *intrinsic qualitative characters* of our sundry sensations, in contrast to the causal/relational/functional profiles that those same sensations may also display. That latter dimension of sensational reality, he is happy to concede, is extremely important for our ongoing explanations of human behavior, and it may well find a successful and exhaustive reductive explanation in terms of physical neuroscience. Indeed, he positively expects this to happen. But the qualitative features of our sensations are a different matter, according to him. Those features are ontological *simples*, he avers, and for that reason, they offer nothing in the way of internal structure that a physicalist theory of the brain might hope to *re*construct, as a successful intertheoretic reduction would require. The qualitative character of my visual sensation-of-red, for example, simply confronts me, or so it seems. It does so in a manner quite distinct from any alternative sensation-of-color, but not because it invites any hope of some signature 'decomposition' into anything else at all, let alone into something physical. Such *qualia*, as they have come to be called, should therefore be counted as something outside the physical order, as something beyond the causal/functional profile in which our sensations are admittedly embedded. Chalmers' dualistic conclusion here is thus one instance of the long-familiar position called *epiphenomenalism*, although he prefers the expression *naturalistic dualism*.

We may open our examination of this argument by noting that the bulk of one's sensational life is characterized, not by simplicity, but by an extraordinary and ever-changing *complexity*. Listening to a conversation, looking around a flower garden, tasting a braised-lamb stew, smelling the aromas in a wood-working shop – our sensations in such cases display intricacies that are amazing. And not always obvious. A young child may not appreciate that the distinctive taste of her first ice-cream cone *resolves* itself into sensations of sweetness, creaminess, and strawberry. And it may take her awhile to learn that such decompositions are both common and useful to keep track of. For the complexities we encounter are indeed composed, quite often, of simpler elements or constituting dimensions. In time, we do learn many of those simpler dimensions. A dinner-table conversation contains my brother's unique voice as an identifiable element; the complex flower-garden displays the striking orange of a typical poppy blossom; the lamb stew displays the distinctive taste of thyme, sprinkled into the mix at the outset; and the smell of yellow cedar stands out from the other smells in the wood shop, at least to a seasoned carpenter. Each of these particular qualitative features of one's inner phenomenological life is certainly a *simpler* dimension of a more complex whole.

But is each of these examples, or any of them, itself an ultimate, undecomposable simple? Perhaps. But how does one tell? I may indeed be *unable* to specify any sub-dimensions whose peculiar concatenation constitutes the sound of my brother's voice, or the poppy's visual orange, or the taste of thyme, or the smell of yellow cedar. But neither could the still-learning child specify, at least at the outset, the taste of sweetness, the taste of creaminess, and the taste of strawberry-ness as constituting sub-dimensions of her taste of the ice-cream cone, even though those elements were undoubtedly there, and even though she subsequently came to appreciate them. How, then, do I know when I have genuinely 'hit bottom' in a given case, as opposed to merely having reached the current limits of my capacity to articulate *how* I manage to discriminate the qualitative feature at issue?

This question has a certain bite to it in the present context, because hitting at least a *local*, *current*, or *apparent* 'bottom,' note well, is absolutely inevitable on *both* a dualist *and* a materialist account of how we discriminate the qualitative characters of our sensations. The only alternative to an apparent or presumptive 'bottom,' *some*where or other, is an infinite sequence of qualities discriminated via a recognizable concatenation of simpler qualities, each of which is discriminated via a recognizable concatenation of still-simpler sub-qualities, where each of those is discriminated in turn via a recognizable concatenation of still-simpler sub-sub-qualities, and so on without end. Qualitative characters that are at least *apparent* simples are thus utterly inevitable on *both* approaches to understanding the mind, dualist and materialist.[2] Their undoubted existence, accordingly, implies nothing one way or the other about their underlying ontological character, despite a widespread presumption that it speaks, somehow or in some degree, in favor of dualism. It doesn't. Every cognitive creature, even in an *exhaustively physical* universe, must display a current limit on how far it can decompose the qualities it can discriminate. Let us not be too impressed in our own case, then, by the mere existence of apparent 'qualitative simples,' however robust their apparent simplicity. Their existence is entailed by *both* of the philosophical approaches here at issue. Such 'simples' simply have to be there, if only to mark the limits of our current understanding. Their existence is not only *consistent* with both of the ontological positions at stake here; it is positively *entailed* by both of them. Accordingly, to infer the ontological simplicity of a given qualitative character from its *apparent* simplicity is to commit the fallacy of Arguing from Ignorance, as in, "I am *unaware* of any constituting elements in this qualitative feature, therefore, there *aren't any* constituting elements."

This *a priori* point looms larger when we reflect on the fact that the domain of *external, objective* things and properties displays exactly the same contrast between complex, decomposable features and (apparently) simple features as is found in the subjective realm. The *objective* red of an apple and the *objective* temperature of warm water, for example, are also *apparent* simples, ontologically; they are without definitional analysis, semantically; and they are 'immediately' accessible, epistemologically. But no one since the eighteenth century supposes that such objective perceptual properties are thereby revealed as genuine ontological simples. The physical sciences of objective color, temperature, sound, and so forth have provided us with decisive analyses of the underlying ontological *complexities* that constitute the (objective!) perceptual qualities here at issue. Those qualities are

[2]This important fact is evident even in the case of Chalmers' 'zombies.' On his own hypothesis, the zombies behave, speak, and argue exactly as we do, and therefore encounter the same decompositional limitations that we do when addressing their own inner states. They, too, *despite being purely physical*, confront what they, too, describe as a family of 'qualitative simples,' and they are no less puzzled by them than we are. Indeed, if they embrace Chalmers' line of argument (and it will be exactly as plausible to them as it is to us), they will end up believing that they, too, have nonphysical qualia, when, *ex hypothesis*, they *don't*. But if *their* argument for that conclusion is manifestly unsound, why is *our* argument for that conclusion any better?

entirely real. But they are also entirely physical and more than a little complex. It just took the physical sciences awhile to learn about their constituting elements. Our *internal* phenomenological qualities may be awaiting a precisely similar fate.

Indeed, the waiting period seems already to be over. But I will return to the matter of the emerging Neuroscience of qualitative states in a few pages. Let us here focus on the earlier and quite independent complaint that nothing of any ontological significance *follows from* either the epistemological opacity of our current sensational discriminations, or from the semantic/analytical simplicity of those qualities as judged by the lights of our current conceptual framework. As we saw, the claim that the subjective qualitative characters at issue are ontological simples is evidently not the outcome of a sound demonstrative argument based on either or both of these two premises. Rather, that ontological claim now looks more like an *explanatory philosophical hypothesis* whose hope is to provide a uniquely compelling explanation of those two premises. After all, if our conscious qualia really *are* ontological simples, wouldn't you expect that our discrimination of them, one from another, would be inarticulable? And wouldn't you expect that our concepts of them would be without internal structure?

Perhaps so, but we must remind ourselves that we can already point to independent explanations of both premises, explanations that do not engage in weakly-motivated ontological profligacy. Moreover, if the postulation of ontologically simple supra-physical qualia is to purchase its plausibility by means of its comparative explanatory virtues, as the above interpretation suggests, then that postulation must be prepared to have its own explanatory virtues explored and evaluated in some detail. To that end, let us look into its actual performance.

4.5 The Explanatory Performance of Epiphenomenal Qualia

Exactly *what are* the phenomena that the postulation of epiphenomenal qualia is supposed to explain? It is hard to see what they might be, for the simple reason that the postulated qualitative simples at issue are held to be *epi*phenomena, phenomena that are caused *by* physical phenomena in the brain, but *which have no causal properties of their own* – not within the physical realm, and not among each other either. They are held to be causally inert: a dynamically impotent sideshow, continuously reflecting the brain's activity, to be sure, but with absolutely no causal effects of their own.

How, then, can they possibly provide systematic *explanations* of anything at all? On the epiphenomenalist's own hypothesis, qualia are precluded from explaining anything about our bodily behavior: that must be done by appealing to the facts about our physical environment and its interactions with our brains. They are precluded from explaining anything about the behavior of our brains themselves: that job is exhausted by the physical neurosciences. And finally, they are precluded from explaining anything about *each other*: that job is exhausted by the idiosyncratic physical activities of each person's brain. Epiphenomenal qualia have no causal effects on one another, nor, indeed, on anything whatever. On the face of it, then, they are explanatorily impotent.

"Well, no," it will be objected, "for they do explain the *existence* of consciousness. Collectively, the complex flux of your epiphenomenal qualia *constitutes* your ongoing consciousness. Without that supra-physical flux, there would be no genuine consciousness. There might be the purely 'functional' form of consciousness displayed by Chalmers' zombies during their 'waking' hours, but there would be no *qualitative* consciousness."

This is the core claim of epiphenomenalism. But there remains a stubborn problem. Indeed, there are at least two of them. The qualitative features at issue cannot constitute someone's consciousness unless they are somehow *apprehended* by that someone, unless their local instantiations are *detected*, *noticed*, *registered*, or *recognized* by that someone. But on the epiphenomenalist's own story, to state the first problem, those qualitative features are wholly without impact or causal effect of any kind on anything. In what, then, does their *apprehension* consist?

And to state the second problem, there is no 'someone' there to do the apprehending or conceptualizing in any case. The epiphenomenalist explicitly eschews any form of substance dualism, and, *ex hypothesi*, the qualitative features at issue can have no causal effects on the physical brain. Who and/or what, then, is 'home' to host, enjoy, or somehow *respond* to this qualitative 'show'? Evidently, no one and/or nothing. To be sure, the physical brain, or some part of it, is the true *subject* of each proposed qualitative feature, on the epiphenomenalist's own account. They are supposed to be supra-physical features of the *brain*. But on that same account, the brain itself is supposed to be totally and eternally blind to the occurrence (and to the *non*-occurrence as well: recall Chalmers' zombies) of any and all such supra-physical features. The price of epiphenomenalism, apparently, is the absence of any*thing* to be aware *of* the supra-physical features that the position itself proposes. Accordingly, those qualitative features themselves disappear from the causal matrix of the world in general, forever undetectable by anything, into an inaccessible metaphysical vacuum, where, beyond merely existing, they do precisely nothing, even to each other.

As an explanation of consciousness, this is a train wreck. Aside from failing to provide *any* positive explanations concerning the qualitative contents of consciousness and their causal role(s) in our cognitive economy and our physical behavior, and aside from leaving it an absolute mystery what these 'ontological simples' are and why they should exist at all, epiphenomenalism is flatly inconsistent with the core conviction of our common-sense conception of mental phenomena, namely, the conviction that our conscious mental states are *causally involved* in the unfolding drama of our conscious mental lives, and *causally responsible* for the unfolding physical behaviors to which it continuously gives rise. The point being made here is that the epiphenomenalist's claim to be faithful to our antecedent conception of our mental states is a five-star fraud to begin with. The allegedly fundamental division the epiphenomenalist draws between our conception of the causal/relational/functional aspects of our inner states, on the one hand, and our conception of the qualitative/introspectible aspects of those states on the other, is not a mutually-exclusive division that our common-sense conceptual framework respects at all. On the contrary, commonsense ascribes *both* kinds of aspects/properties to one and the same internal states, and it portrays their qualitative characters as an integrated *part* of the avowedly causal activities in which those states participate.

To illustrate this point, the state of *pain* is perhaps the first of many hundreds of examples that jump to mind. If a pain is strong enough to register in one's consciousness in the first place, then the familiar and unwelcome *qualitative* character that it displays will prompt one's attention to its possible causes, provoke aversion to its presence, kindle practical reasonings aimed at relieving it, distract one from one's antecedent activities, occasion regret at whatever you did to run afoul of it, and ultimately drive behaviors that one hopes will make it go away. The qualitative character of your pain is not a disconnected bystander to this modest explosion of causal consequences: it is typically what ignites them all in the first place. That is to say, as our current Folk Psychology conceives of things, the qualitative character of pains is a fully integrated *part* of the dynamical profile that pains typically display, not a causally impotent bystander to a causal process that, strictly speaking, does not include it.

The case of pain is typical. The qualitative characters of *all* of our sensational and emotional states are causally potent elements in the dynamical profiles of each of those states. The dynamical profiles vary, of course, across the wide range of such sensational and emotional states, but those diverse causal profiles are just a further reflection of the diverse *qualitative* characters that give rise to them.

The situation here, *within* the narrow dynamical domain of human and animal cognition, is not different from the situation within the much larger dynamical domain of the physical world at large. As we noted, that larger domain *also* displays a great many *qualitative* features such as the objective *pitch* of a sound, the objective *warmth* of the air in an oven, the objective *redness* of a ripe strawberry, and so on at considerable length. (Throughout history, these, too, have often been thought to be 'ontological simples.') And these qualitative features are *also* causally integrated elements in the dynamical profiles that the objective physical world displays. The pitch of a sound – a middle A, or 440 Hz, for example – is causally related to many things: to the wavelength λ of that sound, for one, via the equation $\lambda = \nu/\omega$, where ω is the pitch and ν is the velocity of sound. (The wavelength of that sound must therefore be 340 m/s divided by 440 Hz = .773 m. Change the pitch – the qualitative feature at issue – and you will thereby cause the wavelength to change.)

The warmth of the air in an oven – 300 °F, say – is also causally related to many things: to the fact that a cup of water will eventually come to the boil if placed in that oven, for example. The redness of a ripe strawberry – which has an overall electromagnetic reflectance peak at around .63 μm – will have characteristic causal effects on a spectrometer, and on the angle at which reflected light will be refracted through a prism, and (of course) on the human eye itself. These *external* qualitative sensory characters, familiar to us all, are certainly not causally impotent, supra-physical epiphenomena. On the contrary, they are an integrated part of the world's causal structure, they and thousands of other robustly qualitative objective features as well. And we can see *how* and *why* they are thus integrated when we finally appreciate how they are constituted within the underlying ontological complexities of the physical world.

Why should we think that their inner analogs – the qualitative features of our own conscious states – are any different? Why should the states of the physical brain and nervous system, which even Chalmers agrees are characterized by the causal/functional profiles here at issue – *not* have qualitative features that are just as causally integrated within the relevant dynamical profiles as are their manifold external brethren? Why should being located *inside* the skin introduce such an enormous ontological contrast with qualitative states that are located *outside* the skin? What *motivates* this lack of parity in one's construal of these two classes of qualitative features, especially when it flies in the teeth of the evident convictions of common sense, and of the daily explanatory practices that they make possible for all of us?

This presumptive parity between the semantic, ontological, epistemological, and causal status of the qualitative features of both our inner states and the world's many outer states finds a further parallel when we look at the business of *explaining* their various phenomenological characters and causal profiles in terms of the underlying physical reality that Physics, Chemistry, and Biology have been slowly revealing to us. We all know that the pitch of a sound is the oscillatory *frequency* of a compression wave in the atmosphere. We all know that the temperature of the air in an oven is the mean kinetic *energy* of the molecules that make up the air. We all know that the redness of a ripe strawberry is a peculiar *reflectance-efficiency profile* that leans strongly toward the long wavelengths within the optical range of the electromagnetic spectrum. These 'outer' qualitative characters, and thousands more besides, have all found highly revealing reductive explanations from the relevant sciences, explanations that positively *account* for their causal/functional integration with the rest of the world.

Moreover, those same explanations also account for the *structure* of the mutual *similarity-and-difference* relations among the diverse *qualitative* features within a given qualitative domain. Thus, different *pitches* and different *temperatures* are each arrayed on a one-dimensional similarity continuum, as befits features that vary in only one dimension. (Namely, oscillatory frequency, and mean molecular kinetic energy, respectively.) And different *colors* are arrayed within a three-dimensional similarity space, as befits a feature that varies in three significant dimensions. (Namely, the spectral *location* of its global reflectance peak (its hue), the degree of *concentration* of that reflectance peak (its saturation), and the overall *area* under its energy-reflectance profile (its brightness).)[3] Here we see the underlying physical theories providing systematic explanations of central *qualitative* facts concerning the qualitative features themselves, and not just of their causal/functional profiles.

Finally, and perhaps most importantly, note that the explanations that science now provides for the external, objective qualitative features discussed above reveal that

[3]For an accessible account of the underlying nature of objective colors, see Churchland (2007).

they are not ontological simples at all, despite a fairly convincing first impression. The oscillatory frequency of a compression-wave train in the atmosphere is a modestly *complex* phenomenon. So is the mean of the kinetic energies of the millions of ballistic molecules that make up any gas. And so is the 3-dimensional configuration of the relevant three aspects of the strawberry's electromagnetic reflectance profile. Our native sensory organs are *causally sensitive* to these complex properties, to be sure, which is why we can detect pitch, warmth, and color so reliably, but neither our sensory organs nor our brains have any initial cognitive inkling of the ontological complexities that constitute them. That difficult matter is for the relevant sciences to address and reveal, not for our unaided mechanisms of bare discrimination. And so, in our uninstructed ignorance, we are naturally but wrongly tempted to construe these several properties – the pitches, the temperatures, and the colors – as qualitative and ontological simples, even though they are nothing of the sort.

These external cases provide a clear lesson for addressing our focal case of *internal* qualitative characters. The qualia of our inner states are *also* spontaneously discriminable, one from another, by our native interoceptive mechanisms, whatever those mechanisms might happen to be. Not surprisingly, those native, internal discriminatory mechanisms are *also* cognitively blind to whatever ontological complexities might happen to underlie those internal qualitative characters, and cognitively blind to how those complexities might play a causal role in the discriminations at issue. Just as we found in the outer case. But here also, this does not mean, not for a second, that such underlying ontological complexities are *not there*. Indeed, given that the brain is more complex, by far, than a compression-wave train, or an oven full of gas, or a light-reflecting surface, we should positively *expect* that its internal states will possess extraordinary ontological and causal complexities, complexities *that are initially opaque to our native discriminatory mechanisms and cognitive comprehension*. Those internal states may be spontaneously discriminable by us, one from another, but finding out exactly *what* it is that is being discriminated, and *how*, is a job for the *sciences* of the brain, in strict parallel to the external cases discussed in the preceding paragraph.

That job, as was pointed out earlier, is already well under way. How sounds are processed and represented in the cochlea of the inner ear, so as to send a range of qualitatively distinct (and highly complex) neuronal activation-patterns to the auditory cortex, is a matter that is already understood. The cochlea is wonderfully configured to do a fine-grained energy-profile analysis across the frequency spectrum of any incoming sound. So also – though here the story is still provisional – is the manner in which those peripheral inputs are synaptically transformed and subsequently coded within a multi-dimensional similarity-and-difference 'space' of activation-patterns within the auditory cortex itself. (Pitch, of course, turns out to be only one of many dimension of variation among sensations of sound.)

The same is true for the brain's internal representations of color, both at the retina and in the brain's downstream cortical area V4. The Hurvich-Jameson model mentioned earlier, of how chromatic information is both processed and represented in the brain, gives us a detailed account of the neuronal niceties that underlie subjective human color experience, an account that gives a highly illuminating explanation for the internal *phenomenological* structure of human color-qualia space itself, that is, of the qualitative similarity-and-difference relations that severally unite all of the colors.[4] It further provides predictions of and systematic explanations of the qualitative character of tens of *thousands* of distinct *color after-images* that are produced when one fixates for time on any one of a hundred (different) colored circles, and then relocates one's gaze on any one of a hundred (different) uniformly colored backgrounds. The resulting circular after-image will have a distinct, predictable, and entirely explicable color-quality different from either of the two contributing stimuli. The model even predicts the existence of, and tells us how to produce, color sensations of qualitatively *novel* sorts, such as sensations of a 'red,' or a 'blue,' or a 'green,' each of which is simultaneously as black as the blackest-possible black. I know this description sounds impossible, or even semantically ill-formed, but the predictions turn out to be true and the physical mechanisms involved are straightforward.

Evidently, and as rightly expected, the domain of internal qualitative features is not at all explanatorily impenetrable by the resources of the physical sciences. Just as in the case of the external qualitative features, we already possess some striking explanatory accounts of the nature and contents of our internal qualitative lives, and it would be foolish not to expect more. None of this strictly *entails* that epiphenomenalism is mistaken about the ontological status of our inner qualia, but that position is currently being overwhelmed by an alternative tide of explanatory success, and that position's initial strength derived from nothing more than a highly prejudicial 'analysis,' and a whopping 'argument from ignorance' in any case, both of which missteps are unmasked by the considerations of the preceding pages.

Accordingly, the truth would seem to be that absolutely *none* of the 'apparently simple' qualitative characters that grace our inner lives are genuine *ontological* simples at all. They are, all of them, complex neural and physiological states, states whose qualitative characters are ontologically *embodied* in that precious physical complexity. The dynamical activities of the brain are positively *driven* by those very physical complexities, and so the philosophical claim that these alleged 'simples' are also causally impotent bystanders to the brain's dynamical adventures is flatly inconsistent with the recent insights of Neuroscience, as well as with the antecedent convictions of Folk Psychology. It may be that the overall cognitive profile that characterizes *conscious* brain activity remains to be understood. Indeed it does.[5] But the account of our qualitative conscious states offered by epiphenomenalism holds out no analytical or explanatory virtues to tempt us towards that position, and the competing neurobiological account of those very same states already holds out a broad range of ontological, explanatory, and predictive virtues that pull us in

---

[4] See again Churchland (2005), *Op cit*.

[5] Although, for an opening stab at what such a cognitive profile might look like, and how it might be embodied in the recurrent structure of the brain's global 'wiring diagram,' see Churchland (1995).

precisely the opposite direction. Add up their respective contributions to our current understanding and there is simply no contest. Epiphenomenalism will soon be a museum piece.

The more common forms of Property Dualism – which do not attempt to disconnect our inner qualitative characters from the dynamics of our cognitive activities – are not quite so badly off as is epiphenomenalism, for they do not fly in the face of the constituting convictions of Folk Psychology and the explanatory practices they sustain. But, as has been known for more than 50 years, these less extreme forms of Dualism do fly in the face of basic Physics itself, a rather more damning matter, since any position that includes non-physical elements in the causal dynamics of the brain must violate both the law that energy is neither created nor destroyed, and the law that the total momentum in any closed system is always conserved. In short, you simply can't get a change in any aspect of the physical brain (for that would causally require both energy changes and momentum changes) save by a compensatory change in some other *physical* aspect of the brain, which will thereby lay claim to being the cause at issue. There is simply no room in a physical system for ghosts of any kind to intervene in some fashion to change its dynamical behavior. Any physical system is 'dynamically closed' under the laws of Physics. (Indeed, it was this very difficulty, over a century ago, that initially motivated the desperate invention of Epiphenomenalism in the first place.)

Still, one might choose to simply reject, or somehow to circumscribe, the currently accepted laws of Physics, and contrive to make a case for an 'interactive' Dualism based on its comparative explanatory and predictive successes, relative to the same successes displayed by the physicalistic Neurosciences. This, I propose, is the only possible route by which an honest Dualism of any kind can hope to succeed. Any other route, as we have seen above, will involve nothing but subterfuge and self-deception. But if this honest route is to be taken, it must begin by acknowledging that, to date, "... Dualism is less a theory of mind than it is an empty space waiting for a genuine theory of mind to be put in it."[6] If the 'explanatory successes' of Dualism are to be fairly weighed against those of current Cognitive Neuroscience and of basic Physics, they must first be brought into existence. So far, there is nothing there to permit such a comparative evaluation to even begin. But while we are waiting, we can fairly contemplate the steadily accumulating and highly enlightening explanatory successes produced by our theoretical and experimental probings of the physical brain, even on the topic of its diverse qualitative states. After all, we will need to *know* about those successes, and in great detail, should the prospective contest just imagined ever materialize.

---

[6]This quotation is drawn from a textbook published over a quarter-century ago: Churchland (1984). Little or nothing has changed since then.

References

Chalmers, D. 1996. *The conscious mind.* Cambridge: Cambridge University Press.

Churchland, P.M. 1981. Eliminative materialism and the propositional attitudes. *Journal of Philosophy* 78(2): 67–90.

Churchland, P.M. 1984. *Matter and consciousness*, 19. Cambridge, MA: The MIT Press.

Churchland, P.M. 1985. Reduction, qualia, and the direct introspection of brain states. *The Journal of Philosophy* 82(1): 8–28.

Churchland, P.M. 1995. *The engine of reason, the seat of the soul*, 211–226. Cambridge, MA: The MIT Press.

Churchland, P.M. 1996. The rediscovery of light. *The Journal of Philosophy* 93(5): 5–32.

Churchland, P.M. 2005. Chimerical colors: Some phenomenological predictions from cognitive neuroscience. *Philosophical Psychology* 18(5): 527–560. Reprinted in my (2007) *Neurophilosophy at work.* Cambridge: Cambridge University Press.

Churchland, P.M. 2007. On the reality (and diversity) of objective colors: How color-uqalia space is a map of reflectance-profile space. *Philosophy of Science* 74(2): 119–149. Reprinted as Chap. 10 of Churchland, P.M. 2007, *Neurophilosophy at work.* New York: Cambridge University Press. Also Reprinted in Matthen, M., and J. Cohen (eds). 2010. *Essays in honor of larry hardin*. Cambridge, MA: The MIT Press.

Churchland, P.M. 2012. *Plato's Camera: How the physical brain captures a landscape of abstract universals*. Cambridge, MA: The MIT Press.

Jackson, F. 1982. Epiphenomenal qualia. *The Philosophical Quarterly* 32(127): 127–136.

Levine, J. 1983. Materialism and qualia: The explanatory gap. *Pacific Philosophical Quarterly* 64: 354–361.

Lewis, D. 1983. Postscript to 'mad pain and martian pain'. In *Philosophical papers*, vol. 1. New York: Oxford University Press.

Nagel, T. 1974. What is it like to be a bat? *Philosophical Review* 83(4): 435–450.

Nemirow, L. 1980. Review of Thomas Nagel's *mortal questions*. *Philosophical Review* 89(3): 473–477.

Chapter 5
Churchland on Arguments Against Physicalism

Torin Alter

5.1 Introduction

In "Consciousness and the Introspection of 'Qualitative Simples'" Paul Churchland criticizes a familiar family of anti-physicalist arguments, including Thomas Nagel's (1974) "What is it like to be a bat?" argument, Frank Jackson's (1982, 1986, 1995) knowledge argument, and related arguments developed by David Chalmers (1996, 2010) and others. In Churchland's view, those arguments lead to the pessimistic view that science can shed no light on the qualitative features of conscious experience. He provides good reasons to reject that pessimistic view. However, I will argue, he is wrong to associate it with at least two of the anti-physicalist arguments he considers: the knowledge and conceivability arguments.[1] Proponents of those arguments can share Churchland's more optimistic view about the science of consciousness. Indeed, at least some proponents, including Chalmers, advocate a similar view. Churchland also attacks the anti-physicalist arguments more directly, identifying and criticizing assumptions that he sees as underlying them. But, I will argue, those attacks are at best inconclusive, at least with respect to the knowledge and conceivability arguments.[2]

---

[1]The most widely discussed version of the conceivability argument (Chalmers 1996, 2010) involves the claim that zombies (creatures that lack consciousness but are physically identical to conscious human beings) are conceivable. See Sect. 5.4.

[2]I restrict my claims to the knowledge and conceivability arguments mostly because the other anti-physicalist arguments Churchland discusses have been less well developed. But I suspect that parallel conclusions about those other arguments could be defended on similar grounds.

T. Alter (✉)
Department of Philosophy, The University of Alabama, 325 Ten Hoor Hall,
Tuscaloosa, AL 35487-0218, USA
e-mail: talter@ua.edu

R. Brown (ed.), *Consciousness Inside and Out: Phenomenology, Neuroscience, and the Nature of Experience*, Studies in Brain and Mind 6, DOI 10.1007/978-94-007-6001-1_5,

5.2 Explanatory Impenetrability

According to Churchland, not only *can* the physical sciences shed light on the qualitative features of experiences, but this project is well under way. After describing some relevant results, he concludes, "Evidently, and as rightly expected, the domain of internal qualitative features is not at all explanatorily impenetrable by the resources of the physical sciences" (p. 54).[3] This, he suggests, shows that anti-physicalist arguments such as the knowledge argument must be unsound.

But it is unclear how that is supposed to follow. The knowledge argument does entail that there are aspects of phenomenal consciousness that physical science, as traditionally conceived, cannot exhaustively explain.[4] That claim, though far from trivial, is considerably weaker than the claim that physical science cannot explain any aspect of the phenomenal domain. The considerations Churchland adduces against the latter, stronger claim do not necessarily threaten the former, weaker one.

Consider one of Churchland's examples of the sort of explanation the physical sciences have already provided:

> The Hurvich-Jameson model ... provides predictions of and systematic explanations of the qualitative character of tens of *thousands* of distinct *color after-images* that are produced when one fixates for time on any one of a hundred (different) colored circles, and then relocates one's gaze on any one of a hundred (different) uniformly colored backgrounds. The resulting circular after-image will have a distinct, predictable, and entirely explicable color-quality different from either of the two contributing stimuli. The model even predicts the existence of, and tells us how to produce, color sensations of qualitatively *novel* sorts, such as sensations of a 'red,' or a 'blue,' or a 'green,' each of which is simultaneously as black as the blackest-possible black. (pp. 53–54)

Those results are significant. But why think they conflict with the knowledge argument? Perhaps the implicit reasoning is this: "Pre-emergence Mary would know all about the Hurvich-Jameson model. Thus, she would be able to make all manner of correct predictions about color phenomenology, including predictions about novel experiences. So, she would surely know what it is like to see plain old red!"[5]

This is suspicious. Understanding the Hurvich-Jameson model will enable Mary to understand many structural aspects of seeing red, some of which might be reasonably thought part of the experience's phenomenology. But can she deduce *all* phenomenal aspects of seeing red from the model (in combination with other physical truths)? This is not clear. Or rather, this is not clear unless we assume that the phenomenal character of seeing red is exhausted by structural properties

---

[3]Page numbers taken from Chap. 4 of this volume. Citations refer to that chapter unless otherwise specified.

[4]The qualification "as traditionally conceived" is vital. Physical science might be conceived broadly so as to include, for example, protophenomenal properties (Stoljar 2001). In that case, the knowledge and conceivability arguments would be compatible with the claim that physical science might exhaustively explain consciousness.

[5]Pete Mandick (2009) develops an argument along roughly these lines. I criticize it in Alter (2008).

of the sort the Hurvich-Jameson model describes. Absent such a question-begging assumption, there would seem to be no incompatibility between the sorts of discoveries Churchland describes and the knowledge argument.[6]

Similar reasoning applies to the conceivability argument, whose most prominent advocate is Chalmers. Churchland regards Chalmers as an opponent, but this is misleading where explanatory impenetrability is concerned. On that topic the two philosophers seem largely in agreement. Indeed, the science Churchland describes in connection to the Hurvich-Jameson model would seem to follow an approach Chalmers (1995, 1996, 2004a) expressly recommends as part of a science of consciousness: find structure in conscious experience and correlate that structure with neural or computational structure. For example, he notes that the methods of cognitive science and neuroscience could be used to "*explain the structure of consciousness*" (Chalmers 1995, p. 206). He writes,

> For example, it is arguable that an account of the discriminations made by the visual system can account for the structural relations between different color experiences, as well as for the geometric structure of the visual field . . . In general, certain facts about structures found in processing will correspond to and arguably explain facts about the structure of experience. (Chalmers 1995, p. 206)

The sort of explanation that the Hurvich-Jameson model provides would seem to exemplify what Chalmers has in mind. Unlike Churchland, he denies that the sorts of structural features exhibited by neural and computational processing *exhaust* phenomenality. But other than that, Churchland's view about how science should and does investigate consciousness seems quite congenial to an approach Chalmers favors.[7]

5.3 The Knowledge Argument

According to Churchland, the anti-physicalist arguments that (in his view) have led some to embrace an explanatory impenetrability thesis are fundamentally flawed. Let us examine this charge, beginning with Jackson's knowledge argument. Jackson infers the falsity of physicalism from Mary's post-emergence epistemic progress,

---

[6]If the Hurvich-Jameson model (or the model plus other physical truths) appears initially to explain all qualitative features of seeing red, this may stem from a tendency to conceive of the model in phenomenal terms, for example, in terms of what it is like to see a certain after image. I have been assuming that the model is characterized in entirely physical terms. Otherwise, if the model is specified partly in phenomenal terms, then even if Mary could deduce the phenomenal character of seeing red from the model this would not support physicalism or threaten the knowledge argument.

[7]See also Chalmers' (1995, 1996) discussion of *the principle of structural coherence*. Churchland's view also seems to align well with Thomas Nagel's (1974, p. 449) suggestion that we develop "an objective phenomenology not dependent on empathy or the imagination" the goal of which "would be to describe, at least in part, the subjective character of experiences in a form comprehensible to beings incapable of having those experiences."

that is, from the claim that she learns what it is like to see in color when she leaves the black-and-white room.[8] The inference is indirect. Indeed, it must be: physicalism is a doctrine about the nature of the world, not how we know about it. At least, this is true of the doctrine that concerns Jackson. He characterizes physicalism as the claim that all (correct) information is physical information (Jackson 1982). So, what is the basis for the inference?

Churchland suggests that the inference is based on what I will call *the constitution principle*: the claim that physicalism is true only if scientific knowledge constitutes (as opposed to *describes* or *explains*) subjective knowledge.[9] But, he argues, the constitution principle is false. He writes,

> [Jackson] is expecting that explicit/discursive/scientific knowledge should somehow constitute subjective knowledge of visual experiences. But that expectation is unreasonable on its face. As well expect that your exhaustive discursive knowledge of the micro-organization of a professional golfer's motor cortex would constitute actual practical 'knowledge,' on your part, of how to hit a golf ball 200 yards down the middle of the fairway, even if you have never swung a golf club before in your life. (p. 46)

Churchland adds that his golf analogy "gives a specific voice to a classic objection to Jackson's original argument" (p. 46) based on the Lewis-Nemirow ability hypothesis. On the ability hypothesis, to know what it is like to see in color is to possess abilities such as the ability to imagine in color (Lewis 1983a, b, 1988; Nemirow 1980, 1990, 2007). Ability hypothesis proponents typically argue that Mary's epistemic progress fails to threaten physicalism because it consists in her acquiring abilities but no information. But Churchland does not rest his case on the ability hypothesis per se. Instead, he advances the more general claim that phenomenal knowledge is non-propositional knowledge of some sort, perhaps ability knowledge, perhaps something else. That claim, he suggests, would suffice to undermine the constitution principle and thus the knowledge argument.

But the knowledge argument does not assume the constitution principle. The inference from Mary's progress to physicalism's falsity involves two key claims. One is *non-deducibility*: there are truths about consciousness that cannot be a priori deduced from the complete physical truth. The other is *non-necessitation*: there are truths that are not metaphysically necessitated by the complete physical truth. Mary's progress is used (in conjunction with other premises) to establish non-deducibility; non-deducibility is then used (in conjunction with other premises) to establish non-necessitation, which in turn is used (in conjunction with other

[8]Jackson (1998, 2003, 2007) has since rejected the knowledge argument, but not for the reasons Churchland recommends. For criticisms of Jackson's reasons, see Alter (2007).

[9]I assume that "scientific knowledge" refers to knowledge of the sort Mary learns pre-emergence by watching science lectures on black-and-white television; and that "subjective knowledge" refers to knowledge of what it is like, otherwise known as phenomenal knowledge.

premises) to establish that physicalism is false.[10] Each of those steps depends on controversial assumptions. But none assume the constitution principle.[11]

The knowledge argument does rely on a related claim, which I will call *the propositional knowledge claim*: phenomenal knowledge is at least in part propositional. The propositional knowledge claim seems plausible. There would seem to be something specific it is like to see red: a truth colorsighted folk typically know and Mary learns only after leaving the black-and-white room. Churchland rejects the propositional knowledge claim.[12] But he does not provide a clear, compelling argument against it. He writes,

> . . . it is plain that we are here looking at two distinct kinds of knowledge. Having discursive scientific knowledge of anything requires having a *language*. Knowing what it is like to see red requires nothing of the sort. (p. 46)

If that is supposed to be an argument against the propositional knowledge claim, it is unconvincing. Suppose Noam lacks language but knows what it is like to see red. He might be a parrot, say, or a human being who lacks language but is otherwise normal. He cannot articulate what he knows. It does not follow that his phenomenal knowledge does not consist even partly in knowing truths. Imagine that one day Noam acquires language and remarks, "Here is a truth I knew well before I acquired language: seeing red has a certain phenomenal character *R*." I see no good reason to conclude that Noam's remark could not be true, where *R* is the phenomenal character of typical experiences of seeing red. Churchland's argument might show that scientific and phenomenal knowledge are of distinct epistemic kinds. But that conclusion need not conflict with the propositional knowledge claim.

Perhaps Churchland is working with a narrow conception of propositional knowledge on which such knowledge is necessarily quasi-linguistic (or necessarily symbolic). However, proponents (and most opponents) of the knowledge argument use the term "propositional knowledge" in a broader sense, e.g., for the narrowing down of possibilities, whether or not this involves language or symbols. Under this broader conception, Churchland's argument does not undermine the propositional knowledge claim.[13]

---

[10]That summary omits various details. For example, as Chalmers (2004b, 2010) argues, it is best to formulate the knowledge argument so that the conclusion comes out as a disjunction: either physicalism is false or Russellian monism is true. Also, references to the complete physical truth should strictly speaking be to a conjunction of the complete physical truth, a second-order 'that's all' truth, and the complete indexical truth (loc cit.).

[11]For discussions of these assumptions, see Alter and Howell (2012) and Chalmers (2010). Churchland does not cite any of the post-1996 philosophical literature on this subject, other than his own contributions.

[12]Or if he does not, and he is suggesting instead that knowledge of what it is like is only partly non-propositional, then the propositional part is left over to fuel the knowledge argument.

[13]For arguments defending the propositional knowledge claim, see Lycan (1996), Alter (2001), Stanley and Williamson (2001), Alter and Howell (2009), and Howell (2012).

5.4 Chalmers and the Conceivability Argument

Let us turn to Chalmers' arguments. Here is how Churchland describes Chalmers' reasoning:

> . . . the qualitative features of our sensations are ontological *simples*, [Chalmers] avers, and for that reason, they offer nothing in the way of internal structure that a physicalist theory of the brain might hope to *re*construct, as a successful intertheoretic reduction would require. The qualitative character of my visual sensation-of-red, for example, simply confronts me, or so it seems. It does so in a manner quite distinct from any alternative sensation-of-color, but not because it invites any hope of some signature 'decomposition' into anything else at all, let alone into something physical. Such *qualia*, as they have come to be called, should therefore be counted as something outside the physical order, as something beyond the causal/functional profile in which our sensations are admittedly embedded. Chalmers' dualistic conclusion here is thus one instance of the long-familiar position called *epiphenomenalism* . . . (pp. 46–47)

In response, Churchland begins by noting that most of our experiences are not simple:

> . . . the bulk of one's sensational life is characterized, not by simplicity, but by an extraordinary and ever-changing *complexity*. Listening to a conversation, looking around a flower garden, tasting a braised-lamb stew, smelling the aromas in a wood-working shop – our sensations in such cases display intricacies that are amazing. (p. 47)

But the relevant issue is not whether familiar experiences exhibit such "ever-changing *complexity*". I know of no one, including Chalmers, who denies that they do. The relevant issue – whether phenomenal consciousness is wholly physical – arises for the phenomenal components of such experiences. Those components (or some of them) are the ones that appear simple, if any features of experience do.

Churchland's main criticism is more interesting. He notes that because our cognitive powers are limited, there must be an upper bound on how far we can decompose our experiences. So, that there are apparently non-decomposable phenomenal qualities does not show that any phenomenal qualities are in fact non-decomposable. He writes,

> Every cognitive creature, even in an *exhaustively physical* universe, must display a current limit on how far it can decompose the qualities it can discriminate. Let us not be too impressed in our own case, then, by the mere existence of apparent 'qualitative simples,' however robust their apparent simplicity. . . . Such 'simples' simply have to be there, if only to mark the limits of our current understanding. . . . Accordingly, to infer the ontological simplicity of a given qualitative character from its *apparent* simplicity is to commit the fallacy of Arguing from Ignorance, as in, "I am *unaware* of any constituting elements in this qualitative feature, therefore, there *aren't any* constituting elements." (p. 48)

Those points are plausible.[14] If Chalmers did give such an Argument from Ignorance, then it would provide no reason to doubt physicalism. But he does not.

---

[14] However, Friends of Mary would quibble with the first sentence in the quoted paragraph if it is meant to refer to every *conceivable* cognitive creature.

For one thing, he does not commit himself to the view that "the qualitative features of our sensations are ontological *simples*". He allows for the possibility that phenomenal properties result from the combination of *proto*phenomenal properties. On this view – a version of Russellian monism – the qualitative features of our sensations are far more complex than they seem. Also, although Churchland describes Chalmers' view as epiphenomenalism, that is but one of three options Chalmers leaves open. The other two are interactionist dualism and Russellian monism. And recently (Chalmers 2010, pp. viii, 132) he has distanced himself further from the epiphenomenalist option.[15]

Chalmers does argue for something close to the claim that phenomenal properties, "offer nothing in the way of internal structure that a physicalist theory of the brain might hope to *re*construct". He does not deny that experiences have *phenomenal* structure. But he does hold that (C) the nature of consciousness cannot be exhaustively explained solely in terms of causal/nomic spatio-temporal structure.[16] However, he does not infer (C) directly from the apparent simplicity of experiences. He invokes a variety of arguments for (C). The one that has received the most attention is the conceivability argument – more specifically, the argument from the conceivability of zombies. Does the conceivability argument rest on the fallacious reasoning Churchland rightly rejects?

It does not appear to. It does not seem to rely on a premise about the apparent simplicity of qualia. Chalmers begins with thought experiment: a scenario in which everything physical is held constant but in which there is no phenomenal consciousness, i.e., a zombie-world scenario. He then argues that the zombie-world scenario is conceivable not just at first glance but on ideal reflection. And his inference from the conceivability of the zombie-world scenario to (C) involves several substantive premises, none of which appear to rest on any obvious fallacy.

Chalmers presents several versions of the conceivability argument. The most precise ones rely on two-dimensional semantics, and explaining that apparatus would involve a lengthy digression. For present purposes it will suffice to consider the following simplified version:

1. $P\&{\sim}Q$ is conceivable.
2. If $P\&{\sim}Q$ is conceivable, then $P\&{\sim}Q$ is metaphysically possible.
3. If $P\&{\sim}Q$ is metaphysically possible, materialism is false.

---

4. Materialism is false.

Here P is the conjunction of all microphysical truths about the universe, specifying the fundamental features of every fundamental microphysical entity in the language of

[15]For more on protophenomenal properties and Russellian monism, see Alter and Nagasawa (2012).

[16](C) involves some simplification. Here is a more precise formulation: there are truths about consciousness that are not metaphysically necessitated by the conjunction of the complete truth about causal/nomic spatio-temporal structure-and-dynamics, the complete indexical truth, and a second-order "that's all" truth. See footnote 10.

> microphysics. Q is an arbitrary phenomenal truth: perhaps the truth that someone is phenomenally conscious, or perhaps the truth that a certain individual (that is, an individual satisfying a certain description) instantiates a certain phenomenal property. $P\&{\sim}Q$ ("P and not Q") conjoins the former with the denial of the latter. (Chalmers 2010, p. 142)

Churchland raises some important issues that bear on that argument. One depends on an analogy to "objective perceptual qualities" (p. 48) such as redness and warmth. According to Churchland, such qualities are, like phenomenal properties, apparently simple. However, he observes, "no one since the eighteenth century supposes that such objective perceptual properties are thereby revealed as genuine ontological simples" (p. 48). Such properties are complex and wholly physical. Churchland writes, "It just took the physical sciences a while to learn about their constituting elements. Our *internal* phenomenological qualities may be awaiting a precisely similar fate" (p. 48). Does Churchland's analogy undermine a premise of the conceivability argument, such as the premise that $P\&{\sim}Q$ is conceivable?

This seems doubtful. As Kripke (1972) argued four decades ago, such analogies are problematic. In the case of warmth, for example, we can easily distinguish the objective phenomenon of warmth – high mean molecular kinetic energy – from the sensation that the objective phenomenon typically causes in us. That is, we can easily distinguish the reality (warmth itself) from the associated phenomenal appearance (the feeling of warmth). But in the case of sensations themselves, a corresponding appearance/reality distinction is less easily drawn. Here the phenomenal appearance would seem to be identical to, or part of, the reality. On reflection, the comparison between objective perceptual qualities and phenomenal consciousness seems only to highlight and reinforce the distinctive challenges that the latter poses for physicalism.[17]

Churchland mentions other objections, but they are underdeveloped. For example, he suggests that the apparent conceivability of zombies might derive from ignorance of physical truths that have yet to be discovered. There might be something to this. But to constitute a serious threat to the conceivability argument, good reasons to accept the ignorance line would have to be supplied – reasons considerably better than dubious analogies to successful reductions of objective perceptual qualities.[18] To take another example, Churchland notes that zombies, who are completely physical by definition, might give an argument like Chalmers' argument for nonphysical qualia and asks, "But if *their* argument for that conclusion is manifestly unsound, why is *our* argument for that conclusion any better?" (p. 48, n. 2).[19] That is an excellent question, but Chalmers has provided a plausible answer (Chalmers 2010, pp. 159–60): both arguments depend (as support for premise 3 in

[17]However, for an interesting defense of the analogy, see Pereboom (2011). I criticize Pereboom's argument in Alter (2012).

[18]Daniel Stoljar (2006) develops the ignorance line in detail. For criticisms of Stoljar's arguments, see Alter (2009), Bennett (2009), Chalmers (2010), and Gertler (2009).

[19]See Balog (1999) for a development of this challenge.

the version quoted above) on the assumption that someone is conscious, and that assumption is true of us but false of the zombies. If there are problems with that answer, Churchland does not indicate what they might be.

5.5 Conclusion

The best way to show that something is possible is to show that it is actual. Churchland employs this method to good effect: he shows that the physical sciences can shed light on the nature of conscious experience by showing how it has already done so. This is no mean feat. But he is wrong to suggest that critics of physicalism such as Chalmers would not welcome that result. Indeed, Churchland's naturalistic approach to studying consciousness fits well with the approach Chalmers expressly recommends. So, there is less disagreement here than Churchland suggests.

Not that there isn't any disagreement. Chalmers and other anti-physicalists argue that the sorts of structural truths discovered by the physical sciences do not exhaust the complete truth about consciousness. Churchland rejects that conclusion and the arguments on which it is based. I have argued that his criticisms of at least two of those arguments are unconvincing. But whether or not I am right, the significance of this dispute for his naturalism should not be exaggerated. Endorsing the knowledge and conceivability arguments in no way requires resisting his commendable efforts at showing how the physical sciences are already in the process of providing insight into the nature of consciousness.[20]

References

Alter, T. 2001. Know-how, ability, and the ability hypothesis. *Theoria* 67: 229–239.

Alter, T. 2007. Does representationalism undermine the knowledge argument? In *Phenomenal concepts and phenomenal knowledge: New essays on consciousness and physicalism*, ed. T. Alter and S. Walter, 65–76. New York: Oxford University Press.

Alter, T. 2008. Phenomenal knowledge without experience. In *The case for qualia*, ed. E. Wright, 247–267. Cambridge, MA: MIT Press.

Alter, T. 2009. Does the ignorance hypothesis undermine the conceivability and knowledge arguments? *Philosophy and Phenomenological Research* 79: 756–765.

Alter, T. 2012. Review of Derk Pereboom's *Consciousness and the prospects of physicalism*. *Mind* 121: 1115–1124.

Alter, T., and R.J. Howell. 2009. *A dialogue on consciousness*. New York: Oxford University Press.

Alter, T., and R.J. Howell (eds.). 2012. *Consciousness and the mind-body problem: A reader*. New York: Oxford University Press.

Alter, T., and Y. Nagasawa. 2012. What is Russellian monism? *Journal of Consciousness Studies* 19(9–10): 67–95.

[20]For helpful comments, I thank David J. Chalmers, Robert J. Howell, Rekha Nath, and Derk Pereboom.

Alter, T., and S. Walter (eds.). 2007. *Phenomenal concepts and phenomenal knowledge*. New York: Oxford University Press.

Balog, K. 1999. Conceivability, possibility, and the mind-body problem. *Philosophical Review* 108: 497–528.

Bennett, K. 2009. What you don't know *can* hurt you. *Philosophy and Phenomenological Research* 79: 766–774.

Bickle, J. (ed.). 2009. *The oxford handbook of philosophy and neuroscience*. New York: Oxford University Press.

Chalmers, D.J. 1995. Facing up to the problem of consciousness. *Journal of Consciousness Studies* 2: 200–219.

Chalmers, D.J. 1996. *The conscious mind: In search of a fundamental theory*. New York: Oxford University Press.

Chalmers, D.J. 2004a. How can we construct a science of consciousness? In *The cognitive neurosciences III*, ed. M. Gazzaniga. Cambridge, MA: MIT Press. Reprinted in Chalmers 2010: 37–58.

Chalmers, D.J. 2004b. Phenomenal concepts and the knowledge argument. In *There's something about Mary: Essays on phenomenal consciousness and Frank Jackson's knowledge argument*, ed. P. Ludlow, D. Stoljar, and Y. Nagasawa, 269–298. Cambridge, MA: MIT Press.

Chalmers, D.J. 2010. *The character of consciousness*. New York: Oxford University Press.

Gazzaniga, M. (ed.). 2004. *The cognitive neurosciences III*. Cambridge: MIT Press.

Gertler, B. 2009. The role of ignorance in the problem of consciousness. Critical notice of Stoljar 2009. *Noûs* 43: 378–393.

Harman, G., and D. Davidson (eds.). 1972. *The semantics of natural language*. Dordrecht: Reidel.

Howell, R.J. 2012. *Consciousness and the limits of objectivity: The case for subjective physicalism*. Oxford: Oxford University Press.

Jackson, F. 1982. Epiphenomenal qualia. *The Philosophical Quarterly* 32(127): 127–136.

Jackson, F. 1986. What Mary didn't know. *The Journal of Philosophy* 83: 291–295.

Jackson, F. 1995. Postscript. In *Contemporary materialism: A reader*, ed. P. Moser and J.D. Trout, 184–189. New York: Routledge.

Jackson, F. 1998a. Postscript on qualia. In *Mind, method, and conditionals: Selected essays*, ed. F. Jackson, 76–79. London: Routledge.

Jackson, F. 1998b. *Mind, method, and conditionals: Selected essays*. London: Routledge.

Jackson, F. 2003. Mind and illusion. In *Minds and persons: Royal institute of philosophy supplement,* 53, ed. O'Hear, 251–271. Cambridge: Cambridge University Press.

Jackson, F. 2007. The knowledge argument, diaphanousness, representationalism. In *Phenomenal concepts and phenomenal knowledge*, ed. T. Alter and S. Walter, 65–76. New York: Oxford University Press.

Kripke, S. 1972. Naming and necessity. In *The semantics of natural language*, ed. G. Harman and D. Davidson, 253–355. Dordrecht: Reidel.

Lewis, D. 1983a. Postscript to 'mad pain and martian pain'. In *Philosophical papers*, vol. 1, 130–132. New York: Oxford University Press.

Lewis, D. 1983b. *Philosophical papers*, vol. 1. New York: Oxford University Press.

Lewis, D. 1988. What experience teaches. In *Proceedings of the Russellian society*, 499–518. Sydney: University of Sydney. Reprinted in Lycan 1990.

Ludlow, P., D. Stoljar, and Y. Nagasawa (eds.). 2004. *There's something about Mary: Essays on phenomenal consciousness and Frank Jackson's knowledge argument*. Cambridge: MIT Press.

Lycan, W.G. (ed.). 1990. *Mind and cognition: A Reader*. Cambridge: Basil Blackwell.

Lycan, W.G. 1996. *Consciousness and experience*. Cambridge, MA: MIT Press.

Mandick, P. 2009. The neurophilosophy of subjectivity. In *The oxford handbook of philosophy and neuroscience*, ed. J. Bickle, 601–618. New York: Oxford University Press.

Moser, P., and J.D. Trout. 1995. *Contemporary materialism: A reader*. New York: Routledge.

Nagel, T. 1974. What is it like to be a bat? *Philosophical Review* 4: 435–450.

Nemirow, L. 1980. Review of Thomas Nagel's *mortal questions*. *Philosophical Review* 89(3): 473–477.

Nemirow, L. 1990. Physicalism and the cognitive role of acquaintance. In *Mind and cognition: A reader*, ed. W.G. Lycan, 490–499. Cambridge: Basil Blackwell.
Nemirow, L. 2007. So this is what it's like: A defense of the ability hypothesis. In *Phenomenal concepts and phenomenal knowledge*, ed. T. Alter and S. Walter, 32–51. New York: Oxford University Press.
O'Hear, A. 2003. *Minds and persons: Royal institute of philosophy supplement,* 53. Cambridge: Cambridge University Press.
Pereboom, D. 2011. *Consciousness and the prospects of physicalism*. New York: Oxford University Press.
Stanley, J., and T. Williamson. 2001. Knowing how. *The Journal of Philosophy* 98: 411–444.
Stoljar, D. 2001. Two conceptions of the physical. *Philosophy and Phenomenological Research* 62: 253–281.
Stoljar, D. 2006. *Ignorance and imagination: The epistemic origin of the problem of consciousness*. New York: Oxford University Press.
Wright, E. 2008. *The case for qualia*. Cambridge: MIT Press.

Chapter 6
Response to Torin Alter

Paul M. Churchland

My thanks to Prof. Alter for his careful, kind, and cautionary commentary. I shall try to live up to the standard he has set.

I begin by focusing on the importance that Alter mistakenly ascribes to the *non-deducibility* of the 'phenomenal facts' – the ones that Mary only belatedly comes to apprehend – from the facts of a presumed complete neuroscience. This admitted lack of entailment is of no significance whatever for the issue of the ultimate reducibility of the former to the latter, because such failures of deducibility are absolutely *typical of* successful scientific reductions. And for a very good reason. The successfully reduced conceptual framework typically boasts a lexicon that is *not included* in the (usually quite different) lexicon of the more general theory that sustains the reduction. "Temperature", for example, does not appear in the lexicon of the molecular/kinetic theory of gases. "Light" does not appear in the lexicon of electromagnetic theory. "Pitch" does not appear in the lexicon of the compression-wave theory of sound. And so on, for many other examples. Accordingly, no sentence containing one of these older lexical items – "temperature", "light", or "pitch" – (barring tautologies and other trivial exceptions) will be deducible from *any* sentences of the reducing theory, ever, for purely *formal* reasons.

That is precisely why 'intertheoretic identity statements' or 'bridge laws' or 'correspondence rules' (as they have been variously called) such as "Temperature = mean molecular kinetic energy", or "Light = EM waves", or "Pitch = oscillatory frequency" are proposed so as to *connect* the two lexicons and thereby to make possible the systematic deduction of the specific convictions or postulates of the older theory from the postulates of the new and more general theory. If, in this way, the newer theory succeeds in 'reconstructing' the assembled convictions characteristic of the older conceptual framework, and if the newer theory is independently worthy of our belief, then the relevant identity statements

P.M. Churchland (✉)
Department of Philosophy, UCSD, 9500 Gilman Drive #0119, La Jolla, CA 92093, USA
e-mail: pchurchland@ucsd.edu

R. Brown (ed.), *Consciousness Inside and Out: Phenomenology, Neuroscience, and the Nature of Experience*, Studies in Brain and Mind 6, DOI 10.1007/978-94-007-6001-1_6,

are also rendered beliefworthy. This same pattern, I claim, is well on its way to being followed in the case of our own folk psychology vis-à-vis our emerging computational neuroscience.

In the same paragraph that he points (irrelevantly) to the non-deducibility just discussed, Alter also attempts to impose a further condition on the beliefworthiness of the intertheoretic or interconceptual identities here at issue. (In the context of the mind-body problem, a live candidate identity would be, for example, "A sensation of red = a 50, 90, 50 % activation-triplet across the opponent-process neurons in cortical area V4".) Specifically, to be beliefworthy, he says, such identities must be 'metaphysically necessary' in the (dubious) sense proposed by Kripke some 40 years ago when articulating a novel semantics for modal logic. But in the mind-body case, Alter says, such identities are *not* 'metaphysically necessary', and so materialism must be false.

Not to waste the reader's time, I reject this manufactured flavor of necessity as a measure of intertheoretic identity, or indeed, as a measure of much of anything at all, beyond the accidental scientific and metaphysical prejudices of the philosophers attempting to apply it in any particular case. There is no firm and objective lever here with which to ply the truths at issue.

These points have little to do with Jackson's 'knowledge argument' in any case, since that is a matter of whether our neurally-omniscient Mary learns something new – something she did not know before – when she finally learns *what it is like* to have a sensation of red. Let all us agree that she does. The issue here is the ontological significance of her admitted epistemological novelty. On this matter I am compelled to point out (what Jackson and everybody else seems to have missed over the last 30 years) that Mary would enjoy exactly the same sort of epistemological novelty were she finally to learn *what it is like* to have a *50, 90, 50 % activation-triplet across the opponent-process neurons in cortical area V4*. For this is exactly the kind of *neuronal* state that her color-deprived history has denied her, despite her exhaustive book-learning. Though she knows *about* such states, in a discursive way, she has never been *in* that state. Until now. "So that's what it's like!" she marvels. In *this* case, however, her undoubted epistemological novelty does not tempt us to infer that the relevant *activation-triplet* we have just produced in her is nonphysical! Of course it is physical. Evidently, her epistemological novelty here has *nothing to do* with the ontological status of whatever it was that produced that novelty. Not in this case, and not in the case of her finally having a sensation of red either. The novelty in either case, it emerges once more, lies in the *kind of epistemological apprehension* she enjoys, not in the ontological status of the kind of state that is thus newly apprehended. I am happy to concede that the sensation of red *may*, after all, turn out to enjoy an ontological status quite distinct from the activation-triplet at issue. (This is, after all, a wholly empirical issue.) But Jackson's argument does absolutely nothing to *show* that it does.

After his discussion of Jackson, Alter turns to Chalmers and defends a specific version of the original zombie-argument, a version that once again appeals explicitly to intuitions about what is and what isn't 'metaphysically possible'. I do not begin to share his particular intuitions here, and I reject the integrity of these artificial modalities in any case. I also reject the idea that a successful reduction always requires an 'appearance/reality' distinction of some kind, a distinction that is supposedly hard to draw in the unique case of sensations themselves.[1] What a successful reduction *does* require is that the assembled causal profiles of, and the similarity-and-difference relations between, the target entities/properties be reconstructible from within the resources of the aspirant reducing theory. And that requirement has *already been met* in the case of our internal conscious-sensations-of-color as the target entities/properties, on the one hand, and the Hurvitch-Jameson opponent-process theory of how external colors get coded inside the human brain, on the other. (See again my 2005 paper exploring this success, reprinted in my 2007 collection, *Neurophilosophy at Work*.) We are not looking at a tenuous possibility here. We are looking at a done deal. A successful reconstruction of precisely the kind required is already in place.

I close by accepting Alter's wise advice to keep an open mind on the possibility that a form of dualism that is *not* committed to the 'ontological simplicity' of the states that it posits may yet emerge and come to enjoy widespread explanatory and reductive success. The pursuit of such a worthy aim should be Number One on the research agenda of contemporary dualists. And the emergence of a genuine competitor to the neuroscientific research program, one already savoring its own successes, can only enrich the scientific process. By contrast, attempting to construct purely *a priori* arguments to block, right out of the starting gate, the prospects for either program (as so many have done over the past 40 years) is a waste of everyone's time. Scientific issues aren't settled in that way.

---

[1]The reduction of classical atomic theory to sub-atomic physics, for example, involved observables or appearances at *neither* level of the reduction. And the reduction of the Greek 'planets' (i.e., wandering stars) to the planets of Newtonian physics involved observables or appearances at *both* levels of the reduction. In addition, I agree that a sensation-of-red does not *seem* to be a neural activation-triplet. There is an appearance/reality distinction right there.

Part III
Property Dualism and Panpsychism

Chapter 7
Orthodox Property Dualism + The Linguistic Theory of Vagueness = Panpsychism

Philip Goff

By 'consciousness' I mean *the property of being a thing such that there's something that it's like to be that thing*. The meaning of this rather cumbersome phrase can be illustrated with reference to our commonsense beliefs about what things have the property it denotes. According to common sense, there's something that it's like for a rabbit to be cold, or to be kicked, or to have a knife stuck in it. In contrast, there's nothing that it's like for a table to be cold, or to be kicked, or to have a knife stuck in it. There's nothing that it's like *from the inside*, as it were, to be a table (according to common sense). Consciousness, as I will understand it, is the property of *having an inner life* of some kind or other; a property ordinary opinion supposes to be confined to the biological realm.

There are a number of powerful arguments in the literature – I will focus on the zombie-conceivability argument and the knowledge argument – which have the conclusion that consciousness is a non-physical feature of reality.[1] Call these arguments 'the standard arguments'. A sizeable minority of philosophers (i) accept the soundness of the standard arguments, and so take consciousness to be a non-physical feature of reality, (ii) nonetheless take consciousness to be a property of physical objects rather than immaterial substances, a basic property which arises from physical properties in accordance with fundamental psycho-physical laws of nature. Call such philosophers 'orthodox property dualists'. The purpose of this

[1]Chalmers (1996, 2002) and Jackson (1982).

P. Goff (✉)
Department of Philosophy, University of Liverpool, 7 Abercromby Square,
Liverpool L69 7WY, UK
e-mail: philgoff1@googlemail.com

R. Brown (ed.), *Consciousness Inside and Out: Phenomenology, Neuroscience, and the Nature of Experience*, Studies in Brain and Mind 6, DOI 10.1007/978-94-007-6001-1_7,

paper is to argue that orthodox property dualism, in conjunction with the linguistic theory of vagueness, implies *panpsychism*: the view that consciousness is ubiquitous in nature.[2]

Of course, one might accept this conclusion and go a number of ways with it. Depending on the strength of a property dualist's antecedent commitments, accepting my argument might lead her to embrace panpsychism, or to embrace metaphysical vagueness, or to look hard again for a flaw in the standard arguments. I will not be exploring any of these options in what follows. Nonetheless, I take it to be philosophically significant in itself that the conjunction of two popular views has such a surprising implication.

The argument will proceed in three stages. In Sect. 7.1, I will argue that the orthodox property dualist is committed to two theses concerning the *concept* of consciousness: *conceptual dualism* and *phenomenal transparency*. In Sect. 7.2, I will argue that the orthodox property dualist who accepts the linguistic theory of vagueness, because of her commitment to *phenomenal transparency* and *conceptual dualism*, must accept *phenomenal precision*: the thesis that it can never be vague whether or not a given thing is conscious. In Sect. 7.3, I argue from *phenomenal precision* to *panpsychism*. In Sect. 7.4, I will support my case with some methodological remarks.

7.1 Conceptual Dualism and Phenomenal Transparency

7.1.1 Conceptual Dualism

Each of the standard arguments kicks off with an epistemic premise: zombies are conceivable, Mary learns something knew when she leaves her room. For each of these epistemic premises, accepting its truth commits one to the following principle:

> *Conceptual Dualism*: The physical facts do not entail the phenomenal facts, i.e. there is no way of moving a priori from knowing the kind of things physics has to tell us about the world to knowing what conscious states there are, or indeed whether there are any conscious states.

If the physical facts entailed the phenomenal facts, then zombies would be inconceivable, and pre-liberated Mary would know what it was like to see red. The orthodox property dualist, by definition, accepts the soundness of the standard arguments, and therefore is committed to *conceptual dualism*.

---

[2]Note that my argument is not primarily aimed at forms of anti-physicalism other than property dualism, such as Russellian monism (although I think the argument has some force against Russellian monism as I explain in footnote 14), nor against property dualists who do not take the standard arguments to be sound.

7.1.2 Phenomenal Transparency

Each of the standard arguments begins with an epistemic premise. Each of the standard arguments tries to derive from its epistemic premise a metaphysical conclusion: zombies are possible, the physical description of reality is incomplete. Doing this requires a commitment to the following principle:

> *Phenomenal transparency*: The concept consciousness (I will refer to concepts with underlined words) reveals the nature of consciousness, i.e. it is a priori (for someone possessing the concept consciousness, in virtue of possessing that concept) what it is for something to be conscious.

Spelling out this principle, and why the orthodox property dualist is committed to it, requires a bit of work.

It is plausible that concepts denoting properties come divided up into two categories: transparent and opaque. A transparent property concept reveals the nature of the property it denotes, in the sense that it is a priori (for someone who possesses the concept, in virtue of possessing the concept) what it is for an object to have that property.[3] To put it another way, a transparent property concept reveals what is ascribed in an application of the concept. An opaque property concept reveals nothing of what it is for an object to instantiate its referent[4] (I develop this framework for thinking about concepts in more detail in Goff (2011, MS) and Goff and Papineau (forthcoming)).

The best way to clarify and make the case for this distinction is by giving examples. Suppose David's favourite property is Euclidean sphericity, but I am blissfully unaware of this joyous fact. Now consider two ways in which I might think about Euclidean sphericity. I might think of it as *David's favourite property*, where I use that description as a rigid designator. Alternately I might think of it in geometrical terms, as the property of *being a thing with all points on its surface equidistant from its centre*. There is a clear sense in which, when I think of Euclidean sphericity as *David's favourite property*, I don't understand its nature. I have no idea what it is for something to instantiate 'David's favourite property', or as we might simply put it *I have no idea what David's favourite property is*. In contrast, when I think about the same property in geometrical terms I do understand its nature. I know what it is for an object to be spherical: it's for it to have all points on its surface equidistant from its centre. The concept Euclidean sphericity is transparent; the concept David's favourite property (rigidly designated) is opaque.

---

[3]Concept *C* renders fact *F* a priori if it is metaphysically possible for someone to know *F* in virtue of possessing *C*, without relying on any empirical information beyond what is required to possess *C*.

[4]An opaque concept may (but may not) reveal accidental properties of the property it denotes, e.g. it is plausible to think that the concept *being water* reveals that in the actual world the property denoted realises the property of being the colourless, odourless stuff in oceans and lakes. I call an opaque concept which reveals accidental properties of the referent which uniquely identify it in the actual world 'mildly opaque' (see footnote 5).

If consciousness is taken to be opaque, there is no way of moving beyond the epistemic premise of any of the standard arguments. Consider the zombie conceivability argument. For those physicalists – probably currently the majority – who accept that zombies are conceivable, the challenge is to explain why the conceivability of zombies does not entail their genuine possibility. If it is an option to hold that consciousness is opaque, it is obvious what the physicalist can say:

> The concept consciousness denotes a purely physical or functional property – that's why zombies are impossible – but because consciousness is opaque, it's not a priori that consciousness denotes a purely physical or functional property – that's why zombies are conceivable.

If we allow that consciousness is opaque, the conceivability of zombies has no metaphysical significance.[5]

Consider the knowledge argument. The challenge for the physicalist is to say what Mary learns upon liberation. If it is possible that the concept consciousness, and consequently our concepts of its determinates, i.e. specific modes of consciousness, are opaque, it is obvious what the physicalist can say:

[5]On Chalmers 2D semantic framework (1996, 2002, 2009) the primary intention of each term/concept is a priori evaluable (without this, the move in his two-dimensional argument from the conceivability of a state of affairs to its genuine possibility at some world considered as actual would be implausible). He also holds, as most people do, that the primary and secondary intentions of consciousness are the same (which justifies the move in the two-dimensional argument from the possibility of zombies at some world considered as actual to the possibility of zombies at some world considered as counterfactual). It follows that consciousness has an a priori evaluable secondary intention, which is equivalent to its being transparent, see Nida-Rümelin 2007 for a detailed analysis of this (where I talk about concepts 'revealing the nature' of properties, she talks of concepts 'affording a grasp' of properties). Thus, Chalmers' two-dimensional argument against materialism, at least in its standard form, is dependent on the thesis that consciousness is transparent.

Chalmers does claim that the two-dimensional argument goes through without the premise that the primary and secondary intensions of consciousness are identical, as he believes that the conceivable truth of <P&~Q> — where P is the complete physical truth about our world and Q is some arbitrary phenomenal truth about our world – entails that the primary intention of that proposition is true at some world W, and given that W is a minimal physical duplicate of our world but not a duplicate simpliciter, physicalism must be false. The idea is that Q might be what I have called elsewhere 'mildly opaque', i.e. does not reveal the nature of its referent, but reveals accidental features of the referent which uniquely identify it in the actual world. In this case, although W is physically indiscernible from us, it lacks certain properties, i.e. the properties which uniquely identify Q in the actual world, and hence physicalism is false. Similarly to the case of translucency discussed in *Aside*: *Why not translucency*? (below), if the orthodox property dualist wanted to say that consciousness is mildly opaque, I would focus on the accidental features of consciousness that consciousness does reveal to us the nature of, rather than focusing on consciousness itself, and argue for the conclusion that those accidental features are ubiquitous in nature. However, I don't know of any anti-physicalists who do take this approach; it would mean distinguishing the properties we use to think about consciousness, i.e. the property of being a thing such that there's something that it's like to be that thing, from the essential nature of consciousness itself.

> Liberated Mary gains a phenomenal concept which denotes a purely physical or functional property, but is conceptually novel for her because it is opaque, and hence it is not a priori that it denotes a purely physical or functional property. Therefore, when she leaves the room, Mary does not become acquainted with a new feature of reality, but rather finds a new way of thinking about a feature she already knew about in her room.

In the case of each of the standard arguments, a move beyond the merely epistemic is premised on denying that consciousness is opaque. The orthodox property dualist, i.e. the property dualist who is a property dualist on account of the standard arguments, is committed to *phenomenal transparency*.[6]

7.1.3 *Aside: Why Not Translucency?*

I have divided up property concepts into the transparent and the opaque. But these categories do not seem to be exhaustive. Why think that each concept reveals either *all* or *nothing* of the nature of its referent? There seems room for the category of *translucent concept*, where a property concept is translucent if it reveals *some but not all* of the nature of the property it denotes, i.e. something but not everything of what it is for an object to have that property is a priori knowable (for someone possessing the concept, in virtue of possessing the concept). Is it open to the orthodox property dualist to take consciousness to be translucent: revealing some but not all of the nature of consciousness?

I take it that if a property concept is translucent, then the property it denotes is complex, involving within itself a number of aspects. At least one aspect will be denoted transparently, i.e. its nature will be a priori accessible, and at least one aspect will be denoted opaquely, i.e. it will be denoted, but its nature not a priori accessible. Take for example the concept being a sphere roughly the same size as the Earth. This concept reveals the nature of one aspect of the property it denotes, i.e. *being a sphere*, but does not reveal the nature of another aspect of the property it denotes, i.e. *being roughly the same size as the Earth*; empirical work must be done to discover the nature of the latter but not the former aspect.

We can thus consider a translucent concept as a composite of two 'sub-concepts', one transparent and one opaque. I call the transparent sub-concept the 'window' of the whole concept, and the opaque sub-concept the 'screen' of the whole concept. In the above example, the concept of being a sphere is the window of the whole concept, whilst the concept being the same mass as the Earth is its screen.

---

[6] Without phenomenal transparency, it is impossible to move beyond the epistemic premise of the standard arguments. But with phenomenal transparency, and the epistemic premise of either of the standard arguments, the falsity of physicalism follows pretty quickly. If consciousness reveals the nature of consciousness, and consciousness does not reveal consciousness to have a physical nature (which follows from the truth of either of the epistemic premises), then consciousness does not have a physical nature.

If someone wants to defend the claim that the concept having an inner life/being something such that there's something that it's like to be it is translucent, then they are obliged to give us an account of how that concept divides into window and screen. Which aspects of the property of having an inner life are a priori accessible, and which do we refer to but not understand without empirical investigation?

As far as I am aware only physicalists have even given such an account of our phenomenal concepts, i.e. the concepts we form when we think about conscious states in terms of what it's like to have them. Robert Schroer, for example, claims that we can know a priori certain facts about the internal structure of conscious states, but not the intrinsic nature of the basic elements in that structure (Schroer 2010). This allows Schroer to combine *conceptual dualism* with an account of phenomenal concepts according to which they reveal significant information about the states they denote. For Schroer, physical states do not entail phenomenal states, as although we know a priori the internal structure of phenomenal states, we don't know a priori whether the elements composing that structure are physical or non-physical (and hence don't know a priori whether the entire state which results is physical or non-physical).

On Schroer's account, although the whole concept is priori distinct from the physical facts, the window is not: if we knew all the physical facts, we could see that the internal structures connoted by phenomenal concepts are realised in the brain. But for the standard arguments to have force, the window as well as the whole concept must be a priori distinct from the physical facts. Otherwise the physicalist can simply give the following explanation of why zombies are conceivable but not possible (as Schroer in fact does):

> For each phenomenal concept, both window and screen denote purely physical or functional properties – that's why zombies are impossible – but because the screen is opaque, it's not a priori that the screen denotes a purely physical or functional property, and hence not a priori that the whole concept denotes a physical or functional property – that's why zombies are conceivable.

For the standard arguments to succeed, there must be at least one aspect of conscious experience which is understood a priori, and which is not entailed by the physical facts. It is difficult to see how an orthodox property dualist might divide the concept of having an inner life into an aspect we transparently understand and an aspect we don't, and indeed difficult to see what their motivation for doing so would be. But if they did divide up the concept into window and screen, we could simply substitute the word 'consciousness' in what follows for 'consciousness*', defined as 'that aspect of consciousness we understand the nature of a priori'. I will continue to assume that consciousness is transparent, but we can note that if consciousness turns out to be translucent rather than transparent, then my argument is to be read as aiming to show that consciousness*, rather than consciousness, is ubiquitous in nature.[7]

---

[7]The situation is similar to the case of the imagined – as far as I am aware non-existent – anti-physicalist who wants to claim that we pick out consciousness in virtue of its accidental features, which I discuss in footnote 5.

7.2 Orthodox Property Dualism + Linguistic Theory of Vagueness = Phenomenal Precision

Despite well known contemporary defences of epistemic and metaphysical accounts of vagueness, the 'linguistic theory of vagueness', i.e. the broad spectrum of views which locate the source of vagueness in language rather than the world, remains probably the most popular approach to dealing with vagueness.

According to the linguistic theory of vagueness, vagueness is the result of *semantic indecision*: for any vague predicate there are multiple 'sharpenings' of the predicate, such that the meaning of the predicate does not settle on any of these sharpenings. Consider the vague predicate 'is tall', We could stipulate, somewhat arbitrarily, that anything that is exactly 6 ft or taller counts as 'tall', and anything shorter is not tall. This is one 'sharpening' of the predicate 'is tall', that is, one way of making the predicate precise. Alternately, we could stipulate that anything that is exactly 6 ft and 1 in. or taller counts as tall, and anything shorter is not tall. This is an alternative sharpening of 'is tall', that is, an alternative possible way of making the predicate precise. The predicate 'is tall' is thus associated with a *spectrum of sharpenings*: a range of possible ways of making the predicate precise.[8]

The linguistic theory of vagueness tells us that a vague predicate is vague because no one has bothered to single out one of its sharpenings as the unique meaning of the predicate. To put it metaphorically, the predicate hasn't made up its mind which of those precise meanings it wants to plump for. Suppose John is a borderline case of tallness. According to the linguistic theory of vagueness, it's not that in reality there is some fuzzy, indeterminate state of affairs of John's neither having nor lacking a certain quality. In the world there's just John with some utterly precise height. It's *the predicate* that is indeterminate such that there's no fact of the matter as to whether or not it applies to things with John's exact height. The indeterminacy is in language rather than the world.

The linguistic theory of vagueness explains the vagueness of a predicate in a way that involves the associated spectrum of sharpenings. Clearly if this kind of explanation is to work, then each vague predicate must be associated with a spectrum of sharpenings. However, the predicate 'is conscious' – and hence the concept it expresses – does not seem to be associated with a spectrum of sharpenings, at least not a priori.

This can be a difficult point to get across, because the word 'consciousness' is used in lots of different ways by different philosophers and scientists. Sometimes the predicate 'is conscious' is used to mean *is aware/cognitively sophisticated to a certain level*, perhaps roughly the level we would be inclined to

---

[8]With some vague predicates, as with 'is tall', the sharpenings are determinates of a single determinable. In the case of other vague predicates, e.g. 'is a religion', there is a weighted cluster of properties, *involves belief in a supernatural being*, *involves ritual*, *involves a moral code*, such that each sharpening involves some of those properties but it is not the case that each sharpening involves all of those properties.

call 'self-consciously aware'. This does seem to be a notion of consciousness which is associated with a spectrum of sharpenings, ranging from more to less cognitively sophisticated. But that's not the notion of consciousness we are concerned with. I am using the predicate 'is conscious' to mean *has an inner life of some kind or other*, and this doesn't seem to be a notion of consciousness which is a priori associated with a spectrum of sharpenings. You either have an inner life or you don't. Of course you can have a richer or a less rich inner life, a more sophisticated or a less sophisticated inner life. But the property of *having an inner life* itself does not present itself to us as one that admits of degree: you either have it or you don't.

The physicalist wanting to embrace phenomenal vagueness, at least if she is prepared to deny *phenomenal transparency*, need not worry that the sharpenings of consciousness are not available a priori. She can claim that the semantic workings of the concept, and therefore its spectrum of sharpenings, are determined 'outside the head'. David Papineau, for example, an explicit rejecter of *phenomenal transparency*,[9] denies that the semantic workings of consciousness – constituted of causal or teleological facts – are a priori accessible. Those semantic workings, according to Papineau, leave it indeterminate whether the concept picks out the capacity for higher-order judgement, or the physical basis for that capacity in humans.[10] There is thus a recognisable sense in which consciousness – and hence the predicate 'is conscious' – has (at least) two sharpenings: (A) the capacity for higher-order judgement, (B) the physical basis of higher-order judgement in humans. On sharpening (A) silicon duplicates of humans count as 'conscious', on sharpening (B) they don't. Papineau does not take (A) and (B) to be a priori accessible: the semantic workings of consciousness are not a priori accessible, and so neither are the more referentially precise versions of those semantic workings.[11]

Nothing I have said above casts doubt on Papineau's view, or anything like it. But notice that it assumes the falsity of *phenomenal transparency*, at least if we are assuming the truth of the linguistic theory of vagueness. According to the linguistic theory of vagueness, what is ascribed in the application of a given vague predicate is to be understood in terms of the predicate's indeterminacy over its sharpenings.

---

[9]In his 2006 Papineau gives an explicit denial of *phenomenal transparency*.

[10]Papineau 2002, ch. 7. In fact, Papineau is open to the possibility of conscious states which cannot be thought about, and because of this ends up thinking that the concept consciousness is indeterminate such that on one sharpening it refers to attention, on one sharpening it refers to pre-attention, and on one sharpening it refers to the property of being material! It is an under-emphasised implication of this (I have confirmed with Papineau in conversation that he embraces this implication), that there is no fact of the matter as to whether or not panpsychism is true, just as there is no fact of the matter as to whether I am tall. On one sharpening of consciousness, the table and the pillar of salt are conscious, on another sharpening they are not. It is ironic that Papineau's denial of transparency, which allows him to escape the argument for panpsychism given in this paper, gets him in the end to panpsychism (at least on one legitimate sharpening of consciousness).

[11]It is because of this option, open to the a posteriori physicalist like Papineau, of claiming that the semantic workings of consciousness are outside of what is a priori accessible to the concept user, that I reject the kind of argument Michael Antony (2006) gives for the non-vagueness of consciousness.

Assuming the truth of this view, if what is ascribed in the application of a given vague predicate is a priori knowable, then the sharpenings of that predicate must be a priori knowable. In other words, if the linguistic theory of vagueness is true, then the sharpenings of a transparent predicate must be a priori knowable. If the orthodox property dualist wants to claim that consciousness is associated with a spectrum of sharpenings, whilst remaining faithful to the linguistic theory of vagueness, then, given her commitment to *phenomenal transparency*, she is obliged to hold that these sharpenings are a priori accessible.

There seems to me only one even vaguely plausible proposal as to what the spectrum of sharpenings a priori associated with consciousness is: that which would be offered by the analytic functionalist (even this proposal does not seem very plausible to me, but then that is because I don't find analytic functionalism very plausible). If the predicate 'is conscious' is a functional or behavioural predicate, then presumably it is associated a priori with a spectrum of sharpenings, which can be captured with a fine grained enough functional description.[12] But of course the orthodox property dualist, given her commitment to *conceptual dualism*, is obliged to deny that the predicate 'is conscious' is a functional predicate. The functional and behavioural states of an organism are entailed by the physical facts about that organism. Therefore, if 'is conscious' were a functional or behavioural predicate, then it too would be entailed by the physical facts, contrary to *conceptual dualism*. Putting the analytic functionalist's proposal on one side, there just doesn't seem to be another candidate for being the spectrum of sharpenings a priori associated with consciousness.

Perhaps it might be objected that the sense of consciousness can be sharpened, but that we lack the necessary concepts to grasp the resulting sharper concept. By analogy, it might seem initially plausible that someone might possess the concept phenomenal red, without possessing a concept of any more specific phenomenal shade of red. Such a person would possess a concept, the sense of which can be sharpened, and yet be unable to sharpen it. Perhaps this is the situation we are in with respect to the general concept consciousness; the sense of the concept can be sharpened, but we lack the concepts required to do it.

However, for the reasons discussed above, assuming the truth of the linguistic theory of vagueness, there couldn't be a transparent vague concept which does not allow a priori knowledge of its sharpenings. According to the linguistic theory of vagueness, what is ascribed in the application of a given vague predicate is to be understood in terms of the predicate's indeterminacy over its sharpenings. Given this, for what is ascribed be a priori knowable, it must be a priori knowable what the relevant sharpenings are.

There is a sense in which our inability to find sharpenings of consciousness is not *entirely* conclusive evidence that there are no a priori knowable sharpenings

[12]On Lewis's kind of materialism (see Lewis 1994) there would be a priori associated with consciousness a spectrum of sharpenings of the property of *being consciousness*, but not of consciousness itself (Lewis takes mental concepts to be flaccid designators).

of consciousness. Certain facts which are rendered a priori knowable by concept *C* may be out of the cognitive reach of a given individual possessing *C*, due to that individual's cognitive limitations.[13] Perhaps one might suppose that if we were better reasoners we would be able to see how to sharpen consciousness. This seems to me an implausible leap of faith. We are not dealing with some difficult mathematics which is beyond our cognitive capacities, but which greater beings than ourselves could deal with. We are dealing with the basic semantic structure of a single concept. If our best efforts to find sharpenings of consciousness do not yield them, then we must suppose that there are no such things, at least not accessible a priori.

I conclude, therefore, because of their commitment to *phenomenal transparency* and *conceptual dualism*, the orthodox property dualist is unable to make sense of consciousness having sharpenings. If she wants to remain faithful to the linguistic theory of vagueness, the orthodox property dualist must hold that consciousness is not vague.

7.3 From *Phenomenal Precision* to Panpsychism

In the bible we hear that God turned Lot's wife into a pillar of salt. You get the impression that it happened pretty quickly, but let's suppose that in fact God did it in really small stages: He took Lot's wife, made a slight adjustment to one fundamental particle, a slight adjustment to another fundamental particle, and so on until He had a pillar of pure salt.

Had God gone about it this way, the result would be a temporally continuous series, with Lot's wife at one end, a pillar of salt at the other, and in between a series of objects such that any two objects next to each other in time differ at most by a slight adjustment of a fundamental particle.

Here's a common sense assumption:

Commonsense Assumption: Lot's wife is conscious and a pillar of salt is not conscious.

It follows from *Commonsense Assumption* that we have consciousness at one end of the series but not the other; somewhere along the series consciousness disappears. If it could be vague whether or not a given thing is conscious, then presumably there would be borderline cases along the series, where it is vague whether or not we have case of consciousness. But assuming *phenomenal precision*, the cut off point must be utterly sharp. Somewhere along the line there must be two objects, next to each other in time, differing only by a slight adjustment to a fundamental particle, such that one but not the other is conscious. This leads us to the following implausible consequence:

---

[13]See footnote 3 for my definition of a priori knowability.

Implausible Consequence: The fundamental psycho-physical laws which specify the physical conditions nomologically sufficient for consciousness are utterly precise, in the sense that the slightest adjustment to the smallest particle can make the difference between whether or not a macroscopic object is conscious.

Why is *Implausible Consequence* implausible? Consider the following analogy. Imagine one day I am blowing up a blue balloon. I blow it up about two thirds of the way when suddenly, to my surprise, the balloon turns pink! In shock, I let the balloon deflate. I try blowing it up again, and find that, at exactly the same point, the balloon turns pink. I experiment with a number of balloons from the same packet but find that the effect is not repeated. Much experimenting later, I discover that the following is a fact about our universe:

Random Fact: When a blue balloon is (i) made from three specific kinds of elastic, A, B and C, such that there is 42 % of A, 38 % of B, and 20 % of C, (ii) has a certain thickness, precise to 1,000,0000,000th of a millimetre, (iii) is blown up such that it's diameter has a certain length, precise to 1,000,000,000th of a millimetre, the balloon turns pink.

The hypothesis that *Random Fact* constitutes a basic law of nature is extraordinary. Were we to discover that *Random Fact* obtains in our world, we would be extremely reluctant to take it as a fundamental law, and would try to find a way of explaining its obtaining in terms of more general laws, ones which did not involve such arbitrarily precise values. Of course it is not *inconceivable* that such a law obtains: there is an extremely strange possible world governed by such a law. The hypothesis that such a law obtains is not necessarily false, but is extremely theoretically implausible. It is rational to avoid such a hypothesis it if at all possible.

But if the supposition that *Random Fact* constitutes a basic law of nature is to be avoided, then so much, much more so is *Implausible Consequence*. A law L specifying that physical conditions P are sufficient for macroscopic consciousness, where P are utterly precise down to slightest change in the smallest particle, would involve such arbitrarily precise specifications – many times more so than those involved in *Random Fact* – that it would be crazy to suppose that L was brute. *Implausible Consequence* is to be avoided at all costs.[14]

[14]This argument is aimed at orthodox property dualists, whom I have stipulated to hold that consciousness is a fundamental feature of reality arising in accordance with basic psycho-physical laws of nature. But not all anti-physicalists take consciousness to be a fundamental feature of reality. Many Russellian monists (Russell 1927; Feigl 1958/1967; Maxwell 1979; Lockwood 1989; Chalmers 1996; Griffin 1998; Stoljar 2001) take phenomenal properties be realised in *proto-phenomenal* properties, certain qualities of physical objects which are not themselves phenomenal properties, but are somehow intrinsically suited to constitute phenomenal properties (clearly, our grasp of such qualities is frustratingly meagre). Perhaps the Russellian monist could hold that the conditions sufficient for consciousness are utterly precise, but that this fact is explained in terms of some more fundamental laws involving protophenomenal properties, laws which do not involve such arbitrarily precise specifications. However, even on the supposition that consciousness is not fundamental, it is still pretty implausible to suppose that a slight adjustment to a single fundamental particle – one of countless billions – in the brain could make the difference between the whole brain having or lacking the determinable property of consciousness. So I am inclined to think that

But, as I hope to have shown, *Implausible Consequence* follows from the conjunction of *Commonsense Assumption* and *phenomenal precision*: *Commonsense Assumption* entails that somewhere along the series consciousness disappears, *phenomenal precision* entails that the disappearance of consciousness must be sharp rather than vague. If we want to reject *Implausible Consequence*, then we must reject (at least) either *Commonsense Assumption* or *phenomenal precision*. Given that the orthodox property dualist is committed to *phenomenal precision*, she must reject *Commonsense Assumption*: she must hold either *that neither Lot's wife nor the pillar of salt are conscious* or *that both Lot's wife and the pillar of salt are conscious*. Given her realism about consciousness, the orthodox property dualist is hardly going to go for the former disjunct. Therefore, she is obliged to think that both Lot's wife and the pillar of salt are conscious.

Of course it's not going to end there. We could take any pair of macroscopic objects such that common opinion takes the former but not the latter to be conscious, and do the same thing. To return to the example used at the start of the paper, we might imagine God turning a rabbit into a table, particle by particle, and by a similar chain of reasoning get to the conclusion that the table is conscious. We quickly end up with panpsychism: the view that consciousness is ubiquitous throughout nature.[15]

In setting up the thought experiment I have implicitly assumed unrestricted composition, such that none of the changes God makes to the particles which initially compose Lot's wife results in those particles ceasing to compose anything. Let's entertain the supposition that composition is restricted, such that when the particles are arranged Lot's wife-wise they compose, but when they are arranged pillar of salt-wise they fail to compose; somewhere along the series we cease to have a composite object.

Call the time when the particles stop composing '*C*', and the time when the particles stop *phenomenally composing*, that is composing a conscious object, '*P*'. Let us consider in turn the supposition (i) that *C* precedes *P* (ii) that *C* is simultaneous with *P*, (iii) that *P* precedes *C*.

Supposition (i) is impossible. It cannot be the case that *C* precedes *P*, for any time at which the particles phenomenally compose is a time at which the particles compose.

Supposition (ii) implies that *C* is a precise time, given that *P* is a precise time (assuming *phenomenal precision*). The supposition that *C* is a precise time entails a sharp cut off point between the particles composing and the particles ceasing

---

the considerations outlined here have some force against the Russellian monist, even though the argument is primarily aimed at, and has more force against, the orthodox property dualist.

[15]Throughout the thought experiment I have, for simplicity, assumed that there are fundamental particles, and have spoken of time as though it were ultimately composed of indivisible moments. Neither of these simplifications is essential to the argument. We might instead suppose that God makes a slight adjustment to a sub-atomic particle every 100,000,000,000,000th of a second. Even if there are no fundamental particles, it is still implausible that the fundamental psycho-physical laws of nature are precise such that a slight adjustment of a sub-atomic particle can make the difference between the presence and absence of macroscopic consciousness.

to compose: a slight adjustment of a fundamental particle makes the difference between the particles composing and failing to compose. But this leads to:

> *Implausible Consequence**: The mereological laws which specify the physical conditions sufficient for particles to compose are utterly precise, in the sense that the slightest adjustment to the smallest particle can make the difference between the presence and absence of macroscopic composition.

However, *Implausible Consequence** is just as implausible as *Implausible Consequence*.[16] It is just as implausible to suppose that there are basic mereological laws involving such arbitrarily precise specifications, as it is to suppose that there are basic psycho-physical laws involving such arbitrarily precise specifications. It is just as implausible to suppose that there are sharp cut off points between macroscopic composition and its absence as it is to suppose that there are sharp cut off points between macroscopic phenomenal composition and its absence. We must reject the supposition that *C* is a precise time, and hence the supposition that *C* is simultaneous with *P*.

Finally, let us consider supposition (iii). Given the implausibility of sharp cut off points between macroscopic composition and its absence, we must suppose that *C* is a vague time. So we are supposing that at some precise time *P* the particles stop phenomenally composing, and at some later vague time *C* the particles stop composing altogether. I don't think this is a plausible supposition, as I hope to demonstrate in what follows.

At the first moment after *P*, call it '*P + 1*', there must be a definite fact of the matter as to whether the particles phenomenally compose (assuming *phenomenal precision*).[17] Given that the particles definitely phenomenally compose at *P*, it is implausible to suppose that they definitely do not phenomenally compose at *P + 1* – this would lead to *Implausible Consequence* – therefore at *P + 1* the particles must definitely phenomenally compose. And if the particles definitely phenomenally compose at *P + 1*, then they definitely compose at *P + 1*.

But now consider the second moment after *P*, call it '*P + 2*'. There must be a definite fact of the matter at *P + 2* whether or not the particles phenomenally compose. Given that they phenomenally composed at *P + 1*, they must phenomenally compose, and hence compose, at *P + 2*, on pain of the truth of *Implausible Consequence*. We could keep doing this for every subsequent moment until we get to the particles arranged pillar of salt-wise, which entails that there is no moment along the series at which the particles stop composing, i.e. *C* does not exist. Supposition (iii) cannot be sustained once we have signed up to *phenomenal precision*.

Thus, once we have committed to *phenomenal precision*, we cannot plausibly hold that any of the adjustments God makes result in the particles failing to compose. We now have a complete argument, not only for panpsychism, but also for unrestricted phenomenal composition, and hence for unrestricted composition,

---

[16] A similar claim is argued in Sider (2001), 120–134, a strong influence on this argument.

[17] Those who take time to be infinitely divisible may substitute 'moment' for '100,000,000,000,000th of a second', see footnote 15.

at least regarding macroscopic objects. All combinations of particles numerous enough to be arranged macroscopic-wise phenomenally compose, and hence all such combinations of particles compose.

Why do I make the qualification that phenomenal composition is unrestricted 'regarding macroscopic objects'? I have been implicitly supposing in the above thought experiments that the number of particles remains unchanged in these imagined transformations of woman to pillar of salt, or rabbit to (presumably quite small) table. But what if God took a conscious being and annihilated one particle a time, until only one particle remained? Is the orthodox property dualist obliged to think a single particle has an inner life?

It seems to me that the argument still has force when we are dealing with objects composed of very high numbers of particles. For a conscious object composed of seven billion particles, it is implausible to suppose that the psycho-physical laws are precise such that the removal of a single one of those seven billion particles could render it non-conscious. But it is not clear to me that the argument has force when we are dealing with objects composed of small numbers of particles. The smaller the number of particles required for consciousness, the less implausibly arbitrary the values involved in the psycho-physical laws, e.g. it is not implausible to suppose that the basic psycho-physical laws specify that at least four particles are required for phenomenal composition.[18]

The orthodox property dualist, then, is not obliged to subscribe to *unrestricted phenomenal composition*, but only to *unrestricted phenomenal composition at the macroscopic level*. We thus end up with a very different kind of panpsychism to that defended by contemporary panpsychists such as Galen Strawson[19] and Sam Coleman.[20] These panpsychists warrant the name in virtue of holding that *the fundamental constituents of reality* are conscious, but are reluctant to attribute consciousness to inanimate macroscopic objects. Given the vagueness of the boundary between the animate and the inanimate, and given the commitment to *phenomenal precision* that I would argue the commitment to the soundness of the standard arguments forces upon these panpsychists, the considerations I have outlined above put severe strain on this kind of view.[21]

---

[18]For a similar reason I believe Sider's 'vagueness argument' for unrestricted composition is inconclusive. It does not seem implausible to me to suppose that the basic laws of mereology specify that at least four particles are required for composition. Sider's argument gives us strong reason to think that *macroscopic* composition is unrestricted, but has no force when applied to cases at the fundamental level involving a small number of particles.

[19]Strawson (2006).

[20]Coleman (2006, 2009).

[21]The argument of this section is heavily influenced by the Lewis/Sider 'vagueness argument' for unrestricted composition, see Lewis (1986, 221–213) and Sider (2001, 120–134).

7.4 Common Sense and Serious Metaphysics

I would like to finish by strengthening my case with some methodological considerations. One might think that the case I have made is less than conclusive, as the orthodox property dualist can always avoid panpsychism without giving up on the linguistic theory of vagueness by going for *Implausible Consequence*. *Implausible Consequence* is in itself a very unattractive option, but, when the alternative is conscious pillars of salt, one might be forgiven for suddenly finding it attractive.

Even if this thought is right, we still have an interesting result. We have the orthodox property dualist facing a difficult choice between deeply implausible fundamental laws, metaphysical/epistemic accounts of vagueness, and conscious pillars of salt. But I do want to go further, and to do what I said I would do, which is to argue that orthodox property dualists who are committed to the linguistic theory of vagueness should be panpsychists. In order to do this, I must lessen the theoretical concern regarding panpsychism, which is what I will try to do in what follows.

What is the worry about panpsychism? I don't think it can be a worry about economy. For sure the panpsychist believes in more consciousness than does the average man. But this is at worst a sin of *quantitative* rather than *qualitative* profligacy – postulating more of a kind we already believe in rather than postulating new kinds – and it is generally agreed by metaphysicians that quantitative profligacy is not an especially heinous sin. It is postulating new *kinds* of thing beyond necessity that we need to avoid.

I think the worry with panpsychism is simply that it is so at odds with ordinary opinion. But when you take a step back, it's difficult to see why this consideration should concern the metaphysician. If we're trying to find out the nature of reality as it is in and of itself, why should we care what the average Joe thinks about things? Scientists often tell us weird things about the world. How often do other scientists say, 'Now hold on, Steve, this is getting quite out of kilter with what the average person thinks . . . maybe we should have second thoughts . . . '. Not often. And if fit with common opinion is not a serious consideration in science, it is difficult to see why it should be a serious consideration in metaphysics.

One contemporary metaphysician to have offered an argument for a concern for common sense is David Lewis.[22] I assume that something like Lewis's justification is implicitly guiding the practices of contemporary commonsense-ophile metaphysicians:

> . . . it is pointless to build a theory, however nicely systematised it might be, that it would be unreasonable to believe. And a theory cannot earn credence just by its unity and economy. What credence it cannot earn, it must inherit. It is far beyond our power to weave a brand new fabric of adequate theory *ex nihilo*, so we must perforce conserve the one we've got [i.e. the theory that is implicit in common sense] . . . .It's not that the folk know in their blood

[22]Having 'a concern' for common sense does not render it sacrosanct. Arguably Lewis ends up straying quite far from what would be acceptable to the average Joe.

> what the highfalutin' philosophers may forget. And it's not that common sense speaks with the voice of some infallible faculty of 'intuition'. It's just that theoretical conservatism is the only sensible policy for theorists of limited powers . . . [23]

How do we choose between theories in the sciences? One thing we do is weigh theoretical virtues: where there are empirically equivalent theories, we choose between them on the basis of simplicity, elegance, etc. But of course our primary concern, our starting point for enquiry, is fit with the empirical data. We first turn to the empirical data, and then when we've got everything we can there, we turn to theoretical virtues (no doubt an oversimplification, but it'll do).

How do we decide between theories in metaphysics? Again, one thing we do is weigh theoretical virtues. But, as Lewis says, that can't be the starting point for our enquiry; we can't weave a theory out of elegance, simplicity, etc. We could end up anywhere! So the interesting question is: What constitutes the starting point of metaphysical enquiry? What plays the role in metaphysics that empirical data plays in science?

Lewis, because he doesn't think there's anything better, opts for common sense. The Lewisian method is to start with the theory that is implicit in common sense, and then move beyond that on the basis of theoretical virtues. Crucially, Lewisian metaphysics is built on common sense only because *there isn't anything better*.

But the orthodox property dualist does have something better. The orthodox property dualist claims to have a concept which: (i) transparently reveals the nature of its referent, and (ii) is satisfied. Indeed, I take it that most orthodox property dualists believe that we know with Cartesian certainty that the concept of consciousness is satisfied; each person knows for certain that s/he is conscious. A transparent concept which we know for certain is satisfied amounts to a window onto a bit of the world as it is in and of itself. Much better than common sense!

Unlike Lewis, the orthodox property dualist has no need for common sense; she is able to build metaphysics on much firmer foundations. She might, like Descartes, try to start and finish with *that which cannot be doubted*. However, history is testimony to the failure of Descartes' research project. The orthodox property dualist metaphysician would be better advised to steer a middle way between Descartes and David Lewis. She should follow Descartes in starting with the undoubtable, but follow Lewis in moving beyond the starting point of enquiry by appeal to theoretical virtues. Here's the slogan: Start with *the undoubtable*, then move to *that which the undoubtable renders most probable*[24] (I develop this 'post-Galilean' approach to metaphysics in much more detail in my (MS)).

The only reason a metaphysician need care about common sense is from want of anything better upon which to build metaphysics. But the orthodox property

---

[23] Lewis (1986, 134).

[24] Strictly speaking we have certainty only *that the concept of consciousness is satisfied*. We are not infallible concerning what it takes for the concept to be satisfied (although I take it that we can have strongly justified knowledge about the latter).

dualist has something better: a priori access to the complete nature of a certain feature of reality, i.e. consciousness. The orthodox property dualist should forget about common sense, and embrace conscious pillars of salt.[25]

References

Antony, Michael V. 2006. Vagueness and the metaphysics of consciousness. *Philosophical Studies* 128(3):515–538.

Chalmers, D. 1996. *The conscious mind*. Oxford: Oxford University Press.

Chalmers, D. 2002. Consciousness and its place in nature. In *Philosophy of mind: Classical and contemporary readings*, ed. D. Chalmers, 247–272. Oxford/New York: Oxford University Press.

Chalmers, D. 2009. The two-dimensional argument against physicalism. In *Oxford handbook to the philosophy of mind*, ed. Brian P. McLaughlin, 313–339. Oxford: Oxford University Press.

Coleman, S. 2006. Being realistic: Why panpsychism may entail panexperientialism. *Journal of Consciousness Studies* 13(10–11): 40–52.

Coleman, S. 2009. The mind that abides. In *Mind under matter*, 83–107. Amsterdam/Philadelphia: John Benjamins Publishing Company.

Feigl, H. 1958/1967. The 'mental' and the 'physical. *Minnesota Studies in the Philosophy of Science* 2: 370–497. Reprinted (with a postscript) as *The 'mental' and the 'physical'*. Minneapolis: University of Minnesota Press.

Goff, P. 2011. A posteriori physicalists get our phenomenal concepts wrong. *Australasian Journal of Philosophy* 89(2): 191–209.

Goff, P. MS. *Taking consciousness seriously*.

Goff, P., and D. Papineau. forthcoming. What's wrong with strong necessities? *Philosophical Studies*

Griffin, D.R. 1998. *Unsnarling the world-knot: Consciousness, freedom, and the mind-body problem*. Berkeley: University of California Press.

Jackson, F. 1982. Epiphenomenal qualia. *The Philosophical Quarterly* 32: 127–136.

Lewis, D. 1986. *On the plurality of worlds*. Oxford: Basil Blackwell.

Lewis, D. 1994. Reduction of mind. In *Companion to the philosophy of mind*, ed. S. Guttenplan. Oxford: Blackwell.

Lockwood, M. 1989. *Mind, brain, and the quantum*. Oxford: Oxford University Press.

Maxwell, G. 1979. Rigid designators and mind-brain identity. *Minnesota Studies in the Philosophy of Science* 9: 365–403.

Nida-Rümelin, M. 2007. Grasping phenomenal properties. In *Phenomenal concepts and phenomenal knowledge. New essays on consciousness and physicalism*, ed. T. Alter and S. Walter, 307–349. Oxford: Oxford University Press.

Papineau, D. 2002. *Thinking about consciousness*. Oxford: Oxford University Press.

Russell, B. 1927. *The analysis of matter*. Repr., New York: Dover, 1954

Schroer, R. 2010. Where's the beef? Phenomenal concepts as both demonstrative and substantial. *Australasian Journal of Philosophy* 88(3): 505–522.

Sider, T. 2001. *Four dimensionalism*. Oxford/New York: Oxford University Press.

Stoljar, D. 2001. Two conceptions of the physical. *Philosophy and Phenomenological Research* 62: 253–281.

Strawson, G. 2006. Realistic monism: Why physicalism entails panpsychism. *Journal of Consciousness Studies: Controversies in Science and the Humanities* 13(10–11): 3–31.

[25]I would like to thank everyone who contributed at the 'Mental as Fundamental' conference in Vienna, and the participants in the second Online Consciousness Conference, especially my two respondents Jonathan Simon and William Robinson.

Chapter 8
A Wake Up Call

William S. Robinson

Although I am going to argue that we need not accept the conclusion expressed in Philip Goff's title, I am grateful for the opportunity to comment on his paper. For I believe he has raised an important challenge that property dualism, and some other views, must face up to.

We may begin by locating this challenge within a broader framework. The consequences that Goff rightly labels "implausible" are discontinuities. Rejection of discontinuity is a version of Leibniz' Principle of Continuity, often expressed as the idea that there are no leaps in nature. Acceptance of such a continuity principle leads to a potential problem in any case where we may want to allow for gain or loss of binary properties.

Goff's paper makes it clear that there is no problem with binary *concepts*. We can form these from non-binary concepts by introducing sharp thresholds. But if we propose that things can gain or lose binary properties whose corresponding concepts are not arbitrary divisions of a continuum, then we must accept the burden of showing how we can avoid real discontinuities in nature. *Being a conscious entity* and *being conscious* appear to be properties of this latter kind.

There are two familiar contexts that require those who are not panpsychists to think about gaining and losing consciousness. First, it is *prima facie* plausible that there was no consciousness in any Earthly entity four billion years ago. But there is now. So, it seems plausible that consciousness must somehow have come into being. Against this view, however, we can advance Goff's point in this way: The popping into existence of fully formed consciousness saddles us with an implausible discontinuity, but the gradual coming into existence of consciousness seems to conflict with our conception of what consciousness is. Moreover, as Goff makes clear, it is not easy for the property dualist to hold that our conception of consciousness is mistaken.

W.S. Robinson (✉)
Department of Philosophy and Religious Studies, Iowa State University, Ames, IA 50014, USA
e-mail: wsrob@iastate.edu

R. Brown (ed.), *Consciousness Inside and Out: Phenomenology, Neuroscience, and the Nature of Experience*, Studies in Brain and Mind 6, DOI 10.1007/978-94-007-6001-1_8,

The second familiar context that may puzzle us is that it at least seems that we lose consciousness at least once a day, i.e., when we fall into dreamless sleep, and that we regain it either in dreams or upon waking. Now, it may be that this is a deceptive appearance. We are conscious during the dreams we remember, and conscious during dreams that we have but do not remember. Maybe there is a kind of consciousness we have even when we are not dreaming at all, and that we never remember when awake.

An assertion of unending consciousness, however, should be empirically supported in some way. It would be amazing if we could establish continuous consciousness when we are asleep merely by reflections from the armchair. If we agree with this sentiment, then we should suppose that a viable property dualism must contain a plausible account of the possibility of daily episodes of losing and regaining consciousness. Goff's paper can thus be viewed as a wake up call – a call for the property dualist to develop an account of waking, falling asleep, and the arising of Earthly consciousness.

8.1 Preliminaries

I shall attempt to provide such an account. In order to do this, however, I need to reframe the discussion a little, so that I can state a property dualist view in terms that have an easy fit with the way I believe property dualists commonly think.

The first step in this reframing is to note that creature consciousness is not basic. What is basic is, instead, episodes of consciousness. Being a creature such that there is something it is like to be it is being a creature that has episodes of consciousness of various kinds. So, what has to be accounted for is the arising and disappearing of episodes of consciousness.[1]

A thing can be composed of the same parts while having different kinds of conscious episodes. For example, it might or might not happen that a garbage truck arrives while I am writing these comments, and so it might or might not happen that a certain annoying sound will intrude upon my consciousness. The difference between having and not having such an experience does not depend on my having different parts; instead, it depends on differences in activities in some of my parts.

There is, of course, a relation between processes and parts. You can't have boiling unless you have molecules of a liquid. You can't have an episode of a string vibrating at 440Hz without having a string. Similarly, you can't have a pattern of neural firings without having neurons. However, the natural way for a property dualist to think

[1]Here, and elsewhere in these comments, I rely on some views about consciousness that I have explained and defended in Robinson (2004a). These views, of course, may not be accepted by all property dualists. However, what is needed to dispute Goff's conclusion is only one coherent form of property dualism. It need not be shown that views that include property dualism, but are laden with certain further assumptions, cannot lead to panpsychism.

about consciousness is to regard it as depending on neural processes, rather than depending directly on a set of constituents that are involved in those processes. For this reason, it is a bit awkward for a property dualist to phrase the continuity problem in terms of replacement of particles, as Goff does in his pillar of salt illustration. (Of course, particle replacements in molecules composing neurons can be expected to interfere with their normal activations. So, particle replacements would indirectly affect consciousness, by affecting the processes on which consciousness directly depends.)

The move to processes rather than parts is not a sneaky way of avoiding Goff's problem, for the same problem can plainly be formulated in process terms. Suppose we agree that to be a being for which it is like something to be it is to be a being that has (or can have, if it's in a dreamless sleep) episodes of consciousness, and that episodes of consciousness depend on neural processes. What if we now fiddle with those processes? We might, for example, lower the firing rate of some neuron. Maybe that will merely change the quality of consciousness a little – for example, maybe the change in neural firing rate will make a taste a little bit less salty. But the new taste will still be an episode of consciousness. Let's fiddle a bit more. Again, we will change the quality that's in consciousness, but not the fact of there being some kind of consciousness or other.

These reflections provide background for a revised formulation of Goff's *Implausible Consequence*: It is not plausible that lowering the firing rate of a single neuron from, say, 50–49 Hz, should make the difference between there being an episode of consciousness of some kind or other, and there being no conscious episode at all. Yet, it seems that if we agree that consciousness is an all-or-none property that can be gained or lost, we must also agree that it is possible that some very small difference in neural processes could make the difference between there being and there not being an episode of consciousness.

I am now going to try to explain away this apparent implausibility. To do that, I will first make a key distinction, and then accept the possibility of a real continuity in the gain or loss of consciousness in its most basic form. Next, I will explain both why it seems to us that there can't be a real continuity for gain or loss of consciousness, and why, despite those appearances, such a thing is possible.

8.2 Consciousness *Tout Court* vs. Self-Consciousness

Goff suspects that being conscious is not a complex property. (See end of his Sect. 7.1) I understand what he means, but I think that we do need to make a distinction among phenomena that are properly called "episodes of consciousness". Namely, as I shall put it, there are (a) episodes of qualitative consciousness *tout court*. (These may be either inattentive or attentive.) And (b) there are episodes of self-consciousness. This second class seems to me to embrace a range, from cases where subjectivity of experience is barely registered to full-blown introspection. I shall not, however, attempt a complete account of distinctions within this range, nor will

I say much about the difference between attentive and inattentive consciousness *tout court*. The key distinction is between (a) and (b). I regard qualitative consciousness *tout court* as the most basic form of consciousness.

To explain these distinctions a little, let's suppose that a garbage truck does in fact arrive. As I'm writing, I hear it. I am not thinking about it – I am absorbed in my writing. I may go on to attend to it, but in the first moment, I just hear it. That would be an episode of auditory consciousness *tout court*, and it would be inattentive.

Now, perhaps because I am expecting a package, I may begin to wonder whether it's the garbage truck or the UPS delivery truck. I might attend to the quality of the sound and try to convince myself it's one or the other. That would be attentive qualitative auditory consciousness, along with the activation of a good many other mental abilities – for example, inner speech about the likelihood of the package arriving, imagery of going to the door to sign for it, and so forth.

Finally, I might say to myself that I am hearing a truck. That would be an episode of self-consciousness – a placing of the episode of auditory consciousness in a larger stream of episodes of consciousness of many kinds.

One way to get at conscious episodes *tout court* is to think about cases where, as one says, we are "lost in the moment" or are "in flow". Of course, what we know about such cases is based on remembering them (which we might do quite soon after undergoing them), and when we do that we are likely to be having a form of self-consciousness – we are likely to be thinking about our previous experience as *ours*. But what we seem to remember – the remembered episode itself – is an episode of consciousness just by itself that was not, at the time, accompanied by any thought of its being ours.[2]

8.3 Real Continuity

My remarks in this section about a possible real continuity that property dualists should – and, more importantly, can – accept are offered solely for episodes of consciousness *tout court*. I will return to self-consciousness in the next section.

In this section, I will also suppose that we really do lose consciousness at some point in our sleeping hours, and that, of course, we regain it (both when we dream and upon waking). This does not beg the question at issue: I am merely supposing there is a certain *explanandum* in order to show that there can be an explanation of a possible, very gradual transition to it that a property dualist can consistently offer.

[2]Readers will rightly take this sentence as showing that I am not offering a HOT theory of consciousness. However, HOT theorists can accept the distinction I am making in this section. For them, consciousness *tout court* will be sensory episodes accompanied by unconscious thoughts that one is having them, while self-consciousness will involve sensory episodes accompanied by conscious thoughts that one is having them. For explanation of my views on higher order theories, see Robinson (2004a, b).

We typically have several qualities in our qualitative consciousness at any one time. For example, we may see and hear a train approaching while we shiver on the platform. But we can imagine seeing while not hearing anything, tasting something with our eyes closed, and so forth. And we can imagine that some instance of qualitative consciousness involves only one simple quality. For simplicity of discussion, let us focus on a case of this kind.

Let us suppose, for example, that I am having an episode of auditory consciousness characterized by a single sine tone. Let us imagine that this sound decreases in intensity. It decreases more, and continues to decrease. Now it is faint … fainter … is it still there? yes … maybe … no. In neural terms (in terms of what a property dualist is likely to regard as the causes of our consciousness), what's happening is that some pattern of neural activity is growing progressively less different from absence of pattern. That is, it is growing progressively less different from what we might call "neural noise", i.e., unpatterned neural firing. ("Neural noise" is the proper contrast to having neural causes of qualitative consciousness, since even in what are at least apparently conditions of dreamless sleep, our neurons continue to fire from time to time.)

We have arrived at the possibility of real continuity in loss of consciousness (*tout court* and in a simple case) that I said property dualists can consistently allow. The proposal is that the gradual disappearance of auditory quality is the gradual disappearance of an episode of consciousness. What consciousness in the most basic sense *is*, on this proposal, is occurrence of qualitative events, and when the quality goes, so does the consciousness. The quality goes gradually, and therefore so does the consciousness. When the neural causes of consciousness reach indistinguishability from neural noise, we no longer have a cause of qualitative consciousness, and we have reached absence of consciousness.

As to pillars of salt, they do not have the kind of internal complexity that permits occurrence of activity patterns of the kind that are nomologically related to episodes of consciousness. So, they have no consciousness. They cannot have any; they are not conscious beings at all.

This solution may appear to show too much for its own good, because it may appear to rule out the phenomenological suddenness of some of our episodes of consciousness. Imagine, for example, that I am working on this essay on a warm day in Spring, and that several doors and windows are open. A sudden breeze comes up and I am startled when a door bangs shut. The bang seems like an instantaneous onset event. So, shouldn't we allow for instantaneous onsets of conscious, qualitative events?

We should allow for such onsets, and we can do so by recognizing that Goff's argument concerns the *possibility* of gradual gain or loss of consciousness and does not assert that gradual gain or loss is the only kind of gain or loss that is possible. (To avoid tedium, I will henceforth write only of gain or of loss, assuming that parallel remarks hold for the other.) We can put the point in terms of his example of Lot's wife and the pillar of salt. God *can* replace one particle at a time (at least we are supposing so), but that is not the only way of making a transition from Lot's wife to a pillar of salt. God *could* decide that replacing a million particles at a clip is the best way of making the transition. In neural terms, what is needed to respond to Goff's argument is to allow for the possibility of gradual transition from an episode of consciousness to absence of consciousness. It may very well be that in daily life, many neurons change their activity states simultaneously, and so the transitions between neural noise and activation patterns that cause episodes of qualitative consciousness occur either instantaneously or within a very short (e.g., $\sim$20 ms) time frame.

Before going on, it will perhaps be helpful to note that some physicalists also need to account for transitions to and from consciousness, and may find the account I have given here to be useful. I have in mind those physicalists who are 'conceptual dualists' – that is, those who allow that our concepts of phenomenal qualities (and the events in which they occur) are different from our concepts of neural properties, but that the properties themselves, of which these two sets of concepts are concepts, are identical. For holders of such views, the coming to be of a pain, an afterimaging, a taste, a sound, and so on just are the comings to be of neural events of certain kinds. And now, Goff's challenge can be raised against them in the following form.

> It is an implausible view that the difference between a neural event type that's identical with a phenomenal quality type, and a neural event that is not identical with any phenomenal quality type, could consist in whether some neuron fires at 50 Hz rather than 49 Hz. So, you must allow that, despite common views to the contrary, there is consciousness in all neural processes – and, by similar reasoning, perhaps in all processes everywhere.

To respond to such a line of criticism, I think a conceptual dualist might very well want to have access to a view that allows for the possibility of gradual degeneration of distinctive neural patterns toward neural noise and, correspondingly, gradual degeneration of phenomenal qualities toward absence of phenomenal quality.

8.4 Why Does Consciousness Seem Binary?

In this section, I am going to relax our focus on consciousness *tout court* and consider the role of self-consciousness in our thinking about the issues Goff has raised.

One obvious point to make is that the fading out of one phenomenal quality – e.g., a sharp taste gradually fading to no discernible taste as our saliva dilutes the seasoning, or a sound fading to nothing as its source recedes into the far distance – is not remotely the disappearance of all of our consciousness. That is because (i) we are usually having experiences in several sensory modalities, so even if we are not conscious of any sound or taste, we are still having episodes of visual or olfactory or some other kind of sensory consciousness. Further, (ii) we are often talking to ourselves, which involves auditory imagery, which also consists of episodes of consciousness. (iii) There are also what Mangan (2001) has called "fringe" states of consciousness – such things as feelings of confidence, a sense of rightness or "fit" between what we take ourselves to see and what we expected, a sense of familiarity of a face or of our surroundings. Finally, (iv) there is the pleasantness or unpleasantness of our experiences.

The list just given is a list of ways of being conscious, and if any of them are present, we are conscious. So, if we are aware of the fading of a sensory quality, we do not (and should not) think of that as a fading of consciousness in all its forms – there is plenty of other consciousness that is *not* fading. This fact is part of what accounts for our intuition that consciousness is an all or none affair.

There is, however, more behind that intuition. When we think about the issues raised in Goff's paper, we are reflectively conscious, and that reflection does seem to be all or none. That is, while we may be clear or not so clear about what we think, and we may be confident or not so confident about what we think, we do not ordinarily think of ourselves as being "sort of reflectively conscious" or "half reflectively conscious".

There can, indeed, be cases of being *inattentive* in our reflections. For example, the arrival of a garbage truck may distract me, but not completely, so I can be in a state where I'm still thinking about writing these comments, and even thinking that it is I who am writing them, but where my full attention is not on the issues I'm writing about, or the fact that I am writing. But when I reflect on a state of this kind, it does not seem to be one of "less consciousness" – it is, instead, a complicated state of consciousness that involves partial attention to several items. To put the point as a slogan: Being half-focused is not the same as being half-conscious.

We can also think of sudden onsets and offsets of reflective consciousness. To imagine such a case, think once again of a time when you were "in the moment" or "in flow", where you were completely taken up with some activity – e.g., playing tennis against a well-matched opponent. One can sometimes get distracted out of such states in a particular way – namely, by becoming self-reflective (which typically reduces one's ability). Such onsets of reflection are the bane of actors and musicians – the last thing one wants to happen when performing is to have thoughts intrude about the fact that one *is* performing, or about how one is doing in one's performance. But they do happen, and our memory of such situations makes it seem that at one moment we were not reflective at all, and then suddenly we were. This impression may contribute something to our thinking of consciousness as "all or none".

Finally, if we think of being a conscious entity as being something that can have episodes of consciousness, we are employing a binary concept. That is, it is compelling that a being either can or can't have an episode of consciousness. "Sort of can" and "Half can" do not make obvious sense. "Can to some degree" does not make sense unless it is interpreted as meaning "Can have some episodes of consciousness, but only a few, or only of limited kinds, or only under special conditions". But anything that satisfies that interpretation is a thing that *can have episodes of consciousness*, and so, is a conscious entity.

8.5 Gradual Disappearance of Self-Consciousness

To sum up, there are several aspects of our ordinary conscious life and ways of thinking about consciousness that quite naturally lead us to think of consciousness as a binary property. I shall now argue, however, that a closer look will reveal the possibility of gradual transitions in and out of self-conscious states.

Two sections ago, I explained the possibility of gradual disappearance of an episode of consciousness in a single sensory modality. I take this account to be generalizable to all episodes of consciousness *tout court*.

Much of our self-consciousness consists in referring to ourselves in inner speech. Inner speech, however, can degrade in multiple ways. (i) It may become confused, to varying degrees. (ii) It is often accompanied by related visual imagery, and these accompaniments can be fewer, or less distinct, or less related, again in varying degrees. (iii) It may become gradually less self-reflective in its contents. (iv) It may fade away gradually, just as externally produced sounds can.

We need not express our sense of self in inner speech. For example, finding something to be of burning interest implicitly involves a sense of importance *to us*. Our sense of self is still implicit, though less intensely, if we perceive one thing as being farther away than another – that is, it is farther away *from us*. It does seem that self involvement of this kind could disappear gradually, by lessening in intensity, and by lessening in attention to those aspects of experience that involve relations to ourselves.

Mangan's "fringe" states of consciousness are already recognized as somewhat elusive, which suggests that it may not take much change in our neural activations to cause them to disappear. Further, it is plausible that, for example, a sense of confidence could decrease gradually to a neutral state of neither confidence nor unconfidence. As to pleasure and displeasure, we already recognize that these come in degrees, and we already recognize indifference as absence of both.

What about the reflective states that seem to suddenly come and go? Can we imagine being in such a state, and then reducing it by changing the firing rate of one neuron at a time? I think it is fortunate that we do not have familiar examples of halfway formed states of this kind – our brains are almost always better organized than that (which is why we have the impression that reflective consciousness is all or none) – but I believe we can understand such a possibility. Namely, we can think of confusion setting in, and gradually increasing until it is no longer clear just what we are thinking about. We might gradually lose associations we normally have with the words in our inner speech, and then gradually lose any feeling of frustration of the kind we would ordinarily have if we were confused.

My conclusion so far is that, upon careful consideration, the aspects of consciousness that suggest that it is a binary property turn out to be compatible with its gradual disappearance. I cannot undertake to prove that there is no aspect of self-consciousness that will resist the kind of treatment I have offered, but I believe the aspects explicitly considered here provide an adequate set of models for showing how episodes of self-consciousness can gradually disappear.[3] There remains, however, the last item on my list – the binary character of the property of being able to have a conscious episode. This item can be addressed in two steps.

First, let us imagine that the gradual loss of our usual neural activation states has led us into a state of confusion so deep that our consciousness is nothing more than a mass of inchoate images, sensations, and feelings. Now imagine that, as outlined two sections ago, these gradually fade to nothing. We will then be in a dreamless sleep; we will be having no conscious episodes.

If our neural equipment remains intact, we will remain conscious beings in the sense of being *able* to have episodes of consciousness, if the interference with normal neural operations were removed. However, it would require no further change *in consciousness* for our neural equipment to be disabled – either by divine fiat, or in consequence of replacement of their particles with others that are incompatible with normal neural functioning. After some set of replacements, the remaining neural and formerly-neural equipment will not be able to get into an activation pattern that will cause any conscious episode. The victim of replacement will no longer be such that it *can* have an episode of consciousness, and, as far as consciousness goes, it might as well be a pillar of salt.

8.6 Conclusion

Philip Goff's paper challenges property dualists, and some others, to explain how they can fit together the possibility of gradual degeneration of the causes of consciousness with the apparent binary character of the property of being conscious, without accepting implausible discontinuities. The problem is real but, I have argued, there is a solution that does not lead us to panpsychism.

References

Mangan, B. 2001. Sensation's ghost: The non-sensory "fringe" of consciousness. *Psyche* 7(18). Available at http://www.theassc.org/files/assc/2509.pdf

Robinson, W.S. 2004a. *Understanding phenomenal consciousness*. Cambridge: Cambridge University Press.

Robinson, W.S. 2004b. A few thoughts too many? In *Higher-order theories of consciousness*, ed. R.J. Gennaro, 295–313. Amsterdam: John Benjamins Publishing Company.

Robinson, W.S. 2010. *Your brain and you: What neuroscience means for us*. New York: Goshawk Books.

[3]One might suggest here that a *self* cannot gradually disappear. The text, however, shows how some aspects of a self appear in several guises, and can gradually be lost. My view is that other aspects of self go the same way. For more on the notion of a self, see Robinson (2010).

Chapter 9
What Is Acquaintance with Consciousness?

Jonathan Simon

It is a plausible thought that we are acquainted with our own phenomenal states, and that there are special canonical concepts of those states – phenomenal concepts – that in some sense or another facilitate this acquaintance. Let *Acquaintance* be the claim that our most general concept of phenomenal consciousness – the concept consciousness – is such a canonical concept, facilitating acquaintance with the property of being phenomenally conscious. In 'Orthodox Property Dualism + The Linguistic Theory of Vagueness = Panpsychism', Phillip Goff attempts to put a version of Acquaintance to work. He first argues that Orthodox Property Dualists – those who accept Property Dualism on the basis of arguments from epistemic gaps to ontological gaps[1] – are committed to his version of Acquaintance, which he dubs *Phenomenal Transparency*.[2] He then argues that Phenomenal Transparency implies

[1]Goff defines an Orthodox Property Dualist as one who on the basis of the standard arguments takes consciousness to be "... a basic property which arises from physical properties in accordance with fundamental psycho-physical laws of nature." This loads much into the definition. For example, if one is persuaded by the standard arguments to think that consciousness does not supervene on the physical, but one also thinks that consciousness might be constructed out of some equally non-physical sort of proto-consciousness, then one does not count as an Orthodox Property Dualist. Also, Goff does not say which are the standard arguments, which is frustrating because he goes on to make a universal claim about what acceptance of them entails. For example, is the Max Black argument (discussed in Block 2006; Perry 2001; White 1983) a standard argument?

[2]Goff's notion of Transparency is not to be confused with other notions of Transparency in the literature, for example the notion connected to the notion of Luminosity (Williamson 2000), or the notion connected to Diaphonousness in debates about Representationalism (Tye 2002), or the notion connected to the epistemic status of second order beliefs (Byrne 2005; Barnett forthcoming), or to the notion that deliberation about whether to believe that p gives way to deliberation about whether p is true (Shah and Velleman 2005).

J. Simon (✉)
Postdoc in the School of Philosophy, Centre for Consciousness at ANU, 4506 Elm St, Chevy Chase, MD 20815, USA
e-mail: jas741@nyu.edu

R. Brown (ed.), *Consciousness Inside and Out: Phenomenology, Neuroscience, and the Nature of Experience*, Studies in Brain and Mind 6, DOI 10.1007/978-94-007-6001-1_9,

that consciousness is not a vague concept, and finally he argues that this means Orthodox Property Dualists should be Panpsychists.

If Goff is correct, then Orthodox Property Dualists are committed to much more than they may have thought. I do not think Goff is correct, but the question of how Acquaintance relates to Property Dualism is a fascinating one. With this in mind, I propose to explore in some detail why Orthodox Property Dualists are not committed to anything like Phenomenal Transparency, and why nothing like Phenomenal Transparency commits anyone to the preciseness of consciousness. In what follows, I will first contrast Phenomenal Transparency with a few alternative ways of making Acquaintance precise. I will then argue that none of them are forced upon those who accept standard arguments for Property Dualism. I will then argue that none of them imply that consciousness is non-vague. I will conclude with a remark or two challenging Goff's claim that Orthodox Property Dualists who hold that consciousness is not vague should embrace Panpsychism.

9.1 (One) from Knowledge and Conceivability Arguments to Phenomenal Transparency?

It is quite plausible that we are acquainted with our own consciousness, and that somehow our most general phenomenal concept, consciousness, facilitates this acquaintance. However, it is very difficult to say, in more substantive terms, what this means. One fairly lightweight approach just holds that acquaintance with consciousness is simply a matter of *being* conscious. Accordingly, we might understand Acquaintance as simply the claim that you cannot possess the concept consciousness without being, or having been, conscious. Alternatively we might think of it as involving demonstrative knowledge, or a sort of demonstrative ability: the ability to know, or to truly think "*This* is consciousness". Along related lines we might associate it with the special sort of introspective access that we (allegedly) have to our own phenomenal states – infallibility, or immunity to error through misidentification, or even just a high degree of reliability in ordinary self-reports. But others have more heavyweight understandings of the idea. It is intuitive that if you are acquainted with a property then you know that property, in roughly the way that if you are acquainted with a person, then you know that person. But what sort of knowledge is this? Many languages reserve a distinctive word for it, distinguishing it from ordinary propositional knowledge (*connaitre* rather than *savoir*, *kennen* rather than *wissen*). There is nevertheless a temptation, to which Goff and others succumb, to understand acquaintance with properties in propositional terms. Goff officially defines Phenomenal Transparency thus:

"*Phenomenal transparency*: The concept consciousness reveals the nature of consciousness, i.e. it is a priori (for someone possessing the concept consciousness, in virtue of possessing that concept) what it is for something to be conscious ... To put it another way, a transparent property concept reveals what is ascribed in an application of the concept". He contrasts this with Opaque concepts which reveal "... nothing of what it is for an object to instantiate [the property referred to]." He speaks of translucent concepts which reveal "... *some but not all* of the nature of the property it denotes, i.e. something but not everything of what it is for an object to have that property is a priori knowable." Then, in a pair of footnotes, he concedes that opaque concepts may reveal "accidental" properties of their objects. In other work he explicitly contrasts Transparent concepts with those that do not reveal essential properties of their referents.[3]

All this is to say that Goff seems to have in mind a very propositional understanding of Phenomenal Transparency. He seems to be saying that Phenomenal Transparency means that all of the essential facts about consciousness are *a priori* knowable – indeed, knowable in virtue of possessing the concept consciousness. Accordingly, I shall understand Phenomenal Transparency – the thesis Goff endorses – to be the thesis that all of the essential facts about the property of being phenomenally conscious are *a priori* knowable in virtue of possession of consciousness.

Phenomenal Transparency is a powerful thesis. There is far more to the essence of consciousness than whether or not it is physical. Is it possible for one and the same subject to have disunified conscious experience? Is it possible for consciousness to exist without time? Without space? Is consciousness necessarily relational? If so, is it possible to be related by the relevant relation to uninistantiated universals? Are total phenomenal states metaphysically prior to partial phenomenal states?[4] Is Panpsychism necessary? Is Panpsychism possible?

One might be attracted to Property Dualism (on the basis of standard arguments) yet not wish to commit to the claim that all of these questions have *a priori* knowable answers. One might wish to remain agnostic about, or even endorse, the view that some of them are unknowable, or are knowable but only *a posteriori*. Before we consider Goff's argument, I will present two alternative conceptions of Acquaintance – neither of them lightweight, but both still a few weight classes down from Phenomenal Transparency.

The first is the thesis Chalmers (in his new book, *Constructing the World*) calls *Epistemic Rigidity*: if consciousness is epistemically rigid then this means that we have some fixed conception of what property consciousness is – in every scenario that we can conceive, consciousness picks out the same property. The second is a thesis I will call *A Priori Surveyability*. *A Priori* Surveyability says that, for any purely phenomenal proposition Q, we can know *a priori* that if we can conceive of a scenario in which Q, then it is metaphysically possible that Q.

In what follows I will argue that neither Phenomenal Transparency nor *A Priori* Surveyability nor Epistemic Rigidity are forced on those who accept Property Dualism on standard grounds. I will argue along the way that Phenomenal Transparency

[3]Goff, "*A Posteriori* Physicalists Get Our Phenomenal Concepts Wrong" p. 4.

[4]For a discussion of this and some related issues see Geoff Lee (forthcoming).

is strictly stronger than *A Priori* Surveyability, and strictly stronger than Epistemic Rigidity. I will then argue that none of these three theses entail that consciousness is not vague.

Goff's main argument for the claim that Orthodox Property Dualists are committed to Phenomenal Transparency is that Property Dualists must appeal to Phenomenal Transparency to reply to the usual objections to the standard arguments. Dualists who begin from the premise that zombies are conceivable need Phenomenal Transparency to reply to the objection that zombies might be conceivable but not possible. Dualists who begin from the premise that Mary learns something new must appeal to Phenomenal Transparency to maintain that Mary does not simply re-learn an old fact under a new mode of presentation.

I do not disagree with Goff that dualists *may* appeal to Phenomenal Transparency in order to defend their arguments. But it would be heavy-handed and unnecessary for them to do so. It would be heavy-handed because there is a direct argument from Phenomenal Transparency to Property Dualism: if all of the essential facts about phenomenal consciousness are *a priori* knowable, then if phenomenal consciousness is essentially physical, or were necessitated by the physical, this would be *a priori* knowable. But these things are not *a priori* knowable (this follows from the epistemic premise of the standard arguments, a premise Goff calls *Conceptual Dualism*). Therefore, Phenomenal Transparency implies that phenomenal consciousness is not essentially physical or necessitated by the physical.[5] So if dualists really took their case to hinge on Phenomenal Transparency, the focus on Mary and on Zombies would be needlessly circuitous. At best, these appeals would be useful to support the Conceptual Dualism premise, but that is a premise which many materialists accept anyway.

I will now argue that an appeal to Phenomenal Transparency would also be unnecessary. I will consider two developments of standard arguments for Property Dualism – the Fundamental Scrutability Argument, and the Two-Dimensional Argument. I will argue that neither relies on, or commits us to, Phenomenal Transparency. These arguments are elaborations of the standard arguments Goff mentions – elaborations that reply to the usual objections Goff mentions without appealing or committing to Phenomenal Transparency (or indeed, as I will argue, to any heavyweight Acquaintance thesis). In lieu of an Acquaintance thesis, the proponent of the Mary argument may appeal to (what Chalmers calls) the Fundamental Scrutability thesis. And the proponent of the Zombie Conceivability

[5]Note the important difference between Goff's Phenomenal Transparency thesis, and the similar principles at play in Johnston (1992), Lewis (1995), Byrne and Hilbert (2006) and Stoljar (2006) usually called Revelation principles. Those principles tell you that a token experience puts you in a position to know all of the essential truths about either it or its perceptual object. These theses are about something like perceptual content and perceptual justification, not about concept possession bestowing or enabling *a priori* justification. Nevertheless some of the points I make here in critique of Goff mirror points that Stoljar makes in critique of Lewis.

argument may appeal to the distinction between two dimensions of possibility: (what Chalmers calls) primary and secondary possibility.[6]

Before we move to these arguments, an important distinction. There are two ways that the premises of an argument might imply Phenomenal Transparency. The premises must tell us about some method (not itself dependent on further *a posteriori* knowledge) for coming to know the essential facts about consciousness. How do the premises secure us the *a priority* of the knowledge acquired using this method? The first way is for the premises to explicitly imply that the method gives us *a priori* knowledge. The second way is for the premises themselves to be warranted *a priori*, and for that warrant to transmit.[7] Here, I will only consider possibilities of the second sort. The two standard arguments I will consider (The Fundamental Scrutability argument, and the Two-Dimensional argument) make no relevant mention of *a priori* knowability or anything like it. I take it that the more interesting and more likely possibility is that one of these arguments is such that: its premises give us a method for determining the essential facts about consciousness (perhaps by implying those facts directly), and these premises are themselves *a priori* knowable.

On this basis we are in a position to recognize a first problem for Goff. Phenomenal Transparency is not just the claim that all essential facts about consciousness are *a priori* knowable, but rather the claim that they are so in virtue of possession of the concept consciousness.[8] If what I have just said is correct, this means that Phenomenal Transparency is only a consequence of an argument if each premise of that argument is knowable in virtue of possession of the concept consciousness. But in both of the standard arguments I consider below (the Fundamental Scrutability and Two Dimensional arguments), at least one premise is a general metaphysical claim, having nothing in particular to do with consciousness. Such a premise is unlikely to be justified in virtue of possession of the concept consciousness, even if it is *a priori* knowable. It is therefore unlikely that either of these standard arguments commits anyone to Phenomenal Transparency. However, I do not see that anything hinges, for Goff, on the claim that the relevant knowledge be knowledge

[6] The fundamental scrutability thesis is implicit in the discussion in Chalmers and Jackson (2001), and explicit in Chalmers (2012). The Two-Dimensional argument is comprehensively presented in Chalmers (2010).

[7] An example. Argument X implies, from premises of whatever justificatory status, that it is *a priori* that something is essentially true of consciousness if we can conceive of that thing being essentially true of consciousness. Argument Y implies, from premises knowable *a priori*, that something is essentially true of consciousness if we can conceive of it being essentially true of consciousness. I take it that the question on the table is whether any standard argument for Property Dualism is an argument of the latter sort – it is fairly obvious that none is an argument of the former sort.

[8] And we may presume he means *solely* in virtue, as it basically goes without saying that possession of the concept plays some supporting role here, assuming that concept possession plays a justificatory role at all.

(exclusively) in virtue of the possession of the concept consciousness.[9] Accordingly, in what follows I propose to overlook this problem. The question I will focus on is whether the Kripke-Jackson-Chalmers reasons for favoring Property Dualism – for taking the property of being phenomenally conscious to be a non-material property – lead one to the view that all of the essential facts about this property are knowable *a priori*.

This brings into focus a number of more interesting reasons why Goff's thesis is false. To begin with, both the Fundamental Scrutability and the Two Dimensional arguments may be run without making any explicit appeal to the concept consciousness, but instead only employing concepts of specific phenomenal states such as pain or seeing red. One might accept a form of one of these arguments, while taking the general concept consciousness to be incoherent or defective or not well defined. One might accordingly doubt that there is any such property as phenomenal consciousness *per se*, and doubt that there are any essential facts about it. This already shows that none of the three versions of Acquaintance under consideration – Phenomenal Transparency, Epistemic Rigidity, and A Priori Surveyability – follow from the standard arguments (except perhaps vacuously).

On to the Standard Arguments. Jackson and Chalmers have done much to explore possible replies to the 'old fact, new guise' objection to the Mary argument.[10] One such reply is to hold that Physicalism implies that all facts follow *a priori* from the physical facts. In support of this, one might identify Physicalism with the claim that only physical facts are metaphysically fundamental facts, and then defend the principle that all of the facts follow *a priori* from the metaphysically fundamental facts (a principle that Chalmers calls 'Fundamental Scrutability'). Fundamental Scrutability, together with Conceptual Dualism (here: the claim that there is something Mary learns) entails the falsity of Physicalism so construed. The 'old fact, new guise' objection to the Mary argument fails because it violates Fundamental Scrutability.

We may accept both premises of this argument (Fundamental Scrutability and Conceptual Dualism) without accepting any heavyweight version of Acquaintance. At best, the argument implies that we can know *a priori* that consciousness is fundamental and non-physical. It hardly promises, for example, that we can know *a priori* whether a disunified consciousness is possible, or whether it is possible for consciousness to be adverbial, or whether it is possible for consciousness to exist in a world without time. Indeed, this argument does not even give us a way to determine whether it is possible for there to be worlds with the same physical facts but different phenomenal facts. The argument only tells us that if we cannot derive

[9]At least for the purposes of this paper. Elsewhere he indicates that he believes the metaphysics of mind can be adequately done solely by exploring our concept consciousness (see for example his '*A Posteriori* Physicalists Get Our Phenomenal Concepts Wrong.'). I also suspect that his views on the meaning-based *a priori* here somehow stand behind his faith that the Transparent-Translucent-Opaque distinction carves at the joints.

[10]See for example Jackson and Chalmers (2001).

these facts from other facts, then there must be further fundamental facts from which these facts may be derived. Likewise, the argument does not guarantee that we can know *a priori* what consciousness refers to, and it does not guarantee that we can know *a priori* that a purely phenomenal proposition 'Q' is possible because we can conceive of it.

Another reason that this argument does not lead to Phenomenal Transparency (and still would not, even if its premises implied answers to every question about the essence of consciousness) is that Fundamental Scrutability might not be knowable *a priori*. Fundamental Scrutability is almost certainly not knowable on the basis of possession of the concept consciousness alone, as Goff strictly requires, but it also may not be knowable on the basis of possession of any other non-defective concepts, or on the basis of any other wholly *a priori* method. It is a general metaphysical principle – and for all I have said, a contingent one – which suggests there are methods for raising our credence in it that might be tainted by the *a posteriori*. For example, there might be a special sort of Metaphysical Intuition which is best understood as *a posteriori*,[11] or it might be that the kind of reasoning we employ in Metaphysics to arrive at conclusions whose negations still make sense to us (as the negation of Fundamental Scrutability arguably does) involves general evidence or elegance assessing methods whose justification is at least partially *a posteriori*.[12] Nor need this imply that a thesis like Fundamental Scrutability enjoys no *a priori* justification – only that whatever *a priori* justification there is does not suffice on its own for knowledge without an *a posteriori* boost.[13] If this is correct, then anything we learn about the essence of consciousness from an argument involving Fundamental Scrutability may well only be knowable *a posteriori*.

Let us move to Chalmers' two-dimensional defense of the conceivability argument. Letting 'P&~Q' stand for the conjunction of all of the actual physical truths, and the negation of some actual phenomenal truth, a simplified version of Chalmers' argument runs:

1. I can conceive of P&~Q.
2. If I can conceive of Φ, then there is a possible world that *verifies* Φ.

[11]Though most who speak of Intuition as a legitimate epistemic method take it to be *a priori*, this depends on how exactly we understand Intuition, and how exactly we understand *a priority*.

[12]Though it is implausible that basic epistemic rules be *a posteriori*, derived ones may be. And the rules which make symmetry, elegance and parsimony out to be theoretic virtues may well be derived (from a combination of basic rules and experience). Had we repeatedly experienced the symmetric or parsimonious theory losing out, we might not think of these as theoretical virtues. Alternatively it might be *a priori* knowable that these virtues give a theory some *prima facie* justification, but call for further *a posteriori* evidence in order to get from *prima facie* to all things considered justification. Thanks to Daniel Nolan for discussion. Another possibility is that the relevant rules are *a priori* but the knowledge that a given theory has one of the virtues mentioned by the rules turns out to be *a posteriori* (and, for that matter, contingent).

[13]For example, it might be *a priori* knowable that these virtues give a theory some *prima facie* justification, but nevertheless *a posteriori* evidence might be required in order to get to all things considered justification.

3. Therefore, there is a possible world that verifies 'P&~Q'.
4. But a possible world, w, cannot verify 'P&~Q' unless 'P&~Q' is in fact true at w.
5. Therefore, 'P&~Q' is possible, and therefore, Physicalism is false.

Chalmers' basic move is to reply to the objection that conceivability does not imply possibility by drawing a distinction between two kinds of possibility: primary and secondary possibility. These stand for different ways of evaluating whether a given proposition is true at a world. The way of *Secondary Possibility* corresponds to the standard metaphysical conception of possibility on which, e.g., it is not true at any world that water = XYZ. The way of *Primary Possibility* corresponds to something that makes room for the possibility that water is XYZ. According to Chalmers, a proposition is primary possible at world w (*verified* at world w) if that proposition would have been true had w been actual. Chalmers argues that, though it may not be secondary possible that water = XYZ (i.e. there may not be any world w at which 'Water = XYZ' is in fact true), it is nevertheless primary possible.

The effect of all of this is to equip the Dualist, who wants to argue from the conceivability of zombies (more generally, from the conceivability of phenomenal differences without physical differences) to their possibility, with a reply to the usual objection that conceivability does not imply possibility. Considerations of semantic externalism may show that conceivability does not imply secondary, metaphysical possibility, this response goes, but they do not show that conceivability does not imply *primary* possibility. This is a stable reply because primary possibility does not in general imply secondary possibility, as the example of 'Water = XYZ' shows. It is only because of special features of phenomenal concepts that certain propositions involving them are secondary possible if they are primary possible.

The upshot is that one may employ an argument such as this one to respond to the standard objections, without appealing explicitly to Phenomenal Transparency, or for that matter to Epistemic Rigidity or *A Priori* Surveyability. It is noteworthy that the response enjoys a deal of plausibility even before we delve into the details of Chalmers' analyses of conceivability and of primary possibility. We need only appeal to some suitable distinction between ways of evaluating propositions at worlds, where one of them makes room for the possibility of propositions like 'Water = XYZ' and the other does not.

A natural worry is that the argument nevertheless *commits* us to Phenomenal Transparency. But this is not the case. First, for all the argument says, some of its premises may only be knowable *a posteriori*. This would mean that, even if the premises of the two-dimensional argument imply every essential fact about consciousness, or imply that you may know every single such fact by introspecting in the right way, you still would not know those facts *a priori* if you know them on the basis of this argument. The argument has three premises: steps (1), (2) and (4). It is hard to imagine that (1) might be true but only knowable *a posteriori*. But matters are less clear regarding (2) and (4). (2) Involves a very powerful metaphysical claim linking things minds can do to ways the world might be. (4) Also amounts to a substantive claim about how a possible world may be structured. A dualist who thinks that metaphysical theses, though necessary, involve *a posteriori* justificatory elements, may well hold that one or both of these premises is therefore *a posteriori*, and accordingly that whatever these premises imply about the essence of consciousness is only knowable *a posteriori*. It is worth mentioning that (2), which relates conceivability to primary possibility in general, clearly is not knowable in virtue of possession of the concept consciousness, even if it is knowable *a priori*.

It is also for this reason that the argument does not commit us to Epistemic Rigidity or *A Priori* Surveyability. If the argument were *a priori*, then we could rule out *a priori* that Consciousness were physical. This would make room for the truth of Epistemic Rigidity. However if the premises of the argument are not all knowable *a priori*, then one might hold that it is in fact conceivable that consciousness is identical to some physical property P (but also conceivable that consciousness is a non-physical property), *pace* Epistemic Rigidity. Likewise, if premise (4) is not knowable *a priori* then *A Priori* Surveyability is false.

But even assuming that each premise of the argument is *a priori* knowable (in virtue of possession of the concept consciousness), the argument still does not commit us to Phenomenal Transparency, though it may then entail Epistemic Rigidity and *A Priori* Surveyability.[14] Phenomenal Transparency implies that we can know *a priori* a great number of the things that are possible, and a great number of the things that are not possible, regarding phenomenal consciousness.[15] But this does not follow from any or all of the premises of the argument, even if they are all knowable *a priori*.

First, it might be that there are possibilities that are inconceivable.[16] It might turn out, for example, that Disunified Consciousness is in fact possible, though we cannot make coherent sense of it. This might be a cognitive or conceptual limitation, not a metaphysical one. If we cannot conceive of disunified consciousness we cannot

---

[14]Regarding Epistemic Rigidity this may depend on how we understand the notion of conceivability at play in that thesis. If we understand it in terms of what is knowable *a priori*, then Epistemic Rigidity may follow from the Two-Dimensional argument. But if we understand conceivability in some more psychological and less epistemic way the matter is less obvious. See note 16 below.

[15]I take it that nothing I say in the below will hinge on how exactly we understand 'essential'. My arguments are all compatible with the more restrictive reading of that notion suggested by Kit Fine (1994), where various modal facts about some property are not essential facts about that property. I shall focus on examples of facts that ought to be in the running for counting as essential facts, no matter what one's analysis of essentiality. I also note that the more of a restricted notion of Essence Goff appeals to, the lower the likelihood that Phenomenal Transparency will make all of the facts *a priori* regarding how a concept may be sharpened if it is vague.

[16]This is ruled out by one analysis of Conceivability, the one Chalmers calls 'Ideal Negative Conceivability', which says that something is conceivable if it cannot be ruled out *a priori*. This means something is inconceivable if it can be ruled out *a priori*. If we define 'ruling out *a priori*' as a success term, we establish by definition that every possibility is conceivable. However, if we define 'ruling out *a priori*' as '*a priori* (conclusively) justifying disbelief' then we leave the question open. Alternatively, we leave the question open if we embrace some other conception of Conceivability, for example the one Chalmers calls 'Positive Conceivability.' See Chalmers (2010).

know *a priori* that it is possible. But it is compatible with the argument we are considering that it nevertheless *be* possible. This is also a reason why Phenomenal Transparency does not follow from Epistemic Rigidity or *A Priori* Surveyability.[17]

Second, it might turn out that there are things which appear to be metaphysical possibilities for consciousness but that in fact are not, in roughly the same way that it appears to be metaphysically possible that water be XYZ, though in fact it is not. Some (including Goff) seem to think that premise (4) in the two-dimensional argument implies that the primary possibilities for consciousness are exactly the secondary possibilities for consciousness. This is false.

First of all, in order to defend (4), we only need one true phenomenal proposition 'Q' such that if 'P&~Q' is primary possible then it is secondary possible. Strictly speaking, then, we need not accept *A Priori* Surveyability even if we take every premise of the Two-Dimensional argument to be *a priori*. Admittedly though (barring the worry I outline above that there is no coherent concept like consciousness with which to generalize the point), it is compelling that if (4) is true for one purely phenomenal proposition 'Q' it should be true for very many of them. In sporting spirit, then, let us grant that (4) generalizes. Let us grant that all *purely* phenomenal propositions 'Q' – propositions that only involve purely phenomenal concepts, singular terms, and logical terms – are only primary possible if they are secondary, i.e. metaphysically, possible.[18]

This still does not get us to Phenomenal Transparency. Again, Phenomenal Transparency implies that all essential facts about consciousness are *a priori* knowable. And we are now allowing that it is *a priori* knowable that every purely phenomenal proposition 'Q' is primary possible only if it is metaphysically possible. And we are allowing that it is *a priori* knowable that conceivability implies primary possibility. It follows that it is *a priori* knowable that if 'Q' is conceivable, then it is metaphysically possible. However, not all essential facts about consciousness are expressible as purely phenomenal propositions. Rather, many essential facts about

[17]Incidentally, Phenomenal Transparency also does not imply Epistemic Rigidity or *A Priori* Surveyability so long as we make room for things to be both conceivable but *a priori* knowably false, and it does not imply *A Priori* Surveyability provided that the possible truth of at least one purely phenomenal proposition is not an essential fact about consciousness.

[18]Chalmers sometimes refers to this as the thesis that the 1-intension of consciousness is identical to its 2-intension. Chalmers points out that his argument does not rely on this claim. He has in mind the possibility that 'Q' turns out to rigidly designate a physical property, but one which itself has (second order) non-physical properties. Then a possible world might verify 'P&~Q' even though strictly speaking 'P&Q' is true there. Nevertheless, on this picture in order for that world to verify 'P&~Q', the physical property rigidly designated by 'Q' must not instantiate the second order non-physical properties that it actually instantiates. Hence Physicalism is still falsified. My point in this paragraph of the text illustrates a different way that the argument against Physicalism may succeed even if the 1-intension of consciousness is not identical to the 2-intension. But my central claim is that even granting that the 1-intension is the 2-intension, Phenomenal Transparency does not follow – for one thing because (as I argue above) the identity of 1- and 2-intensions might not be *a priori* knowable, for another thing because (as I argue below) even if this is *a priori* knowable, it does not follow that all essential facts about consciousness are.

consciousness may pertain to structural elements of consciousness which are not strictly phenomenal, or which relate consciousness to other things, for example to Time, to Space, or to Perceptibles. Propositions of this sort are not covered by the basic insight that stands behind (4). If something looks and feels like pain, it is pain. But it does not follow that if something looks and feels to be relational, it is relational, or that if something looks and feels to be disunified, it is disunified, or that if something looks and feels to be atemporal, it is atemporal. Indeed, if something looks and feels to be a world where a conscious being is near water, it does not follow that it is a world where a conscious being is near water. In the latter case, we may appeal (on Lewisian combinatorial principles, perhaps) to the claim that there is in fact an H2O world of the relevant sort, so that the primary possibility is also a secondary possibility. But it is not obvious how this strategy generalizes. This is a reason to think that Phenomenal Transparency does not follow from the Two-Dimensional argument, and it is also a reason to think that Phenomenal Transparency does not follow from Epistemic Rigidity or *A Priori* Surveyability.

The specificities of Chalmers' analysis of primary and secondary possibility – especially as they are systematically developed in his new work *Constructing The World* – may of course impose further constraints here, possibly even constraints leading us to something like Phenomenal Transparency. Chalmers certainly indicates that he believes consciousness is both Epistemically Rigid and *A Priori* Surveyable. But though a rigorous analysis of the sort Chalmers has developed is important in order to conclusively show that the Two-Dimensional strategy of response to the 'Conceivability doesn't imply Possibility' objection is a success, we need not follow Chalmers in all particulars in order to be persuaded by some version of the argument. Whether or not Chalmers himself accepts any of the heavyweight Acquaintance theses, there is a basic but still attractive version of his Two-Dimensional argument that does not commit us to any of them. Whether on the whole it is more dialectically efficacious to arrive at Dualism by direct appeal to one of these principles is another matter – though I doubt it.[19] I conclude that Goff is incorrect. There is certainly something to the idea that phenomenal concepts *acquaint* us with the properties they present. But even Property Dualists may not wish to cash this intuition out in one of the heavyweight senses that we have considered. Property Dualists – even those who accept Property Dualism for standard reasons – are not committed to Epistemic Rigidity or *A Priori* Surveyability, and they certainly are not committed to Phenomenal Transparency. They may hold that none of the essential facts about consciousness are knowable *a priori* (for example, if they hold that some of the needed justification for one of the premises in their preferred argument is *a posteriori*) or they may hold that some but not all of the essential facts about consciousness are knowable *a priori*.

---

[19]An audience that is not hostile to Phenomenal Transparency is most likely an audience that already accepts Dualism anyway (cf. Lewis 1995; Stoljar 2006). For an example of an argument from something like Phenomenal Transparency to Dualism see Nida-Ruemelin (2007).

Goff offers a quick argument against this latter possibility. Consciousness may not be translucent (in the sense that some but not all essential facts about it are knowable *a priori*) because this would imply, he claims, that the concept is "... a composite of two 'sub-concepts', one transparent and one opaque." But this is almost certainly false. Many would say that the concept red is in Goff's sense translucent: it is *a priori* that crimson is a shade of red, and that red is more similar to orange than to green, but it is not *a priori* whether or not red is a surface reflectancy property. Yet it is doubtful that red factors into two concepts, for example primitive red and surface reflectancy. The whole point is that we do not know *a priori* what to say about the relationship between red and surface reflectancy. We may say with Goff that a concept is translucent if it has aspects which are *a priori* knowable and other aspects which are not *a priori* knowable. But this is just to say that some propositions involving the concept are knowable *a priori* while others are not. It does not follow from this that there exist concepts corresponding perfectly to these distinct sets of propositions, let alone that our ordinary concept has these other concepts as components.[20]

9.2 (Two) from Transparency to Non-vagueness?

In the previous section, I argued that Orthodox Property Dualism does not commit one to Phenomenal Transparency or to either of the other heavyweight Acquaintance theses that I have considered. But it is independently worth asking whether the non-vagueness of consciousness follows from any of these Acquaintance theses. I argue here that it does not.

Goff argues that it does. He argues as follows. If a concept is vague then it has a set of associated sharpenings. If a concept is vague and transparent then its set of associated sharpenings must be *a priori* knowable. But we have no idea what a sharpening of consciousness might be, let alone *a priori* knowledge of what one might be. It follows that if consciousness is transparent, then it is not vague.

---

[20]Goff's reasoning here seems to be influenced by two illicit assumptions: that translucent concepts denote complex properties, and that these factor into simple properties for which there are transparent and opaque concepts. Goff is perhaps assuming (without argument) that all *a priori* knowledge is knowledge had in virtue of possession of transparent concepts. This claim is implausible, and begs some of the central questions here. I conclude that the partition of concepts into Transparent, Translucent and Opaque, at least as Goff understands it, is not very dialectically effective – in particular, it does not carve at the battle lines in the Metaphysics of Mind. I also note that for all we have said here, there may well be some concept which is such that grasp of it puts you in a position to know all of the essential facts about consciousness – for example, some infinitary concept that encodes all of those facts. The implausible claim Goff makes is that our ordinary, everyday concept of consciousness has this feature.

My own view is that, though heavyweight Acquaintance theses may help us to establish that we have no *a priori* knowledge of what a sharpening of consciousness might be, they have no role in establishing that the concept is vague only if we have such knowledge.[21] The latter is a thesis that must flow either from general considerations about the nature of vagueness, or not at all. In my dissertation I advance such an argument for the non-vagueness of consciousness flowing from general principles about the connections between a vague concept and other concepts to which it is related.[22] I also take it to be far more difficult to establish that we can have no *a priori* knowledge of sharpenings than Goff seems to think, but that is another story for another day.[23]

The thesis that consciousness facilitates acquaintance with consciousness says nothing at all about vagueness, nor do any of the specifications of that thesis that we have considered. Phenomenal Transparency is a claim about *the* property denoted by the concept consciousness. Epistemic Rigidity as I have characterized it also is, and *A Priori* Surveyability is only a claim about *purely* phenomenal truths. But if consciousness is vague then there is no single property that it denotes (assuming as Goff does that vagueness is semantic indecision), and there may be truths such that it is indeterminate whether they are purely phenomenal truths. Each version of the Acquaintance thesis may be elaborated in different ways, some stronger than others, to take vagueness into account.

Phenomenal Transparency may be extended either to the claim that all of the facts which are determinately essential facts about consciousness are *a priori* knowable, or to the (far stronger) claim that determinately essential facts are *a priori* knowable and also it is *a priori* knowable that indeterminately essential facts about

---

[21]For example, we might be able to argue that the sharpenings of a transparent concept must themselves be transparent. This would narrow the pool of candidates significantly.

[22]Simon (2012).

[23]Very briefly: Conceptual Dualism ensures that no ordinary material concepts give us an adequate conception of a sharpening. But this leaves room for the wide range of what we might call 'Protophenomenal' concepts – concepts that are not material, but that give us conceptions of sharpenings, by showing us how clear cases of consciousness may be located on some sort of continuum with clear cases of non-consciousness. Many views that countenance protophenomenal concepts will not count as Orthodox Property Dualist, in Goff's sense, even if they believe that phenomenal and protophenomenal reality do not supervene on material reality. Nevertheless, that is a view to which one might be led by the standard arguments, and it is puzzling that Goff does not spend more time addressing it. But Orthodox Property Dualists might also avail themselves of something like protophenomenal concepts: for example, concepts of properties that are related to consciousness the way red is related to green, even if these other properties are uninstantiated. There might still be vagueness in exactly how far up the hierarchy from infima species to maxima genera our concept consciousness lies, since there is no reason for its reference to be determined by the exact range of actual instances. Compare: our concept red would be vague even if there were no actual orange things. Alternatively, there might be some conceptual analysis in neutral terms that entailed some deep commonality between the physical and the phenomenal, while still respecting the claim that the phenomenal is every bit as fundamental as the physical. Goff points out that no one has yet successfully carried out the sort of conceptual analysis that these approaches require. But it is hard to see why that means we are entitled to assume no one will.

consciousness are indeterminately facts about consciousness. The former extension of the principle is the most that the face value reading of the original principle gets us. From:

6. F is an essential fact about consciousness if F is an *a priori* knowable fact about consciousness
7. It is indeterminate whether F is an essential fact about consciousness

   We get:

8. It is indeterminate whether F is an *a priori* knowable fact about consciousness

Not that F is *a priori* knowably indeterminate. And (6) is already stronger than Phenomenal Transparency, which does not imply that *only* essential facts about consciousness are *a priori* knowable.

Similar reasoning applies to the other versions of Acquaintance. Epistemic Rigidity may be extended either to the claim that we have a fixed conception, at any conceivable scenario, of what it is for a property to be determinately identical to consciousness, or to the stronger claim that we have a fixed conception, at any conceivable scenario, of what it is for a property to be determinately identical to consciousness or of what it is for a property to be only indeterminately identical to consciousness. *A Priori* Surveyability extends either to the claim that we know *a priori* of any determinately pure phenomenal truth that if it is conceivable then it is metaphysically possible, or to the claim that adds to the latter that we also know *a priori* of any indeterminately pure phenomenal truth that if it is conceivable then it is metaphysically possible (or indeterminately possible?).

As I have argued, one might be a Property Dualist on the basis of standard arguments and not accept even the weakest of these principles. But even if we were committed to one of them, it is unclear that our reasons would have anything to do with vagueness, and so unclear how we might come to be committed to the extension of that thesis that makes explicit provision for the case of indeterminacy, rather than to the one which only talks about the determinate case. We have as yet no argument from any of the Acquaintance theses to the claim that the sharpenings of consciousness would have to be knowable *a priori*.

For independent reasons, I agree with Goff that consciousness is not vague, and I welcome his attempt to show that there are extra reasons for Property Dualists to think so. But I worry that he has not found a dialectically efficacious ground for holding that the sharpenings of this concept would have to be knowable *a priori*. Goff seems to be drawn to the view that this is a special constraint on consciousness: there are many vague concepts whose sharpenings need not be known *a priori*. I urge Goff to reject this claim and join me (and Michael Antony) in arguing that we would be able to know the sharpenings of consciousness *a priori*, if it were vague, because we can know the sharpenings of any vague concept *a priori*.[24]

---

[24]Though I do not think we need to accept Supervaluationism, or the language of Sharpenings, to make the point. The Semantic Indecision theory does not imply Supervaluationism – especially if

9.3 (Three) from Non-vagueness to Panpsychism?

Goff concludes his paper with an argument that Property Dualists committed to the non-vagueness of consciousness should accept Panpsychism. His argument centers on the claim that the Panpsychist Property Dualist may give a more elegant account of the Psychophysical Laws than may the Emergentist Property Dualist. I do not think this is obvious. Both views must say that very fine grained physical changes correlate with phenomenal changes. The Panpsychist who can derive the phenomenal facts about macroscopic things from the phenomenal facts about microscopic things may have an advantage, but Goff does not endorse that sort of Panpsychism. I also am not sure what vagueness has to do with anything: it is unclear how the vagueness of consciousness would make for more elegant ultimate psychophysical laws. I will conclude with a sketch of one easy way to draw out the problems with Goff's argument here: it has potentially absurd consequences, even by Goff's lights.

It is plausible that consciousness is not the only precise phenomenal concept. Just as the most general phenomenal concept may be precise, so too the most specific phenomenal concepts may be precise. For example, there may be a precise concept specifying the exact phenomenal state that Phillip is in right now. Let us call this concept P. The principles Goff invokes seem to imply that it would be implausibly arbitrary if some infinitesimally precise physical difference – the movement of an electron infinitesimally to the left – should make the difference between Phillip's being in the state specified by P, and his not being in this state. But Goff's reasoning then implies that no microphysical difference can make the difference. It seems to follow that Phillip's phenomenal state will never alter!

References

Barnett, D.J. forthcoming. First person transparency.

Block, N. 2006. Max Black's objection to mind-body identity. In *Oxford studies in metaphysics*, vol. II, ed. Dean Zimmerman, 3–78. Oxford: Oxford University Press.

Byrne, A. 2005. Introspection. *Philosophical Topics* 33: 79–104.

Byrne, A., and D. Hilbert. 2006. Color primitivism. In *Perception and the status of secondary qualities*, ed. R. Schumacher. Dordrecht: Kluwer.

Chalmers, D. 2010. The two dimensional argument against materialism. In *The character of consciousness*. Oxford: Oxford University Press.

Chalmers, D. 2012. *Constructing the world*. Oxford: Oxford University Press.

---

the latter is taken to be the view that sharpenings are supposed to *extend* the meanings of vague terms while preserving what they already mean, where this means rejecting that the vagueness of a vague term is bound up with what it means. For difficulties with this idea see Schiffer (2003). For example, it is unclear how this picture could apply in intensional contexts, where the vagueness is inherently a part of what is meant e.g. 'She only likes bald men'. I defend my own proposal in my dissertation, *The Sharp Contour of Consciousness*.

Chalmers, D., and F. Jackson. 2001. Conceptual analysis and reductive explanation. *Philosophical Review* 110: 315–361.
Fine, K. 1994. Essence and modality. In *Philosophical perspectives,* vol. 8, ed. J. Tomberlin as the *Nous* Casteneda Memorial Lecture, 1–16.
Johnston, M. 1992. How to speak of the colors. *Philosophical Studies* 68: 221–263.
Lee, G. forthcoming. Unity and systematicity in Chalmers' theory of consciousness
Lewis, D. 1995. Should a materialist believe in qualia? *Australasian Journal of Philosophy* 73: 140–144.
Nida-Ruemelin, M. 2007. Grasping phenomenal properties. In *Phenomenal concepts and phenomenal knowledge, new essays on consciousness and physicalism*, ed. T. Alter and S. Walter. Oxford: Oxford University Press.
Perry, J. 2001. *Knowledge, possibility and consciousness*. Cambridge, MA: MIT Press.
Schiffer, S. 2003. *The things we mean*. Oxford: Oxford University Press.
Shah, N., and J.David Velleman. 2005. Doxastic deliberation. *Philosophical Review* 114(4): 497–534.
Simon, J. 2012. *The sharp contour of consciousness*. Doctoral dissertation. New York University.
Stoljar, D. 2006. The argument from revelation. In *Naturalism and analysis*, ed. David Braddon-Mitchell and Robert Nola. Cambridge, MA: MIT Press.
Tye, M. 2002. Representationalism and the transparency of experience. *Noûs* 36(1): 137–151.
White, S. 1983. The curse of the qualia. In *The nature of consciousness: Philosophical debates*, ed. N. Block, O. Flanagan, and G. Güzeldere, 695–718. Cambridge, MA: MIT Press, 1997.
Williamson, T. 2000. *Knowledge and its limits*. Oxford: Oxford University Press.

Chapter 10
Reply to Simon and Robinson

Philip Goff

Jonathan and Bill's pieces certainly rattled me, but after much thought I feel confident about my responses. There are many points in both pieces, but I will focus below on what I take to be the key objections.

10.1 Response to Simon

Consider the following distinctions between property concepts:

> *Transparent* – A concept C of a property P is transparent if and only if C reveals what it is for P to be instantiated.
>
> *Translucent* – A concept C of a property P is translucent if and only if C reveals something but not everything of what it is for P to be instantiated.
>
> *Mildly opaque* – A concept C of a property P is mildly opaque if and only if C reveals nothing of what it is for P to be instantiated, but does reveal an accidental feature of P which uniquely identifies it in the actual world.
>
> *Radically opaque* – A concept C of a property P is radically opaque if and only if C reveals neither what it is for P to be instantiated, nor an accidental feature of P which uniquely identifies it in the actual world.

I think that the concept consciousness is transparent. Call this thesis 'phenomenal transparency'. I claimed in 'Orthodox property dualism + the linguistic theory of vagueness = panpsychism' that any decent argument against physicalism is implicitly or explicitly reliant on phenomenal transparency. I now think I should have been slightly more careful and said that any decent argument against physicalism implicitly or explicitly relies on the denial of the thesis that consciousness is radically

P. Goff (✉)
Department of Philosophy, University of Liverpool, 7 Abercromby Square,
Liverpool L69 7WY, UK
e-mail: philgoff1@googlemail.com

R. Brown (ed.), *Consciousness Inside and Out: Phenomenology, Neuroscience, and the Nature of Experience*, Studies in Brain and Mind 6, DOI 10.1007/978-94-007-6001-1_10,

opaque (call this thesis 'radical phenomenal opacity'). The reason that any decent argument depends on denying phenomenal opacity is that, if the physicalist can claim that consciousness is radically opaque, no metaphysical conclusions can be drawn from the epistemic premises of the standard arguments against physicalism.

Simon disputes this on the grounds that we might argue against physicalism on the basis of the Chalmers' two-dimensional argument against physicalism, which he suggests is not reliant on accepting phenomenal transparency. However, the entire two-dimensional setup is premised on the denial of radically opaque concepts, as radically opaque concepts do not have primary intensions. If a concept has a primary intension, then a thinker with idealised rational faculties can determine a priori its extension across possible worlds considered as actual. But if a concept is radically opaque, then it reveals neither the essence of its referent, nor a property which uniquely identifies it in the actual world. It just doesn't, therefore, provide the concept user with enough information to allow her to locate the referent in a possible world considered as actual. The two-dimensional framework collapses if consciousness is radically opaque.

Simon also suggests that we might argue against physicalism from the premise that all facts are a priori scrutable from the fundamental facts. We could argue against physicalism in this way, but, without some way of ruling out radical phenomenal opacity, this would not be a very good argument. If there are non-fundamental facts couched in radically opaque concepts, then these facts will certainly be counterexamples to the thesis that all facts are a priori scrutable from the fundamental facts. Radically opaque concepts are just blind pointers, which reveal neither essential nor accidental features of their referents (or at least not accidental features that uniquely identify the referent). No matter how much information we have about the fundamental nature of the world, we're not going to be able work out whether it instantiates *that property*, where reference to *that property* is determined wholly outside of what is a priori accessible.

The point is that if consciousness is radically opaque, then it reveals nothing substantive about either the essential nature of consciousness, or its defining nature in the actual world. How could we possibly demonstrate, then, that that nature is not physical? But if we can rule out radical phenomenal opacity, then we do know something substantive about consciousness a priori. If consciousness is not transparent then the complete nature of consciousness is not a priori accessible, but the complete nature of a substantive property closely related to consciousness will be a priori accessible. If consciousness is translucent, then we have a priori access to an aspect of the nature of consciousness, call that aspect consciousness*. If consciousness is mildly opaque, then we have a priori access to a property of consciousness which uniquely identifies it in the actual world, call it consciousness**.

In either case, we have the basis for arguing against physicalism, as there is a property we completely understand the nature of which does not reveal itself to be physical (assuming conceptual dualism). Even if we can't show that consciousness is not physical, we can show that consciousness* or consciousness** is not physical.

Similarly, if phenomenal transparency is false, we cannot argue that consciousness is non-vague, and on that basis argue that consciousness is ubiquitous in nature.

But it seems that we will be able to argue that consciousness*/consciousness** is non-vague, and on that basis argument that consciousness*/consciousness** is ubiquitous in nature (at least if no other aspects of my argument fail).

Simon suggests in a number of ways that phenomenal transparency is too strong. It would make *every single essential truth* about consciousness a priori, and surely we don't have to commit to this just to argue against physicalism!

> Phenomenal Transparency is a powerful thesis. There is far more to the essence of consciousness than whether or not it is physical. Is it possible for one and the same subject to have disunified conscious experience? Is it possible for consciousness to exist without time? Without space? Is consciousness necessarily relational? If so, is it possible to be related by the relevant relation to uninistantiated universals? Are total phenomenal states metaphysically prior to partial phenomenal states? Is Panpsychism necessary? Is Panpsychism possible?

There seems to be an implicit argument here of the following form:

1. Phenomenal transparency entails that we can know a priori whether consciousness is necessarily relational, whether consciousness is necessarily unified, whether panpsychism is possible, etc.
2. It is clearly not the case that these metaphysical questions are answerable a priori.
3. Phenomenal transparency is false.

Maybe some will find this implausible, but I am more than happy to accept that all the above metaphysical questions are answerable a priori (although that doesn't mean it's easy to answer them!). I take our immediate acquaintance with conscious experience to be the most important source of data in metaphysics; consciousness is the one bit of the world as it is in and of itself that is directly revealed to us. In my recently completed monograph, 'Taking Consciousness Seriously', I develop a 'post-Galilean' conception of metaphysical enquiry which takes careful reflection on one's own conscious experience to be a fundamental source of data.

Simon also criticises my move from phenomenal transparency to the non-vagueness of consciousness. Given phenomenal transparency, I can know a priori what it is for the property of consciousness to be instantiated. Simon objects that, if the linguistic theory of vagueness is true, there is no single property picked out by consciousness, rather it picks out a number of properties corresponding to the various sharpenings of the concept.

This will of course be true on some metaphysically robust notion of property. But in this case we can turn to my other formulation of the definition of a transparent concept: a concept C is transparent just in case C reveals what is ascribed in the application of a predicate. And as I say in the paper, 'According to the linguistic theory of vagueness, what is ascribed in the application of a given vague predicate is to be understood in terms of the predicate's indeterminacy over its sharpenings. Assuming the truth of this view, if what is ascribed in the application of a given vague predicate is a priori knowable, then the sharpenings of that predicate must be a priori knowable.' I'm not clear what Simon's response to this argument is.

(I reply below to Simon's final objection).

10.2 Response to Robinson

I like William S. Robinson's interpretation of my argument as a continuity argument. A theory's credibility is dependent on its theoretical virtue, and it is theoretically vicious to suppose that there are sudden jumps in nature. If consciousness in an all or nothing property, and a very slight neurological change can the make the difference between its presence and its absence, then this would seem to constitute an unacceptably inelegant discontinuity in nature; coherent but improbable.

Robinson, then, accepts much of my argument, and, as a non-panpsychist property dualist, he takes from my argument the obligation to remove this appearance of discontinuity. His primary means of doing this is to move the focus from *consciousness as such* to *specific episodes of consciousness*. Specific modes of consciousness, thinks Robinson, plausibly admit of degree:

> Let us suppose, for example, that I am having an episode of auditory consciousness characterized by a single sine tone. Let us imagine that this sound decreases in intensity. It decreases more, and continues to decrease. Now it is faint ... fainter ... is it still there? yes ... maybe ... no. In neural terms (in terms of what a property dualist is likely to regard as the causes of our consciousness), what's happening is that some pattern of neural activity is growing progressively less different from absence of pattern.

If specific episodes of consciousness can fade out in the way Robinson suggests, then their gradual appearance or disappearance does not constitute a discontinuity in nature. And if specific episodes of consciousness are more basic than consciousness in general, as seems plausible, then it might seem that the smooth appearance/disappearance of episodes of consciousness entails the smooth appearance/disappearance of consciousness itself.

Of course one specific state of consciousness can gradually change into another specific state of consciousness. As Simon points out, if I didn't accept this, I would face a parody of my argument to the conclusion that my current specific state of consciousness cannot give way to another. The determinable property of *consciousness* – the property of being a thing such that there's something that it's like to be that thing – takes many determinate forms, and a particular conscious thing can gradually move from one of those determinate forms to another, e.g. when a sad mood slowly melts into happiness. Compare with shape. The determinate property of *being shaped* takes a number of determinate forms, and a particular shaped thing can gradually move from one of those determinates to another, for example, we can imagine a plasticine cube gradually being moulded into a sphere.

However, this does not entail that the determinable property itself can gradually disappear. It would be a massive change in the world if a thing suddenly went from not having a shape to having a shape, or from having a shape to not having a shape (at least on the assumption that shape is irreducible). Similarly, it would be a massive change if things went from not having an inner life to having an inner life, or vice versa (at least on the property dualist assumption that consciousness is fundamental). In both cases an entire determinable way of being is gained or lost. Respect for theoretical virtue impels us to avoid such massive changes.

Robinson's example certainly constitutes a gradual change in consciousness, a continuous alteration in the way some subject's inner life is modified. But I don't think the example captures a sense in which there can be gradually *less consciousness* in a subject. The property of consciousness, of having an inner life, does not admit of degree. For this reason, the gain or loss of that property would constitute a momentous change in an object, just as the gaining or losing of shape would (again given the assumption that both of these properties are irreducible). Property dualists with a taste for continuity should take consciousness to be everywhere or nowhere.

Part IV
Naïve Realism, Hallucinations, and Perceptual Justification

Chapter 11
It's Still There!

Benj Hellie

11.1 The Big Picture

The view concerning perception developed in 'There it is' (Hellie 2011) involves, most centrally, the following theses:

I. A. One brings *a* within the scope of attention only if *a* is an aspect of one's perceptual (or sense-perceptual) condition;
   B. If one sees veridically, one ordinarily brings within the scope of attention such an *a* partly constituted by the condition of the bodies surrounding one;
   C. The perceptual condition of a dreaming subject is never partly constituted by the bodies surrounding them;

II. One brings *a* within the scope of attention just if it is *situatedly analytic* for one that *a* is genuine (where this is partly constitutive of the character of one's rational position);

III. If two subjects are in distinct rational positions, what it is like for them differs.

Section 11.2 defends thesis (III). Section 11.3 lays the groundwork for the defense in Sect. 11.4 of thesis (II). Thesis (I) is obvious so I don't bother defending it.

B. Hellie (✉)
Department of Philosophy, University of Toronto, 170 St George Street, Toronto, ON M6J 1N6, Canada
e-mail: benj.hellie@utoronto.ca

R. Brown (ed.), *Consciousness Inside and Out: Phenomenology, Neuroscience, and the Nature of Experience*, Studies in Brain and Mind 6, DOI 10.1007/978-94-007-6001-1_11,

11.2 A Hallucination Puzzle

The view runs headlong into a sort of 'hallucination puzzle', discussed in Sect. 11.5 and extended to other troubling phenomena of perceptual epistemology in Sect. 11.6. Consider Sam, an ordinary 'veridically perceiving' subject. The following principle, in the spirit of an internalism about the phenomenological, is alluring:

PI There is a possible subject 'Dreaming Sam' who is dreaming but for whom what it is like does not differ from what it is like for Sam.

Those giving in to the allure of (PI) will then argue as follows:

1. By (I-B), Sam brings within the scope of attention an aspect *a* of her perceptual condition partly constituted by the condition of the bodies surrounding her;
2. So, by (II), it is situatedly analytic for Sam that *a* is genuine;
3. By (III) and (PI), Sam and Dreaming Sam are in the same rational position;
4. So, by (2) and (3), it is situatedly analytic for Dreaming Sam that *a* is genuine;
5. So, by (4) and (II), Dreaming Sam brings *a* within the scope of attention;
6. But, by (I-C) and (I-A), Dreaming Sam does not bring *a* within the scope of attention

In contradiction with (5). So if we go with the alluring phenomenological doctrine (PI), we have to get rid of one of the theoretical hypotheses (I-B), (II), or (III).

I have discussed this sort of hallucination puzzle before (Hellie 2006, 2007, 2010). At its core is a three-way incompatibility among something like the *externality* of the object or content of (veridical) perception—in the current presentation, (I-B)—the *relationality* or *factivity* of the perceptual stance or attitude toward this object or content (II), and an alleged *inner supervenience* of perception (PI). This structure was brought to my attention by Michael Martin (see in particular his 2002); Martin has also argued convincingly (Martin 2000) that this structure is at the heart of the twentieth-century analytic dispute over perception. Like Martin,

I think the best resolution of the puzzle jettisons the 'inner supervenience' claim in (PI): accordingly, I am on the side of the view variously known (unhappily) as 'disjunctivism' or 'naive realism' or (more happily) as 'direct realism'.[1]

[1]Or at least I think the puzzles for the theorist are maximally difficult if direct realism is as- sumed at the outset. My experience has been that when faced with an aporia, the maximal amount of grain is revealed in the phenomena when we stay on the vehicle heading off the cliff for as long as possible before hitting the eject button. In the present case, we might say that there would be no real problem about perception if it were in fact as the intentionalists say: false belief puzzles, if they exist at all, are far less compelling than hallucination puzzles, so presumably the answer is different. More generally, philosophers should be hesistant to think solutions consist in the rephrasal of everyday worries in high-tech locutions. Here I think Martin would agree: Martin (2000).

11.3 Phenomenology Meets Epistemology

In 'There it is', I defend (III) on the grounds that it accounts for the significance of 'simulation' (Heal 2003): that it explains the inextricable role of the 'second-person perspective' in real-life rationalization of the reactions of the other. What is this principle doing in a discussion of the hallucination puzzle? Answer: linking the best case for externality—found in the phenomenologically-oriented literature on perception—with the best case for relationality/factivity—found in the epistemologically-oriented literature.

In these two largely independent strands of literature, phenomenal internalists occupy closely corresponding positions:

- Among the more phenomenologically-oriented, we find
  - (i) *Sense-data theorists* (Robinson 1994), who require existing objects or factive contents of perception, and correspondingly locate these objects or facts within an internal realm;
  - (ii) *Intentionalists* or *representationalists* (Crane 2007; Chalmers 2004), who allow the objects or contents of perception to be as in or as concerning external bodies, and correspondingly sometimes require the objects to be nonexistent or the contents false.
- While among the more epistemologically-oriented, we find:
  - (i) *Evidentialists* (Carnap 1932; Lewis 1973), who think our basic justification must be infallible, and correspondingly locate its subject-matter as concerning an internal realm;
  - (ii) *Fallibilists* (Pollock 1974; Pryor 2000), who think our basic justification typically concerns an external realm, and correspondingly allow basic justification for false claims.

These evidentialist and fallibilist views assume something like (III). This principle is rejected in the epistemological literature by *classical externalists* (Goldman 1976; Williamson 2000), according to whom basic justification *concerns the external* and *must be infallible.* Obviously no analogue to this position could exist in the phenomenologically oriented literature: discuss the nonphenomenological and the subject is changed. Notably, the *modern externalist* analogue to direct realism advocated in 'There it is' is largely absent in the epistemological literature.

In the phenomenologically-oriented literature, externality is better off than factivity: the sense-data theorist is in a weak position relative to the direct realist and the representationalist. Obviously, ordinary perception is 'transparent' in at least the sense that we find no sense-data there to which to turn attention (Harman 1990): this is part of why 'There it is' is friendly toward (I-B).

By contrast, when we pit the representationalist against the direct realist, the outcome is more mixed. Why can't we say that in a dream, I'm (to use a slightly perplexing locution occasionally arising in conversation) 'turning attention to a mere intentional object'? Well: phenomenologically, 'there it is', whatever it may be.

My inclination is that when I focus my attention on Pirate, his in-my-face, no-doubt-about-it presence is something like a baseline of certainty around which other bits of uncertainty and error are 'wrapped'. My feeling here is that this demands factivity or relationality, but other (perhaps more sensitive) compatriots of mine remain unconvinced and surely not in bad faith; so if the issue is not yet settled, exactly what it would take to do so emerges as entirely nebulous.

It is the opposite over in the more epistemologically-oriented literature, where factivity is better off than externality: here it is the fallibilist who is in a weak position relative to the evidentialist (and the classical and modern externalists). This is largely for programmatic reasons, but the program is extremely deeply rooted and far-reaching. The core idea of the analytic tradition, perhaps, is that inquiry should be modeled on proof. It is easy to see why we should accept theorems proven from analytically true axioms—but why would we accept anything 'proven' from something that might be true, might be false? To insist that this is the best we can do is to consign us to 'frictionless spinning in the void' (McDowell 1994). 'There it is' generates friction through a model of focusing attention on *a* as the 'tokening' of a 'sentence' in a 'Lagadonian' (Lewis 1986) language (in essence, the story says about attention to the external what Chalmers (2003) says about attention to the internal). In a Lagadonian language, grasp of one of its sentences requires recognizing that the semantics is part of the orthography. My proposal is that focus of attention on *a* is tokening of a sentence with *a* as a part and meaning that *a* exists: since the sentence cannot be tokened unless it is true, it is 'analytic'—but *situatedly* analytic in the sense that this sentence is difficult to utter: that's (II). On this proposal, perception provides 'axioms'—sentences that are implicitly known to be, and the contents of which are phenomenologically presented as, infallible: an suitable basis for the rest of a picture of the world, surely.[2] This Lagadonian story is available to the evidentialist and the classical and modern externalists, unavailable to the fallibilist: too bad for the fallibilist, in my view, and good for the modern externalist.

But, pitted against the modern externalist, the evidentialist can mount a strong defense. Why not say that we have an 'implicit' understanding or posit of statistical connections between internal evidence and the external world which leverage internal evidence into external belief? Without some further articulation of what

[2]The technical implementation of this idea in 'There it is' is not fully adequate. A better approach builds on a quasi-Stalnakean approach to self-location (Stalnaker 2008). One's overall picture of the world is represented by a set of doxastically possible worlds *plus* an object of attention *a; a* is rigidly designated across the doxastic possibilities; the act of attention to *a* is the tokening of a Lagadonian judgement that *a* exists with *a* as a part. In a doubly complex-demonstrativejudgement ('this tomato is that color'), the semantic values of the subject and predicate are extracted from *a* as the constituents of *a* satisfying the restrictor predicates. If there are any: which is a presupposition of the doubly complex-demonstrative judgement.

On this model, the act of attention is infallible and—if the presupposition is true—entails the doubly complex-demonstrative judgement: that is how perception justifies belief. (I develop this story in more detail in *Semantics, Self, and World:* in preparation).

this 'implicit' posit consists in—not at all an easy task if the subject-matter of epistemology is understood on an all-too-common hydraulic metaphor, as a sort of flow of normative juice—it is hard to say what the problem is supposed to be.

So the external-content theorist is strong on the phenomenology but weak on the epistemology, while the factive-attitude theorist is strong on the epistemology but weak on the phenomenology. If only there were some connection between the phenomenological and the epistemological! Fortunately there is: namely (III). This means we can assemble the transparency argument for externalism and the programmatic case for factivity: in combination, these yield direct realism on the phenomenological side aka modern externalism on the epistemological side (henceforth I will use these labels interchangeably).[3]

11.4 Lucid Dreaming

It would be nice if modern externalism could be made to work: the big challenge is addressing the allure of the phenomenological internalist doctrine (PI)—an allegedly phenomenologically manifest datum.

The alleged datum can be contested. Obviously epistemology goes a lot better if (PI) is false. Perhaps (as I argue in the first sections of Hellie 2007) (PI) just cannot be made consistent with the genuine phenomenological presentation of perception ('there it is'). Perhaps (PI) cannot be explained in a way that makes sense: in the second half of Hellie (2007) I sketch out a multiplicity of candidate meanings, while in the first half of Hellie (2010) I attempt to undermine the sense that there is a clear meaning to any phenomenological internalist claim fit to conflict with modern externalism.

Still, alleged data is most convincingly tackled head-on. So 'There it is' argues that it is *false* that what it is like for Sam and Dreaming Sam is the same. The wedge is *lucid dreaming:* what it is like to lucidly dream of seeing a tomato differs from what it is like to knowingly see a tomato and what it is like to be taken in by a dream of seeing a tomato. But to be lucidly dreaming could be like a case of seeing a tomato while somehow under the mistaken impression that one is lucidly dreaming.

I suggest that this class of phenomena, not to my knowledge really explored in either the phenomenological or epistemological literature, is best explained as follows. 'What it is like for one' is what the *world* is like for one. And there is

[3] To be explicit, modern externalism gives the following verdicts on its competitors: evidentialism is right to ally the rational and the phenomenological, right to require basic justification to be infallible, wrong to require the content of basic justification to be internal; fallibilism is right to ally the rational and the phenomenological, right to allow the content of basic justification to be external, wrong to allow basic justification to be fallible; classical externalism is right to require basic justification to be infallible, right to allow the content of basic justification to be external, wrong to dissociate the rational and the phenomenological—and all three positions (like sense-data theory, like intentionalism) are wrong to insist on phenomenological internalism.

more to what the world is like for one than which *a* (or even: which kind of *a)* is taken up within the scope of attention: other beliefs also matter, especially one's presuppositions concerning *a*.

So in particular, we analyze our four cases as follows:

S/C This is the normal seeing case Sam inhabits (seeing/correct presuppositions). When one is seeing a tomato, the phenomenological contribution of attention to the red color of the tomato—what it is like to be attending to that color—consists of the fact being 'for one' that that particular state of redness exists as a target of attention. When, against this background, one presupposes correctly that one is seeing, there is no conflict between this fact and one's broader sense of how one interacts perceptually with the world.

Accordingly, a complete story about what it is like for one can simply mention the fact that one is focusing attention on a particular state of redness.

D/C This is the lucid dreaming case (dreaming/correct presuppositions). When one is dreaming of seeing a tomato, the target of one's attention is something other than a token state of color of any tomato: it is rather a qualitative *red-like* state of, perhaps, some dreamy tomato-simulacrum, part of the brain, imagined or recollected previous tomato-encounter, neural image, or something else. Then, the phenomenological contribution of that act of attention consists of the fact being 'for one' that that particular state of red-likeness exists as a target of attention. When, against this background, one presupposes correctly that one is dreaming, there is no conflict between this fact and one's broader sense of how one interacts perceptually with the world.

Accordingly, a complete story about what it is like for one can simply mention the fact that one is focusing attention on a particular state of red-likeness.

D/M This is the case Dreaming Sam inhabits, in which she is taken in by a dream (dreaming/mistaken presuppositions). When Dreaming Sam is dreaming of seeing a tomato, the phenomenological contribution of attention to the red-likeness of whatever is given—what it is like to be attending to that feature—consists of the fact being 'for Dreaming Sam' that that particular state of red-likeness exists as a target of attention. But her presupposition that she is seeing is incompatible with any such dream quality being a target of attention. So when Dreaming Sam affirms this presupposition, the fact that is 'for Dreaming Sam' in attention conflicts with Dreaming Sam's broader sense of how Dreaming Sam interacts perceptually with the world.

So what then *is* it like for Dreaming Sam? The question admits of no coherent answer, because the condition the world would have to meet in order for it to be faithful to how the world is 'for Dreaming Sam' is unsatisfiable. As a result, the best we can say is that Dreaming Sam's overall position is 'fragmented' (Lewis 1982): the view Dreaming Sam adopts in attention and the view Dreaming Sam adopts presuppositionally cannot be put together into a unified view. In one fragment, that particular state of red-likeness exists as a target of attention. But in another, some particular state of color exists as a target of attention; by Dreaming Sam's reckoning, a state of redness. The incoherence is not obvious because the Lagadonian language of attention and whatever 'language' Dreaming Sam's presuppositions are carried in 'code' the incompatible content in a way that obscures the incompatibility. At bottom, then, the hallucination puzzle is a Frege puzzle.

That is the story that would be given by a sympathetic 'second-person' observer armed with the apparatus of 'There it is'. The subject in that position would put things differently: the fragment according to which Dreaming Sam sees a state of redness is in charge of *articulation* of things. So Dreaming Sam would articulate what it is like for her by saying that she turns attention on a state of redness.

In that sense, what it is like for Dreaming Sam is 'indiscriminable' from what it is like for Sam (compare Martin 2004). But this indiscriminability should not be taken as a mark of identity. For although what it is like for Dreaming Sam is indiscriminable from what it is like for Sam *for Dreaming Sam,* for others it is *discriminable.* In particular, these are discriminable for the sympathetic external observer armed with my apparatus. Why? If I attempt to make sense of what it is like for Dreaming Sam, I want to answer in two stages: focusing *just* on the perceptual side of things, I find a 'dreamy' property; but bundling this together with Sam's overall view of things, I find redness. The discrimination from Sam's position consists in the absence of any available 'dreamy' property in Sam's case.

Why trust the external observer above Dreaming Sam? Ordinarily when someone makes a mistake about a situation and when someone else does not, we trust whatever conclusions the latter subject comes to above the conclusions of the former subject when both are in otherwise equally good positions to understand what is going on. Why are Dreaming Sam and I in otherwise equally good positions? Because simulation is roughly as good as the real thing (when the simulation is based on genuine experience rather than mere speculation, and I have been in Dreaming Sam's position before).

S/M This is the odd case in which one is seeing but thinks one is lucidly dreaming (seeing/mistaken presuppositions). Here the story is analogous to the previous case, (D/M): out of respect for trees, I will leave the 'plugging-and-chugging' as an exercise to the reader.

11.5 Bedrock

We might wonder what *makes* Dreaming Sam's position indiscriminable from Sam's? In particular, *why,* given her presuppositions, does she interpret the state of red-likeness she sees as a state of *redness,* rather than a state of *greenness?* I have attempted to provide a 'happy-face' answer to this question before,[4] but 'There it is'

[4]In the second half of Hellie (2010) I provide a precise account of what the indiscriminability could consist in. Some problems: the account presupposes an 'epistemic two-dimensionalism' (Chalmers

argues for the 'unhappy-face' response on which there is no hope for an articulation of the internal cognitive mechanics generating the data.[5]

The grounds for despair are these. Dreaming Sam's picture of the world is incoherent. So if what is wanted is an answer pitched at the level of rational psychology, *we can't give one.* Rational psychology runs out of steam as soon as someone loses coherence: all that can be done at that point is to break the subject into multiple 'fragments', each comprehensible through rational psychology. But rational psychology concerns the doings of individual coherent subjects; interaction among fragments is out of bounds, to be explained at the physiological or ecological level if at all. That is why, in the McDowellesque slogan of 'There it is', if what is wanted for Dreaming Sam is 'justification', too bad: all that can be given is 'exculpation'. Obviously we can use ourselves as instruments to see what we would think in Dreaming Sam's position. But that would be, again, to offer only exculpation: it would 'make sense' of her reaction to her situation as recognizably *human,* but it would do so without providing any *rational* basis for the reaction. Ultimately the task here is not for philosophers: we are good at rational psychology, bad at empirical psychology. If we want answers we should pass the file on to someone else.

'But can't we just say that Dreaming Sam thinks she sees something red rather than green because the dreamy simulacrum *looks red*'? No. On a correct analysis, 'that looks red' expresses one's sense, concerning the object one is looking at and arrived at by looking at it, that it is red (I develop this in *Semantics, Self, and World).* For Dreaming Sam, this sense *just is* her thinking that she sees something red, because she has no further basis for thinking she sees something red beyond going by looking. This is the *product of* her attending to a state of red-likeness while under the impression that she is awake, and not any distinctive further fact about the relation of red-likeness to red in Sam's view. If there is explanatory power to 'I think it is red because it looks red', it derives entirely from the presuppositional content of 'it looks red': what it conveys is 'I arrived at my belief that it is red by looking at it'. That is a *causal* claim, and not a *rationalizing* claim. Similarly, 'I think it is not green because it looks red' conveys 'I arrived at a belief that it is red by looking at it and what is red is not green'. So the purported explanation merely restates what we already know: namely, that Dreaming Sam arrived at the belief concerning the simulacrum that it is red rather than green by looking at it, and that a normal person

---

2003) which 'There it is' rejects (140); the account does not explain how lucid dreaming could seem different from seeing; the account is not at all easy to distinguish from a 'qualia' account.

[5]If I understand his position correctly, Soteriou (2005) assumes that we must face dreams with the assumption that they are real if we are to characterize them at all, and argues that the inevitable incoherence blocks any understanding of the nature of dreams. The assumption seems to be based on an overly demanding understanding of the 'transparency' of perception which would rule out lucid dreaming. But the prospect remains that our only strategy for understanding the nature of perception involves making no false assumptions, in which case we in effect merely recapitulate something more determinate than what we already knew. Soteriou's important insight is that while we respond intellectually to perception, this does not involve in any way our analyzing it.

would have the same reaction in her position (namely, focusing attention on red-likeness while presupposing that one is seeing). It offers no added clarification of why it is a good idea to react as she does.

So, I conjecture, the best we can say in response to what makes Dreaming Sam's position indiscriminable from Sam's is that *this is just how she reacts:* given her habits of translating perceptually coded vehicles into articulately coded vehicles, she is disposed to translate Lagadonian red-likeness into 'red' rather than 'green'. Not because that helps to best make sense of things; not because she has introduced a meaning postulate which makes the transition analytic. Rather, that is just how she is 'wired up'—at a physiological or ecological level.

11.6 The Importance of the Second-Person Perspective

A philosophical theory of perception should explain why there is any need for a philosophical theory of perception—again, a point emphasized by Martin (2002). We wouldn't find the need for such a theory if we didn't focus on cases in which we are misled. After all, the theory (I)-(III) is completely self-consistent, and upon superficial reflection is just plain obvious. The monkey in the wrench is (PI): as stated explicitly, the principle concerns a case of being misled; as generalized to the 'phenomenal internalist spirit', it highlights one among the many 'pathologies'—odd cases beloved by modern thought in which things ordinarily going together get teased apart—serving as data for the philosophy of perception.

On the theory of 'There it is', these pathologies are episodes of incoherence. When Fred is incoherent, he is in the worst possible position to understand what is going on with him: noticed incoherence collapses to one or another variety of coherence, so incoherent Fred must be self-ignorant. So what it is like for Fred can only be understood from the second-person view. Unfortunately, a distinctive element of the modern philosophy of mind is the isolation of the subject, the paradigm of which is the Cartesian 'soul-pellet' ontology. Granting the isolation of the subject, it is natural to think that the route to understanding the phenomenological would be from the first-person if from anything. But get rid of the isolation of the subject and this assumption vanishes—which, for the philosophy of perception, unties the knot.

Perhaps the main overarching moral of 'There it is' is the importance of the second-person perspective. Reflection on the role of the second-person motivates the principle (III), which in turn is the keystone in the case for direct realism aka modern externalism: without a doubt, the natural phenomenologico- epistemological view. Accord the appropriate level of respect to the second- person perspective, and it ceases to be clear why we would resist this natural view.

References

Carnap, Rudolf. 1932. Psychology in physical language. *Erkenntnis* 3: 107–142. Reprinted in ?
Chalmers, David J. 2003. The content and epistemology of phenomenal belief. In *Consciousness: New philosophical essays*, ed. Quentin Smith and Alexander Jokic. Oxford: Oxford University Press.
Chalmers, David J. 2004. The representational character of experience. In *The future for philosophy*, ed. Brian Leiter. Oxford: Oxford University Press.
Crane, Tim. 2007. Intentionalism. In *Oxford handbook of philosophy of mind*, ed. Ansgar Beckermann and Brian McLaughlin. Oxford: Oxford University Press.
Goldman, Alvin. 1976. Discrimination and perceptual knowledge. *Journal of Philosophy* 73: 771–791.
Harman, Gilbert. 1990. The intrinsic quality of experience. In *Action theory and the philosophy of mind, vol. 4 of philosophical perspectives*, ed. James Tomberlin, 31–52. Atascadero: Ridgeview.
Heal, Jane. 2003. *Mind, reason, and imagination*. Cambridge: Cambridge University Press.
Hellie, Benj. 2006. Beyond phenomenal naivete. *The Philosophers Imprint* 6(2): 1–24.
Hellie, Benj. 2007. Factive phenomenal characters. *Philosophical Perspectives* 21: 259–306.
Hellie, Benj. 2010. An externalist's guide to inner experience. In *Perceiving the world*, ed. Bence Nanay. Oxford: Oxford University Press.
Hellie, Benj. 2011. There it is. *Philosophical Issues* 21: 110–164.
Lewis, David. 1973. Why conditionalize? In?
Lewis, David. 1982. Logic for equivocators. *Noûs* 16: 431–441.
Lewis, David. 1986. *On the plurality of worlds*. London: Blackwell.
Martin, Michael G.F. 2000. Beyond dispute: Sense-data, intentionality, and the mind-body problem. In *History of the mind-body problem*, ed. Tim Crane and Sarah Patterson. London: Routledge.
Martin, Michael G.F. 2002. The transparency of experience. *Mind and Language* 17: 376–425.
Martin, Michael G.F. 2004. The limits of self-awareness: Disjunctivism and indiscriminability. *Philosophical Studies* 120: 37–89.
McDowell, John. 1994. *Mind and world*. Cambridge, MA: Harvard University Press.
Pollock, John. 1974. *Knowledge and justification*. Princeton: Princeton University Press.
Pryor, James. 2000. The skeptic and the dogmatist. *Nous* 34: 517–549.
Robinson, Howard. 1994. *Perception*. London: Routledge.
Soteriou, Matthew. 2005. The subjective view of experience and its objective commitments. *Proceedings of the Aristotelian Society* 105: 193–206.
Stalnaker, Robert C. 2008. *Our knowledge of the internal world*. Oxford: Oxford University Press.
Williamson, Timothy. 2000. *Knowledge and its limits*. Oxford: Oxford University Press.

Chapter 12
Perceptual Justification Outside of Consciousness

Jacob Berger

12.1 Introduction

It is often assumed that rationality and consciousness share some sort of essential connection. Thus some theorists build rationality into their accounts of consciousness. Ned Block, for example, famously claims that a mental state exhibits what he calls access consciousness if it "is poised for free use in reasoning and for direct "rational" control of action and speech" (2007, p. 168).[1] In his (2011) paper "There It Is" and his (2014, this volume) précis "It's Still There!" Benj Hellie develops a complex account of how perceptions justify beliefs—an account which effectively builds consciousness into rationality. As Hellie puts it, he "advances a picture of the nature of rationality and rational explanation in which consciousness plays a central role" (2011, p. 110).

Hellie develops his account of perceptual justification against the backdrop of the view known as direct realism. Hellie notes that direct realism involves a cluster of commitments, but that a central feature is the recognition of the distinction between so-called good cases wherein perception is accurate and bad cases such as hallucinations. And Hellie develops a sophisticated semantics for perceptual justification according to which perceptions in good cases can justify beliefs and can be explained by intentional psychology. In bad cases of perception, by contrast, Hellie argues that one's perceptions are in an important way defective and so rational explanations do not apply in those cases. Adapting John McDowell's (1994)

[1] All references to Block are from his paper "On a Confusion about a Function of Consciousness," originally published in (1995) and reprinted in his (2007) collection. Page references to Block are from the (2007) version.

J. Berger (✉)
The Graduate Center, City University of New York, Philosophy Program, 365 Fifth Avenue, New York, NY 10016, USA
e-mail: jfberger@gmail.com

R. Brown (ed.), *Consciousness Inside and Out: Phenomenology, Neuroscience, and the Nature of Experience*, Studies in Brain and Mind 6, DOI 10.1007/978-94-007-6001-1_12,

well-known expression, Hellie claims that in bad cases "we cannot offer a theory of justification; we must content ourselves with exculpation" (2011, p. 111).

Though there is much to say about Hellie's rich and challenging papers, I'll focus this commentary on Hellie's view of the relationship between perceptual justification and consciousness. It is undoubtedly true that some conscious perceptions justify beliefs: If I consciously perceive that there is a red apple, my conscious perception justifies my belief that there is a red apple. However, there is increasingly good evidence that perceptions can occur outside of consciousness, as in cases of so-called subliminal perception in normal individuals and blindsight in people with damage to visual cortex. I'll argue that such perceptions can justify beliefs and rationalize behavior, even though these states are not within consciousness. I will reserve judgment regarding Hellie's treatment of the difference between good and bad cases, but I'll argue there can be what he views as good cases of perceptual justification outside of consciousness.[2]

12.2 Perceptual Justification Outside of Consciousness

At the outset of "There It is," Hellie glosses the notions of rationality and justification in the following way:

> [T]he core notion of rationality is something like manifest coherence of the stream of consciousness. If so, the most basic interpretation of the claim that A justifies B means something close to: from the first-person perspective, B was required to maintain coherence of the stream of consciousness in light of A (2011, p. 111).

Though Hellie does not explicitly mention consciousness in his précis, he does commit to what he calls thesis (III): "If two subjects are in distinct rational positions, what it is like for them differs" (2014, p. 127). Assuming, as most do, that there is something that it is like for one only if one is in a conscious state, it would seem to

[2]To be more precise, Hellie restricts good cases of perceptual states only to "what one is 'attending to'" and he claims that anything else "has no direct presence within one's stream of consciousness and therefore cannot be rationally significant" (2011, p. 131). But Hellie offers no reason to hold this position in his papers, and it is not clear why one would hold it. Even if Hellie were right that rationality requires consciousness (a view which I'll challenge), whether consciousness requires attention is a vexed issue because of the considerable debate about how to understand attention. At first blush, however, it seems clear that there are many conscious perceptions that do not involve attention, such as those involved in the periphery of one's consciousness. Such states may be less rich informationally than states that do involve attention, but it is not as though these peripheral perceptions need be illusory and hence bad cases. There thus seems to be no reason to deny that a conscious perception without attention can justify a belief. Additionally, there is mounting evidence that we can attend to stimuli in the absence of consciousness (see, e.g., Koch and Tsuchiya 2007; van Boxtel et al. 2010). For brevity's sake, I will not review this evidence here and I acknowledge that some dispute it (e.g., de Brigard and Prinz 2010). But if attention can occur nonconsciously, even if Hellie were right that rationality requires attention (which is doubtful), it would not provide a reason to deny that rationality can occur outside of consciousness.

follow that one's position cannot be rational in virtue of a nonconscious state. Hence Hellie's thesis effectively holds that perceptual justification cannot occur outside of consciousness.

I'll begin by offering a theoretical reason to be open to the possibility that perceptual justification does not require consciousness. Hellie may be correct that rationality involves maintaining a certain kind of coherence, but it need not be the coherence of one's stream of consciousness. The first thing to note is that, even if all perceptions and beliefs do occur within the stream of consciousness, the property of being within consciousness is distinct from the property of having a particular content. Many states that exhibit distinct contents can all occur within consciousness. Furthermore, if a perception justifies a belief, whatever rational connection holds between those states holds in virtue of their respective contents. If I perceive that there is an apple, then I am justified in forming the belief that there is an apple—and that justificatory relationship holds because those states exhibit relevant contents involving the apple. Thus the fact that some states occur within consciousness is independent of whether their relationships are rational (cf. Rosenthal 2008, p. 832).

As a result, the core notion of rationality is better glossed as coherence within one's stream of psychological states, whether or not those states occur within consciousness. On this revised view of justification, if A justifies B, B was required to maintain coherence of the stream of psychological states—conscious or otherwise—in light of A.

Hellie himself does not say much in his papers about the nature of psychological states, consciousness, or their relationship to one another, but he does briefly characterize the stream of consciousness as "the sequence of experiences (understood as token occurrences) one undergoes" (2011, p. 115). The term 'experience' is often taken to imply a state that is within consciousness and many do make the Cartesian assumption that all psychological states occur consciously. But Hellie offers no reasons to think this, and there is substantial evidence that psychological states such as perceptions and beliefs can, and often do, occur outside of consciousness. Nowadays it is commonplace to talk about nonconscious intentional states such as beliefs and desires that guide or influence behavior. There are many ordinary cases of, for instance, knowing what others believe or desire before those people know themselves.

There is also evidence that qualitative states such as perceptions can occur outside of consciousness. Consider, for example, the remarkable experimental work with the blindsight patient TN. TN suffered bilateral damage to visual cortex. As a result, under typical conditions reports TN that he cannot see anything and generally behaves as though he cannot. But experimenters recently found that TN was nonetheless able to navigate successfully a corridor which included many barriers (de Gelder et al. 2008).

Arguably, what happened in TN's case is that, during his walk down the corridor, TN nonconsciously perceived the barriers, which justified him in forming beliefs that there were obstructions in his environment. Additionally, because of his goal to walk down the corridor, these newly formed beliefs interacted rationally with states (such as the belief that one cannot walk through barriers) to produce the rational behavior of moving skillfully around the barriers. In other words, on the basis of TN's nonconscious perceptions, certain beliefs about his environment were required in order to maintain coherence—not in his stream of consciousness, but in the stream of his psychological states generally.

TN's case is certainly striking and doubtless calls for explanation. One might worry, however, that blindsight is a special phenomenon. After all, TN suffered brain damage and we lack an exhaustive understanding of these kinds of conditions. To the extent that Hellie is attempting to provide an account of perceptual justification in normal individuals, perhaps he is warranted in withholding judgment regarding such cases.

But there are many everyday instances of perceptual justification outside of consciousness in normal individuals as well. For example, it is often the case that while you are in a crowded place engrossed in a book, at some point you may happen to look up to immediately lock eyes with someone who has been staring at you for a period of time. A sensible explanation is that, though you were not at first aware that you saw the person stare at you, the fact that you subliminally saw the person justified you in believing there was such a person. Because of your standing interest in investigating people that stare at you, the confluence of these states rationally caused you to look up directly at the person.

There are in addition experimental examples of similar phenomena. In one study (Castiello et al. 1991), participants were asked to grasp one of three targets as fast as possible when one of the targets was illuminated. In some trials, as soon as participants began reaching for one target, the light was switched to another target. Despite the changes in the targets, participants grasped for the illuminated targets in fluid and uninterrupted ways. Participants were also asked to report if and when they noticed that the targets changed. Remarkably, participants typically reported noticing that they needed to change their movement around 300 ms after they had corrected their behaviors and began grasping for the illuminated target. It is thus arguable that at first the participants subliminally perceived the changes.

Again, though the participants were for some time unaware that they had perceived the changes, given their prior goals of reaching for the illuminated targets, these subliminal perceptions rationally caused them to alter the trajectory of their reaches. So even though these perceptions lay outside of consciousness, the behavioral adjustments that they caused were arguably rational. Indeed, correcting one's behavior in light of new incoming perceptual information to achieve one's goals is the paradigm of rational activity.

Importantly, these nonconscious perceptions do not seem to be bad cases of the sort that Hellie discusses in his papers. To navigate the corridor successfully, for example, TN's states had to register information about his environment accurately. Put another way, TN did not hallucinate a barrier, consciously or otherwise. Hellie may be right that rationalizing explanations only apply in good cases, but at first blush it appears there is no good reason to deny that some nonconscious perceptions are good cases that can justify beliefs and rationalize behavior.

12.3 Potential Replies

There are several reasons, however, why Hellie might deny that these are genuine cases of perceptual justification outside of consciousness. As noted above, Hellie is committed to thesis (III), which holds that if two subjects are in distinct rational positions, what it is like for them differs. Hellie notes in his précis that he defends (III) in "There It Is" "on the grounds that it accounts for the significance of 'simulation' (Heal 2003): that it explains the inextricable role of the 'second-person perspective' in real-life rationalization of the reactions of the other" (2014, p. 129). In other words, Hellie argues that, in order to determine whether someone else's activity is rational, one must "'push into', 'take up', 'project [oneself] into', or 'simulate' [her] point of view, and rehearse the narrative she advanced to herself" (2011, p. 121). Hellie's account of what it is to take up another's perspective in this way is complex, but, roughly, a narrative is a set of sentences that characterize the states of one's consciousness and one rehearses a narrative by simulating that narrative to determine whether it is coherent.

Thus Hellie might argue that one cannot find, for example, TN's activity rational because one is unable to adopt his point of view to find his narrative coherent or not. Since the states of TN that register information about his environment are not conscious, there is nothing that it is like for TN to be in them. One might thus think TN does not have a point of view with regard to those states. If thesis (III) were true, it would seem TN cannot be in a rational situation in virtue of his nonconscious states.

But since rational connections hold in respect of content and not consciousness, as I've argued, we ought not to characterize a narrative in terms of consciousness. Instead, a narrative is better understood as the set of states of one's psychology, whether or not those states are conscious. And though one might assume that one only has a point of view with regard to one's conscious states, this is arguably unsupported. In order to navigate the corridor successfully, TN's nonconscious perceptions must represent his environment egocentrically and are thus from his point of view—he is simply unaware of the states that are from that point of view.

If so, then it is arguable that one can evaluate TN's narrative. Though we cannot imagine what it is like to be in TN's perceptual states—for there is nothing that it is like to be in them—we can reflect on the narrative of TN's psychological states characterized in terms of their content. On that basis, we do not regard TN's behavior as irrational, even though many of the contents involved in issuing in that behavior are nonconscious. Hellie's argument therefore fails to establish that perceptual justification requires consciousness.

During an exchange at the Third Annual Online Consciousness Conference, Hellie considers several other sorts of replies to these kinds of cases. Hellie proposes that, whatever the states outside of consciousness that register information about one's environment might be, it may be that they are not properly called 'subliminal perceptions' because they are not genuine psychological states. On this view, nonconscious states are, to use Daniel Dennett's (1969) term, subpersonal—that is, not personal-level psychological states such as beliefs, desires, and perceptions. These states may be able to be explained in terms of biology or physics, or perhaps by some sort of nonmental computational explanation, but they are not within the purview of intentional psychology. These states would thus be akin to bad cases insofar as they cannot enter into rationalizing explanations.

This reply mirrors Hellie's discussion of having one's attention captured, which he claims is an example of what he calls "arational update in the stream of consciousness" (2011, p. 131). Hellie avers that the processes that underlie one's shifts of attention cannot be explained in terms of rationality. Likewise, Hellie suggests that, in cases such as looking directly up at someone who has been staring at you, whatever nonconscious processes give rise to one's behavior are not rationally explicable. This seems to be a commonly held position. For example, in discussing his characterization of access consciousness, Block hastens to add that "[t]he "rational" is meant to rule out the kind of control that obtains in blindsight" (2007, p. 168).

But this reply is unconvincing. These sorts of nonconscious states play the same functional roles as their conscious counterparts—they are simply not within one's stream of consciousness. TN's states enable him to respond differentially to a range of stimuli during his walk down the corridor. And, to repeat, these nonconscious states interact with psychological states such as beliefs and desires in ways that are rational. Had TN previously believed that there were no barriers in the corridor, his registering visual information about them would doubtless have resulted in a conflict of some sort. TN might not have immediately reported a feeling of confusion, but there is good reason to suppose that there would have been evidence of such a conflict such as delays in his ability to act on that information. For these reasons, it is natural to describe these kinds of nonconscious states with the same intentional and qualitative vocabulary that we use to characterize conscious states.

Hellie does not offer any reasons to think that nonconscious states are not psychological in his papers, but there are several reasons why one might find this claim inviting. First, many assume that the paradigms of psychological states are those that occur within consciousness. This may appear to be the case because the only states of which we seem to be directly aware are conscious states. TN is certainly not aware that he sees any barriers and would deny that he does. If we are in states that are not in consciousness, we only know about them in ways that seem indirect—such as by being told that we are in them or through conscious inference. This may seem to suggest that nonconscious states are no more psychological and thereby open to rational explanation than the states of one's stomach.

Similarly, without sufficient prompting, TN would not verbally cite his states as justifications for why he behaved as he did, which might suggest that these states cannot function as the reasons for his actions. Thus Block writes that "although the information that there is an X affects [a person with blindsight's] "guess", it is not available as a premise in reasoning ... or for rational control of action or speech" (2007, p. 172). If by 'available' Block means capable of being verbally cited as a reason, Block holds—and Hellie may agree—that the capacity to verbally cite one's justifications for action is essential to rationality.

But even if one is not aware of being in a state, it does not entail that the state is not a genuine psychological state such as a perception or a belief. Though TN is not aware that he sees any barriers, this alone does not show that he does not see any barriers. The only reason to hold that one is always aware of one's psychological states is the dubious Cartesian assumption that the mind is always and fully known to itself. Likewise, the fact that TN cannot verbally cite his perceptions as the reasons for his behavior does not show he does not have those reasons. Serving as a premise in reasoning and rationally guiding behavior is distinct from being available to be reported as a reason for one's behavior. It therefore seems that we are often not aware of the reasons for our actions—and stipulating that one must be aware of one's reasons begs the question against the possibility of nonconscious perceptual justification.

Perhaps more fundamental to Hellie's project is that it is a central tenet of direct realism that the only states that can enter into rationalizing explanations are those which put us into direct contact with the world. Hellie unpacks the way in which perception is direct in part in terms of the so-called transparency of experience. As Hellie observes, direct realism is committed to the idea that "ordinary perception is 'transparent' in at least the sense that we find no sense-data there to which to turn attention (Harman 1990)" (2014, p. 129). Direct realism denies that people perceive external objects indirectly by, for example, perceiving mental intermediaries such as a sense data and, on that basis, inferring that those external things exist. But since we are never aware of being in nonconscious states in ways that do not seem indirect, one might think that those states cannot put us into the kind of direct contact with the world that the direct realist emphasizes.

It is hard to see, however, how the fact that we are not aware of being in nonconscious states could make a difference as to whether they put us in direct contact with the world. No version of indirect realism is entailed by the idea that some perceptions occur outside of consciousness. TN may be directly aware of barriers, even if he is not conscious of his direct awareness of them.

Crucially, insofar as direct realism is committed to a distinction between good and bad cases, there is evidence not only that accurate perceptions can occur outside of consciousness, but also that illusions can too. For example, there is a visual phenomenon known as the simultaneous brightness-contrast illusion, wherein a gray object on a dark background is typically illusorily perceived to be brighter than the same gray object on a lighter background. Recently, Marjan Persuh and Tony Ro (2012) used a technique known as metacontrast masking to determine whether this perceptual illusion can take place outside of consciousness.

In a typical metacontrast-masking study, participants are briefly presented with a stimulus, which is immediately followed by a non-overlapping mask that renders the stimulus invisible to consciousness (see, e.g., Breitmeyer and Öğmen 2000). Though participants report not seeing the stimuli, they can be primed by them in various ways that can be behaviorally detected, which suggests that they subliminally perceive such stimuli. In their study, Persuh and Ro found that gray stimuli on dark backgrounds primed as though they were perceived to be brighter than the same stimuli on lighter backgrounds, even when those stimuli were masked. That is to say, this perceptual illusion can occur even if that perception occurs outside of consciousness.

So one can endorse the direct realist's distinction between good and bad cases, even if one denies that perceptual justification requires consciousness. The question of whether rationality takes place only in good but not in bad cases is independent of the question of whether some perception takes place outside of consciousness. If Hellie is correct about when rationalizing explanations apply, then perhaps no rational explanations can be given of these nonconscious illusions.

In sum, there are no good reasons to deny that nonconscious states are genuinely psychological and that they can, at least sometimes, justify beliefs.

In light of these considerations, Hellie instead might endorse the possibility that the states that register information about TN's environment may be in his stream of consciousness. On this view, TN consciously perceives the barriers, even though he cannot report that he perceives them. If TN has any other conscious states, such as a conscious desire to walk down the hall, these other states may simply be encoded in TN's consciousness differently than his conscious perceptions. Thus TN's walking down the corridor is wholly rationally explicable, even though TN may not be able to verbalize some of the rational activity that generates his behavior.

But there seems to be no reason to regard TN's perceptions as conscious, especially in the face of his fervent denial that he consciously sees anything. Indeed, common sense as well as most recent experimental work holds that one's report that one is not aware of being in a state is taken as excellent evidence that the state is in not within the stream of consciousness. This is why most agree that if any states are outside of consciousness, TN's perceptions are paradigm cases. Since it clearly fits better with both folk and experimental psychology to regard some psychological states as being within consciousness and others as not, this reply is unmotivated.

12.4 Conclusions

I have argued that perceptual justification can occur outside of consciousness. But insofar as one of Hellie's main goals was to develop a semantics for perceptual justification according to which rationalizing explanations apply in good but not in bad cases of perception, Hellie's account of that difference could hold even if such cases can and often do occur outside of consciousness.

Acknowledgement I thank Myrto Mylopoulos, David Pereplyotchik, and Marjan Persuh for help with this commentary. I also thank David Rosenthal for his thoughtful comments on the online version of this commentary, which was presented at the Third Annual Consciousness Online Conference. I especially thank Benj Hellie for his detailed replies to that online commentary and Richard Brown for organizing the conference as well as this volume.

References

Block, N. 2007. On a confusion about a function of consciousness. In *Consciousness, function, and representation: Collected papers*, vol. 1, 159–213. Cambridge, MA: MIT Press; originally published as N. Block (1995), *The Behavioral and Brain Sciences*, 18 (2): 227–247.

Breitmeyer, B.G., and H. Öğmen. 2000. Recent models and findings in visual backward masking: A comparison, review, and update. *Attention, Perception, & Psychophysics* 62(8): 1572–1595.

Castiello, U., Y. Paulignan, and M. Jennerod. 1991. Temporal dissociation of motor responses and subjective awareness: A study in normal subjects. *Brain* 114(6): 2639–2655.

De Brigard, F., and J. Prinz. 2010. Attention and consciousness. *Wiley Interdisciplinary Reviews* 1(1): 51–59.

de Gelder, B., M. Tamietto, G. van Boxtel, R. Goebel, A. Sahraie, J. van den Stock, B.M.C. Stienen, L. Weiskrantz, and A. Pegna. 2008. Intact navigation skills after bilateral loss of striate cortex. *Current Biology* 18(24): R1128–R1129.

Dennett, D.C. 1969. *Content and consciousness*. London: Routledge & Kegan Paul.

Harman, G. 1990. The intrinsic quality of experience. *Philosophical Perspectives* 4: 31–52.

Heal, J. 2003. *Mind, reason, and imagination*. Cambridge, UK: Cambridge University Press.

Hellie, B. 2011. There it is. *Philosophical Issues* 21: 110–164.

Hellie, B. 2014. It's still there! In *Consciousness inside and out: Phenomenology, neuroscience, and the nature of experience*, ed. Richard Brown, 127–136. Dordrecht: Springer.

Koch, C., and N. Tsuchiya. 2007. Attention and consciousness: Two distinct brain processes. *Trends in Cognitive Science* 11(1): 16–22.

McDowell, J. 1994. *Mind and world*. Cambridge, MA: Harvard University Press.

Persuh, M., and T. Ro. 2012. Context-dependent brightness priming occurs without visual awareness. *Consciousness and Cognition* 21(1): 177–185.

Rosenthal, D.M. 2008. Consciousness and its function. *Neuropsychologia* 46: 829–840.

van Boxtel, J.J.A., N. Tsuchiya, and C. Koch. 2010. Consciousness and attention: On sufficiency and necessity. *Frontiers in Psychology* 1(217): 1–13.

Chapter 13
Some Thoughts About Hallucination, Self-Representation, and "There It Is"

Jeff Speaks

Benj's "There it is" is a characteristically original and wide-ranging exploration of the relationship between certain direct realist theories of perception and the nature of perceptual justification—with some formal semantics thrown in for good measure.[1] Here I'll focus on just one of the many topics about which Benj has something to say: his remarks on the topic of the relationship that must obtain between a perceptual state and a belief in order for the former to immediately justify the latter.

Setting some of the subtleties of Benj's story to the side, and focusing for now only on the case of veridical experience, the basic picture is as follows[2]: being in a veridical perceptual state involves accepting a sentence. This sentence, speaking loosely, represents the subject of the experience as having an experience of the type he is having—so, to use Benj's example, the sentence accepted in virtue of Sam's looking at the red color of a widget would represent Sam as having the property of looking at the red color of a widget. The relationship between veridical experiences of this sort and the corresponding sentences accepted has some interesting and unusual features: (i) whenever a sentence of this language which ascribes the property of having a veridical experience of the right sort is accepted, it is true; and (ii) everyone who has a veridical experience of the right sort accepts a sentence which self-ascribes the property of having just that sort of veridical experience. As Benj puts it, the sentences in question are infallible, and the properties they ascribe to subjects are self-intimating.

---

[1]A previous draft of this paper was give in response to "There it is" at the 3rd annual Online Consciousness Conference. "There it is" is now published as Hellie (2011).

[2]See Hellie (2011), §3.

J. Speaks (✉)
Department of Philosophy, University of Notre Dame, 100 Malloy Hall, Notre Dame, IN 46556, USA
e-mail: jspeaks@nd.edu

R. Brown (ed.), *Consciousness Inside and Out: Phenomenology, Neuroscience, and the Nature of Experience*, Studies in Brain and Mind 6, DOI 10.1007/978-94-007-6001-1_13,

One might wonder: how could a language have these features? How could it be impossible to accept a sentence of a language without it being true? Benj's answer is that the language in question is a "Lagadonian language" in which (at least some of) the expressions are objects and properties which refer to themselves. If Sam himself, and the property of looking at the red color of a widget are expressions in this language, and if accepting the sentence which represents Sam as looking at the red color of a widget just is a matter of the name for Sam (namely, Sam himself) instantiating a predicate which expresses the property of looking at the red color of a widget (namely, that property), then we can see how the Lagadonian language could be infallible and self-intimating.

Now, to be sure, this Lagadonian language raises some further questions. Surely Sam can instantiate some properties—like the property of gaining 2 lb—without representing himself as instantiating these properties. So it must be that some of Sam's properties are predicates of the Lagadonian language, and some aren't. But what explains this distinction between two sorts of properties? In the standard case, we explain the distinction between things which are and things which aren't expressions of a language in terms of the use to which the expressions are put by a certain community of language users—whether use is specified in terms of Gricean intentions, Lewisian conventions of truthfulness in trust, or in less psychologistic terms. But in the case of the Lagadonian language, "use" seems to just be a matter of property instantiation—which won't give us the wanted contrast between the property of looking at the red color of a widget and gaining 2 lb, since Sam instantiates each.

Now, at this stage Benj is, I take it, just sketching a framework for thinking about these issues rather than giving a fully worked out theory of the Lagadonian language. The present worry is less an objection to the framework itself than a question which, it seems to me, a fuller development of Benj's theory should be able to answer.[3] So let's set worries about the Lagadonian language to the side and press on to the account of perceptual justification.

To explain how accepting a sentence of the Lagadonian language can rationalize a belief, Benj suggests that we have to understand how sentences of this language might be related to sentences of the distinct language which does underwrite beliefs. The mechanism for this is the subject's regarding a sentence of the Lagadonian language as equivalent to a sentence of the belief language. Roughly, if $\mathcal{R}$ is a sentence of the Lagadonian language and B is a sentence of the belief language, this requires that the subject have a certain cluster of attitudes toward the biconditional $\ulcorner$ $\mathcal{R}$ iff B $\urcorner$—one must regard it as trivially true and its negation as incoherent, and one must take questions about why it is true to be unintelligible. When these

[3]Some initially plausible answers won't work. For example, one might try to draw the distinction in terms of availability of the relevant properties for reasoning; the proposition that I am looking at the red color of a widget is immediately available to affect my beliefs and actions, whereas the proposition that I have gained 2 lbs might not be. But of course the proposition that I am looking at the red color of a widget might be similarly unavailable, as can be seen from cases in which I'm unsure whether I'm having a veridical or hallucinatory experience.

conditions are satisfied, this is sufficient for B to have the same content as $\mathcal{R}$, which in turn is sufficient for (to continue with the example above) Sam to believe that Sam is looking at the red color of a widget. Since sentences of the Lagadonian language are infallible, a belief formed in this way will always be true.[4]

Again, one might raise some questions here about how the central terms of the theory are to be understood. It is fairly clear what it means to regard two sentences of a language like English as equivalent, in Benj's sense; but it's not quite as clear when one of the sentences is, like $\mathcal{R}$, a subject instantiating a certain property. As far as I can tell, I have never had any attitudes at all toward a biconditional one of whose constituent sentences is my instantiating the property of looking at something red, mainly because it never crossed my mind that my instantiating such a property could even be a sentence in a biconditional. But, if Benj's theory is to explain the rational status of my beliefs about me looking at red things, this must be something which I've done many times—and it is very puzzling how I could have adopted the cluster of attitudes described in the preceding paragraphtoward the relevant biconditionals without noticing. This suggests, I think, that we need something more than the suggested interpretation of 'regarding $\mathcal{R}$ and B as equivalent' if it is to do the work assigned to it by Benj's theory.

There are also some worries about the view of the individuation of contents implied by this story, according to which two sentences have the same content for a speaker if the speaker (in the sense sketched above) regards the two sentences as equivalent. This makes certain sorts of mistakes about equivalence impossible: if someone regards a biconditional as trivially true, its negation as incoherent, and its truth as inexplicable, it follows that that person is correct. This is in a way parallel to a familiar consequence of other coarse-grained views of content, like the view that propositions are sets of worlds, which entails that no one believes any necessary falsehoods. And, it seems to me, one might object to Benj's theory using just the same sorts of examples standardly used to argue against those coarse-grained views of content, like mathematical mistakes. Suppose that a mathematician regards a pair of formulae as equivalent, the negation of their biconditional as incoherent, and their equivalence as inexplicable (perhaps the mathematician thinks that all mathematical truths are inexplicable)—does this really entail that the formulae are synonymous out of that mathematician's mouth? Now, there are things that can be said here—roughly, the sorts of explanations that proponents of possible worlds semantics give of apparent cases of believing necessary falsehoods. But those who are unconvinced by these explanations will regard this consequence of Benj's theory

---

[4]See Hellie (2011), 138 and following. Strictly, what follows is just that the belief is true at the moment at which it is formed, presuming that this moment is the same as that at which the relevant experiential property is instantiated by the subject. The belief might quickly be falsified by a change in the veridicality of the subject's experience. (Or, if we think of beliefs as having their truth-values eternally, ordinary mechanisms of 'belief maintenance' might quickly lead to a false belief if there's a change in the veridicality of the subject's experience).

as an unwelcome one—and it's not one that Benj can avoid so long as he maintains the explanation of the way in which perceptual beliefs acquire their contents from perceptions.

So far I've only been talking about Benj's approach to veridical experiences which the subject takes to be veridical; in the later sections of the paper, Benj provides an extensive taxonomy of the different ways in which experiential episodes might fall short of this norm. Here I'll just focus on what Benj has to say about the familiar case in which a subject is having a hallucinatory experience which she mistakenly takes to be veridical.

Suppose that our subject is Sam, and that Sam is dreaming that he is looking at the red color of a widget. In this case, Sam will, by virtue of so dreaming, accept a sentence $\mathcal{R}_8$ which represents Sam as instantiating the property of dreaming that he is looking at the red color of a widget—since, as above, Sam and this property are both terms which represent themselves in our Lagadonian language. What perceptual belief will Sam form in this case?

Benj's idea is that the way to answer this question is by looking at Sam's conditional evidential policies, which we can suppose to include the following two:

(A) Regard $\mathcal{R}$ and "I am looking at the red color of a widget" as equivalent if I am looking at the red color of a widget.
(B) Regard $\mathcal{R}_8$ and "I am dreaming that I am looking at the red color of a widget" as equivalent if I am dreaming that I am looking at the red color of a widget.

The question is then how these policies are related to the content of the belief actually formed. It can't be that adopting policies (A) and (B) is sufficient for one to form, in a particular situation, whichever belief (A) and (B) dictate—for, if this were sufficient, one would always believe that one is veridically perceiving when one is, and always believe that one is dreaming when one is. And of course we don't always do this, since we can be mistaken about whether we are dreaming or veridically perceiving.

And, in a way, the fact that we don't always do this can be used to pose a problem for Benj's theory. Recall that, in the veridical case, the content of a belief is determined by the proposition associated with the perceptual experience via the subject's regarding the experience as equivalent to the sentence in her "belief language." As noted above, one might worry about what, exactly, it means to regard a sentence of the Lagadonian perceptual language as equivalent to a sentence of the belief language. But, *whatever* it takes for a subject to regard these sentences as equivalent, it seems that a subject might take exactly the same attitude toward an episode of dreaming and a sentence of the belief language. And if the subject can do this, it is hard to see, on Benj's picture, why this should not be sufficient for the subject to believe that she is dreaming. But one simply can't, in this way, form true beliefs about whether we are dreaming or having a veridical experience—it isn't that easy! This, I think, casts some doubt on Benj's explanation of how belief formation works in the case of veridical experience.

To press this point for just a moment: consider the veridical case, in which I instantiate the property of looking at the red color of a widget, and the dreaming case of type D/M, in which I instantiate the property of dreaming that I am so looking but don't know that I do, and let's stipulate that in each case I'm equally convinced that my experience is veridical. (My dispositions to act are the same, in each case I assert that my experience is veridical, I am disposed to take just the same bets about the veridicality of the experience, etc.—add in whatever seems required.) What I think needs some explanation is why in the veridical case I manage to regard my instantiation of the relevant experiential property—in that case, the property of looking at the red color of a widget—as equivalent to some belief sentence, and that in the dreaming case I don't manage to that with myinstantiation of the relevant dreaming property. This looks mysterious to me because it seems that in the two cases my attitude toward the experiential episode I'm undergoing is exactly the same.

It's natural to try to answer this challenge by appealing to the conditional evidential policies (A) and (B); but it's not obvious to me that this helps. Even if it is my policy to regard S and S* as equivalent only when p is the case, it does not follow that I will regard them as equivalent when p is the case, or that I won't when it isn't. But if this is granted, then it should be possible, in the dreaming case just described, that I regard $\mathcal{R}_8$ and "I am looking at the red color of a widget" as equivalent. (I am, after all, as certain as I ever am that this is just what I am doing.) But then, given Benj's claims about regarding as equivalent, it follows that $\mathcal{R}_8$ and "I am looking at the red color of a widget" are synonymous for me. Since $\mathcal{R}_8$ is a sentence of a Lagadonian language, I take it that it cannot change it's meaning; which implies that, for me, "I am looking at the red color of a widget" means that I am dreaming that I am looking at the red color of a widget. And this, in turn, means that I believe that I am looking at the red color of a widget. But I plainly don't believe this in the above case.

We're thus forced to the conclusion that it is impossible for a subject who does not have the correct beliefs about which experiential property she is instantiating to regard her instantiating that property as equivalent with any sentence—since, otherwise, she would, contra our supposition, have the true beliefs about, e.g., whether she is dreaming or veridically perceiving. My problem is that I don't quite see what "regard as equivalent" could mean which would secure this result.[5]

Returning to policies (A) and (B), it's clear that neither policy has anything to say about the case in which I am dreaming that I am looking at the red color of a widget, but believe that I am looking at the red color of a widget. So what should we say about this sort of case? Benj says:

---

[5]One might say: this is impossible because 'regard as equivalent' is sufficient for synonymy, which makes it impossible that a subject should ever regard as synonymous $\mathcal{R}_8$ and "I am looking at the red color of a widget." But I think that this gets Benj's preferred order of explanation backwards: "I am looking at the red color of a widget" and other sentences of the language of belief are supposed to get their contents from being regarded as equivalent to the relevant Lagadonian sentences; they don't have meanings independently which are available to constrain the objects which the subject is able to regard as equivalent. Otherwise, I think, we'd lose Benj's explanation of the truth of the beliefs formed in the ordinary veridical case.

> The question is not easily posed from the first-person perspective. If one is under the impression one is looking, then from the first-person perspective things are this way: I am looking. Fixing this, the question of what to do if one tokens $\mathcal{R}_8$ is then a question of what to do in an incoherent situation. Rationalizing policies and rules provide answers about what do if things are this way or that way; given a way things can't be, such policies are silent.....
>
> At this point, we see what I take to be the root of philosophical perplexity about perception. In a delusive case, *one's perspective is incoherent*: the perceptual aspects of one's perspective affirm a certain hypothesis; the doxastic aspects affirm a certain incompatible hypothesis. In such circumstances, all bets are off from the point of view of intentional psychology . . . .[6]

The idea is that, just in virtue of dreaming that he is looking at the red color of a widget, given our remarks about the Lagadonian language above, Sam represents himself as dreaming that he is looking at the red color of a widget. But he believes himself to be looking at the red color of a widget; since it is impossible to be both looking at the red color of a widget and dreaming that one is looking at the red color of a widget, the proposition which is the content of Sam's belief is inconsistent with the proposition associated with his perceptual state.

This leads Benj to say two surprising things about Sam. The first is that there is "no coherent answer" to the question of what it is like for Sam. The second is that, for the reasons just given, Sam's perspective is incoherent, and that for this reason, in Sam's case, "considerations of rationality do not apply." I find both of these conclusions hard to accept; I'll discuss them in turn.

About what it's like to be Sam, Benj says

> So what then is it like for Dreaming Sam? The question admits of no coherent answer, because the condition the world would have to meet in order for it to be faithful to how the world is 'for Dreaming Sam' is unsatisfiable.[7]

But this seems to me to be a *non sequitur*. Even if (and here I agree with Benj) there is a certain kind of equivalence between what it's like for Dreaming Sam and the condition which the world would have to meet to be faithful to Sam's experience, we can't infer from the fact that the world could not satisfy this condition that there is no such condition—any more than we can infer from the necessary falsehood of a mathematician's belief that there is nothing that that mathematician believes. If we can coherently describe incoherent beliefs, why not say the same about Sam?

Further, it seems to me that there must be at least some coherent things that we can say about what it is like for Sam. Consider Sam', who is like Sam but for the fact that he is dreaming that he is looking at the green color of a widget. Surely what it's like to be Sam' is different than what it's like to be Sam; if we deny this, then it seems that we've lost track of the notion of 'what it's like' which made it seem

---

[6]Hellie (2011), 153.

[7]Hellie (2012), 132.

interesting in the first place. But if we accept this, then, contra what Benj says, it seems to me that there must be facts about what it is like for Sam and Sam'.[8]

Let's turn now to Benj's claim that "considerations of rationality do not apply" to Sam. One way to bring out just how surprising this claim is to imagine Sam and Sam*, each of whom are dreaming that they are looking at the red color of a widget and each of whom mistakenly takes themselves to be looking at the red color of a widget. On the basis of this experience, Sam comes to believe that there is a red widget before him, and Sam* instead forms the belief that there is a blue widget before him. Surely there is a straightforward sense in which Sam's response to his dream is more rational than Sam*'s—even if we want there to be a sense in which Sam's response is less fully rational than the response of a subject who forms this belief on the basis of a veridical experience of the red color of a widget. Benj tries to capture this intuition by saying that it is indeed more natural to form Sam's belief than Sam*'s—but while this is no doubt true, I don't think that this succeeds in capturing the intuition, which I find quite compelling, that Sam was rational to form his belief, and Sam* (bizarrely) irrational to form his.

This might just boil down to a battle of intuitions, and Benj might fairly point out that this bullet might well be worth biting to preserve the sort of direct realist picture to which he is drawn. But I wonder whether one could preserve much of that direct realist picture without having to say these surprising things about Sam and Sam*.

Even if we grant that in Sam's case the proposition associated with his dream state is inconsistent with a proposition he believes to be true, this fact doesn't by itself show that Sam is now wholly outside the realm of rationality. Even proponents of coarse-grained views of contents think that we have to say something about the rationality of subjects with inconsistent commitments, if only because inconsistency is so common. Indeed, it is especially common when the subject's inconsistent commitments are such that the subject herself fails to recognize their incompatibility. And the sort of inconsistency which arises in the dreaming case seems—given the aspects of Benj's framework sketched above—to be a case of compartmentalization of just this sort, since the subject who believes that she is perceiving veridically is apparently in no position to know that she is correctly representing herself (in the Lagadonian language) as dreaming that she is looking at the red color of a widget. Given that we should have something to say about subjects whose commitments are globally inconsistent the claim that "considerations of rationality do not apply" to subjects like Sam and Sam* seems like an overreaction.

It's also worth noting that the idea that Sam's dream and his belief are inconsistent is not an essential part of the direct realist picture; the alleged contradiction is generated by (i) the self-representational aspect of Benj's theory, on which the proposition which a subject accepts is not just about the red color of the widget apparently before him, but also about the subject's relation to that widget and

[8] I think that Benj makes these claim about Dreaming Sam in order to deny PI. But one could deny PI, and admit the existence of indiscriminable but genuinely distinct "what it's likes", without denying that there is anything that it's like to be Dreaming Sam.

(ii) the claim that dreaming, just as much as veridically experiencing, has this self-representational aspect. But one might wonder—especially from the perspective of a direct realist who is unafraid to think of veridical experiences and matching hallucinations as belonging to very different categories—about the motivation for (ii). Remember that, so long as we want to avoid the conclusion that every subject represents himself as having every property which he has, that we have to find some way of distinguishing between those properties which are expressions of the Lagadonian language and those which are not. So why not think that the property of looking at the red color of a widget is one of the properties in the former category, and the property of dreaming that one is looking at the red color of a widget is not? Why not say that the mechanism by which we form true beliefs in the veridical case is radically different than the mechanism by which we form true beliefs about our hallucinatory and illusory experiences? This would avoid the conclusion that subjects who are dreaming represent themselves as such, and hence would avoid the conclusion that subjects who are "taken in" by a hallucination are thereby inconsistent as well as simply mistaken about the scene before them.

The availability of this option is important even if the theory which results from taking it ends up not being attractive. It is important because it shows that the claim which Benj takes to be the "root of philosophical perplexity about perception"—namely, that, "[i]n a delusive case, *one's perspective is incoherent*"—is not generated by Benj's direct realism. Quite the opposite: it is generated by Benj's commitment to there being a certain kind of *commonality* between veridical and hallucinatory experience: namely, that both involve accurate self-representation, in the Lagadonian language, of one's current experiential state.

References

Hellie, Benj. 2011. There it is. *Philosophical Issues* 21: 110–164.

Hellie, Benj. 2012. There it was. In *Consciousness inside and out: Phenomenology, neuroscience, and the nature of experience*, Studies in brain and mind, vol. 6, ed. R. Brown. Springer press.

Chapter 14
But Where Is a Hallucinator's Perceptual Justification?

Heather Logue

Sam sees a tomato on the table before her, and sees its redness. Call this situation 'the good case'. In virtue of seeing the tomato and its redness, Sam is justified in believing that there is a red tomato before her—at least, that's what we ordinarily think. However, it is possible for Sam to have a subjectively indistinguishable experience in which she sees a tomato, but doesn't see its color (e.g., an illusory experience in which the subject sees a white tomato bathed in red light), or in which she doesn't see anything in her environment at all (e.g., a total hallucination "as of" a red tomato). Call these situations 'the bad cases'. In light of such possibilities, it's hard to resist the conclusion that Sam's good case experience bestows no more justification on the claim that there is a red tomato before her than it does on (e.g.) the claim that there is a white tomato bathed in red light before her. So how can Sam's belief that there is a red tomato before her be perceptually justified? (As is well known, a structurally similar problem can be raised about whether Sam's perceptually-based belief amounts to knowledge.)

Hellie's theory of perceptual justification affords a solution to this problem. As I interpret him, Hellie holds that in the good case, Sam has perceptual justification for beliefs about her environment that she *lacks* in the bad cases. In this commentary, I will frame his theory in somewhat different terms than he does (in order to highlight its relation to a certain kind of anti-skeptical strategy), and I will focus on the epistemological (rather than the phenomenological) aspects of his theory. In particular, I will outline Hellie's take on the idea that Sam has perceptual justification for beliefs about her environment that she lacks in the bad cases. Then, I will outline another plank of Hellie's theory that he takes to follow from this claim, given certain background assumptions. Finally, I will argue that this second plank

H. Logue (✉)
School of Philosophy, Religion, and History of Science, University of Leeds, Woodhouse Lane, Leeds LS2 9JT, UK
e-mail: heatheranne@gmail.com

R. Brown (ed.), *Consciousness Inside and Out: Phenomenology, Neuroscience, and the Nature of Experience*, Studies in Brain and Mind 6, DOI 10.1007/978-94-007-6001-1_14,

of the theory is unacceptable, and thus that we ought to reject at least one of the background assumptions that lead to it.

As I understand it, Hellie's theory of perceptual justification is a version of a kind of view called 'epistemological disjunctivism' (the term was coined in Snowdon 2005; for examples of the view, see McDowell 1982, 2008; Williamson 2000: Ch. 8, Pritchard 2008; Neta 2008). This view aims to exonerate perceptual experience from the charges of epistemological impotence outlined above by claiming that the good case experience has some epistemological power that the bad case experiences lack.

As I interpret Hellie's particular version of epistemological disjunctivism, the content of Sam's good case experience is that she is seeing a red tomato (p. 132). The vehicle of this content is simply the subject seeing the tomato and its redness—the language of perceptual experience is Lagadonian, in that the vehicle of representation just is what is represented (p. 130). Thus, this type of vehicle is what Hellie calls "contextually analytic", in that it's guaranteed to be true whenever it's tokened (since it is what it represents).

The experiences in the bad cases have different contents. The subject of a subjectively indistinguishable illusion sees the tomato, but not its redness (e.g., if the lighting conditions are such that a non-red tomato looks red, the tomato has no redness for the subject to see). And the subject of a subjectively indistinguishable hallucination doesn't see anything in her environment at all. So there are no Lagadonian vehicles fit to express the content that Sam is seeing a red tomato. So given that the language of perceptual experience is Lagadonian, the contents of the bad case experiences must be different.

It's a short step from here to the desired conclusion about perceptual justification. The content of the good case experience (that Sam sees a red tomato) justifies the belief that there is a red tomato before her. By contrast, the bad case experiences don't have this content, and so if Sam has any justification for the belief that there is a red tomato before her at all in these cases, it's not of this sort. (More on this issue at the end.)

In summary, the first plank of Hellie's theory of perceptual justification is that Sam has a source of perceptual justification for the belief that there's a red tomato before her in the good case that she lacks in the bad cases, and this idea is cashed out in terms of experiences involving Lagadonian representation. The second plank of Hellie's theory concerns the relationship between the contents of experience and the contents of beliefs about one's experiences.

In the bad cases, Sam may well *falsely believe* that she is seeing a tomato and its redness (i.e., that she is in the good case). And in the good case, Sam may well falsely believe that she isn't seeing a tomato and its redness (i.e., that she's in one of the bad cases). Let us call such cases 'mismatch cases', to reflect the fact that there is a mismatch between the subject's experiential situation and what she believes about it. According to Hellie, a subject in a mismatch case is psychologically *incoherent*. On his view, such a subject has an experience with a content that is *incompatible* with the content of her belief about her experience. For example, a bad case experience has a content that is incompatible with the belief that one is seeing a tomato and its redness (pp. 132–133).

Now, what exactly is the content of a bad case experience, such that it is incompatible with the proposition that one is seeing a tomato and its redness? If the language of perceptual experience is Lagadonian, the contents of bad case experiences have to concern what's actually going on in one's perceptual situation. And one thing that is going on in the subjectively indistinguishable illusions and hallucinations is that the Sam is "focusing attention on a particular state of red-likeness" (p. 132). That is, Sam isn't focusing attention on a particular state of *redness*, since she doesn't perceive an instance of redness in such cases. So whatever she's focusing her attention on, it's something distinct but perceptually indistinguishable from a particular state of redness. (Hellie wants to remain neutral on exactly what focusing attention on a particular state of red-likeness consists in—at least in principle, it might consist in a "... qualitative red-like state of, perhaps, some dreamy tomato-simulacrum, part of the brain, imagined or recollected previous tomato-encounter, neural image, or something else" (p. 132). For his purposes, exactly how we cash out this talk isn't of primary importance.)

So here's where we're at: the content of Sam's good case experience is that she sees a red tomato (i.e., that she sees a particular state of redness instantiated by the tomato she sees). The content of the bad case experiences is something along the lines of the following: Sam is focusing attention on a particular state of red-likeness.

Of course, we get incompatibility between the content of the bad case experiences and the belief that one is seeing a red tomato only if focusing attention on a particular state of *red-likeness* is incompatible with seeing a particular state of *redness*. On the face of it, it might seem that these are perfectly compatible—after all, redness is a special case of red-likeness. But this is to forget that this talk of focusing attention on a particular state of red-likeness in cases of illusion and hallucination must be cashed out in some way or other—focusing attention on a particular state of red-likeness is a determinable, of which seeing a particular state of redness is a determinate. Cases of illusion and hallucination don't involve *this* determinate, but they involve some *incompatible* determinate (e.g., seeing a red' sense-datum). Hence, the content of a bad case experience (that one is focusing attention on a particular state of red-likeness *in a way that isn't seeing a particular state of redness*, whatever that may be) is incompatible with the belief that one is seeing a red tomato.

So, according to Hellie, in a mismatch case, Sam's overall psychological state is incoherent in that the content of her experience is incompatible with the content of her belief about her experience. Moreover, he claims that "rational psychology runs out of steam as soon as someone loses coherence" (p. 134)—e.g., that in a mismatch case, rational psychology is inapplicable to Sam, and therefore she has *no reason at all* for believing that there is a red tomato before her. Echoing McDowell, Hellie says that "... if what is wanted ... is 'justification', too bad: all that can be given is 'exculpation'." (p. 134). That is, although we can give a *causal* explanation of Sam's belief that there is a red tomato before her in a mismatch case, there is no *rational* explanation of this belief. Incoherence carries the penalty of deportation from the space of reasons, and so there is no rational explanation of a belief that is inconsistent with some of one's pre-existing mental states.

Although I'm sympathetic to the first plank of Hellie's theory (that the good case experience affords a sort of perceptual justification that the bad case experience cannot), I'm skeptical of the claim that rational psychology doesn't apply to subjects in mismatch cases. For it seems that subjects in mismatch cases are subject to standards of rationality. Consider a bad mismatch case in which Sam hallucinates a red tomato but believes that she is seeing one. It seems like there are some things it would be rational for her to believe, and some things it wouldn't be rational for her to believe. For example, given that it perceptually appears to her that there is a red tomato before one, and her (mistaken) belief that she's seeing one, it's rational for her to believe that there is a red tomato before her, and irrational for her to believe that there is a giant purple plum before her (given that it doesn't *also* perceptually appear to her that there is some such thing before her). She has no reason whatsoever to believe that there is a purple plum there, but at least *some* reason to believe that there is a red tomato there.

Hellie would reply that we can satisfactorily re-describe this case solely in terms of what *causes* Sam's beliefs rather than in terms of what *rationalizes* them (pp. 134–135). For example, when I say that it would be *rational* for Sam to believe that there is a red tomato before her, one might think that what I'm really getting at could be equally well captured by saying that a hallucination as of a red tomato in conjunction with the belief that one is seeing a red tomato *normally causes* the belief that there is a red tomato before one. Similarly, when I say that it would be *irrational* for Sam to believe that there is a purple plum before her, one might think what I'm getting at could be equally well captured by saying that a hallucination as of a red tomato (and nothing else) in conjunction with the belief that one is seeing a red tomato (and nothing else) normally *doesn't* cause the belief that there is a *purple plum* before one.

We can see what is unsatisfactory about such re-descriptions in the context of another mismatch case. Suppose now that Sam is seeing a red tomato, but falsely believes that she's hallucinating on the basis of trustworthy testimony. She's been told by her honest friend that's she's accidentally taken a drug that generates realistic total visual hallucinations, but this isn't the case—her friend believes this because the type of headache pill Sam just ingested looks unfortunately similar to the hallucinogenic pills sold around their neighborhood. Now, suppose that Sam goes on to believe that there is a red tomato before her, *despite* her belief that she's merely hallucinating one. Since this is a mismatch case, Hellie would say that Sam is incoherent and hence "[r]ational psychology runs out of steam" (p. 134). However, I submit that it would be *irrational* for Sam to believe that there is a red tomato before her, and that it wouldn't do just as well to say that the belief that one is hallucinating a red tomato doesn't normally cause the belief that there is a red tomato before one. For Sam is *epistemically blameworthy*—she believes that there is a red tomato before her, in spite of the fact she has no undefeated rational basis for this belief whatsoever (it visually appears to her that there is a red tomato before her, but whatever evidential force this fact might have is undermined by her friend's testimony). Thus, she is subject to *criticism*, which suggests that her belief isn't merely a causal aberration but also irrational.

In order to support the connection between Sam's being subject to criticism and her belief's being irrational, let us suppose that Sam's belief is just a causal aberration for the sake of argument—that all that's going on is that her overall mental state (including the belief that she is hallucinating a red tomato) causes a further mental state that it normally wouldn't (the belief that there is a red tomato before her). The mere fact that Sam's overall mental state causes something that it normally wouldn't isn't *in itself* grounds for criticizing Sam. We might be surprised by the abnormality, but there isn't anything intrinsically *wrong* with it. Sam's belief is subject to criticism only if *norms of rationality* apply to the causal connections between her mental states. In short, if Sam's belief is subject to criticism (which it seems to be), then it is irrational, and so rational psychology must apply to her after all.

It is worth noting that the mismatch cases are importantly different from paradigm cases in which rational psychology goes out the window—e.g., cases in which the subject's beliefs are brought about by being hypnotized, or being hit on the head. If one acquires a belief in one of these ways, one isn't subject to criticism for having it, even if it's contradicted by one's experience and other beliefs. Given the way it was formed, it isn't subject to standards of rationality. All we care about is what *caused* it; the question of whether it is *rationalized* by the subject's other mental states is beside the point. But the beliefs in the mismatch cases discussed are not like this. There is a sense in which *Sam* determines whether or not she believes that there is a red tomato before her (by taking into account or discounting how her experience presents her environment as being, her pre-existing beliefs, the relevant evidence, and so forth)—this isn't settled by a hypnotist or a hammer striking her head in just the right way. This is why Sam's beliefs are subject to criticism, and hence why rational psychology applies to her in the mismatch cases.

I've argued that rational psychology is applicable to the subject of a mismatch case. But Hellie holds that if the subject of a mismatch case is incoherent, then rational psychology doesn't apply. So if we're to maintain that rational psychology is applicable in such cases, we must argue that incoherence doesn't undermine the applicability of rational psychology to a subject, or that the subject isn't incoherent. In my view, both options are promising, but I won't explore either here. Suffice it to say that if we want to hang on to the claim that rational psychology is applicable to subjects in mismatch cases (which I think we should), we have to take one of these two routes. (It is worth briefly noting that if we reject the claim that the language of perceptual experience is Lagadonian, we can secure versions of epistemological disjunctivism on which the subjects in mismatch cases are not incoherent—see those described in Logue 2011, section 2.)

I will bring this commentary to a conclusion by briefly addressing a different lingering issue. I've implied that a subject in a bad case who believes that she's seeing a red tomato has at least some reason for believing that there is a red tomato before her (in claiming that there are certain beliefs about her environment it would be rational for her to form, and others it wouldn't be rational for her to form). For the sake of argument, let us assume along with Hellie that the subject in the bad cases has experiences with something along the lines of the following content: that one is focusing attention on a particular state of red-likeness. If that's right, then it's not immediately obvious what the subject's reason for believing that there is a red tomato before her could be. Recall that this talk of focusing attention on a particular state of red-likeness is a placeholder for something *incompatible* with seeing a particular state of redness. And the fact that one is in an experiential state that is incompatible with seeing a particular state of redness definitely isn't a reason for believing that there is something red before one (indeed, it's a reason to be skeptical about whether there is). So what reason could the subject of the bad cases have for believing that there is a red tomato before her?

The subject's reason could simply be that it *perceptually appears* to her that there is a red tomato before her. Of course, it perceptually appears to her that there is a red tomato even though there isn't. But we can say that this fact provides *some* justification for the belief at issue—just not justification that is sufficient for *knowledge*. More precisely, we can say that the fact that it perceptually appears to the subject that there is a red tomato before her gives her at least a *little* justification for believing that any of the scenarios that could have given rise to it obtain—e.g., a little justification for believing that she is seeing a white tomato bathed in red light, a little justification for believing that she is having a drug-induced total hallucination of a red tomato, and a little justification for believing that she is seeing a red tomato (and so on for any other scenario of this sort). Hence, it can be rational for the subject to believe that any such scenario obtains (depending on what the subject's background beliefs are, of course). It's just that reasons of this sort are far from sufficient for knowledge that any such scenario obtains. In short, given that a proposition can afford *partial* justification for a belief that *p* even though it doesn't support the proposition that *p* over incompatible alternatives to it, we can hold that the fact that it perceptually appears to the subject of the bad cases that there is a red tomato before her affords partial justification (in other words, a rather weak reason) for the belief that there is a red tomato before her.

In conclusion, while I think we should explore and attempt to defend the idea that the subject in the good case has more and better perceptual evidence for claims about her environment than she does in the bad cases, I don't think we should accept that rational psychology doesn't apply to her in the bad cases—nor do I think we have to, since there are ways of elaborating the first claim that don't entail the second. So in my view, Hellie's version of epistemological disjunctivism isn't its most attractive variant.

References

Logue, H. 2011. The skeptic and the naïve realist. *Philosophical Issues* 21: 268–288.

McDowell, J. 1982. Criteria, defeasibility, and knowledge. *Proceedings of the British Academy* 68: 455–479.

McDowell, J. 2008. The disjunctive conception of experience as material for a transcendental argument. In *Disjunctivism: Perception, action, knowledge*, ed. A. Haddock and F. Macpherson. Oxford: Oxford University Press.

Neta, R. 2008. In defense of disjunctivism. In *Disjunctivism: Perception, action, knowledge*, ed. A. Haddock and F. Macpherson. Oxford: Oxford University Press.

Pritchard, D. 2008. McDowellian neo-Mooreanism. In *Disjunctivism: Perception, action, knowledge*, ed. A. Haddock and F. Macpherson. Oxford: Oxford University Press.

Snowdon, P.F. 2005. The formulation of disjunctivism: A response to Fish. *Proceedings of the Aristotelian Society* 105: 129–141.

Williamson, Timothy. 2000. *Knowledge and its limits*. Oxford: Oxford University Press.

Chapter 15
Yep—Still There

Benj Hellie

15.1 Berger

15.1.1 Why Did I Do It?

That is one kind of question. An entirely different one is 'why did it happen to me?' What is the difference? Though there are many layers to be peeled back from this particular onion, my view is that any answer to the former must contain, at bottom, some variant of *it made sense in light of blah blah blah*—whereas in the latter case, no element of sense-making need be involved.

Answers of the former sort are *rationalizing explanations*; and when a certain rationalizing explanans explains a certain rationalizing explanandum, we might perhaps say that the former 'justifies' the latter.

The notion of *sense-making* is at least a bit elusive. But I am inclined to think it is a first-person notion: something which does not impinge upon my stream of consciousness cannot make sense of anything I do, nor can it be made sense of by any aspect of my condition; by contrast, the character of the stream of consciousness is uniformly available for sense-making—and also in almost all cases (perception and life being the exceptions at the extremes, alongside perhaps certain shifts of attention) for being made sense of.

What goes for me goes for you. And what goes for you goes for 'pseudo-you': you as from within the 'second person perspective'—from within my simulations

Thanks to Berger, Logue, and Speaks for this thoughtful set of comments.

B. Hellie (✉)
Department of Philosophy, University of Toronto, 170 St George Street, Toronto,
ON M6J 1N6, Canada
e-mail: benj.hellie@utoronto.ca

R. Brown (ed.), *Consciousness Inside and Out: Phenomenology, Neuroscience, and the Nature of Experience*, Studies in Brain and Mind 6, DOI 10.1007/978-94-007-6001-1_15,

of you. The core, paradigmatic, fundamental notion of rationalization is making sense from the first- or second-person perspective; where this is tied in the ways gestured at to the stream of consciousness.

This much was philosophical common currency up through the 1930s (Carnap's great 'Psychology in physical language' takes it as a starting point), and is a core message of certain core texts of post-war philosophy (including, as I read it, Ryle's monumental but perplexing *The Concept of Mind*, and also perhaps Anscombe's essential text *Intention*). Unfortunately, these (to my mind, rather obvious) points were washed out of the mainstream of Anglophone philosophy starting in about the mid-1950s (long story).

Berger's assertion that what rationalizes is not consciousness but 'content' is in this vein. Unfortunately, it is really not plausible that content has anything by itself to do with rationalization. My computer is a content engine, but has no interiority, and cannot be literally brought under rationalizing explanation (though, of course, the 'intentional stance' is always available). The same for this thermostat. The same for this lectern—which does not, after all, desire more than anything to be at the center of the universe and believe that it is at it: rationalizing its just sitting there.

What is rationalizing and rationalized is the contour of the stream of consciousness. That suggests that the character of the stream of consciousness is exhaustively characterizable in content-theoretic terms. That would also suggest that there is something to Brentano's distinction between 'original' and 'derived' intentionality: perhaps even that there is no intentionality outside of consciousness, but only an 'intentional stance'.

Berger's various examples deserve attention: I will focus on the most emblematic. The blindsight patient who makes it down the hall: what is it like for them? Is it like anything? Maybe so: maybe it is simply hard for them to describe. I have seen suggestions that blindsight patients navigate by echolocation. Having navigated by echolocation, I can say that it is like something—albeit something hard to describe. And I would say that going this way rather than that way *makes sense:* something looms up that way but not this way, and so I go this way rather than that way.

But perhaps it is like nothing. If so, then why do they go this way rather than that way? Why is *going this way* something they do, rather than *going that way?* Or is this a poorly-posed question? Would it be better to ask why *going this way* is something they simply find themselves engaged in? I am inclined to think that if there is absolutely no difference for the blindsight patient to be in one sort of corridor or another, there is no answer of the former sort: at best answers of the latter sort.

More concisely: it is an open question whether blindsight is like something. If so, no problem for the consciousness-rationality link because consciousness is present. If not, then also no problem, because rationalization is absent.

15.2 Logue and Speaks

We would like to be able to say that in good cases, perception provides indefeasible justification for beliefs about the external world. That would be the holy grail of perceptual epistemology. Why can't we say that?

Because not all cases are good, of course: for example, sometimes we are misled by dreams. Without getting fancy, philosophers have exactly two options in regard to such cases: good and bad cases are the same, providing indefeasible justification at most for beliefs about the *internal* world; bad cases are different.

Let's not give up yet: let's say bad cases are different. This seems to leave exactly two options: bad-case perception is intrinsically bad; badness is due to a bad mixing of bad-case perception and bad-case cognition.

I'm inclined to wonder how anything in the mind could be intrinsically bad. Moreover, it seems obvious that (a) good-case perception can mix badly with cognition (as when we mistakenly think we are dreaming); (b) sometimes bad- case perception isn't misleading (as when we lucidly dream).

What could the bad mixing be? The difference between lucid dreaming and bad-case dreaming is over whether we think we are asleep. If we do, no problem; if we don't, problem. And the difference between good-case being awake and mistakenly thinking we are dreaming is over whether we think we are awake. If we do, no problem; if we don't, problem. To account for this, we could say that waking perception implies we are awake, while dreaming implies we are dreaming. Then the bad mixing would be contradiction between what perception implies and what thought implies.

So bad cases are bad because we contradict ourselves. We don't notice the contradiction because perception involves a different mode of presentation than thought. So bad cases are Frege cases. Frege cases generally seem like not a big problem; so bad cases shouldn't seem like a big problem either. That's the line in 'There it is'.

Now in more detail. In 'mismatch cases', though one is in fact presented with an *F*, one's general background assumptions about one's situation entail that one is not presented with an *F*. Two examples:

1. (a) Fred is under the impression he is dreaming, and therefore not presented with any surface-color tropes but only with pseudo-color tropes
   (b) Fred is in fact seeing, so that what is presented is a redness trope
   (c) Fred thinks to himself 'that is pseudo-red' and thereby judges (incorrectly) that he is presented with something pseudo-red
2. (a) Ro is under the impression she is seeing, and therefore not presented with any pseudo-color tropes but instead with color tropes
   (b) Ro is in fact dreaming, so that what is presented is a pseudo-redness trope
   (c) Ro thinks to herself 'that is red' and thereby judges (incorrectly) that she is presented with something red

Central to my view is that if one is presented with a certain trope, one is certain that one is: it exists and is presented to one at every 'doxastic possibility'—every world compatible with what one believes—and this certainty tracks which quality the trope is an instance of. *There it is:* I may be uncertain about much, but this uncertainty is 'wrapped around' a basis of certainty in the reality of what is presented. That seems to be the epistemological core of direct realism.

If so, (1b) requires that in all Fred's doxastic possibilities, he is presented with a redness (and therefore surface-color) trope, while (2b) requires that in all

Ro's doxastic possibilities, she is presented with a pseudo-redness (and therefore pseudo-color) trope. But by (1a), in all Fred's doxastic possibilities, he is presented only with pseudo-color tropes, while by (2a), in all Ro's doxastic possibilities, she is presented only with surface-color tropes. So no world satisfies all the requirements for Fred, and no world satisfies all the requirements for Ro.

So when it comes to explaining the judgements in (1c) and (2c), what are we to say? Can we use the following 'good case' as a template?

1. (a) Sam is under the impression she is seeing, and therefore not presented with any pseudo-color tropes but only with surface-color tropes
   (b) Sam is in fact seeing, so that what is presented is a redness trope
   (c) Sam thinks to herself 'that is red' and thereby correctly believes that she is presented with something red

A prior question then arises: what would an explanation of (1c) look like? I suggest a family of 'evidential policies' which implicitly underlie one's strategy for recoding perception into thought. Roughly, a surface color term like 'red' is a 'recognitional concept' in the sense that one is guided in one's applications of it (and other color terms like 'green' and 'blue') by policies to be used when one is seeing: these policies create something like analytic equivalences between such color terms and the 'language' underlying perceptual presentation (in another chapter of the story, this language is said to be 'Lagadonian'). So: Sam accepts as definitional of 'red' its equivalence with some perception-language expression. She thinks she is seeing, and opens the drawer containing the policy for cases of seeing—which includes this definitional equivalence. Out of all the policies in that drawer (including those for 'green', 'blue', and the rest), the best fit is the policy for 'red'. So she carries it out.[1] As it happens, she is in luck: the policy she carries out does in fact apply (because the expression defined as equivalent to 'red' is in fact the expression she is producing), and she carries that policy out because she correctly believes it applies, so everything is great: full rational explanation has been produced.

---

[1] Speaks raises various concerns about the 'implicit definition' story: one, in particular, seems to involve the misconception that 'regard as equivalent' applies at the token level rather than the type level (obviously any attempt to appeal to something like syntactic derivation as a source of justification requires the definienda to be types rather than tokens).

Back to Fred and Ro. Does the story for Sam apply? No. Sam opens up the drawer of policies to be used when seeing, because, in her view, she is seeing. Is that why Ro opens that analogous drawer—applies concepts appropriate to seeing? Well, what does Ro believe—how are things, in Ro's view? There is no full, transparent, coherent answer.[2] There are two partial coherent transparent answers: she is seeing; she is presented with a pseudo-redness trope. These combine to make one full coherent nontransparent answer: in one compartment, she is seeing, and in another compartment, she is presented with a pseudo-redness trope. The kind of answer we might have wanted to 'why does Ro apply concepts appropriate to seeing' is a full, transparent, coherent story about the world from Ro's point of view—and there just isn't one.[3]

OK, so something has resulted in Ro's applying concepts appropriate to seeing. Why then does she apply 'red' rather than 'green'? Presumably, doing the former would be appropriate just if one is presented with a redness trope; the latter just if one is presented with a greenness trope. Ro isn't presented with either; so she does something inappropriate.

So: why (2c)? Not for reasons having to do with how the world is for Ro or with the 'definitions' governing the use of the expressions in which she encodes how the world is for her. Perhaps considerations of this sort exhaust the abstract backbone of (the relevant portions of) rational psychology. (Logue gestures in her closing remarks at the prospect of some alternative to or amendment of these resources, but I do not fully understand what she has in mind.) If so, there is no *rational* explanation of (2c).

This conclusion bothers both Logue and Speaks.[4] They argue:

(A) It would be worse for Ro to judge that she is presented with something purple (Logue, 3; Speaks, 8).

- Reply: it would certainly be *weirder.* If I put myself in Ro's position, I find that I too would believe I am presented with something red. That gives a sort of 'humanizing' aspect to Ro's actual reaction, whereas the alternative would be alien. But that only shows her reaction to be rational if the position

[2]Speaks seems in places to read me as meaning, implausibly, that there is *no* answer: not my claim.

[3]Perhaps the 'compartment' associated with articulate thought typically gets to carry the day here, as Speaks suggests. Perhaps so; but appeal to compartments is not transparent; more to the point, when we speak about compartments, we do so with the intention that rational psychology doesn't apply to the facts on the ground, but only to a certain abstraction from them.

[4]Logue and Speaks are also bothered by the claim that 'rational psychology is inapplicable to the subject of a mismatch case'/'the subject is wholly outside the realm of rational psychology'. I am also bothered by that claim, insofar as I understand it. I am inclined to think the fundamental subject-matter of rational psychology is the update: the transition from neutrality to belief, or the commencement of an action. A subject might perform both rationally explicable and rationally inexplicable updates.

I imagine myself into is rationally unimpeachable. It doesn't strike me from within as rationally impeachable, of course. But only internalists think that has overwhelming dialectical force.

On the counteroffensive: what exactly is the world like for Ro such that she judges herself to be presented with something red rather than something purple? No doubt a non-direct realist answer appealing to various philosophical inventions (sense-data, edenic representational states, structured propositions containing uninstantiated universals) could be concocted. But because no one but the philosopher understands this answer, it does not characterize what the world is like for Ro, and therefore cannot be a correct explanation of her judgement.

(B) It would be worse for Fred to judge he is presented with something red (Logue, 4).

- Reply: it certainly would. In that case, the 'conceptual' or 'articulate' fragment of Fred's psyche would harbor an obvious incoherence. An incoherence between aspects of one's view coded in the same way seems much less human than an incoherence between aspects coded in different ways. Maybe some would even find it worthy of reproach, Logue suggests: though in my view rational psychology is about coherence and incoherence—sense and nonsense—rather than reproach and approbation.

(C) We can explain (2c) by saying 'it looks/perceptually appears to Ro as if something red is before her' (Logue, 5).

- Reply: that would be an explanatory solecism. 'That looks *F* to me' means 'going by looking, that is *F*'. So 'Ro judged that to be red (rather than green) because it looked red (rather than green) to her' would mean 'Ro judged that to be red (rather than green) because going by looking, here's how things were for her: *that is red (rather than green)*'. That seems hard to distinguish from 'Ro judged that to be red (rather than green) because, upon looking, she judged *that is red (rather than green)*'. That is obviously equivalent to 'Ro judged that to be red (rather than green) because she went by looking'. We already knew Ro was going by looking, so that provides no further illumination. (Compare 'There it is', 159; 'It's still there', 11.)

  Counteroffensive: I wonder also how to extend this story to explain (1c): does it perceptually appear to Fred as if he is presented with a pseudo-red trope? If so, does it also perceptually appear to him as if he is presented with a redness trope? And does it perceptually appear to Ro as if she is presented with a pseudo-redness trope? If the perceptual appearances are supposed to be the justifiers, the suggestion is of a common factor view rather than of a version of 'epistemological disjunctivism' superior to mine.

(D) Perhaps we can explain (2c) using resources beyond my direct realist apparatus, in order to avoid pronouncing Ro incoherent (Speaks, 9).

- Depending on the details, I foresee several complications. (i) In order to avoid pronouncing Fred incoherent, we would need to explain (1c) using the alternative apparatus—in which case there goes direct realism. (ii) If there is *nothing* presented to Ro, we are left wondering why she goes for red rather than purple or green; and either way, surely when I am dreaming *there it is,* whatever it may be. (iii) If something is presented to Ro that is determinable as between pseudo- red and surface-red, then that is also presented to Fred (seeing mismatch) and Sam (good case). But this sort of 'moderate view' drains the life out of direct realism, and threatens the good case with 'mission creep'. The dialectic over mission creep is intricate: I discuss it elsewhere ('The multidisjunctive conception of hallucination', in Fiona MacPherson, editor, *Hallucination:* MIT, 2013). Finally, the moderate view makes cases of lucid dreaming inexplicable.

15.3 Summary

Put crudely, my position is that consciousness and rationality are one and the same; and that because consciousness is complex, so is rationality—so complex even that sometimes when we do what we think consciousness rationalizes, it yet does not. That's the problem in perceptual bad cases: we are in a rational pickle.

Logue and Speaks seem tempted by the idea that whenever we do what we think consciousness rationalizes, we are right. Consciousness and rationality are one and the same, and consciousness is simple, so we never really get in a rational pickle.

Berger thinks that consciousness doesn't have much to do with rationality at all. So there is no question of whether we can be mistaken about what consciousness rationalizes.

I would say that my sympathies lie closer to the view of Logue and Speaks than to the view of Berger. But both views are widespread in contemporary philosophy. I used to accept both, perhaps because of some picture-thinking I absorbed in my education. But having thought my way out of both, I have a hard time seeing the appeal of either.

Part V
Beyond Color-Consciousness

Chapter 16
Black and White and Colour

Kathleen A. Akins

16.1 Part I

16.1.1 Introduction

This paper is the result of a chance remark made by a colleague: "So, you are writing about luminance vision. What exactly *is* luminance vision?" "Really?!" I thought. "How could *anyone* in neuroscience not know *that*?" Now as a child I was always mystified by a common proverb: "Pride goeth before a fall". If you tried *very* hard and managed to reign in your pride—if pride genuinely '*goeth*'—why were you destined to fall? That seemed rather harsh even by the standards of the Old Testament. It was only a few years ago that I suddenly realized that 'goeth' does not mean 'go away' a discovery that brought absurd relief. Still, it was not until I wrote this paper that I gained a more robust understanding of the proverb. 'Luminance vision' is a phrase as common as mud in the vision sciences. But getting a grip on luminance vision (both on what it is and what it is not) is difficult.

The central topic of this paper, as the reader will have guessed, is the nature of luminance vision and the difference between luminance vision and its close cousin, chromatic vision. *Prima facie*, the topic is not very interesting, certainly not to readers whose central interest lies in the phenomenology of colour vision. Why should we learn about 'black and white' vision when our research concerns colour vision and its phenomenology, topics of greater philosophical interest (objectively speaking)?

K.A. Akins (✉)
Department of Philosophy, Simon Fraser University, 4604 Diamond Building, Burnaby, BC V5A 1S6, Canada
e-mail: kathleea@sfu.ca

R. Brown (ed.), *Consciousness Inside and Out: Phenomenology, Neuroscience, and the Nature of Experience*, Studies in Brain and Mind 6, DOI 10.1007/978-94-007-6001-1_16,

The answer is this: On this point, Wilfrid Sellars (1956, 1962) was prescient. When we try to understand the neural processes of visual perception, those that eventuate in our conscious perceptions of the world, we (often) start with an analogy, one taken from our everyday experiences of the physical world. The analogy in this instance is that of 'black and white versus colour' a very engrained notion indeed. Since the days of cave paintings, quite literally, we have known that images can be rendered in two ways: In black charcoal, *outline and shade in* your favorite wild beast with a gang of stick men in hot pursuit; now *colour in* the animal with ochre and sienna. Contemporary photography divides up the same way. A 'black and white' or monochrome photograph is a rendering of an image 'in one colour', usually along a greyscale, although one could certainly use any other single hue to create a monochrome photograph. Thus old-fashioned sepia-toned photographs are monochrome prints rendered in an orange hue at various levels of darkness. What makes such a photograph useful, however, is that it takes the pattern of *light intensity* in a reflected image—at some level of spatial grain, across some range of illumination, and with a certain degree of contrast—and renders it in a single colour. (Or at least what appears to be a single colour to normal trichromatic viewer. Remember that a black and white printer can use black ink or a triplet of ink colours. A tri-colour printer produces 'monochrome' prints as long as the trichromatic viewer is unable to see any colour differences or specific colours within the printed image.) In a monochrome photograph, we see what is portrayed in virtue of a rendering of image intensity. A *colour* photograph, on the other hand, also renders light intensity but to this is added *hue*, wavelength contrast within the image that *is* accessible to the human visual system. Prior to colour film and printing, for example, black and white photographs were often hand-tinted by painting over them with translucent coloured pigments. Digital photography programs have similar functions. You can 'paint' with virtual translucent colour over the black and white image using a virtual brush. So in all cases, from cave paintings to digital photo advertisements, colouring a black and white image adds, well, *colour*—from which one can infer the colours of objects in the world. This is how the black and white or colour distinction became so embedded in the cognitive psychology of most citizens of the contemporary world.

Like Sellars, I suspect that the distinction between 'black and white' and 'colour' was applied, in the first instance, to physical media—illustrations, photographs and what not—and that the same terminology was then borrowed to describe visual sensations, to one aspect of our visual experience when we inspect black and white or colour images or when we find our way around at night. This view is highly controversial, of course, but fortunately it is not a point on which much hangs at the moment (I hope). What *is* clear, however, is this. If we think of the neurophysiological distinction between luminance and chromatic systems of vision as one of 'black and white' and 'colour', this *is* an analogy. No one thinks that there is literally a black and white image that is passed from retina to cortex with a quick stop at the LGN. There is no parallel system that conveys the coloured pigment, spatially arranged, to be paired with the monochrome image wherever it ends up. This is an analogy, one meant more or less literally depending upon who uses it—and, given the particulars of its usage, it will turn out to be more or less apt. It is also clear that the analogy is deeply entrenched. It is very difficult to imagine the workings of the visual brain along any other lines except the division between the 'black and white' and the 'colour' of public images. When one first learns that the ganglion cells in the retina are of two types, 'chromatic' or 'luminance' cells, it is natural to think that here too 'black and white' and 'colour' is the essence of the divide: luminance cells encode light intensity (i.e. brightness or darkness) and chromatic cells encode, well, the *other* dimension of light, wavelength or hue. If not that, what would the nature of the division be?

The central task of this essay is to pry the reader (not to mention, the author) out of the analogy's firm grip. To do so, we will look at the case of the rod achromat, a person who has only one type of photoreceptor, the rods, and whose visual experience depends upon luminance information alone. Although the normal trichromat also has a rod-based visual system—night vision—we will be looking at the pure case of luminance vision, a person who has, and has always had, only luminance vision. By the end of the essay, it should be clear why this was a good place to begin: The rod achromat's experience is quite different from what most people would imagine it to be. Hence, our own experience of night vision may not be quite what we imagine either.

As the reader will have guessed, the nature of luminance vision is only the *prima facie* topic of this paper. The hidden agenda is a philosophical one, about the nature of visual experience. The picture of luminance and chromatic processing that emerges, with a restructuring of the 'black and white or colour' divide, is of two systems that function in analogous and complementary ways to discern the multiple features of the visual world. The claim is not that luminance and chromatic systems work in exactly the same ways, i.e. they instantiate the same algorithms—but that the two systems use comparable (and often common) mechanisms to perform the multitude of visual functions that comprise human vision. They are intertwined systems, both of which are concerned with the broad goal of *seeing the distal world*. The point of dismantling the analogy, then, is *to make room for chromatic processing*. If a sharp black and white photograph shows you more or less what you would see if you looked at the same scene in person—if *that* is what you get from luminance processing—then there is only one thing left for chromatic processing to contribute: The colours. In other words, this common analogy, between black and white or colour, is a hindrance to understanding the natural fault lines of human visual processing. And we cannot understand the phenomenology of vision if we do not have these fault lines firmly in place.

16.1.2 Luminance Vision in the Rod Achromat

What is it like to be a human rod achromat?

A rod achromat is a person who lacks all three of the cones in a normal human trichromatic retina (a 'complete' achromat) or a person who lacks these cones functionally if not anatomically (i.e. the retinae of some ‘achromats’ contain cones but they do no contribute to normal vision.) In short, the rod achromat lacks the human system for ‘daylight’ vision, trichromatic vision.

What the rod achromat retains, however, is a virtually normal rod system for low light or night vision. In the human trichromatic retina, there are two main pathways for luminance information, one from the rods and one from the cones. (This often comes as a surprise to people who have been taught since grade school that “rods are for ‘black-and-white’ and cones are for ‘colour’”.) Both luminance systems use the same outgoing pathway from the retina, the magnocellular pathway. But because the rods and cones function under different levels of illumination, their use constitutes a sort of ‘timeshare’ arrangement, depending upon the light level: In daylight, the cones send luminance signals to cortex via the magnocellular pathway, while at night, the magnocelluar pathway carries luminance information from the rods. (It’s a bit like students who ‘double bunk’ for lack of money: the ‘day crew’ studies by day and sleeps at night; the ‘night crew’ sleeps during the day and occupies the desks at night.) In the retina of the rod achromat, the rods still use the magnocellular pathway despite the fact that there are no cones present to take over the magnocellular pathway in bright lighting conditions. So, surprisingly, given all the different ways that achromatic rod vision could have been organized, an achromat’s retina has roughly the same arrangement as our own minus the cones.

The central visual problem for rod achromats is that the rod system does not function well under daylight conditions. For one, the rod system ‘saturates’ under normal daylight conditions. Each rod absorbs so many photons in bright light that the photoreceptors are bleached of all pigment—and without adequate pigment, the rods no longer respond to light. For the rod achromat, then, sight under daylight conditions is very much like the experience the average human trichromat has when someone suddenly flicks on the bedroom light in the middle of the night. Certainly this *hurts* but it also renders the newly awakened subject entirely blind. This is more or less the constant state of the rod achromat in bright sunlight. The rod achromat’s photosensitivity explains why they prefer darkened rooms and deep shadow, and why even under low light conditions rod achromats wear sunglasses.

Second, the rod system is a highly convergent system: it pools together the signals from many different rods in order to maximize photon catch. To see anything at all at night, one needs a system that makes the best possible use of the miniscule amount of available light. However, this pooling of rod signals also decreases the spatial resolution of the system, i.e. the ability to distinguish two distinct but nearby points. (The higher the visual acuity or spatial resolution of a system, the closer together two points can be and still be seen as being distinct.) So, the rod system, in both achromats and trichomats, has far lower spatial resolution than the cone or daylight system of the trichromat. Just as a myopic trichromat is aided by large print or magnified illustrations, so too is the rod achromat.

The difference in spatial acuity between the trichromat and the achromat is actually a bit more complicated than this. As all mothers know, visual acuity is relative to the ambient light level. This is why, when as a child you sat in the dark reading, your mother probably said (sweetly) "TURN ON THAT LIGHT!" Spatial acuity gets better with increased illumination and this is true for both the trichromat and the rod achromat. The difference between these two systems is the absolute light levels at which rods and cones saturate. For the trichromat, during the day, visual acuity increases with light level and suddenly drops off when the photoreceptors saturate in intensely bright light. The same holds true of the rod system but saturation is reached at much lower light levels. The net effect is that the spatial acuity of rod vision is maximized under 'mesopic' conditions, during dawn and dusk or under the illumination of a full moon at night. Under mesopic conditions the rod achromat has the greatest spatial acuity, indeed the same visual acuity as the trichromat in similar circumstances. During daylight hours, however, the rod achromat is 'all but blind'. But this is not a result of the spatial acuity of the rod system.

So what is it like to be a rod achromat? Dr. Knut Nordby was both a rod achromat and one of the vision scientists responsible for understanding the physiology of rod achromacy (Skottun et al. 1982; Hess and Nordby 1986; Greenlee et al. 1988). He described his visual experience as follows:

> Trying to explain to someone with normal, or nearly normal, colour-vision what it is like to be totally colour-blind, is probably a bit like trying to describe to a normally hearing person what it is like to be completely tone-deaf, i.e. not possessing the ability to perceive tonal pitch and music. My task, though, is probably a bit simpler than the case of the tone-deaf, since practically everyone has had experiences of achromatic (i.e. colour-less or black & white) or monochrome pictures and renderings, and certainly must have witnessed the gradual disappearance of colours when darkness sets in.
>
> A first approximation, then, in explaining what my colour-less world is like, is to compare it to the visual experiences people with normal colour-vision have when viewing a black & white film in a cinema or when looking at good black & white photographic prints (good here meaning sharply focused, high contrast with a long grey-scale, as in crisp, high quality, glossy, technical prints).
>
> This, however, is only part of the story because I have so far only dealt with the achromatic aspect of my perception. To get a fuller understanding of my visual world one must, in addition to my colour blindness, also take into account my light aversion (i.e. hyper-sensitivity to light) and my reduced visual acuity. (Nordby 1996)

This description sums up Knut Nordby's view of his own visual experience, one that accords well with commonly made inferences about achromatic experience from the third person point of view. First, given that both a trichromatic retina and the achromat's retina have very similar rod systems, what the achromat sees at night is probably very close to what the trichromat sees at night: in both cases, the subject sees the world via input from the rods alone. Second, the night vision of a trichromat is commonly described as seeing 'in black and white' or like looking at a monochrome photograph. Making allowance for the lack of illumination at night and the relative 'fuzziness' of rod-based vision even under optimal (mesopic) lighting conditions, what trichromats see at night is very much what we see when looking at a monochrome photograph or watching an old black and white movie. By transitivity, then, the rod achromat sees just what you, a slightly myopic trichromat, would see were you to watch an old black and white film without your glasses but while experiencing a photosensitive headache. And speaking for myself, this is something I *can* easily imagine. (Surprise! A rod achromat is not an alien-life form. Who would have guessed?)

In the next section, I will begin to explain why neither of the above two inferential steps are good ones. The trichromat's experience of the visual world at night is not like the rod achromat's experience of the world during the day; nor is trichromatic night vision like looking at a black and white movie or photograph.

16.1.3 Luminance Information

Like many terms commonly used in science (e.g. 'electron' 'energy' 'power'), we often use 'luminance' without remembering (if we ever knew) its explicit definition. 'Luminance', we all know, has something to do with 'the amount of light'. But if you look up the definition of 'luminance', you will find the following puzzling statement: luminance is the radiant intensity of light as filtered by the human photopic luminance function. In turn, if you look up 'photopic luminance function', you will learn that this is a model (for the normal trichromatic human observer, adapted to photopic conditions) of the probability that an individual photon will be absorbed, expressed as a function of the wavelength of the stimulus. Soooo . . . .after suitable rumination, it seems that 'luminance' refers to the amount of light, at each wavelength, your visual system will absorb. Uh huh. While technically correct, there is something unsatisfying about this definition. We want to know what a luminance system *does*. But this definition does not explain the *function* of a luminance system or distinguish it from the obvious alternative, the chromatic system. Nor does it tell us what luminance vision represents. It appears to say only that a luminance system absorbs photons—presumably, for the purpose of seeing although even this is not a part of the definition. However unsatisfying this definition may seem, I've come to realize that it has profound consequences.

Consider first that all photopigments on earth function in fundamentally the same way. Importantly, each photopigment responds to light across a restricted range of wavelengths, what is commonly known as the receptor's spectral range. Within this range, the response of each photopigment is wavelength sensitive: a light stimulus of the same intensity will be absorbed with a greater or lesser probability depending upon the wavelength of the stimulus. Figure 16.1 illustrates the response curves of the three human cones. Relative to a fixed intensity or amplitude of light—and that is the part to always remember—the graph illustrates the probability that a photon will be absorbed at each wavelength across its spectral range. At the apex of the curve is photopigment's 'preferred' wavelength, the wavelength of light that will result in the greatest light absorption. For example, peak light absorption for the human long wavelength (or L) cone occurs at 470 nm. Importantly, though, this is only the receptor's preferred wavelength; it responds across the entire spectral range although to lesser extent. What makes one photoreceptor different from another is the photopigment it contains, and this in turn means a difference in the spectral range over which each photoreceptor responds.

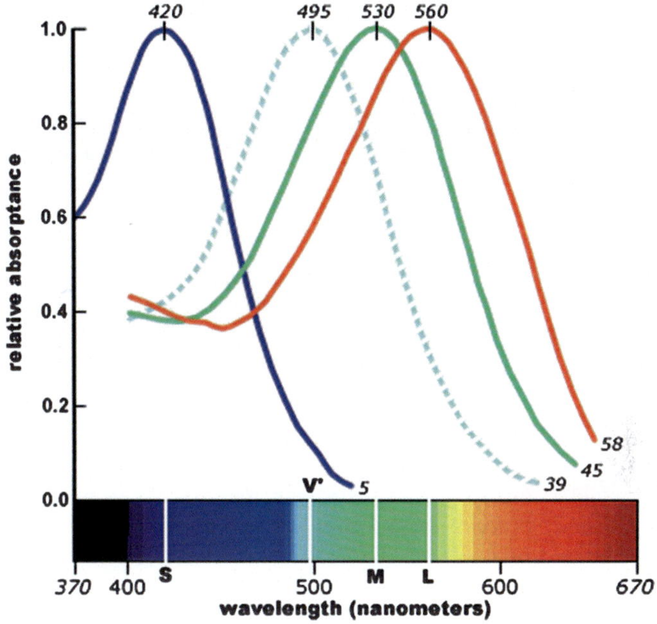

Fig. 16.1 Absorption spectra of the *three* human cones, *S, M & L*, plus the *rods* (*dotted blue line*)

What differentiates cones from rods is primarily their sensitivity to light, how much light is required to affect a response. Again, if one thinks in the old terms of ‘rods are for black and white’ and ‘cones are for colour’, this might come as a surprise. Above, in the case of the achromat, we saw that cones are responsible for luminance signals under photopic conditions. But does it follow that rods are equally capable of producing colour vision? Yes and no, replied the philosopher. Rods, just like cones, respond within a specific spectral window; rods are wavelength sensitive in exactly the same way as cones. The primary difference between rods and cones, as I have said, is the energy required for photon absorption: rods require far less energy and are thus ideal for low light conditions. However, despite their greater sensitivity, the absolute photon catch of rods is still markedly lower than that of cones. This is why rod systems are convergent: they must pool the signals of multiple rods in order to achieve a good signal-to-noise ratio. If yet another type of rod were added uniformly throughout the retina—and one must have at least two types of receptors to discriminate wavelength—this would *halve* again the already poor spatial resolution of night vision. In the dark of night, it is thus the low photon catch of the rods that disqualifies rods for participation in colour vision.[1]

[1]Note that this does not rule out a specialized area of the retina in which rod input is used for chromatic processing. This could be used much like the ‘bucket’ function in painting programs in which colour is added to an object, not pixel by pixel, but given the outlines of the object to be painted.

Still, there is nothing in the function of rods that intrinsically precludes them from chromatic processing and the question of whether (and what) rods might contribute to chromatic vision has been an active one since the 1960s (Stabell and Stabell 1965; Stabell 1967a, b). It used be thought that rod and cone systems were functionally segregated by light level, i.e. the cones 'shut off' at precisely the level of illumination at which the rods 'come on' and vice versa. We now realize that this is false. At dawn, dusk and under the light of a full moon (under *mesopic* conditions), the mammalian visual system contains a sub-population of rods that contribute to chromatic vision (Buck et al. 2006; Cao et al. 2008, 2011; Li et al. 2010; Pang et al. 2010). So the world continues to look coloured even when the photon absorption of the cones is compromised. Moreover, very recent experiments suggest that even under scotopic conditions (in the dark of night without starlight), rods feed into the S cone chromatic pathway—which explains why the predominant colour of night, for the trichromat, seems to be blue (Field et al. 2009). In dim lighting, we *do* see colours partially as a result of *rod* processing. In retrospect, given the similarities between rods and cones, it is not surprising that rods and cones work together under mesopic and scotopic conditions to encode colour. But this co-operative function is predictable only against a general understanding of photoreceptor function.

So human vision has (at least) two major luminance systems, a photopic (bright light) system that depends upon cone input and a scotopic (low light) system that sums rod signals. *Importantly, neither of these luminance systems—indeed no biological luminance system—encodes light intensity per se*. This is in stark contrast (sorry) to a black and white photograph (at least one printed from colour-corrected black and white film) in which the intensity value of light at each point in the photographic image is represented using the greyscale. (One source of confusion for the reader may be that monochrome images, which represent light intensity, are sometimes called 'luminance images' in computer science and artificial intelligence circles.)

This point is the flip side of one familiar to all researchers of colour vision: A visual system with a single receptor cannot discriminate between two stimuli that differ only in wavelength. In the above graph of cone function, recall that the graph plots the probability that a photon will be absorbed against the wavelength of light relative to *a set intensity of light*. Alter the intensity of the light stimulus and the probability that a photon will be absorbed (at a specific wavelength) is altered as well. So the photon absorption of any receptor is a function of both wavelength and intensity. Receptor response does not indicate or provide information about either property independently of the other. As I said above, for colour researchers, this is a well-known—and one might even say 'shopworn'—fact: A visual system with a single receptor (or, what comes to the same thing, without the ability to compare different photoreceptor signals) is 'colour blind'. But what is sauce for the goose is sauce for the gander. The same moral holds, *mutatis mutandis*, for the *intensity* of the light stimulus. If a single photoreceptor conflates the wavelength and intensity of the light stimulus, then each photoreceptor is *intensity blind* as well. This is a fact we don't hear repeated nearly as often. Without a signal comparison between two different types of photoreceptors, intensity cannot be distinguished from wavelength. Thus, a luminance system, which has only one kind of receptor, is just that: intensity blind.

In 2010, there was an art exhibition in Berlin by the design firm Carnovsky that provided a brilliant demonstration of this fact—and of the nature of luminance vision in general. For this exhibit, entitled 'RGB', Carnovsky produced several different wallpapers covered in 19th century illustrations of various species (some of which are even recognizable as the species which they represent). Each animal is rendered by a line drawing—this is important as we'll see—and printed in one of three colours from the standard printer's RGB palette of cyan, yellow and magenta. Although the wallpaper is printed digitally, it looks like a screen print with three colours of creatures layered one upon the other. Under natural illumination or any light source that approximates a uniform spectral power distribution ('white' light), the coloured figures are clearly visible to the human trichromat. The exceptions are the yellow figures on the wallpaper which can be quite difficult to see especially when overlaid with other creatures. a common problem with yellow figures. While the wallpaper is pleasant enough in daylight, the interest of the exhibit really lies in what happens when the wallpaper is illuminated by one of three coloured lights (Figs. 16.2b and 16.3a, b). When a filtered light is switched on, the entire room is suffused with colour and the illustrations themselves now appear as monochrome images rendered in red, green or blue —or what one might call 'black and red', 'green and black' or 'blue and black' (as opposed to black and blue). Some creatures simply disappear, while those that remain appear in very dark shades of the illuminant colours, almost black. Switching between the coloured lights produces dramatic differences in the visibility of the various creatures. For example, under the red light, the blue creatures are visible but the magenta and yellow ones disappear. Under the blue light almost all of the animals are visible but the yellow creatures, previously invisible, now pop out (as black!); the other creatures appear as more misty grey background figures.

Let's take a close look at what is going on in this exhibit. (In figuring out this exhibit, I found it helpful to pick out three figures, one in each colour of ink, from the original wallpaper and then to compare their appearance under each of the three coloured lights. So let the fox be our magenta figure, the alligator be cyan or blue, and the large cockroach be yellow. I know. What cockroach? But it is there, overlaid upon the elephant, visible only under the blue light.)

There are two central 'effects' that create the magic of the RGB exhibit. First, the display uses *spectral filtering* to its best advantage, a 'trick' that every natural system of vision 'learns' to employ over the course of evolution. In the case of the rod achromat, this 'filter' is on the receiver end of things: with only one photoreceptor, the rods, visible light is limited to the spectral range of that single receptor. In the RGB exhibit, the normal trichromat observes a room illuminated by one highly filtered light source, by the red, green, or blue light. Here, visible light is limited to an artificially small window by the filtered light source. That is, for us as trichromats, visible or 'effective' light ranges from 370 to 660 nm, a spectral range of roughly

Fig. 16.2 (**a**) Shows the RGB wallpaper, with figures rendered in the standard three colours of printer's *ink*, *magenta*, *cyan* and *yellow*. (**b**) Shows the same walls illuminated by *narrow-band red light*. For further photographs, see http://www/carnovsky.com

300 nm. But under the filtered lights of the exhibit, all light within the room is restricted to a narrow band of light about 60 nm in width. For the trichromatic viewer, then, spectral bandwidth is restricted by the sender not the receiver. Under the red, blue or green lights, whatever the trichromat sees is made visible by a single narrow spectral band of light, be it red, blue or green, reflected from the surfaces within the room. In effect, then, trichromatic observers have reduced spectral range very much like the restricted range of the rod achromat. In fact, one can think of the

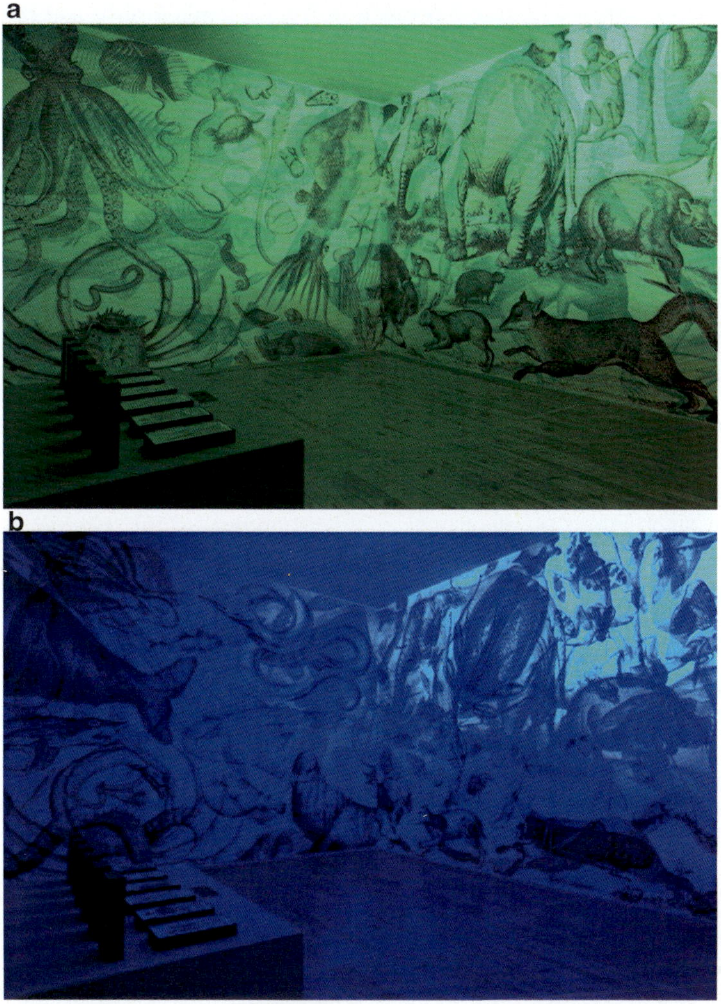

Fig. 16.3 (**a**) Shows the RGB wallpaper illuminated by *narrow-band green light*. Note the high visibility of the *red figures* (which appear *black*) and the low visibility of the *yellow figures* (which appear *yellow*) and *blue figures* (which appear *gray*). In (**b**), the *narrow-band blue light* highlights the *yellow figures*, such as the *cockroach*, which now appear *black*

three lights as producing (very roughly) *functional monochromats*, each with only a short (blue), medium (green) and long (red) cone/photoreceptor.

From this perspective, what we'll call the perspective of the 'Carnovsky monochromat', it is clear that the strength of the returning signal is strongly *colour dependent*. Each of the three inks (red, blue or yellow) absorbs and reflects light continuously across the normal spectrum of visible light; their surface spectral reflectance or SSR is a continuous function. Yet each ink absorbs and reflects light in a wavelength selective manner. Blue objects appear blue because they reflect *more* short-wavelength light given a light source that emits light at a uniform intensity across the wavelength spectrum. Of course, if the light source does not emit 'blue' light, then there is no light for a blue object to reflect. In such a case, say given a predominantly 'red' light source, a blue object will appear black. This is one of the central principles used in RGB exhibit. Each ink is colour selective: there is a certain range of light wavelength that it reflects preferentially. Hence the colour of the light can be chosen so as to maximize or minimize the visibility of each colour of ink and, by extension, the visibility of the creatures rendered in each ink colour.

Take, for example the blue alligator. Even though the alligator's cyan pigment absorbs light continuously across the visible spectrum light, cyan pigment *reflects* far more blue light than it absorbs (which is why it appears blue). Conversely, it *absorbs* far more red light than it reflects (why it does *not* appear red). The blue alligator, illuminated with red light, will thus appear black for it reflects almost no red light (and there is no other light to reflect). In this exhibit, blue light produces the most interesting effects. Neither the yellow nor the red figures will reflect much blue light, so both will 'pop out' under the blue light, a particularly good effect for the all-but-invisible yellow creatures (in daylight). But the yellow figures will also reflect a bit more blue light than the red ones. For this reason the red figures will appear both darker and closer, while the yellow ones will appear as more hazy background figures. The cockroach is the exception that proves the rule: it appears *in front of* the elephant under blue light. This is because darkness is also a function of the level of detail in the line drawings and thus how much pigment is used. The more yellow pigment per unit of area, the closer the lines of drawing, the darker the yellow creature will appear. Here the cockroach is much more finely rendered than the elephant, hence the cockroach appears to be in front, the darker elephant behind.

More succinctly, the RGB exhibit uses narrow bandwidth filters to re-create the monochromat's world, a world in which perceptual 'lightness' is a function of *both* intensity and the predominant wavelength of the reflected image. With only one photopigment, rhodopsin, the rod achromat's visual world varies along a single visual dimension. So too do the perceptions of the 'functional monochromat' who views the Carnovsky world of illustrated figures under coloured light. Still, there is a crucial difference between a trichromat who views a Carnovksy exhibit under filtered light and a monochromat who views the natural world under sunlight. There is no escaping the fact that, for the trichromat, the RGB exhibit appears in shades of red (or green or blue.) The trichromat sees the light and the wall *as coloured*, as having a particular hue, even if the light and every surface are monochrome, i.e. even though they have the *same* hue. This is not information that the rod monochromat could possibly have, that the illuminant has a particular predominant wavelength as does the light reflected from every surface. We must assume therefore that the monochromat's experience is not 'coloured' red *or* blue *or* green and that, in all likelihood, it differs from our experience in this crucial way. This brings us to the second reason why the RGB exhibit works so well.

The second reason why the RGB exhibit is so effective is that the combination of each coloured light and the three ink pigments are designed to enhance (or diminish) *luminance contrast*. When the wallpaper is bathed in red light, for example, only red light can be reflected back from the illustrations. Similarly the white wall, which normally reflects light equally across the entire spectrum, reflects only red light. So, to the trichromat, the wall appears red. As we said, for the trichromat, under a red illuminant, every thing that is visible appears in shades of red from bright red to red-black. But what is *visible* against a bright red wall? A magenta figure (e.g. the fox) will reflect a large percentage of red light. A red fox does not contrast with a red wall. The same holds true for all of the magenta figures. Paradoxically, under the red illuminant, figures rendered in the *blue* ink will be the most visible. A blue figure reflects very little red light under any lighting conditions, hence it will now reflect very little light *at all*. The blue alligator thus appears as a *black* figure against a red wall. Finally, the yellow figures will now be entirely invisible. We are not told the spectral power distributions (SPD) of the coloured lights used in the exhibit. But suppose that the red light source contained some 'yellow' light and that the yellow pigment reflects a bit of red light in addition to yellow light. This lack of visual contrast would render the yellow figures invisible.

The two principles used by the Carnovsky exhibit—of spectral filtering and luminance contrast—mirror two of the most important principles of vision. In fact, this is why the RGB exhibit works so well on us. First, from above, every known photopigment acts like a wavelength filter, responding to light as a function of both wavelength and intensity. Two different pigments may produce profoundly different levels of excitation in response to one and the same reflected figure. In the evolution of any visual system, the type of photopigments/filters in place will have had a direct effect on visibility within the environment and hence on the species ability to *see* its predators, find sustenance, determine the fitness of mates and so on. (In fact it is hard to imagine many physiological facts that would play as important a role as photopigment sensitivity in the general fitness of a species.)

Figure 16.4 (Gegenfurtner 2003) demonstrates the human case, the outcome for the majority of our species. Each illustration shows how the S, M or L cones filter a natural image, the photon catch for each of the three cones given the same reflected image. The original colour photograph depicts a group of fruits and vegetables (Fig. 16.4a), most of which reflect light predominantly from the middle- to long-portion of the spectrum, the yellows and oranges. (Blue vegetables are rare, not to mention somewhat unsettling.) Note, for example, the banana and orange bell pepper in the original colour photograph. Now compare their luminance images as filtered by the M and S cones (Fig. 16.4e, d respectively). The M cone is preferentially sensitive to middle wavelengths with its peak preference in the yellow range. So the brightest objects in Fig. 16.4e, which illustrate net photon absorption by the M cones, are the banana and a lemon. In Fig. 16.4d shows the intensity image as filtered by the S cones: here the orange pepper is black and the banana is dark gray. This is because S cones are highly sensitive to 'blue' light and insensitive to the longer 'red' wavelengths. So in the S cone luminance image, the redder the

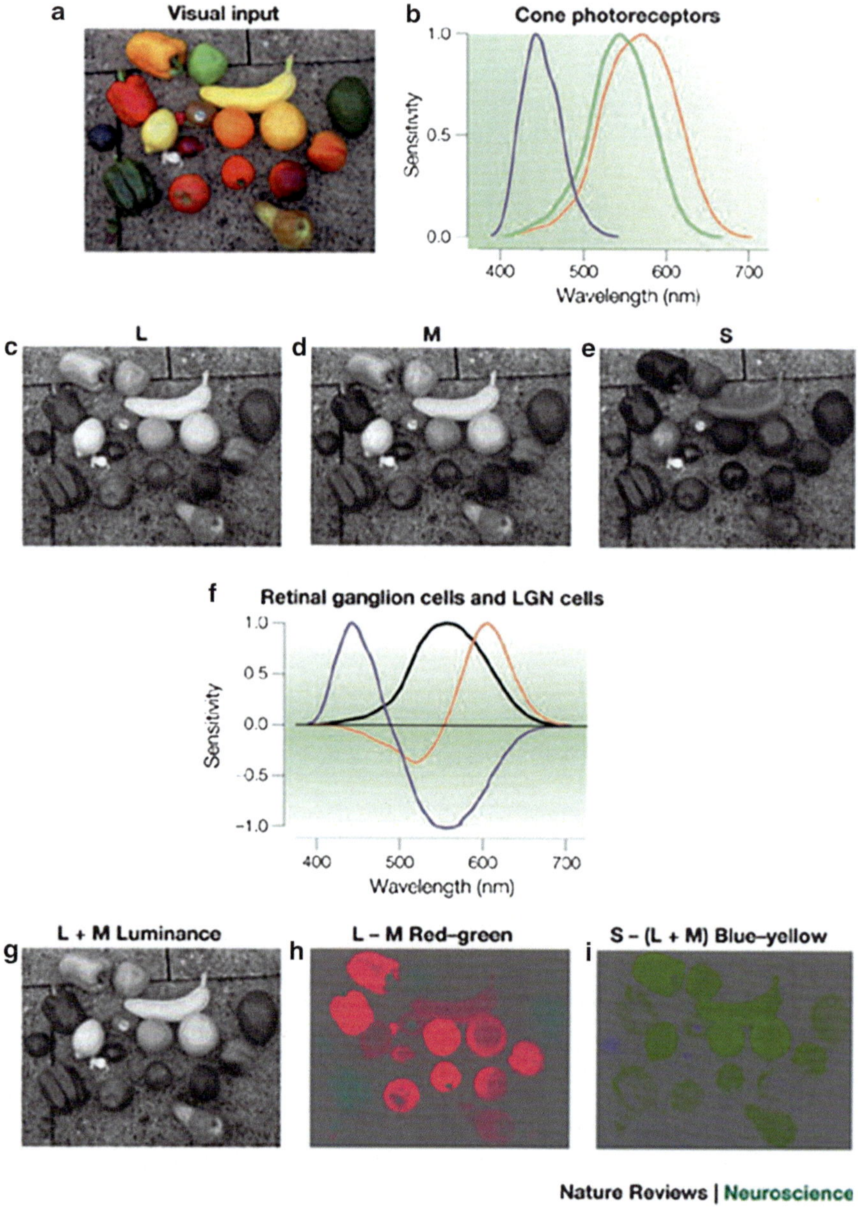

Fig. 16.4 (**c**), (**d**) and (**e**) Illustrate the different measures of luminance, for the *L*. *M*. and *S* cones that would result from a single a *colour photograph* of common fruits and vegetables. (**g**) Shows the luminance image for the $L + M$ luminance channel of the same scene. (**h**) and (**i**) Illustrate the reaction of chromatic ganglion and LGN cells. In (**h**) *magenta signals* a positive response and *green signals* a negative response of an *L-M* or *red-green cell*. In (**i**), *yellow signals* negative response of the $S - (L + M)$ channel. *Grey signals* a lack of response in both images. Note that, in (**i**), there is no positive response shown because none of the fruits or vegetables are predominantly *blue* (Gegenfurtner 2003)

fruit or vegetable, the darker it will appear. In sum, the amount of 'effective' light reflected from an object, the intensity of light as filtered by one photoreceptor or another, depends upon the *spectral* facts of the environment—*the colour of both the object and of the light source*—and on the spectral sensitivity of the photoreceptor at issue. This is why the definition of luminance provided at the outset ('luminance is the radiant intensity of light as filtered by the human photopic luminance function') while seemingly empty carries such weight. The information available to any visual system is as much a function of the wavelength of light reflected from distal objects, as it is a function of its intensity. For a luminance system, object colour matters just as much as object lightness or darkness.

Second, the primary concern of evolution in vision—i.e. what natural selection hinges upon—is *what the organism can see*, the *visibility* of relevant objects, not which objects reflect the greatest or least amount of light. Whatever else, the viewer must segregate an item of interest from its background. So at bottom luminance vision requires the registration of luminance *contrast* between an object and its background. It does not matter whether, for this particular visual system, the object has positive or negative contrast with its background—or whether the luminance contrast arises as a function of genuine intensity differences between the object and its background or because, while the figure and ground reflect the same intensity of light, the spectral sensitivity of the cones 'creates' luminance contrast given their difference in colour. 'It's *all* good!' as they say, as long as we are able to distinguish between an object and its background, as long as there is contrast.

A more recent Carnovsky exhibit nicely illustrates this principle of 'visibility by contrast'. In their second large installation, the space to be 'papered' contained two mirror-image rooms, a design feature that the Carnovsky designers wanted to exploit. Both rooms were papered with a jungle scene in which its inhabitants were obscured by dense foliage when viewed under an even SPD illuminant. One room, designated the 'positive' room, was papered with coloured figures on a white ground just as in the RGB exhibit. The second 'negative' room portrayed coloured figures on a black background when viewed under normal lighting conditions. Looking at one part of the display, say the hidden elephant in Fig. 16.5a, one further difference is apparent: the 'positive' room has a blue elephant, hence a blue-on-white image, and the 'negative' room (Fig. 16.5b) has a red elephant hence a red-on-black image. Once the red light is turned on, the rooms demonstrate their intended yin/yang nature. The first room reveals a *black* elephant on a bright *red* wall (a positive image) while the second room shows a bright *red* elephant on a *black* wall (a reverse image of the other room). Of course, in some sense, no one should be surprised to see a red elephant on black in the negative room. After all, it *is* an illustration of a red elephant drawn on a black background. The surprise is the positive room: the blue elephant on a white wall appears to be a *black* elephant against a *red* background. Added to this 'reversal' is a very nice feature of both rooms. In both the positive and negative configurations, the elephants now stand in plain sight without one bit of flora to hide them, a very nice illusion. Note that both of the rooms, under red illumination, produce monochromatic visual images, one the reverse polarity of the other, both of

Fig. 16.5 Figure (**a**) *above* is the 'positive' rendition of the wallpaper, with *coloured figures* and fauna against a *white* background; Figure (**b**) is the 'negative' rendition, with *fauna* in *white* and *animals* in *colour* on a *black* background

which allow us to see the elephant without difficulty despite the reversal of contrast. Again, it's luminance contrast that makes visible the figures, not absolute luminance (Fig. 16.6).

Before ending this section, let me explain in slightly different terms, *what* the Carnvosky monochromat looses under the narrow-band illumination. Above, I explained the disappearance of certain figures (in the wallpaper) under coloured lights as the result of decreased wavelength contrast. Obviously, if a red figure is

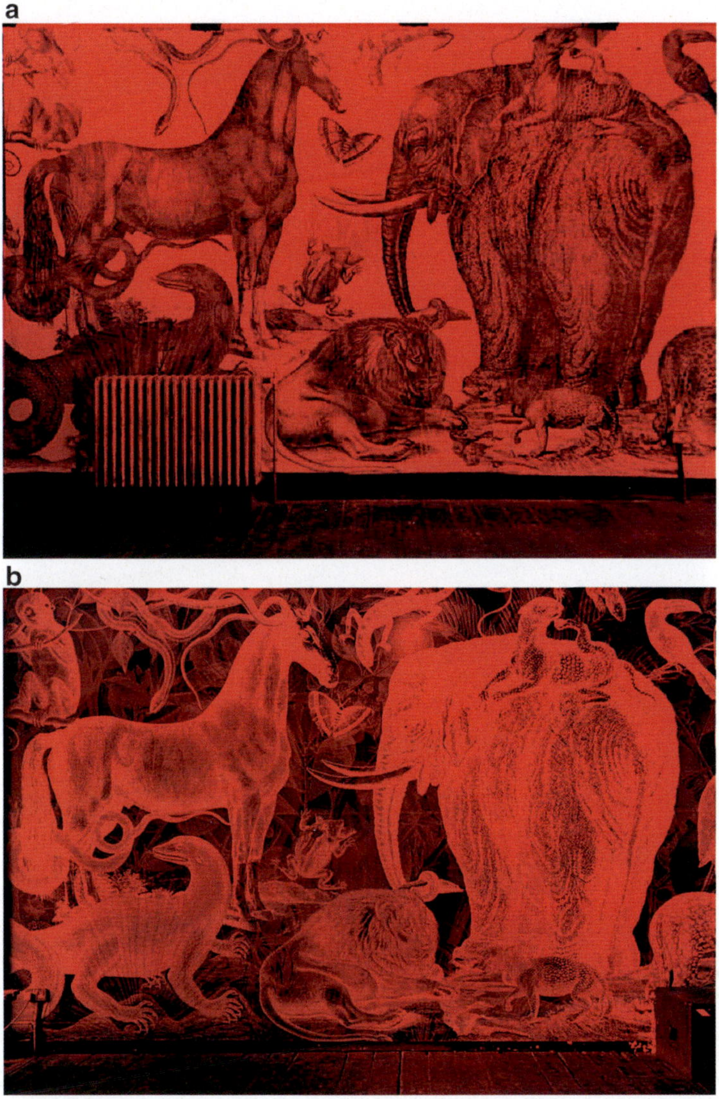

Fig. 16.6 In (**a**), the positive scene is shown under *red* illumination, which yields *black* figures on a *red* background. In (**b**) the negative scene is show under *red* illumination, which yields *red* figures on a *black* background

against a red background, one cannot see it. But it is important to realize that a Carnovsky monochromat also looses *intensity information*, what we think of as the 'black and white' of the original wallpaper. Compare three images derived from the Carnovksy exhibit (Fig. 16.2a) A *full colour* photograph of the wallpaper under daylight (Fig. 16.2a) (ii) An *intensity image* of that same wallpaper, i.e. an image of a

b

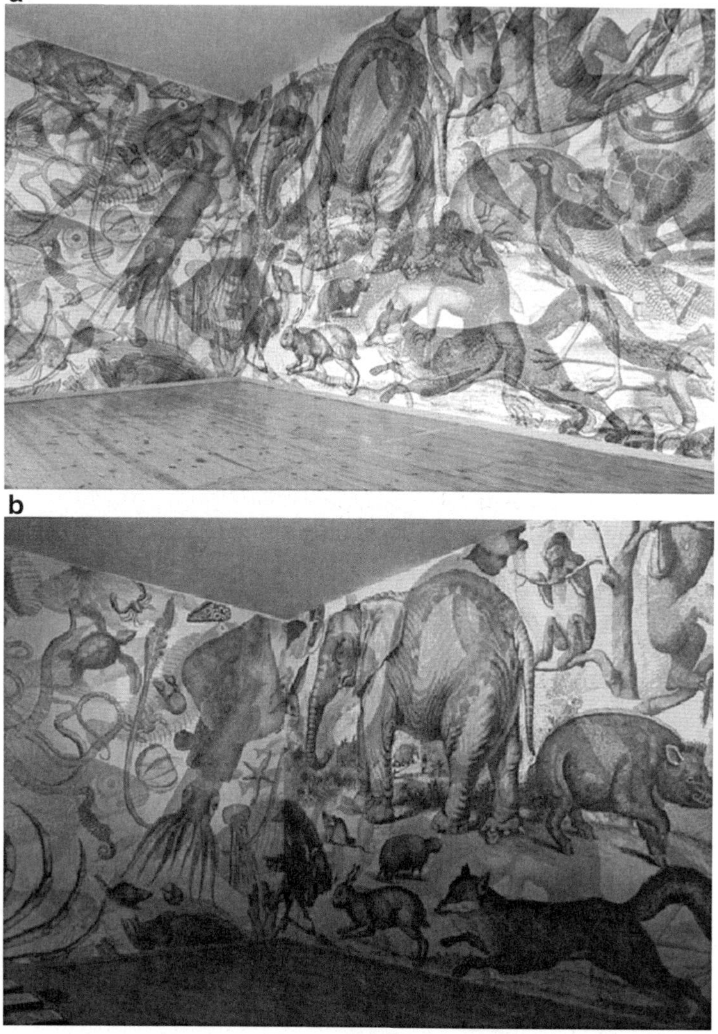

Fig. 16.7 (**a**) A monochrome image of the RGB wallpaper. Note that the *three* inks used to render each of the colours, *cyan*, *magenta* and *yellow*, appear roughly in intensity. In (**b**), the intensities of the figures is now variable, some *lighter* and some *darker*

the wallpaper rendered in greyscale (Fig. 16.7a); and (iii) A *luminance image* of the wallpaper or an photograph of the wallpaper illuminated by the *green* light source which has been rendered as a monochrome black and white image (Fig. 16.7b). In the original colour image, all three ink pigments—cyan, magenta, and yellow—reflect roughly the same overall amount of light. So under daylight, all of the figures have roughly the same intensity, i.e. appear equally dark to the trichromatic viewer in the greyscale intensity image. (Some of the creatures *do* look darker than others because they are rendered using many more lines. See the paragraph below for an explanation.) In the luminance image, however, the figures have a range of luminance values, from very dark to very faint images. There are also figures that have disappeared, which no longer have any luminance value at all. You can see this clearly if look just to the left of where the walls meet in the luminance and intensity images. In the *intensity* image, there are so many figures rendered one on top of the other that it is difficult to extricate, visually, any one of them. But in the same area of the *luminance* image, there is one predominant figure: a giant squid. You can also clearly see several surrounding figures: a conch shell, a sea star, the lion's tail and, in shadow, a large fish with a spikey dorsal fin. By looking at the *coloured* wallpaper, you can also see what creatures have been 'disappeared'—i.e. a huge coiled snake and, by the floor, a large mammal akin to a walrus. Above, I explained why certain figures disappear in terms of the *colour* contrast: A red figure against a red wall is invisible. But this point can also be put in terms of light intensity: *there is less luminance* contrast (*in the luminance image*) *than intensity contrast* (*in the intensity image*). Any narrow-band spectral filter will eliminate all but a small range of wavelengths in the image. But if a spectral filter, like the rods, leads to a loss of light, it leads to a loss of *intensity information as well*[2]. This is why a rod achromat also looses 'black and white', the intensity information of the retinal image.

It is worth emphasizing in the present context how difficult it is 'simply imagine' the achromat's point of view—the effect of a spectral filter on a natural image and the probable consequences for achromatic visual experience. The results are too complex *and* unintuitive. Even people who work with colour and light for a living, the graphic designers at Carnovsky, could not have imagined exactly what would happen when, with their wallpaper newly affixed to the wall, they turned on the first of the coloured lights. In fact, the first RGB images required extensive experimentation with different ink pigments and filtered lights even though the scene was highly artificial, a series of uniformly pigmented line drawings against a uniformly white background (*personal correspondence with Francesco Rugi of Carnovsky Designs*). This same failure of imagination is no less likely for vision scientists or (dare I suggest) philosophers of vision science. A seasoned psychophysicist of colour vision would not be able to predict, accurately and in detail, the appearance of an arbitrary natural scene relative to a specified filter. This is why Gegenfurtner (2003) was allowed to include three images of a fruit and vegetable arrangement in an article in *Nature Reviews Neuroscience*. It is one thing to understand the theoretical principles of spectral filtering and 'visibility by contrast', yet another to imagine the concrete, particular results of their application to a natural scene. So a realistic exploration of achromatic vision would begin with detailed image analysis: To see what the achromat can see, that is, we would start with a series of natural images, apply the rod luminance function to each image

[2]I do not mean to suggest that the luminance image in Figure is an accurate depiction of the rod achromat's situation—either the luminance information available to the rod achromat much less what the rod achromat would see when looking at the Carnovsky exhibit under natural daylight.

pixel-by-pixel, and then do a statistical analysis of the resulting set of images. And this would give us only the *starting state* of achromatic vision, the receptor input, prior to retinal or cortical luminance processing.

That said, my own best guess is that the achromat will not loose whole snakes and walruses, or even medium-sized objects like bowls and buttons. Instead, the achromat will find it difficult to see *surface detail*. Again, examine the Gegenfurtner (2003) images particularly the S cone image (Fig. 16.4e). Of course, we don't expect oranges and apples to look *black* and this makes the fruit look rather peculiar in the S image. Strangeness aside, the fruit also appear somewhat plastic and featureless. Looking at the apple, its 'plasticity' results from the darkening of the apple image (by removing the 'red' light for the light source), an event that serves to highlight, by contrast, each white 'glint' of light from apple's surface. The apple appears 'plastic' because its surface looks too shiny to be real. However, the S-filter also makes it impossible to see the kinds of shading and shadowing that are so useful in normal trichromatic perception. Shading, which is the self-shadowing of an object given directional lighting, is a central cue for shape perception. For example, a round object, lit from the right, has both a circular boundary around its periphery plus a curved pattern of shading, from light to dark, on the left side of the image—two cues for a single property, roundness. We see surface texture by means of patterns of small shadows on the surface of an object: A dimpled orange peel produces a regular array of dimple-shaped shadows against the background of a bright orange surface. But if the image of the orange itself is very dark, as it would be when filtered by an S filter, it would very difficult to distinguish the dark dimples against their black background. (That is why a tight dress looks best in black at least for women with any surface 'texture'.) It is exactly this kind of contrast—patterns of low luminance contrast on reddish or blue-ish objects—that will be absent in rod achromatic vision. Hence surface texture and detail will be invisible on all but blue objects for an S-cone achromat.

In fact, this is why the designers at Carnovsky always wear black dresses. Well, not really. Rather this is why the designers chose line drawings over photographs for the RGB wallpaper. In a line drawing, all of the surface features of an object, its texture and the shading from which we determine the shape of objects, is rendered with just line and empty space. By choosing three colours of ink with the same intensity, the surface features of the animals are rendered at a single level of intensity contrast, the difference between ink and paper. The designers could then heighten or 'disappear' this constant contrast through the judicious choice of light filters. But importantly, if a creature is visible in the RGB exhibit, so too are its features and surface texture—eyes, fur, feathers or scales. The careful choice of pigment and lights, plus the format of line drawings, accounts for the dramatic effects of the exhibit, the appearance and disappearance of whole creatures under different illuminants. But if natural images had been used, with a range of intensity contrasts typical of a high-resolution greyscale image, it is the surface detail that would have gone missing. A trichromat who views a set of natural images filtered by the rod sensitivity function might not find the loss very interesting. No creatures would suddenly vanish. But the cumulative loss of contrast information would amount to a (statistically) dramatic loss of information. *The rod achromat's losses are no less real—and in all likelihood no less substantial—than the information loss that a trichromat suffers in the Carnovsky world.*

16.1.4 A Scenic Detour: Chromatic Processing

In the last section, the central lesson was that the rod achromat does not see 'in black and white' if by that one means that the achromat has access to the intensity information represented by a black and white photograph. Of course, the achromat does not have access to wavelength/colour information but neither does the achromat have access to the other dimension of a visual image, light intensity. Instead, an achromat has a very specific form of luminance information—image intensity filtered as function of wavelength by the rod photopigment, subsequently encoded as differences in photon catch, i.e. luminance contrast. And *that*, as should now be clear, is a different kettle of fish, the explicit consequences of the 'empty' definition of 'luminance' with which Chap. 17 began.

In this section, I want to explain how chromatic systems arise and why, once in existence, they turn out to be highly effective partners for luminance systems. On the one hand, the chromatic cells, which arise post-receptor in the retina, do not encode colour *per se*, neither the colour of objects and various media in the distal world nor what is called 'image colour', the wavelength composition of the retinal image and its spatial arrangement. Rather chromatic cells arise by chance, the inevitable outcome of the genetic variation and the wiring of existent luminance cells.[3] It is this chance composition that endows chromatic cells with highly complex informational properties, as opposed to the properties that a wavelength 'detector' would need to have. On the other hand, once chromatic cells are added to a luminance system, the informational 'reach' of the combined system greatly extends the informational reach of either component. Computational tasks that were beyond the capacity of either luminance cells *or* chromatic cells alone are made possible by the partnership of these two complementary systems. As the reader will have guessed, this partnership makes possible the perception of colour in the distal world, the colour of opaque surfaces, transparent media such as water, and of light itself. But as we will see in the next section, chromatic and luminance processing is also needed for the perception of lightness and darkness, to see coal as black and snow as white. Without chromatic encoding, the rod achromat lacks the veridical perception of surface lightness or darkness. In this second sense, in terms of darkness perception, the rod achromat does not see 'in black in white' either.

---

[3]Note that receptors are neither luminance cell nor chromatic cells. Rather, the signals from various receptors are used as the inputs to luminance and chromatic cells, a distinction that will be explained more fully in a few pages.

To see why chromatic and luminance systems make such good partners, back up a few steps. A central problem with a 'pure' luminance system is that it is rather primitive *qua* a source of information about the distal world. As light travels through the atmosphere, its interactions with bulk matter will affect, in specific and law-like ways, both its wavelength and intensity. For example, the intensity of light diminishes as it travels further and further from its source or as it makes contact with bulk matter; transparent substances act as wavelength filters thus changing the wavelength composition of the emerging light; opaque objects cast shadows, resulting in areas with diminished light intensity (shadows) but of the same wavelength composition. And so on through the laws of optics applied to a natural environment. Thus, each dimension of the light stimulus, if encoded separately, would act *as an independent source of information about distal bulk matter*. By definition, a luminance system conflates these two dimensions of light. So a luminance system cannot make use of the full informational resources of the light *qua* multi-dimensional stimulus except serendipitously. So a luminance system alone is at a functional disadvantage relative to one that registers wavelength and intensity individually.

However handy it would be to have an independent encoding of wavelength and intensity, evolution does not strive *towards* anything. Physiological variations appear, they work or they don't. Once genetically entrenched, these chance variations tend to stay the course unless their retention results in positive harms. Now in our own case, for the evolution of our three cone photopigments, we can chart their paths in the lineage of Old World primates and the mammals that preceded them (Jacobs and Rowe 2004; Jacobs 2008, 2009). For our purposes, the general story of how photoreceptors evolve (opposed to the specific evolutionary history of the three cones of Old World Primates) is the relevant one. Through genetic drift, random links in the amino acid chains of existing photopigments are altered. When a change occurs at a key location, a receptor with a new spectral sensitivity arises. In effect, existing photopigments mutate into new ones through chance substitutions. In each case, the result is a new spectral filter. Very occasionally, on the order of .01 duplications per gene per million years, an additional photopigment gene is created. And *that* photopigment will itself become subject to mutation and drift as time goes on. Over time, then, the existing photoreceptors of each species change their sensitivities, and on rare occasions, a species may gain an entirely new class of photoreceptor. As a direct result of the genetics of photoreception, each species 'auditions' a series of new photoreceptors, each with an individual sensitivity to light.

Above, in the section on luminance processing, I said that one of the central principles of vision is visual contrast: In order to see anything at all, the system must encode a difference in some property of light or another between two spatially adjacent areas of visual space. In the vertebrate retina, this requirement finds its expression in the centre-surround cells of the retina, the LGN and primary visual cortex. As its name suggests, a center-surround cell compares the total photon catch between two regions of visual space, between a central circular region of visual space and the circular area immediately surrounding it (a sort of donut-shaped configuration). It is here, with the formation of centre-surround cells, that chromatic and luminance cells first emerge.

In a retina with exactly one kind of photoreceptor, a centre-surround cell will have—*can* have—only one configuration: it compares the absolute photon catch of the *centre* region of visual space with the total photon catch of the *surrounding* area. This is an *achromatic or luminance* cell by definition because it signals, for two distinct regions of space, the difference in photon catch (i.e. luminance contrast) *for one receptor/filter type*. One way to think of this arrangement—what a luminance cell does best—is that a luminance cell responds most strongly to any intensity changes that occur against a uniformly coloured background. In the terrestrial world of mammals, every visual scene contains numerous instances of this arrangement. It occurs whenever a shadow falls upon a uniformly coloured surface or whenever directional lighting produces shading. The spatial arrangement of a centre-surround cell insures that the background colour, whether it is green, blue or brown, makes no difference to the cell response. Because both the centre and the surround regions are 'filtered' by the same photopigment, both center and surround will react in the same way to the colour of background. The center and surround have the same 'colour' filters. So the colour of the stimulus is factored out and it is the intensity contrast that drives the cell response. A luminance cell is thus maximally sensitive to intensity contrast assuming a uniform background colour (and a colour within the spectral range of the luminance cell).

The addition of a new kind of photoreceptor to a retina—an event that is destined to happen regularly—makes for some interesting variations on this theme. The first option, the 'if it worked once, why not try it again?' option, is the creation of a new and distinct luminance system. These new centre-surround cells would signal the difference in photon catch, for the center and surround regions, by this new photoreceptor. This 'same old' option is actually more interesting than if first seems. If you look at the difference between the luminance images in Fig. 16.4c, e for the S and L cones, you can see why this new luminance cell constitutes a different 'take' on the world. Because the S and L cones are different spectral filters, they will almost always differ in their total photon catch—and so too will their deliverances about luminance contrast. Looking again at Fig. 16.4c, e, it is clear that the two types of luminance cells, centre-surround cells driven by either S or L receptors (but not both), would yield clearly different measures of luminance. A new photopigment equals a new luminance measure.

Note that the question "But which measure of luminance system is *correct*?" is not a coherent question. Indeed, one might say that it misses the whole point of multiple luminance systems. As we know from the Carnovsky exhibit, each type of luminance cell is maximally sensitive only within a certain narrow range of wavelengths. So a new luminance system can extend the range of luminance contrast processing out beyond the spectral boundaries of a single photopigment already in place. Think of it this way. Fine-grained luminance processing is possible only within the small window of response for each photoreceptor. This is why a luminance system, with one receptor, results in such drastic information loss, relative to the intensity information of the image. As each new kind of luminance cell is added, the contrast range of the system as a whole is extended. At the limit, the system as a whole approaches a detector for *intensity contrast*. This is why multiple luminance systems are found in all diurnal mammals, to extend the range of fine-grained contrast encoding.

The addition of a second receptor also makes possible a quite different and equally interesting option. A center-surround cell could compare the photon catch of two *different filters types*, one for the centre region and one for the surround. This is a *chromatic* cell by definition a cell that compares the total photon catch of two different spectral filters or receptor populations. This new arrangement yields a cell with surprising properties.

Once again, look at Gegenfurtner's S, M and L luminance images of common fruits and vegetables, here at the banana and the grapefruit (just under the banana and abutting it). Now imagine how the visual brain might go about the basic task of scene segmentation, of identifying the boundaries between objects, here between the grapefruit and the banana. In the L and M luminance images, the two fruits are a light grey; in the S image, the grapefruit is black and the banana is dark grey. To distinguish the banana from the grapefruit, any of the three images *could* be used: centre-surround luminance cells, fed by just one type of receptor for both the centre and surround would respond to the banana/grapefruit boundary. But if you could compare the banana *in the L image* with the grapefruit *in the S image*, the difference in their luminance values would be striking. Instead of comparing two shades of grey (as in the L and M images), the banana (L image) and grapefruit (S image) have a boundary defined by black on one side, and white on the other, a high contrast boundary. This is why an S—L chromatic cell would be so effective (if it existed—this is a fictive example at least for human vision) (Fig. 16.8). Passing the cell over the border, it would compare the total photon catch of the S cones with the total photon catch of the L cones at the border between the two fruits. And *that* comparison would yield a very strong, highly reliable contrast signal—a *chromatic* contrast signal yet one demonstrated for us, quite admirably, with black and white photographs.

Importantly, the encoding of chromatic contrast does not depend upon any prior categorization of the distal surfaces or of the image areas into *colours*. A chromatic cell merely compares two different *luminance* measures. Nor does a chromatic cell *detect* wavelength contrast per se. For one, there are other stimuli apart from colour contrast to which chromatic cells respond (see below). But even so, a chromatic cell does not provide an objective measure of wavelength contrast. A yellow banana and a pinkish-yellow grapefruit are very similar vis-à-vis their surface spectral reflectance, the percentage of each wavelength of light that the two fruits reflect. So the colour (or more neutrally, 'wavelength') differences between a banana and a pink grapefruit are not very large. The response of a (fictive) S-L cell to their colour differences in the image, however, signals *high* chromatic contrast. An M-L cell, an equally fine example of a chromatic cell, would also respond to this boundary. But an M-L cell would yield a much lower chromatic signal. So a better way of thinking

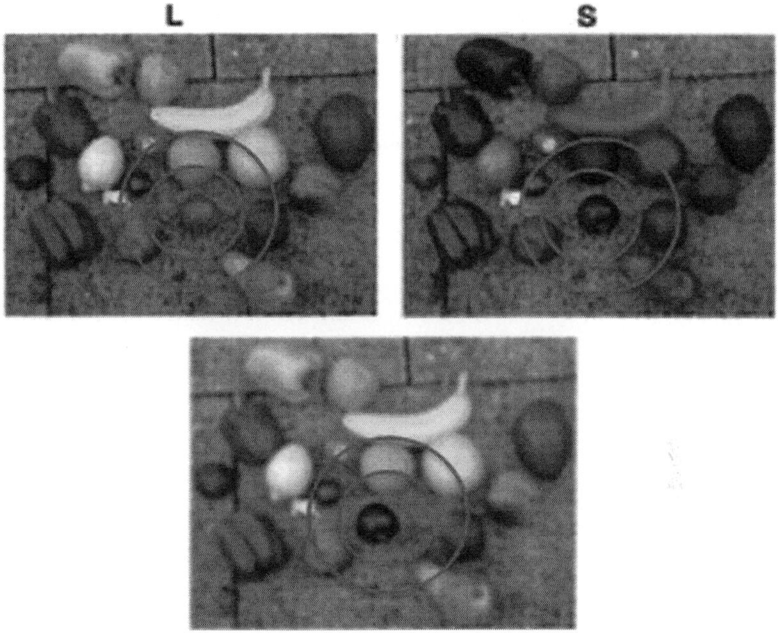

Fig. 16.8 A fictitious L-S cell. *Top*: These two show the visual fields of center-surround luminice cells that would result from a single cell, *L* or *S*. The figure *below* illustrates the nature of a chromatic cell, composed of the *L* surround and L-S center-surround cell that would result from combining the surround of the *L* cell and the center of the *S* cell

of a chromatic cell is as one that takes advantage of a fortuitous state of affairs, the difference in photon catch between two distinct spectral filters, one of which prefers short wavelength light, the other of which prefers long. At its best, a chromatic cell *highlights* a genuine wavelength contrast that exists in the world.

With the designation of ‘chromatic’ and ‘luminance’ cells in hand, however, it is easy to overlook the fact (given the human penchant for orderliness and simplicity) that both chromatic and luminance cells have complex informational properties. A chromatic cell is often called a ‘colour’ cell because it meets the formal neurophysiological definition of a ‘colour’ system, the capacity to respond to wavelength independently of intensity information. Given two contiguous coloured stripes, a red stripe and a green stripe of equal lightness (intensity), an L-M ganglion cell will react to their common border. Hence a chromatic cell fulfills the formal definition of colour vision. Shine a flashlight on only the centre region of an L-M cell, however, and the L-M cell with produce a sustained response to this difference in light intensity. With a broad-band light stimulus (the flashlight) targeted on the center region, the center L receptors receive more ‘red’ light than the surround M receptors receive ‘green’ light (because the surround region is in the dark!). In other words, a chromatic cell does not *detect* wavelength differences but it will respond, vigorously, to certain wavelength differences. The same holds of luminance cells, *mutatis mutandis*. Within the range of greatest wavelength sensitivity for a luminance cell, it will respond vigorously to a difference in light intensity alone. Yet if one looks at the response curve for any photoreceptor, it is clear that particular, large differences in wavelength alone between two stimuli would make a difference in photon catch for receptors of the same type. So a difference in photon catch between a centre and surround, as registered by a luminance cell, *could* be the result of a wavelength difference alone. In sum, both chromatic and luminance cells respond to wavelength differences *of a certain kind* and intensity differences *of a certain kind*. Given their physiology, chromatic cells are more sensitive to wavelength contrast (along a particular colour axis) while luminance cells respond more actively to differences in pure intensity (within a certain spectral range). But neither a chromatic cell nor a luminance cell is specialized for the "if and only if" task of wavelength or intensity detection.

In sum, the essential difference between luminance and chromatic cells comes down to the filters involved in contrast processing—whether the comparison involves one filter or two. This way of describing the distinction represents it as primarily a distinction in 'wiring' or anatomy. In effect, I asked this question: If contrast processing is a hard constraint on the evolution of vertebrate vision, how many ways are there to wire-up a retina with more than one photoreceptor? And no matter how many more receptors are added, whether a species has two photoreceptors or ten, the answer is this: *Just two*. There are only two ways to make a luminance comparison: between filters of the *same* kind and filters of *different* kinds, luminance and chromatic cells respectively. Thus the distinction is one of 'chance and wiring', a division that occurs inevitably given the genetics of photoreception and rules of combinatorics. It is no mystery, then, that the world contains both chromatic and luminance cells, that both forms of cells exist. Rather, the interesting questions concern the widespread integration of chromatic cells into the eyes all diurnal creatures and why the anatomy of biological vision respects this distinction throughout the visual system. If the species is diurnal and the environment contains light across a broad range of stimuli, it is overwhelmingly likely that the species will have both chromatic and luminance visual cells which comprise anatomically separate but often physiologically interactive systems. This universal phenomenon suggests that the luminance/chromatic divide has general informational consequences. But what would those be? What is it about chromatic and luminance systems that make them such good partners?

To answer this question, let us look at edge processing. Take a standard colour photograph. We can represent the separate contributions of wavelength and intensity to a colour photograph (or retinal image) with two separate illustrations, an isochromatic image and an isoluminant image (meaning, literally "all the same colour" and "all the same intensity"). Because light is a transverse wave and, by definition, every wave has amplitude and wavelength, every retinal image can be divided into these two components. So there are potentially two sources of information in a retinal image and at least on visual inspection they appear quite distinct (Fig. 16.9). For example, in the isochromatic (or monochrome) image (Fig. 16.9b), object edges are

Fig. 16.9 These photographs illustrate the separation of a single image (full colour, *top middle*) into two contributing components of light to the image, the Luminance Image and the Spectral Image

demarcated by a difference in contrast with their backgrounds. As nature would have it, objects almost always differ from their backgrounds in intensity. In addition, the isochromatic photo also shows both shadows and object shading. In comparison, in the isoluminant image one sees only ‘colour without intensity’. Now although we often think of shadows as ‘coloured’ in the natural world, most shadows create only a difference in light intensity. Shadows on a green lawn create merely darker areas of (that same green) lawn. (Blue shadows on snow are the exception that proves the rule.) In the isoluminant image, the shadows and object shadings are no longer present (Fig. 16.9c). What we see, in the isoluminant image, are object boundaries and expanses of surface colouring. When one looks at these two types of images, side-by-side, you can see that the object boundaries (and surface colour boundaries) are visible in both illustrations but the shadows are visible in only the monochrome image. Thus in a full colour image (Fig. 16.9a)—in the retinal image—objects (and surface markings) are demarcated by a combined edge of intensity and wavelength contrast, while shadows and shading are demarcated by intensity contrast alone. *One way to differentiate object boundaries from shadow edges, then, would be to distinguish between the intensity and wavelength dimensions of the stimulus in the visual image*. Or at least, that would be a good way in principle. In practice, the tools at hand are chromatic and luminance cells not wavelength and intensity detectors. So the question concerns which features of the world luminance and chromatic cells

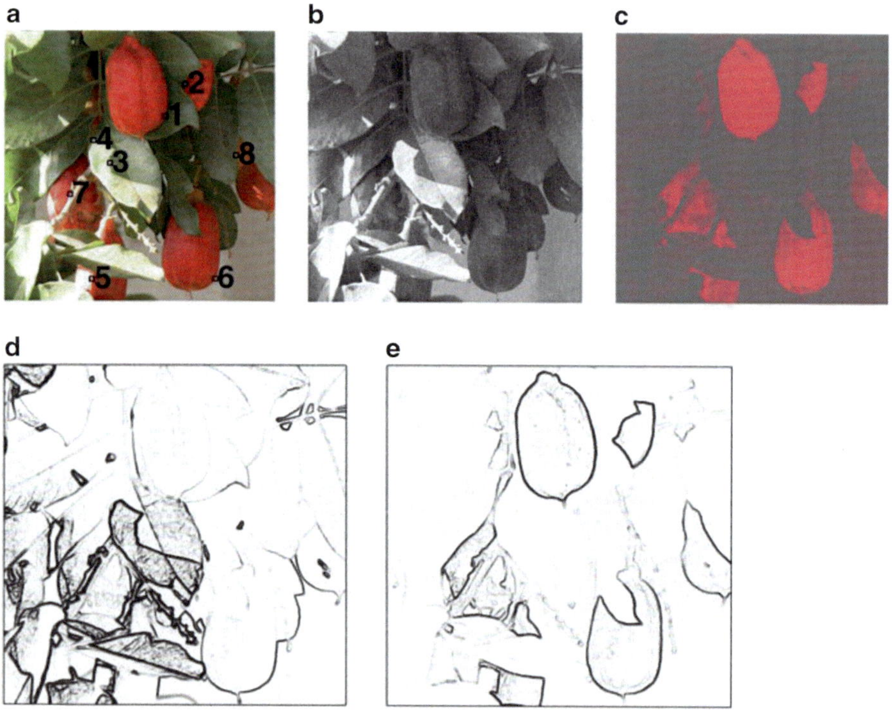

Fig. 16.10 (**a**) Original colour or 'input' image. (**b**) Luminance Image or photon catch of the *S*, *M*, and *L cones* (**c**) *Red-Green image* of the response of the L-M channel. (**c**) Edges derived from the Luminance Image in (**a**). (**d**) Edges derived from the L-M channel in (**c**) (From Hansen and Gegenfurtner 2009)

encode: what would a chromatic system paired with a luminance system make of the objective facts of object and shadow boundaries?

In an experiment by Hansen and Gegenfurtner (2009) they examined 700 images of the natural and artificial world to find out which edges, defined by either chromatic or luminance contrast, human vision can see. The image in Fig. 16.10a depicts the original image or what they called the 'Input Image', a full colour image; the second image, 16.10b, shows the 'Luminance Image' or the total photon catch by the system, calculated by adding together the absorption of the S, M and L cones; Fig. 16.10c shows the M-L Image, namely the contrast between the responses of the L and M cones at each point in the image (using a continuum between red and grey to illustrate the relative responses). Finally 10(e) and 10(f) show the edges that can be computed from these two types of contrast data. Here you can see that the edges determined using luminance contrast were often distinct from—and different strengths than—the chromatic edges.

As the reader may already suspect, a test of a red-green (M-L) chromatic cell to determine what edges it can detect, is hardly likely to fail given an image of *red* fruit against *green* leaves. When the joint edge histograms for all 700 images were computed, however, the result was a general one. *Chromatic edges and luminance edges are statistically independent of one another in natural visual images.* So the addition of an M-L chromatic system to an M + L + S luminance system represents a huge leap in first-order information about edges: chromatic cells encode *the location and strength of edges in the image that luminance cells do not.* Equally importantly, a visual system with both a chromatic and luminance component gains *second-order information* about the *relative* locations and strengths of luminance and chromatic edges. With this comparison in hand, it is possible to differentiate objects from shadows using two simple rules: To find an object edge, look for a discontinuity that triggers both chromatic and luminance cells or for which the chromatic response predominates; to find a shadow, look for contrasts to which *only* luminance cells (but not chromatic cells) respond. These are two rules that will work *most of the time* given the judicious selection, by evolution, of the different chromatic and luminance systems, and given a scene in line with the statistics of images of that species natural environment.

More generally, the addition of chromatic systems to luminance systems has proven useful because they are *complementary systems*. Chromatic systems respond vigorously to spectral discontinuities to which luminance systems have little response; luminance systems respond most vigorously to discontinuities in intensity in which chromatic cells are uninterested. In mammalian vision, this fact has been co-opted in the service of object vision. The anatomical segregation of the chromatic and luminance systems allows mammalian vision to utilize these two independent sources of contrast information *as* independent. Of course, scene segmentation through edge processing is one of the earliest and most essential capacities of any visual system. *But once luminance and chromatic edges are determined, they can be used, either together or independently, for virtually any visual task.*[4]

[4]That said, some visual tasks *are* easier to solve using one source or the other—chromatic or luminance—information. For example, if the task is to track linear motion, a population of luminance cells (each cell with a 'zippy' onset and transient signal) will do this both more quickly and accurately than a population of chromatic cells (which have a sluggish onset and sustained response). But in the information game, *something*—some information—is almost always better than *nothing*. On a foggy, rainy day, you are outside your tent about to fry bacon in a pan (a typical day in Vancouver for any philosopher). A vaguely Grizzly-shaped object wanders into the middle distance. Under conditions of visual haze and scattering, luminance signals are often weak, noisy or often absent while chromatic signals are not as readily affected.. But it doesn't matter that the more robust chromatic signal is a sub-optimal source of motion information. That the Grizzly bear has started to move (motion onset), that it is moving towards the viewer from a leftwards direction (direction of motion) and that it is rapidly increasing in velocity (2nd order velocity information) are bear-inhering-properties that it would be beneficial to perceive. The task is to use whatever information is ready-to-hand in the most optimal way, not to use the ideal source of information. This is where a chromatic system, paired with a luminance system, can provide clarifying information.

Exactly how luminance and chromatic systems work together (and apart) is something that recent experimentation is beginning to explore (Gegenfurtner and Kiper 1992; Cropper et al. 1996; Baker et al. 1998; Mullen et al. 2000; Mullen and Beaudot 2002; Gegenfurtner 2003; Kingdom 2003, 2005; Kingdom and Kasrai 2006; Kingdom et al. 2006; Gheorghiu and Kingdom 2007; Michna et al. 2007; Garcia-Suarez and Mullen 2010). Unfortunately, given the complex informational properties of luminance and chromatic cells/populations, predictions from first principles have limited utility without supporting models. Our best bet is to expect the unexpected. That is, we often assume, even when we know better, that the *point* of a chromatic or a luminance cell is to encode wavelength or intensity. We treat the informational complexity of chromatic and luminance cells as if this were a deficit or hindrance to be surmounted as soon as possible. ("Thank you God, for now the laws of optics are finally within reach!") But the real story—the 'there-is-no-simple-story' story—is that the chromatic/luminance distinction was never 'meant' to track the dimensions of intensity and wavelength. It is a distinction of chance and wiring, as I said above, and one with which evolution has been grappling since the advent of the first center-surround cell. We can be confident, I think, that primate vision stumbled upon any number of interesting ways to use the informational complexity of chromatic and luminance cells to visual advantage.

16.1.5 Albedo Perception: Perceiving Surfaces as Light or Dark

One of the cortical functions for which luminance information is almost certainly used is for seeing opaque object surfaces as light or dark. We see coal or briquettes as dark, copier paper and snow as light, and natural concrete as somewhere in between. As trichromats, of course, we also see these surfaces as having *colours*, what are known as the *achromatic colours*: briquettes are black, paper and snow are white, and untreated concrete is a medium grey. Given that one dimension of the trichromatic colours, both chromatic and achromatic, is darkness/lightness, there seems to be a relation between trichromatic *colour* perception and trichromatic *albedo* perception. Exactly what this relation might be, if any, is the subject of debate within the vision sciences. But the issue for the achromat, who does not see the colours, is much clearer. If the achromat sees surfaces as being light or dark, this must the result of albedo perception proper.

The relevance of albedo perception to the current topic might well seem opaque to the reader: what does the perception of surface lightness by the achromat have to do with the question of whether he or she sees the world 'in black and white'? This very fact, that the relevance of albedo perception to an achromat's visual phenomenology is not clear, is symptomatic of a deeper problem, a common misunderstanding about what is involved in seeing surfaces as light or dark. Let me take a moment, then, to discuss albedo perception before we wind our way back to the main point.

Jonathan Cohen, in a paper on colour properties (Cohen 2004), provides a good example of how we commonly think about albedo perception. In this paper Cohen's main concern is to argue for a certain view about colour properties, the relationalist view. The topic of albedo perception arises only in passing in the context of his 'Master Argument' about colour constancy, how we decide the colours of objects in a complex scene containing directional illumination. Cohen's example is a now famous photograph of a red coffee mug on a table, illuminated from the side by sunlight. Pointing first to the dark red handle of the mug (in shadow) and then to the bright cherry red body of the mug (lit by sunlight), Cohen asks how we choose the correct colour: Is the mug dark red or cherry red? Now, as Cohen notes, although the original photograph is in colour, readers with only a greyscale reproduction can ask a parallel question. If we examine the greyscale version, we will see a mug with a dark grey handle (in shadow) and a glossy light grey barrel (in sunlight), two distinct achromatic colours. Yet we still see the cup as having a uniform surface colour despite the variations of shadow and sunlight. We can thus ask a parallel question about *achromatic* colour: Which grey is the mug's correct colour? Cohen poses the question of colour constancy, then, as a question about a shade of perceptual grey—almost black or light grey?

In computational vision, the problem of albedo is described somewhat differently. Below is a clear (and standard) definition (Anderson and Winawer 2005).

> The amount of light projected to the eyes (luminance) is determined by a number factors: the illumination that strikes visible surfaces, the proportion of light reflected from the surface and the amount of light absorbed, reflected and deflected by the prevailing atmospheric conditions (such as haze or other partially transparent media). Only one of these factors, the proportion of light reflected (lightness) is associated with an intrinsic property of surfaces and hence is of special interest to the visual system. To accurately recover lightness, the visual system must somehow disentangle the contributions of surface reflectance from the illumination and atmospheric conditions in which it is embedded. (pp. 79–80)

An illuminant (light) shines upon a three dimensional object. The intensity of light that falls upon each surface—the object's *illuminance*—is a function of the brightness of the light source and the particular shape of the object. Each object, in virtue of its surface qualities, absorbs and reflects a certain percentage of the light cast by the light source, a property known as *surface reflectance*. Thus, the total light reflected from the object, its *radiance* is a function of both the intensity of light that shines on each point of the object's surface and the percentage of that light which is absorbed. This light then travels to the retina and en route meets with certain media. Perhaps it is dispersed by smoke or haze in the air, transmitted through the coloured sunglasses or through the ordinary transparent lenses of correctional glasses. Even in the "normal" case, however, the light must travel through the atmosphere between the eye and object, then through the cornea and the lens plus the aqueous and vitreous humours of the eye. All of these media are filters, collectively known as *atmosphere*, that absorb, reflect and refract the light of the luminance image before it reaches the retina. Thus, conceptually, the problem of lightness perception for human vision concerns the disambiguation of the contributions of albedo from those other physical factors (illuminance and atmosphere) that result in the proximal stimulus, the retinal image (or luminance image in the parlance of

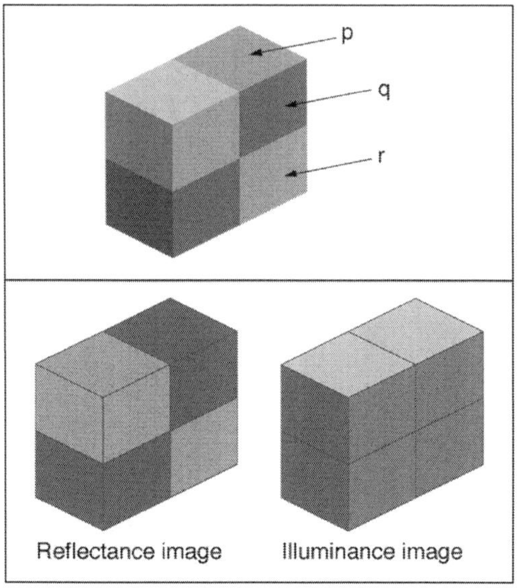

Fig. 16.11 On the standard interpretation, the problem of albedo is the problem of disambiguating the Surface Reflectance Image (*lower left*) from the Illuminance Image (*lower right*), given the Luminance Image (*upper cube*) (From Adelson 2000)

computer science). In albedo research, however, researchers often choose to ignore atmosphere and concentrate on how to disambiguate the contribution of surface darkness from those of the light source. The diagrams in Fig. 16.11 show this simplified problem in schematic form (Adelson 2000). Here, the three components of the computational problem are referred to as 'images' or layers, and the task is to disambiguate the three layers. In order to avoid terminological confusion, let us call what the schematic refers to as the 'luminance image' the '*retinal image*'.

Unfortunately, there is no easy solution to the problem of albedo perception even if one sets aside the vexed effects of atmosphere. Let L(x, y) be the luminance (retinal intensity) image, R(x, y) be the reflectance image, and E(x, y) be the illuminance image. Then:

$$L(x, y) = R(x, y)\, E(x, y)$$

Given only the luminance image alone, the problem is ill-posed or undecidable in the computational sense. For any value of L(x, y), no unique value of R(x, y) can be computed without knowledge of the value of E(x, y). Without E(x, y), an indefinite number of images are possible. Exactly how the human visual system overcomes this problem is the subject of much debate. But for our purposes, the facts are not as important as the principles involved. What matters here is the general form of the various methodologies that have been proposed. Here, then, are two examples, followed by a summary of their shared assumptions.

One prominent solution to the problem of albedo, proposed by Adelson (1993) holds that the visual system uses various "tricks" and short-cuts based upon the properties of the retinal image. Adelson's computations for lightness make use

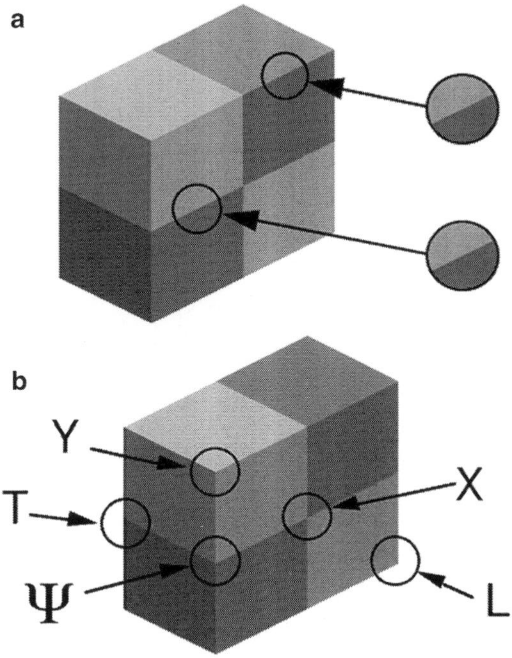

Fig. 16.12 (**a**) In this figure, an identical contrast is caused by a difference in illuminance (*above*) and (*below*) by a difference in albedo or surface reflectance. (**b**) Illustrates a number of different junction types that are typical of changes in surface reflectance and illumination (From Adelson 2000)

of the geometry of the image, how the light and dark areas of the image meet—what he calls "junctions". Both the geometric form of the junction plus the relative luminance of each bounded area at a junction provides important clues about the causes of those edges. For example, Fig. 16.12a demonstrates how two identical edges could have two distinct causes in the world: the upper edge is caused by illumination differences while the lower edge is the result of differences in surface darkness. Figure 16.12b illustrates the various types of junctions used as clues in Adelson's model—X, Y, L, T and junctions—and Fig. 16.13 shows us how the junctions we can affect our interpretation of the scene. In Fig. 16.13, the dotted rectangle, viewed alone, appears to be composed of stripes. But when the rectangle is viewed with one end or the other obscured (i.e. when we see different the contextual cues) the stripes within the rectangle are interpreted in two different ways. If you cover the right side of the illustration, the stripes will appear to be painted or the result of surface reflectance differences; if you cover the left side of the illustration, the dark stripes are appear to be shadows on the risers of steps. Adelson posits that we see the two sides as different because to the left of the rectangle, the junctions are arranged vertically, with their spines connected, while to the right, the dark stripes form horizontal junctions. This arrangement of junctions combined with the arrangement of light and dark areas defined by their edges, determines our interpretations of the stripes. Adelson's theory is considerably more complicated than this, but the above gives the reader the basic flavour of his view.

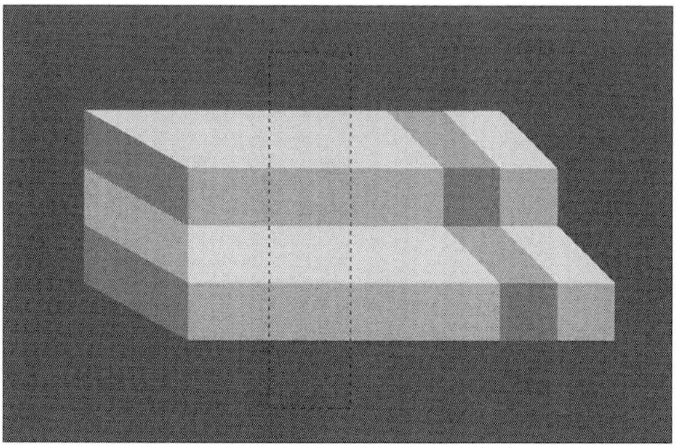

Fig. 16.13 This illustration show how two quite different configurations of the world, on the one hand, a set of stairs with the same surface reflectance, and on the other hand, a solid figure with differences in surface reflectance, nonetheless can cause an identical visual image (in *dotted rectangle*). Context is needed to resolve the ambiguity (From Adelson 2000)

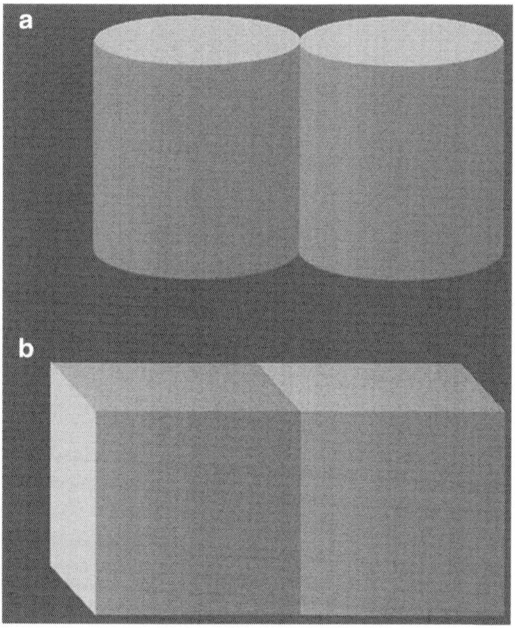

Fig. 16.14 Although (**a**) depicts *two cylinders* and (**b**) shows two adjacent *cubes* of different albedo (surface reflectance) both (**a**) and (**b**) are identical (From Knill and Kersten 1991)

A second popular theory of albedo perception focuses upon properties of the scene that are more directly tied to intensity contrast in the retinal image. For example, Knill and Kersten (1991) have demonstrated that shape information is critical to distinguishing whether an image area with a pattern of diminishing intensity is the result of illumination (on a curved surface) or surface reflectance (of a flat surface) (Fig. 16.14). More recently, Anderson and Winawer (2005, 2008), leading proponents of this second view, demonstrated that the visual system appears to use both local cues (intensity contrast reversal across borders) and more global cues (occlusion and shape information) to separate a retinal image into "layers", what are essentially depth planes corresponding to the plane of objects' surfaces,

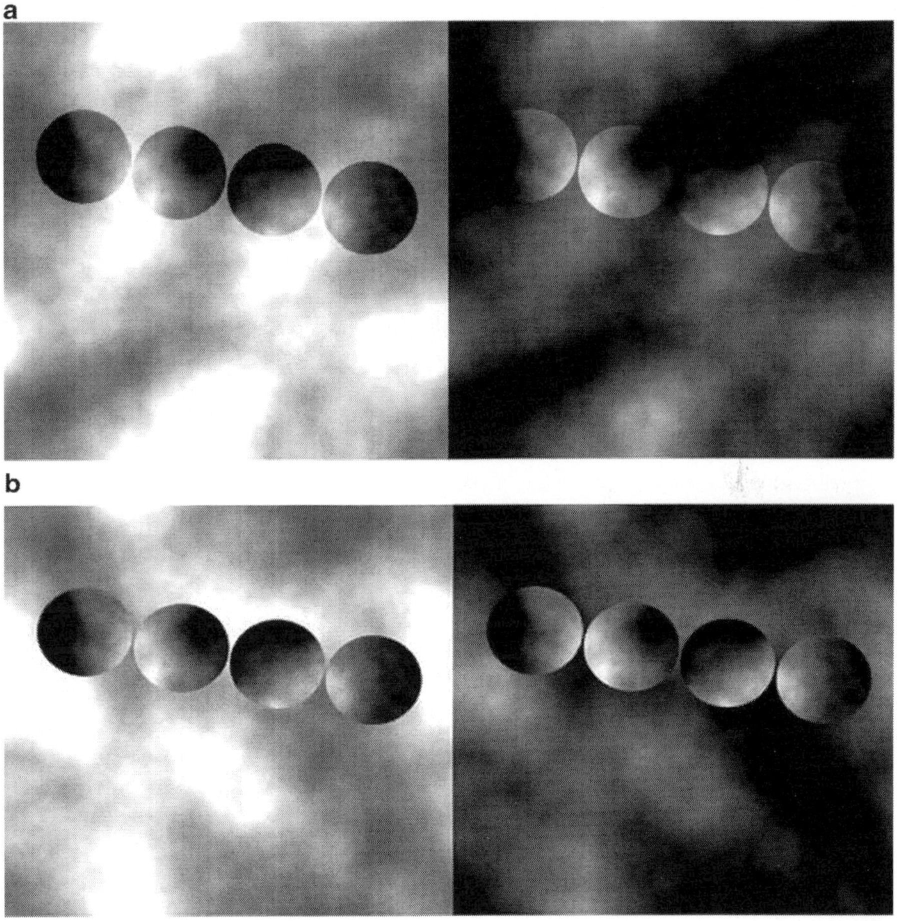

Fig. 16.15 In (**a**) identical disks on the *left* and *right* appear as *dark* disks obscured by dark atmosphere, while on the *left*, the disks appear as *light* disks occluded by dark clouds. *Below* in (**b**) the disks have been rotated, thereby destroying the contrast relations with the background. Here, the disks look identical but the illusions of atmosphere are destroyed (From Andersen and Winawer 2005)

the background and the foreground. In Fig. 16.15, the disks on the left and right are physically identical. However we interpret the two figures very differently: The disks on the left appear to be dark discs partly covered by a transparent white haze or fog; on the right, the discs look like white moons, obscured by dark clouds. When the disks are turned 90°, however, much of the illusion is lost. Anderson argues that the rotation destroys the pattern of contrast reversal at the edges of the disks that is necessary for dividing the image into depth planes. Without depth information, there is no clear disambiguation of surface reflectance (light or dark surface) and the atmospheric factors (fog or dark clouds). We also loose the strong illusion that the two sets of identical disks are entirely different. Insofar as the illusion remains, this is attributable to the overall illumination of the two backgrounds, one of which, as a whole, is statistically darker than the disks, the other of which is lighter. On Anderson's view, then, intensity contrast across a figure-ground border is essential to our perception of lightness and darkness as are the global scene characteristics. The sort of local and global 'clues' that Anderson suggests are of a different kind than Adelson's image junctions.

The reason for the above comparison was to give a sense of how the problem is conceived in computer science and what would constitute a solution relative to that conception. Clearly, there is significant disagreement among computer scientists about how our visual system deals with what is a theoretically intractable problem: each posits different sets of cues and, as a result, each theory would predict different occasional failures and illusions of albedo perception (Adelson 2000; Corney and Lotto 2007; Poirier 2009; Anderson and Khang 2010; Spillmann et al. 2010). On the other hand, there is also significant agreement about the nature of the problem and hence about what a system for albedo perception must do. All researchers agree that:

1. Luminance, illuminance and surface reflectance are physical properties.
2. Surface reflectance (lightness) is a constant property of object surfaces and hence is useful for identifying objects, distinguishing one object from another, and tracking them.
3. In systems of natural vision, albedo perception is a computational process that uses intensity contrast data from the retinal image to determine surface reflectance values. (Intensity contrast could be used to see different kinds of junctions or it could be used to see more large-scale patterns in luminance contrast itself.) The result is the perception of an object surface *as* having a specific lightness or darkness, an intentional property of the object.
4. Albedo perception for surfaces within complex natural scenes depends upon numerous cues within the image. As the number of natural cues diminishes—that is, as the scene/luminance image becomes less complex—mistakes in lightness perception, *lightness illusions*, occur more frequently.
5. Given that there is no decidable procedure for solving the problem of albedo perception, human lightness perception is not 100 % reliable. Yet through the use of multiple cues, human lightness perception approaches a reasonable standard of accuracy.

Turning back to the central problem of this paper, the nature of achromatic experience, let me make explicit why this conception of the problem of albedo is at odds with what I'll call, for the lack of a better name, Cohen's Conception. Going back to Cohen's example of the coffee mug in shadow, observers agree that the coffee mug is a uniform achromatic colour despite the numerous greys caused by the directional illumination. Cohen then asks which shade of grey is the true (achromatic) colour, a question that seems to make sense. But if one goes back to Adelson's three figures that illustrate the problem, Cohen's question appears suspect. The first illustration, what we are calling the Retinal Image, is a representation of the intensity values of light at each point in the image reflected from the distal scene. In the second illustration, the Illuminance image represents the effects of a directional light source on a block. Finally, the Reflectance Image represents the surface reflectance of that same block, the constant proportion of light reflected from the four cubes of the block, the result of its inherent surface properties. Thus in each representation of the block, the greyscale is used to 'stand in for' a different property: image intensity, illuminance of the object, and surface reflectance. Suppose then, that using these illustrations, we point to two different locations in the Retinal Image, say the endpoints of the arrows p and q and we ask 'which shade of grey, medium grey or dark grey, is the small cube *really*?' In the Retinal Image, those two different greys, indicated by the arrows, represent different values of *light intensity*; in the Reflectance Image, the single shade of grey of the uppermost right cube represents a dispositional property of the cube's surface, *surface reflectance*. Neither shade of grey 'is' the correct albedo of the small cube: the two greys do not represent albedo. The question is incoherent.

Cohen's original question about the correct shade of grey initially 'plays' much better because of the dual nature of photographs. The photograph of the cup is an object which itself has albedo, the many variegated greys that make up its surface; the photograph *qua* representation of a scene also portrays a coffee cup, a cup which has a certain surface lightness or darkness. When we are asked whether the cup is 'this grey' or 'that grey', we are looking at two areas of the photograph with different albedo, one near black and the other light grey. Thus it seems to make sense to ask *of the cup* which of the two greys 'matches' the cup's surface. But this is a false dichotomy. There is no reason why the surface reflectance of the *photograph*, at any point, must be the same as the surface reflectance of the *mug* it represents. If the coffee mug had been photographed entirely in dark shadow, its photographic image would have been a montage of very dark greys. Yet we would still *see* the cup as medium grey.

A few more general conclusions can now be drawn. On Cohen's view, the problem of albedo perception is primarily a question of phenomenology: Is the mug *this* grey or *that* grey? As such, it is problem in the first person: the observer must select the correct phenomenal grey. This question seems to make sense as we view the photograph before us with its stable properties of surface albedo. The problem of albedo, as construed by the vision sciences, is the computational question of how a visual system disambiguates two properties of the visual scene conflated within the retinal image, surface reflectance and the effects of the light source upon that surface. Thus construed, the albedo problem is a question of sub-personal processing, of how to account for what we as observers see, namely surfaces as light or dark. It is the essential question of how we gain intentional perceptual representations of the distal world, of how we see the multiple properties of the world from the first person point of view. On this view, the question 'how do we see the lump of coal as dark?' is the same type of question as 'how do we see the chair *as* being behind the table?' (a question about depth perception) or 'how do we recognize a certain grapheme *as* being an 'F'? (a question about object recognition). For the computationalist, then, our very ability to ask the question posed by Cohen—'which is the correct grey?'—is a question that can be asked only *after* all the hard work of viewing a natural scene is done. We gaze at the photograph of the mug on the kitchen table and interpret it as we would the Retinal Image of a natural scene: we see a mug with a uniform surface lightness with its handle in shadow, its body illuminated by bright directional light. We can then examine the photograph and determine *its* albedo qua surface of a paper photograph, there discounting any effects of the illuminant in the actual world, the one in which you, the observer, examines the photograph (does your shoulder cast a shadow over the image in front of you?). With these two processes of albedo perception behind us, we can then ask Cohen's question about achromatic colour—'which is the correct grey?' Looking at the photograph before you, you can ask this first person question only after two problems of albedo perception—for both the mug and its medium of representation—have already been finessed.[5]

This way of thinking about albedo perception may seem a bit odd until one remembers that light itself is a perceived property of any distal scene. We see the dark shadows in the forest (this is what frightens us), white 'glints' of sunlight off water (why we reach for sunglasses), and the fiery sunset reflected in a wall of windows (why we reach for our cameras). Given a retinal image, we could not see the distal scene unless, sub-personally, the visual system was able to distinguish properties of illumination from properties of objects (Arend and Spehar 1993; Kingdom 2008, 2011). This is why the problem of albedo could not be a question of phenomenology *per se*, of comparing, from the first person, the different grays in an image. One of the first and most important tasks of mammalian vision is the division of the scene into illumination and object properties, a task that must occur *prior to* our intentional perceptions of objects and illuminant properties. For example, to see a cube as a cube in a scene with directional lighting, there must be a way to reconcile the proposed object shape, being cube-shaped, and the pattern of illumination. Such a reconciliation would not involve a *precise calculation* of a measure of *absolute albedo* for each visible surface. But it would involve *rough and ready* assumptions about *relative* albedo, e.g. that all of the visible sides have roughly the same surface reflectance, or that the front-facing facet must be lighter than the top surface in order for the object to be cube-shaped. In other words, when, from the first person, we peer at the scene (or image) before us, it is part

---

[5]I haven't discussed how the brain figures out a sensitive measure of absolute surface reflectance as opposed to an assessment of a relative lightness and darkness within a given scene—e.g. the coal is much darker than the table on which it sits. I am myself somewhat skeptical about the utility of such a representational process, whether the brain bothers to asses lightness or darkness in the relevantly fine-grained way that would be necessary to assign each surface a place on an absolute scale from dark to light. Nonetheless I believe that we think of lightness and darkness as an objective property, a continuum along which each surface has an objective place, whether or not our on-line assessment of albedo is, in practice, merely 'good enough' for whatever task is at hand. Whatever the answer to this question, however, albedo perception does not come down to choosing the right grey 'chip'. It's not a question of phenomenology.

of the content of our perception that objects have a certain shape consistent with the conditions of directional lighting, perceptions that presuppose at least some assumptions about relative surface lightness (e.g. that this surface is darker or lighter than that one).

Finally, we can return to the achromat. If we think of the problem of albedo perception following Cohen, as a choice between two phenomenal greys, then an achromat, who sees coal as dark, also chooses among some set of phenomenal greys. Thus there is no real question about an achromat's visual experience as a result of albedo perception: it *is* like looking at a monochrome photograph. *It is the same, by definition.* But if we treat the problem of albedo perception as a computational conundrum, then there is a substantive question to be asked about the achromat and his or her experience: Is the achromat capable of seeing surface reflectance given the luminance information of rod achromatic vision? Note that the computer scientist starts with a bird in hand, an intensity image of the distal scene. Any biological system for object vision starts with a luminance image that, as we have seen, can be of many types. It is this fact that puts the achromat at a true disadvantage for albedo processing—a rod-only luminance system. Without wavelength information, the achromat cannot compute the intensity values of the retinal image and with only one photoreceptor the achromat cannot even approximate image intensity (by combining multiple receptor outputs). So the achromat has no chance to regain this intensity information at any stage of vision.

It is worth pausing to consider the extent to which this puts the achromat at a disadvantage for lightness perception. If one were to give the achromat seven or eight paint chips of different colours side by side, each of which was equally bright, the achromat could not see that this was so. Each would appear to have a different lightness value. Presented with a series of coloured paint chips of varying brightness, an achromat could not order them from light to dark. With only a single receptor, the rods, with a spectral sensitivity centered in the 'green' range of light, the reflections of red and blue objects will be very dark, the reflections of green objects, very light. An achromat might be told that his favorite tie is composed of bright magenta and turquoise stripes but the achromat will see neither the pink nor the blue stripes as bright. His distinctly cheerful tie will appear as somber attire to the achromat, a tie with an overall pattern of low contrast dark stripes—a good tie to wear to, say, a job interview or, better, a funeral.

So the rod achromat suffers a large deficit in lightness perception. But however bad at the task the rod achromat may be, he or she nonetheless perceives the surfaces of opaque objects as having a *constant* property. Suppose you were asked to view the world using a virtual reality mask. The camera, from which your mask receives images, uses a narrow-band green filter and it transmits this information in the form of greyscale images. The camera is pointed at a real table covered in blocks of many colours and your task to arrange them in various ways. With no knowledge of the actual colours of the objects, you would see each cube as being light or dark. If told to build a tower, you would be able to stack and move the cubes using this constant surface property even though, were you to think of the blocks as light or dark, you would be almost certainly wrong in your perceptual judgments. This is the position of the achromat. He sees surfaces *as* having a constant property, ‘pseudo albedo’, which is distinguished from the effects of any particular illuminant on the scene; he knows that this property is not what others would call ‘lightness’ or ‘darkness’, and; he knows that his judgments of surface lightness will usually be wrong. There is a very robust sense, then, in which the achromat’s perception of albedo is relevant to achromatic phenomenology and in which it widens the gap between the achromat’s and the trichromat’s visual phenomenology.

16.2 Conclusions

This paper began with the claim that the common division between black and white and colour leads us astray when we think about the nature of human visual phenomenology and the neural processes that support it. The running example has been that of the achromat whom Nordby claims sees in black and white. My reconstruction of that argument was based on the commonalities between three types of experience: the prototypical case of ‘seeing in black and white’ in which the trichromat looks at a black and white photograph; the trichromat’s visual experience at night when only the rod luminance system is active, and finally; the visual experience of the rod achromat who has only a system for night vision, the rods and the magnocellular system. By transitivity, if the trichromat sees in black and white, so too does the rod achromat. I expect that most readers will now see some of the flaws in this argument but let me walk through the three different cases in order to make explicit their differences and commonalities. The question at issue is whether there is a sufficient overlap, in physiology, to justify the conclusion that the achromat sees ‘in black and white’.

Let’s start first with the trichromat who views a black and white photograph. When a trichromat looks at, say, Arthur Sasse’s famous portrait of Einstein (with protruded tongue), she looks at a physical object that reflects light continuously across the spectrum of visible light. This object also meets the following condition: *either* it reflects each wavelength of light equally, relative to a set intensity of light, *or* the photograph reflects light such that, for any given point in the image, the photon catch of the three cones is the same. This strange disjunction exists because a light wave has both wavelength and amplitude. So an image that conveys intensity information necessarily reflects light of some wavelength or other. The convention of black and white photography attempts to render the image such that the light it reflects has no discernible predominant wavelength—discernible by the trichromat. A black and white illustration can be printed with black pigment/ink or with the three standard colours for printing, cyan, magenta and yellow. If the three inks are each combined in the correct intensities, the three colour ‘filters’, the S, M and L cones, absorb photons in equal measure at each point in the photograph.

The portrait of Einstein, viewed by a trichromat, produces a reaction in all three cones and thus produces signals in visual channels of all types, luminance and chromatic. In the luminance channels, each luminance contrast cell will respond to *only* intensity differences. There are no wavelength differences in the photograph, and thus any differences in photon catch between the center and surround area of a luminance cell indicates intensity differences. Now, in the trichromat, there will be many different luminance channels, each of which uses a distinct spectral filter. In the normal case, when the trichromat views the world, each type of luminance cell would produce a different measure of luminance contrast (for luminance is relative to a filter type). Here, though, light reflected from Einstein's photograph (as opposed to Einstein) produces the same absolute photon catch in each spectral filter. So all luminance contrast cells will produce the same signal, an *intensity* signal, relative to any point in the visual image. That is the luminance side of the equation. In the two chromatic channels of the trichromat, the chromatic contrast cells, with input from two different spectral filters, will *not* 'highlight' wavelength contrast: There is *no* wavelength contrast in the portrait. In effect, a black and white photograph does not silence chromatic processing. Rather it neuters it. The signals of chromatic cells, of whatever type, also indicate stimulus intensity because each type of cone responds with the same photon catch *by definition* (of a black and white photograph). The upshot is that all contrast cells, luminance and chromatic alike, encode intensity contrast across a spatial border—an unsurprising result given that this is the function of modern black and white photographs. It is the recognition of this fact, at some higher level of visual processing, which *may* give rise to the perception of the photograph *as* black and white.

When the trichromat sees the photograph of Einstein *as* a black and white photograph, this would normally involve two different types of intentional perception, of the photograph *as* having only *achromatic colours* (the colours from black to white) and *as* having areas that are *light and dark*. If you go to the paint store and choose 'Ripe Aubergine' as the new colour for your dining room, you are well aware that the colour is a dark one. That is what will make it a cozier and more intimate space. But you do not think that 'Ripe Aubergine' is an *achromatic* colour, a shade of grey. At least in some circumstances, the perception of surface chromatic *colour* and the perception of *albedo* come apart and so too must the processes that produce these perceptions.

Suppose then the trichromat sees the portrait of Einstein as having light and dark areas. Even though a black and white photograph produces, in the retina of the trichromat, a unique 'signature' of retinal encodings (i.e. the 'neutering' of the chromatic systems, etc.), albedo perception nonetheless requires higher-level processing. (This is why I said, in the concluding sentence of the paragraph before the last, that such retinal signals *may* result in albedo perception.) When we look at the portrait of Einstein, the chromatic and luminance responses of the trichromatic retina indicate that there is neither a predominant wavelength nor any wavelength contrast in the retinal image. But albedo perception is the perception of the surface lightness of distal objects, not of the retinal image itself. This is the difficult part to take on board: Albedo perception, even for a black and white photograph, requires a complex computational process. The very same considerations that apply to seeing a 3D block figure, like the one in Fig. 16.11, apply to Einstein's portrait. For example, if the photograph faces towards the light source, it will be brightly illuminated; if the light source is directly behind it, the portrait will be darkened. In addition, nearby objects, such as your own body, can shadow the portrait in any number of ways. Any of these conditions will change the retinal image, reduce or increase its brightness overall, or have a selective effect on the intensity of its various parts. So the problem of albedo still stands: in computational parlance, the observer must separate the Illuminance Image from the Reflectance Image of Einstein's portrait, given the retinal image. During this process of albedo perception, the 'signature' chromatic and luminance signals produced by the black and white portrait will be used to infer the complex pattern of light and dark areas that defines its surface. But the albedo perception is not a process of phenomenal reconstruction—a systematic mapping of the intensity values of the retinal image into a visual area of the phenomenal greys. For one, such a process would not yield a veridical perception of the albedo of the photograph. For another, the process of albedo perception is one of complex *inference* not one of *selection*, in which the observer, from the first person, selects the correct shades of grey from a reproduction, in phenomenal grays, of the retinal image.

In the usual case, the trichromat will also see the Einstein photograph *as* black and white, as containing only achromatic colours. Clearly, a discussion of the perception of colour, achromatic or otherwise, is beyond the scope of this paper. But I think that one point can be made here, given the discussion that has come before: The perception of achromatic colours is unlikely to rely upon entirely low-level visual process. This is why trichromatic observers are often *wrong* about black and white photographs, why we sometimes see a subtly tinted photograph as black and white and why we sometimes see a black and white photograph as containing chromatic colours. For example, there is a contemporary Slovak photographer, Peter Župnik, who applies very faint pigment to black and white photographs, often to the lighting within the scene as opposed its objects. Watching a person examine a Župnik photograph is very interesting: the viewer leans into the photograph, scanning it repeatedly, attempting to discern whether the photograph is black and white or 'coloured', a surprisingly difficult task for most of Župnik's photographs.[6]

The reverse effect—seeing a black and white photograph as coloured—can occur as well. In a now famous experiment, Gegenfurtner and colleagues (Olkkonen et al. 2008) showed subjects a series of photographs, one after the other, each containing the same image of a banana. In the first image, the banana is coloured a saturated blue; each successive photograph involves a change in hue along the blue/yellow colour axis; in the last image, the banana is coloured a saturated yellow. A single frame, in the middle of the series, contains the null point, when the image is neither blue nor yellow but entirely grey. Subjects were told to press a button when the banana first appears yellow. Invariably, subjects choose the null point, the grey image, as the first yellow image. So our knowledge of and memory for surface

[6]You can view Zupnik's photographs at http://www.zupnik.eu. The photograph in my dining room is one of the Day's Dreams series, 'The Private Investigation'. In this photograph, the steam rising off the pig (yes, it is a pig) is tinted blue—the colour of night, of course.

colour can affect our current experience of a black and white image. Note that in both examples given, of the Župnik photographs that appear to be black and white (but are not) or of the banana that appears to be yellow (but is not), the problem is not one of *illusion*. An indefinite number of photographs could have been used in the Gegenfurtner experiments as long as they depicted prototypically coloured objects. Nor is the colouring of the Župnik photographs too subtle for trichromats to see. On the contrary. The cognitive problem is that we both expect to see colour (given daily vision) and expect not to see colour (in a black and white photograph), and this makes it very difficult to discern *what* we are actually seeing. It would seem that whether we see an image as black and white or as coloured is also a matter of sophisticated perceptual *inference*, not a function of the brain's information that wavelength contrast is absent in primary visual cortex.

Finally, the normal trichromatic sees the *objects represented* in the black and white photograph, here Einstein and his tongue, as light or dark. This is albedo perception applied to the objects of representation. With intensity information in hand, adult trichromats are seasoned interpreters of visual images including the surface lightness of represented objects. We trichromats will not be fooled by an image of a white egg sitting in dappled shadow. We will not see a dark egg with darker leopard spots, but a uniformly white egg in dappled shadow. (Indeed, even if we viewed the same image under a piece of translucent red film, we would still see the egg as white and the dapples as a feature of illumination (as long as we could see the red film as a transparent colour filter.)) On the other hand, adult trichromats do have trouble 'toggling' between the lightness/darkness of objects as depicted and of the surface properties of the representation itself. This fact is nicely illustrated by Adelson's Checker-shadow Illusion (Fig. 16.16). In the Checker-Shadow Illusion a green cylinder is depicted as sitting on a black and white checkerboard, seemingly lit by a light source on the right (the cylinder 'casts' a shadow over the checkerboard to its left). When we look at the illustration we see the checkerboard as one normally does, composed of a pattern of black and white squares. But if we are asked about the albedo of the illustration itself, that is more difficult. There two squares, A and B, that are represented as black and white which in fact have the same albedo in the illustration—the white square in shadow and the black square that is (represented as) fully illuminated. If you place a mask over the image that reveals only the squares in question while obscuring the rest of the scene, it is easy to see that these two areas, A and B, of the illustration have exactly the same surface lightness. We cannot see the two areas of the illustration as they are, as opposed to how they are represented as being: This information is lost to the first person point of view in the very process of the interpretation of the scene. It is no surprise, then, that when the scene is obscured we can have a veridical perception of the areas A and B qua surfaces of the illustration itself. Presumably, the infant must learn to see represented albedo while the adult trichromat must try use visual aids *not* to see it, once the capacity for albedo is learned.

What happens to the trichromat when he or she views the world at night? Perhaps the most important difference between night vision and the perception of an achromatic photograph is that in night vision, the trichromat views *the natural*

Fig. 16.16 The checker-shadow illusion. The *squares* marked *A* and *B* both have the same surface reflectance, darkness, in the image, but we see *A* as a *black square* and *B* as a *white square* because this consistent with the objects represented and the lighting of the scene (From Adelson 2000)

world, a world of coloured objects, not an artificial image designed to convey intensity information. We do not see the colours of objects in the dark, but whether the world is bathed in sunlight or moonlight, all surfaces reflect and absorb light as a function of wavelength and intensity. And of course, on the receiver end, rod luminance is still a function of the intensity and wavelength of the stimulus, with a spectral response very much like the Carnovksy Green Monochromat. So, the reflected light of red and blue objects will result in low photon catch; green objects will cause high absorption. In other words, colour matters even at night. At the level of the retina, the rod achromat and the dark-adapted trichromat have the same visual access to the world.

What makes trichromatic night vision interesting, and different from rod achromatic vision, is trichromatic visual memory, both personal and sub-personal, of object properties as seen in daylight, i.e. trichromatic visual knowledge. When the dusk falls on the trichromatic system, the photon absorption of the cones will gradually lessen until their catch is so low that it becomes impossible to distinguish signal from noise. Below a certain level of illumination, chromatic cells respond with idle or random chatter. (Remember that the rods will feed into the magnocellular luminance pathway, thus creating a luminance signal at night.) Tellingly, this is not the 'state of the union' of the trichromat who views a black and white photograph. In that case, the chromatic and luminance cells—all of them—detect *contrast*. A comparison of these different signals would yield the conclusion that the proximal stimulus, the retinal image, does not have (discernible) wavelength contrast or a predominant wavelength. It is this information that allows the trichromat to see the photograph as 'black and white' in either sense (as having only achromatic colours or containing a pattern of light and dark areas) in daylight. But at night, no such information is available. Rather, the trichromatic visual system registers that the robust luminance signals are accompanied by a marked *absence* of any signal from the chromatic pathways. With this information in hand, the trichromatic infant will learn *that the world is dark*, the most likely conclusion consistent with the signal pattern and with the overall reduction in illumination. (It is also possible that the lights have gone out *and* that world's surfaces have simultaneously changed to achromatic colours but this is less likely, as an infant will learn.)

So how does the trichromat see surfaces given visual memory? The trichromat, like the rod achromat, cannot perceive albedo from the rod luminance signals alone. Like the rod achromat, the incoming signals provide the basis for seeing only the strange property of 'pseudo albedo', a measure of surface *luminance*, not surface *reflectance*. These perceptions of pseudo-albedo are inconsistent with the trichromat's memories of object surfaces under daylight. They are also inconsistent with her memories of familiar and prototypically coloured objects—e.g. the cherry blossoms as pale pink as opposed to a dark colour of some kind. Here we arrive at *terra incognito*: we do not know whether or how past experience of albedo and colour influences trichromatic *night* vision. Does the trichromat merely discount her daytime perceptions of albedo and colour when she views a familiar scene at night? Is the trichromat's belief that we cannot see colours at night extended to her perceptions of albedo? Or does daytime knowledge of albedo and colour taint ('tint'?) her nighttime perceptions? This is a question that would reward careful investigation. Certainly if you ask a class of students whether they find their cars by colour at night in a parking arcade illuminated by low energy sodium lights (which produce only 'orange' light), many students will claim that they do, that they can see colours under 'orange' light. It is also clear that our perceptions of the world—even colour perceptions—can be influenced by our memories. The Gegenfurtner experiments (Olkkonen et al. 2008) with the blue/yellow images of the banana suggest that prototypical colours influence our perception of an achromatic photograph. Trichromats are also well aware of which properties of the world are stable and which are not. As the light grows dimmer at dusk, we become less and less able to see the colours of the world. But we do not see the colours themselves disappearing, chromatic colours turning into their achromatic cousins, much less previously light or dark objects assuming a new albedo as Nordby so picturesquely suggests ("certainly (all trichromats) must have witnessed the gradual disappearance of colours when darkness sets in"). But this is not the experience of the trichromat: we are aware that darkness hinders visual perception. Our knowledge of the visual world cuts both ways: An adult trichromat knows the colours of familiar and prototypical objects and that these relatively stable properties may be difficult to see at night. What a trichromat *does* see at night is an open question.

Finally, we can return to the rod achromat. Much of this paper has been concerned with showing just how different the rod achromat's experience must be relative to that of the normal trichromat. For example, a trichromat has a broad range of visible light and hence a broad range of fine-grained luminance contrast information. A rod monochromat, with but one receptor, has a narrow spectral range of visible light and fine-grained luminance contrast is confined to that narrow window. The rod achromat sees the world 'through a glass darkly', with less light given only one receptor and with spectrally biased light in the bargain. For another, the trichromat has a multiplicity of chromatic and luminance channels. So the trichromat is sensitive to edges within naturals images, the chromatic edges, to which the achromat has no access. Through the use of both chromatic and illuminance contrast, the trichromat can perceive both the achromatic and chromatic colours, and the inherent lightness of opaque surfaces. The trichromat can also use these types of information as independent sources of knowledge of edges in the world, a fact that opens the doors to numerous new processing strategies. The achromat, in contrast, cannot make an accurate judgment about surface lightness and cannot see any colours at all, chromatic or achromatic. Finally, the trichromat enjoys excellent spatial resolution in daylight conditions, and hence has the kind of high acuity information necessary for depth perception by stereopsis, the perception of fine-grained surface patterns, for the control of fine motor skills, and so on. The achromat never, under any conditions, achieves comparable spatial information. It is clear that the achromat suffers a general impoverishment of information compared to the trichromat.

Now, *prima facie*, these differences may not seem relevant to the problem at hand. Nordby's argument assimilates achromatic visual experience with trichromatic *night* vision not with normal daytime vision in the trichromat. Surely, one thinks, Nordby is on safe ground here. But the background assumption that the trichromat and achromat have the 'same' luminance system, based on the anatomy of the rods and the magnocellular pathway, holds only if human visual development follows a fixed and unalterable path, if there is no plasticity in the development of vision based upon experience. And this, we know, is false. Visual development in mammals is thought to proceed via a developmental cascade of neural function. Incoming information in the first days or months after birth determines the nature of low-level visual mechanisms, the cells responses and their topographical organization in the LGN and V1. Each subsequent step depends upon the system's current state (hence on past input) and upon current incoming signals. This developmental cascade—the sequential extension of visual function in response to input—continues until maturity. Thus both the type and timing of information determine a system's end state. It also follows that the trajectories of two systems, however similar they may be at the start, can diverge during the course of development as a result as asymmetric input.

In this case, the visual development of the human trichromat and achromat, we know that the visual input is very different from the first moments after birth onwards. Although the human $M + L$ luminance system develops fastest, followed by the $S - (M + L)$ system and then the M-L chromatic channel, by 5 months of age the trichromatic infant is approaching an adult-like sensitivity to contrast in all channels. So with the exception of the first few weeks post birth, a trichromatic infant has access to achromatic and chromatic signals of many kinds. We also know that the learning conditions for vision favor the trichromatic infant. Assuming that most visual learning occurs when the infant is awake, in the presence of an adult trichromat who prefers light to dark conditions, the visual input of trichromatic infants will occur under optimal visual conditions for the trichromat (i.e. bright light) while achromatic infants must learn under the worst possible visual conditions for rod vision (i.e. bright light).

Finally, one of the central arguments in this paper was that adding chromatic function extends the informational reach of any visual system. Chromatic cells highlight one range of contrasts within natural images, luminance cells capture a different one. The two types of cells, and their various sub-types, tap into independent sources of information about edges. It is this difference that makes chromatic and luminance systems such excellent partners. So the presence of multiple chromatic and luminance signals in the trichromat makes possible, both logically and empirically, *different strategies* for the representation of distal properties. This fact, that there are two independent sources of information, physically shapes the trichromatic visual system to accommodate the requisite parallel and/or joint processing of these signals. In other words, the dual encodings both make possible different strategies *and* change the physiology of human vision as it develops. It is thus very unlikely that the achromat and the trichromat will have magnocellular systems with the same functional capacities. Without chromatic input, the achromatic system must either solve the visual problems differently, using only narrow band luminance input, or not all. Indeed, we can expect the achromat's magnocellular channel to be specialized for the processing of rod signals alone, with the all of the attendant differences in information of rod vision. Indeed it may be the case that while trichromat's have a huge visual advantage under daylight conditions, the very innovations for chromatic and luminance interaction, may turn out to be a hindrance to night vision. This is not unlikely, that two developmental cascades with distinct resources, may have specialized in divergent ways. In any event, it seems unlikely in retrospect that the trichromat and achromat 'share' a luminance system in any substantive sense.

We come then, back to the initial question of this paper: Is there any basis for saying that an achromat sees 'in black and white'. The reader will have noticed that I did not, at the outset, offer a definition of this central term. I chose to simply adopt the prototypical case of seeing in black and white and then looked for an overlap in physiology between the prototypical case and the rod achromatic vision that would explain why both observers had a shared phenomenology. At this juncture, there would seem to be very few options for this common property. 'Seeing in black and white' cannot be matter of viewing a scene on the basis of luminance information alone. A trichromat uses chromatic information to see a black and white photograph as black and white. So the prototypical case does not meet this condition. Or perhaps the suggestion is that the luminance content alone explains their shared phenomenology. Here, an argument would be needed to explain in what sense precisely an achromat and a trichromat have systems with a shared content. Recall that luminance is a measure of information that is always *system relative*. If you have a luminance system that depends upon an S cone and I have luminance system that depends upon an L cone, in what sense do our perceptions have 'the same' content? One cannot claim that both systems encode *intensity* information about the scene for they do not. Yet there has to be more to this claim of common content than the bare fact that both systems access the world via a single (but different) receptor suited for photon absorption under daylight conditions. What exactly is meant by the same content? Second, seeing in black and white cannot be a matter seeing a scene as having only achromatic colours, or seeing a scene as being devoid of wavelength differences or a predominant wavelength. All of these abilities require chromatic information that the achromat does not have. So again, seeing the achromatic colours could not be the requisite common property.

Finally, we come to albedo perception, a capacity that both types of observer *do* have in common at least if one includes the 'pseudo-albedo' perception of the achromat. Of all the visual capacities that an achromat and trichromat might share, this seems the most likely. But would a shared capacity for the perception of surface lightness account for the wholesale phenomenology of seeing in black and white? If you think of albedo perception àla Cohen, as choosing the appropriate shade of grey for each surface in a scene, this suggestion makes intuitive sense Once each surface (or translucent body) is so coloured—or at least achromatically coloured—everything that can be seen now has some greyscale colour. The whole world is now coloured in black and white. And *that* would seem to be a good candidate for the phenomenology in question.

Albedo perception as described above, however, is not a matter of choosing the correct phenomenal grey and projecting it upon each surface/interior of each opaque/translucent object. Like shape perception, albedo perception is the perception of a property based upon a complex computational or other neural process. It begins with low-level luminance contrast information as its input and ends with a systematic representation of surface reflectance independently of atmosphere. It is the ability to see lightness, e.g. to see whether the egg is white independently of whether the egg is in dappled shadow or in direct sunlight. So capacity to see albedo (or pseudo albedo) is just one of the capacities for intentional perception that an achromat has. This is a mysterious capacity to be sure, but it is not a problem particular to albedo perception. Importantly, it is a capacity, like the perception of non-linear motion, that can be absent in a perceiver without rendering that subject *blind*, devoid of *all* visual phenomenology, here 'black and white' visual phenomenology. We can imagine a person who has a deficit in motion perception, who sees that a ball has moved from here to there without seeing the ball *move* and indeed such people exist, albeit rarely. It is no less imaginable that a person might suffer a deficit in albedo perception. He or she would see a scene, filled with objects of a certain shape in specific positions of various kinds, but without being able to make judgments of absolute albedo. Similarly a trichromat might see everything that a trichromat can see at night and yet *not* see the surfaces as light or dark. Yes, I made it from the bed to the bath without turning on the light; no, I did not see whether the hotel carpet was darker than my robe. If this is possible, then seeing albedo is not the basis of seeing 'in black and white'. An achromat or the trichromat at night still *sees* the world, but need not have accurate albedo perception per se.

We end up, then, just where we might have predicted (had we donned our Sellars' hats) in the first place. In retrospect, Nordby's suggestion is quite odd. It is the view that a trichromat who looks at an achromatic photograph (one without chromatic colours) has the same visual experience as an achromat who views a world of many colours. But without prior bias—unless one assumes that our neural processes follow the divide of external images—why would anyone believe that this suggestion *must* be right?

Acknowledgments I would like thank the James S. McDonnell Foundation through for the Foundation's generous support through a James S. McDonnell Centennial Fellowship, without which this research would not have been possible. I would also like to thank Martin Hahn for reading and commenting on many drafts of this paper; Pete Mandik for his comments and helping to make this paper intelligible to the New York crowd, and; for their support and time, past and present graduate students, Jason Leardi, Lyle Crawford, Simon Pollon, Emma Esmaili, Nicole Pernat, Gerry Viera, and Robert Foley. Thanks to Richard Brown for organizing and facilitating the Consciousness Online conferences.

References

Adelson, E. 1993. Perceptual organization and the judgment of brightness. *Science (New York)* 262(5142): 2042–2044.

Adelson, E.H. 2000. Lightness perception and lightness illusions. In *The new cognitive neurosciences*, ed. M. Gazzaniga, 339–351. Cambridge, MA: MIT Press.

Anderson, B.L., and B.-G. Khang. 2010. The role of scission in the perception of color and opacity. *Journal of Vision* 10(5): 26.

Anderson, B.L., and J. Winawer. 2005. Image segmentation and lightness perception. *Nature* 434(7029): 79–83.

Anderson, B.L., and J. Winawer. 2008. Layered image representations and the computation of surface lightness. *Journal of Vision* 8(7): 18.1–22.

Arend, L.E., and B. Spehar. 1993. Lightness, brightness, and brightness contrast: 1. Illuminance variation. *Perception & Psychophysics* 54(4): 446–456.

Baker, C.L., J.C. Boulton, et al. 1998. A nonlinear chromatic motion mechanism. *Vision Research* 38(2): 291–302.

Buck, S.L., L.P. Thomas, et al. 2006. Do rods influence the hue of foveal stimuli? *Visual Neuroscience* 23(3–4): 519–523.

Cao, D., J. Pokorny, et al. 2008. Rod contributions to color perception: Linear with rod contrast. *Vision Research* 48(26): 2586–2592.

Cao, D., J. Pokorny, et al. 2011. Isolated mesopic rod and cone electroretinograms realized with a four-primary method. *Documenta Ophthalmologica Advances in Ophthalmology* 123(1): 29–41.

Cohen, J. 2004. Color properties and color ascriptions: A relationalist manifesto. *Philosophical Review* 113(4): 451–506.

Corney, D., and R.B. Lotto. 2007. What are lightness illusions and why do we see them? *PLoS Computational Biology* 3(9): 1790–1800.

Cropper, S.J., K.T. Mullen, et al. 1996. Motion coherence across different chromatic axes. *Vision Research* 36(16): 2475–2488.

Field, G.D., M. Greschner, et al. 2009. High-sensitivity rod photoreceptor input to the blue-yellow color opponent pathway in macaque retina. *Nature Neuroscience* 12(9): 1159–1164.

Garcia-Suarez, L., and K.T. Mullen. 2010. Global motion processing in human color vision: A deficit for second-order stimuli. *Journal of Vision* 10(14): 20.

Gegenfurtner, K.R. 2003. Cortical mechanisms of colour vision. *Nature Reviews Neuroscience* 4(7): 563–572.

Gegenfurtner, K., and D.C. Kiper. 1992. Contrast detection in luminance and chromatic noise. *Journal of the Optical Society of America* 9(11): 1880–1888.

Gheorghiu, E., and F.A. Kingdom. 2007. Chromatic tuning of contour-shape mechanisms revealed through the shape-frequency and shape-amplitude after-effects. *Vision Research* 47(14): 1935–1949.

Greenlee, M.W., S. Magnussen, et al. 1988. Spatial vision of the achromat: Spatial frequency and orientation-specific adaptation. *The Journal of Physiology (London)* 395: 661–678.

Hansen, T., and K.R. Gegenfurtner. 2009. Independence of color and luminance edges in natural scenes. *Visual Neuroscience* 26(1): 35–49.

Hess, R.F., and K. Nordby. 1986. Spatial and temporal limits of vision in the achromat. *The Journal of Physiology (London)* 371: 365–385.

Jacobs, G.H. 2008. Primate color vision: A comparative perspective. *Visual Neuroscience* 25(5–6): 619.

Jacobs, G.H. 2009. Evolution of colour vision in mammals. *Philosophical Transactions of the Royal Society of London. Series B, Biological Sciences* 364(1531): 2957–2967.

Jacobs, G.H., and M.P. Rowe. 2004. Evolution of vertebrate colour vision. *Clinical & Experimental Optometry: Journal of the Australian Optometrical Association* 87(4–5): 206–216.

Kingdom, F.A.A. 2003. Color brings relief to human vision. *Nature Neuroscience* 6(6): 641–644.

Kingdom, F.A.A. 2005. Chromatic properties of the colour-shading effect. *Vision Research* 45(11): 1425–1437.

Kingdom, F.A.A. 2008. Perceiving light versus material. *Vision Research* 48(20): 2090–2105.

Kingdom, F.A.A. 2011. Lightness, brightness and transparency: A quarter century of new ideas, captivating demonstrations and unrelenting controversy. *Vision Research* 51(7): 652–673.

Kingdom, F.A.A., and R. Kasrai. 2006. Colour unmasks dark targets in complex displays. *Vision Research* 46(6–7): 814–822.

Kingdom, F.A.A., K. Wong, et al. 2006. Colour contrast influences perceived shape in combined shading and texture patterns. *Spatial Vision* 19(2–4): 147–159.

Knill, D.C., and D. Kersten. 1991. Apparent surface curvature affects lightness perception. *Nature* 351(6323): 228–230.

Li, W., S. Chen, et al. 2010. A fast rod photoreceptor signaling pathway in the mammalian retina. *Nature Neuroscience* 13(4): 414–416.

Michna, M.L., T. Yoshizawa, et al. 2007. S-cone contributions to linear and non-linear motion processing. *Vision Research* 47(8): 1042–1054.

Mullen, K.T., and W.H.A. Beaudot. 2002. Comparison of color and luminance vision on a global shape discrimination task. *Vision Research* 42(5): 565–575.

Mullen, K.T., W.H. Beaudot, et al. 2000. Contour integration in color vision: A common process for the blue-yellow, red-green and luminance mechanisms? *Vision Research* 40(6): 639–655.

Nordby, K. 1996. Vision in a complete Achromat: A personal account. Retrieved from http://consc.net/misc/achromat.html

Olkkonen, M., T. Hansen, et al. 2008. Color appearance of familiar objects: Effects of object shape, texture, and illumination changes. *Journal of Visualization* 8(5): 13.11–13.16.

Pang, J.-J., F. Gao, et al. 2010. Direct rod input to cone BCs and direct cone input to rod BCs challenge the traditional view of mammalian BC circuitry. *Proceedings of the National Academy of Sciences of the United States of America* 107(1): 395–400.

Poirier, F.J.A.M. 2009. The Anderson-Winawer illusion: It's not occlusion. *Attention, Perception, & Psychophysics* 71(6): 1353–1359.

Sellars, W. 1956. *Empiricism and the philosophy of mind*, 1st ed, 253–359. Minneapolis: University of Minnesota Press.

Sellars, W. 1962. Philosophy and the scientific image of man. In *Frontiers in science and philosophy*, ed. R. Colodny, 35–78. Pittsburgh: Pittsburgh Press.

Skottun, B., K. Nordby, et al. 1982. Temporal summation in a rod monochromat. *Vision Research* 22(4): 491–493.

Spillmann, L., J. Hardy, et al. 2010. Brightness enhancement seen through a tube. *Perception* 39(11): 1504–1513.

Stabell, B. 1967a. Rods as color receptors in scotopic vision. *Scandinavian Journal of Psychology* 8(2): 132–138.

Stabell, U. 1967b. Rods as color receptors in photopic vision. *Scandinavian Journal of Psychology* 8(2): 139–144.

Stabell, U., and B. Stabell. 1965. Rods as color receptors. *Scandinavian Journal of Psychology* 6(3): 195–200.

Chapter 17
What Is Visual and Phenomenal but Concerns Neither Hue Nor Shade?

Pete Mandik

17.1 Introducing Akins's Problem

Though the following problem is not explicitly raised by her, it seems sufficiently similar to an issue of pertinence to Akins's "Black and White and Color" (Akins 2014) to merit the moniker, *Akins's Problem*[1]:

> Can there be a visual experience devoid of both color phenomenology and black-and-white phenomenology?

The point of the present paper is to draw from Akins's paper the materials needed to sketch a case for a positive answer to Akins's Problem. I am unsure about how much of what follows Akins will want to endorse, but I hope this helps move us forward in our collective pursuit of a theory of visual consciousness.

Many philosophers of mind familiar with Jackson's (1982) Mary thought experiment may feel confident that they know both what color phenomenology and black-and-white phenomenology are. Prior to her release from her achromatic captivity, Mary's visual experiences have black-and-white phenomenology, but no color phenomenology. Or so the story goes.

Readers lacking either familiarity with or a taste for the Mary thought experiment may nonetheless feel that they have a grasp on this alleged contrast in visual phenomenology. Such readers arrive at this seeming grasp by appeal

[1]But not "Akins' Problem." Regarding the rule followed here on apostrophes for proper nouns ending in "s," see p. 354 of *The Chicago Manual of Style*, 16th Edition, Chicago University Press, 2010.

P. Mandik (✉)
Department of Philosophy, William Paterson University of New Jersey, 300 Pompton Road, Wayne, NJ 07470, USA
e-mail: mandikp@wpunj.edu

R. Brown (ed.), *Consciousness Inside and Out: Phenomenology, Neuroscience, and the Nature of Experience*, Studies in Brain and Mind 6, DOI 10.1007/978-94-007-6001-1_17,

to a contrast between two main kinds of photographs and other visual media (paintings, movies, etc.). A normal, that is, non-colorblind, viewer of black-and-white visual media during daylight conditions enjoys correlative black-and-white phenomenology. In contrast, colored phenomenology accompanies normal vision of colored media, and, of course, normal vision of a colored world.

One way to get a feel for Akins's problem is by contemplating the visual phenomenology of the genuinely colorblind. Take, for example, the rod achromat. It is overwhelmingly plausible that, in never seeing colors, their visual experiences lack color phenomenology. However, they still have visual phenomenology, right? They can still consciously see things, so there has to be something it's like when they do so. But what *is* it like? A negative answer to Akins's problem goes along with saying that the visual phenomenology of the rod achromat must be black-and-white phenomenology. A positive answer to Akins's Problem goes along with saying that the rod achromat's phenomenology need not be black-and-white phenomenology.

17.2 Undermining the Nordby Argument

One line of thought favoring a negative answer to Akins's Problem is a line that we can reconstruct from remarks made by the rod achromat Knut Nordby (1996), a colorblind vision scientist (who was hip to Jackson's Mary and other philosophical topics in the vicinity (Nordby 2007)). Nordby's line of thought is pertinent to Akins's Problem because it can be interpreted as an argument for the conclusion that if a visual experience is devoid of color phenomenology, then it must have black-and-white phenomenology. The argument toward such a conclusion has two main components. The first component is a claim that achromatic experience is like trichromatic night vision. The second component is a claim that trichromatic night vision is like trichromatic day vision of black-and-white pictures. Presuming the transitivity of *being like*, it would seem to follow that achromatic experience is like trichromatic day vision of black-and-white pictures.

Akins presents considerations against both (1) the analogy between trichromatic night vision and trichromatic day vision and (2) the analogy between achromatic experience and trichromatic night vision. Undermining these two analogies serves to undermine the Nordby-inspired argument for a negative answer to Akins's problem. Of course, undermining an argument for not-P is one thing. Providing reasons for P is another. Nonetheless, I think we can find in Akins's paper (ingredients for) a case for a positive answer to Akins's problem. Further, I think we can find in Akins's view the resources for spelling out *what it would be like* to have a visual experience that had neither color phenomenology nor black-and-white phenomenology.

17.3 What Akins's Problem Isn't

To further clarify what I take Akins's Problem to be, it will be useful to clear out of the way a problem that *isn't* Akins's Problem. One approach to visual phenomenology embraced by certain qualiaphiles is the view that visible properties of worldly items—worldly properties like red, blue, and gray—appear to consciousness via phenomenal properties—mental qualities or qualia like red* (pronounced "red star"), blue*, and gray*. Further, since objects in the world have visible properties besides those concerning their hue and shade, properties such as their size and shape, a qualiaphilic aficionado of the property-star notation may (perhaps) be comfortable making assertions about phenomenology in terms such as tall* and square*. With such terminology in hand, we can formulate a question that is decidedly *not* the same as Akins's Problem. Call this one *Not Akins's Problem*:

> Are there visual qualia other than the hue and shade qualia of, for instance, red* and gray*? Are there additionally, for instance, *spatial* visual qualia, such as big* and round*?

A qualiaphile can answer "yes" to Not Akins's Problem and "no" to Akins's Problem. The imagined qualiaphile defends this pattern of answers by appeal to the following dependency thesis: Just as no visible object can have a visible size and shape without being colored (more specifically, visibly differing from the background in hue or shade), so can no visual experience have space* properties without color* properties (e.g. red*, gray*). So, according to this qualiaphile, even though there *are* non-color qualia, no visual experience can have *only* non-color qualia.

The analogy between visual objects and visual experiences appealed to in articulating the above dependency thesis is part of the package one embraces in holding that visual experiences are picture-like. Literal pictures depict nothing at all without doing so in virtue of the spatial distributions of hues and/or shades in the picture itself. This is a key contrast between pictorial representations and non-pictorial, language-like representations: A linguistic representation of a shape can be totally silent as to its shade or hue in a way that a pictorial representation cannot.

Anyway, I'll say more about this in later sections. For now, the key is that there are two points of contrast between Akins's Problem and Not Akins's Problem. The first point is that Akins's Problem is about *experiences* themselves. It is not about any putative *elements* of experiences (qualia, or whatever). The second point is that Akins's Problem is formulable in a way that is neutral about whether there *are* any qualia, or anything else worth denoting with the property-star notation (e.g. sensations).

17.4 Orthodox Philosophy of Mind and the Negative Answer to Akins's Problem

While I've not conducted anything remotely resembling a formal survey, I'm pretty confident that most philosophers of mind will answer "no" to Akins's Problem. In the present section I want to lay out what I take to be the background driving contemporary philosophy of mind orthodoxy. I will also make some remarks in order to contrast this orthodoxy with Akins's own approach.

According to contemporary philosophy of mind orthodoxy, call it the *Orthodoxy*, when the world makes its impingements upon the mind (instead of the other way around), one or both of two sorts of mental state may be involved: sensations and judgments. There are various contrasts the Orthodoxy appeals to in contrasting sensations and judgments. Perhaps not every adherent of the Orthodoxy will go along with all of them. However, the contrasts are: low level vs. high level, phenomenal vs. cognitive, nonconceptual vs. conceptual, and determinate vs. indeterminate. (The relevant determinate/indeterminate contrast is perhaps best illustrated by a contrast between a hen or hen photo that has a specific number of speckles and a description of the hen as being speckled that is non-committal about which number is the number of speckles.)

Further elaborating the Orthodoxy: There are two sorts of property that these mental states may have in virtue of which they count as mental. The first sort of property is phenomenality or the having of a quale—a property, perhaps intrinsic, in virtue of which it is true of a mental state that there is, in the unwieldy and uninformative parlance of the Orthodoxy, "something it's like". The second sort of property is intentionality or aboutness—a property, perhaps relational, in virtue of which it is true of a mental state that it represents or is about something. One typical sort of account of the relation between the two kinds of mental state and the two kinds of property goes like this: Phenomenality goes more with sensations whereas intentionality goes more with judgments.

There are two ways in which Akins's view of vision departs from this Orthodoxy. The first and main departure is the denial of a role for sensations (and qualia, those properties in virtue of which sensations have their phenomenality). The second is to reserve a use for notions of *what it's like* and *phenomenology* whereby there need be no sensations or qualia for these notions to have an application.

These departures are very much in the spirit of Dennettian qualia-quining (Dennett 1990). While it may be true, maybe even platitudinous, that there is something it's like to be impinged upon by the world via our visual processes, that there is a way our visual mental life appears to us, the processes involved are some combination high-level, cognitive, conceptual, and indeterminate. Contra Block (2003), there is no mental paint. Contra Sellars (1956) and Rosenthal (2005), there's nothing mental worth regarding as an impression that serves as an intermediary between the worldly impingements upon our sensory surfaces and our eventual judgments about what's out there doing the impinging. Neither is there anything red* in your mind when you see something red in the world.

Akins's departure from the Orthodoxy is driven by a close analysis of the neurophysiology of both luminance vision and chromatic vision. This analysis leads to a version of *conceptualism* about the character of conscious experience. As I understand conceptualism for present purposes, it is the view that . . .

> . . . conscious perceptual states have conceptual content, and the mental aspects distinguishing various perceptual states, aspects such as the phenomenal character or sensory qualities of the states, are exhausted by these conceptual contents. Focusing on conscious experience of color, . . . the difference between a conscious experience of red and a conscious experience of blue just is the difference constituted by deploying the concept of red in the one experience and the concept of blue in the other (Mandik 2012, p. 620).

Akins's conceptualism will be key for supplying a positive answer to Akins's Problem. I turn now to briefly sketch what I take Akins's account to be.

17.5 Akins on How Luminance Vision and Chromatic Vision Work

For a quick sketch of Kathleen's view of how luminance vision and chromatic vision work, it helps to spell this out in terms of commonalities and differences between the two kinds of vision. It won't do, it must be noted, to say that chromatic vision is for detecting colors and luminance is for detecting lightness and darkness. Similarly, it won't do to say that chromatic vision receives only wavelength information as input and luminance vision receives only intensity information as input. Part of the problem is that the main respective receptors, cones for chromatic and rods for luminance, are both responsive to intensity within limited wavelength ranges. It is true of each individual receptor, rod and cone alike, that it cannot distinguish wavelength from intensity. So it's not receptor types that will distinguish chromatic from luminance systems, but the way the receptors are wired together and the computations that such wirings enable that do the trick. Chromatic systems involve comparisons between different kinds of receptor, for instance, comparisons between short wavelength cones and medium and long wavelength cones in the blue-yellow opponent system. Luminance systems involve summations across similar kinds of receptor.

One upshot of this way of thinking about luminance and chromatic systems is that having cones as inputs does not alone suffice to make a chromatic system. In fact, systems with only one kind of cone population may be regarded as luminance systems unto themselves, albeit luminance systems with a preferred wavelength range. Such different luminance systems might be usefully analogized to the different uses that distinct color filters can be put to in black-and-white photography. The filtering of different wavelengths results in different luminance contrasts. A clear luminance contrast revealed in one filtration scheme may be invisible in another. An advantage conferred by having multiple cone types isn't so much to see the colors, but instead to have multiple sources of luminance contrast and thus effect better discrimination of objects from backgrounds. Chromatic systems, by comparing activations between populations of different kinds of receptor, are able to disambiguate wavelength from intensity, and thus effect an additional range of contrasts: chromatic contrasts in addition to luminance contrasts. The main point of having these additional sensitivities to contrast is to enable different means for seeing the edges that demarcate objects from their backgrounds.

Now, chromatic and luminance systems do not serve simply to differentiate object from ground. They also underwrite the visual perception of the objective properties of worldly objects. One objective property of objects is albedo or surface reflectance, roughly the objective basis of the perceptible lightness and darkness of objects. The computational problem of discerning albedo is quite difficult, given that the amount of light hitting the eye by itself underdetermines albedo. This underdetermination may be circumvented if decent information is at hand about the current illumination and its interaction with other parts of the scene, but this in turn is likely to involve the contribution of high-level processes sensitive to information about, among other things, spatial structure and material composition. A similar high-level circumvention of stimulus underdetermination can be expected for chromatic systems and the objective basis of object color, spectral surface reflectance.

It is at this point—the point where high-level contributions are appealed to for the circumvention of stimulus underdetermination—that we see Akins's view of vision as a species of conceptualism about consciousness. The high-level contributions tap knowledge about the external world that is encoded in one's conceptual repertoire. One of the key features of conceptualism is the way that it posits representation schemes that aren't picture-like. There are several key features of these non-picture-like representational schemes.

One key feature of non-picture-like representational schemes is their *indeterminacy*. A worldly object, such as my cat Mary, has a determinate size and a determinate shape. It is impossible for Mary to merely have determinable properties like being sized or being shaped. If she is shaped, there must be some particular shape that she has. A key feature of imagistic representational schemes is the representation of determinates by determinates. The blob in the photograph that represents Mary itself has a determinate size and determinate shape, and further, which determinate size and shape the blob has helps determine which determinate size and shape Mary is represented as having.

Another key feature of non-picture-like representation is its *sparseness* or lack of *lavishness*. Pictorial representations are lavish—size can't be represented without also representing shape and much else besides. In contrast, nonimagistic schemes are sparse. A language-like scheme can represent Mary as being the same shape as my other cat, Ernest, while being noncommittal as to which shape they both have. And it can represent Mary as having a shape while being noncommittal about her size, color, etc.

The indeterminacy and sparseness that go along with conceptualism will be key in making coherent how we can motivate a positive answer to Akins's Problem. But before we can proceed to that answer, we need to say a bit about how conceptualism handles phenomenology.

17.6 Conceptualism and Phenomenology

Conceptualism explains phenomenology by way of a two step-line of thought. The first step involves identifying "what it's like" to be in such-and-such conscious state with the way things seem to one when one is in such-and-such state. What it's like, for instance, when one sees a rose as red is explicable without residue by appeal to the ways in which things seem to one in virtue of seeing a rose as red. The second step is to account for the ways things seem in terms of the concepts deployed in having the states in question. Our primary model for this second step comes from the ways in which things appear to us in virtue of thinking about them. If George thinks of the dark thing leaning against the wall as an umbrella and not a walking stick, then, in virtue of his so thinking about it, it will thereby seem to him like an umbrella and not a walking stick. Whether it actually is an umbrella is irrelevant to its seeming as such. Instead, what's most directly relevant here concerning the way things seem to George is the concept thereby deployed, namely, George's concept of an umbrella. And if George and I share an umbrella concept, then there's no bar to my coming to know what it's like to be George thinking that there's an umbrella nearby, since, in possessing the relevant concepts, I grasp how the world would appear to me were I to deploy those concepts.

The conceptualist need not take a stand on whether perceptions are species of thoughts. However, the conceptualist does hold that the account of perceptual appearance is largely the same as the account of cognitive appearance. The account in both cases will largely be spelled out in terms of the concepts deployed in having the relevant conscious states.

The conceptualist allowance of sparse phenomenology allows for a positive answer to Akins's Problem. Just as one can think that the mat on the floor is rectangular without thinking that it differs from the floor in hue or shade, so can one see the mat as rectangular without seeing it as differing from the floor in hue or shade. And all of this is consistent with the fact that the visibility of the mat's shape depends on the mat differing from its background in either hue or shade.

That visual phenomenology can actually be so sparse is evidenced by certain surprising breakdowns of normal functioning. Akins mentions one sort of example when she writes that "[w]e can imagine a person who has a deficit in motion perception, who sees that a ball has moved from here to there without seeing the ball *move* and indeed such people exist, albeit rarely." Evidence more directly pertinent to Akins's Problem comes from studies of cerebral achromatopsic patient, M.S., who is able to see shapes defined only by hue contrasts with their backgrounds even though he is not able to see hues (he cannot visually discriminate, e.g., red from green) (Heywood et al. 1994).

So, then, what *is* the phenomenology of someone exemplifying the positive answer to Akins's Problem? That is, what would it be like to see something without seeing it as having some shade or hue? One example would be simply seeing a mat as rectangular. A reader understanding the previous sentence has the relevant concepts, in particular, the concepts of seeing and of rectangularity, and thus, there's no real bar to the reader's understanding what it would be like to be the hypothesized seer of the mat.

There are two potential lines of objection to the conceptualist's proposal that one might find tempting, but I think they are ultimately unpromising. One line is to suggest that the proposed case is not *visual*. The second grants that it is visual, but suggests that this is not a case of *experience* (as in the nonconscious visual processing of blindsight).

The first line might be articulated like this: We do have a firm grip on the possibility of experiences of shape that are silent about hue and shade, but such a grip comes from familiarity with nonvisual sensory modalities. For instance, I can feel that a game piece in my hand is round without thereby feeling its color. On this line of thought, this absence of color awareness is one of the main features distinguishing tactile awareness from visual awareness. However, if this line of thought is correct, then it would seem that we should predict that someone who had the proposed sparse phenomenology would not be inclined to report that they had it by seeing. They shouldn't report, for instance, that they came to be aware of the mat's rectangularity by seeing it. However, this prediction is unlikely to be correct. It is highly implausible that the cerebral achromatopsic M. S., in making the aforementioned shape-discriminations, is unaware that he's accessing shapes by *seeing*.

This point bears on the second line of objection as well, for, in exhibiting awareness of *seeing* the shape, the seeing of the shape cannot be plausibly regarded as *nonconscious*. One need not be a full-blown adherent of higher-order theories of consciousness (e.g. Rosenthal 2005) to accept that a mental state of which the subject is conscious (in this case, the seeing) is itself a conscious state. In any case, there are other reasons to regard M.S's access to visual shape as conscious. For instance, the availability of shape information is not evident only by implicit means (as in the forced-choice guessing associated with blindsight research). M.S. is able to indicate the shapes in spontaneous verbal reports.

Of course, it remains to be spelled out exactly what does suffice to make the hypothesized sparse experience both visual and conscious, but that's beyond both the present paper and the present state of its author. So, let us end things on a positive note. Here's the correct answer to Akins's Problem: Yes.

Acknowledgements Most of what I know about the neurophilosophy of color I learned while under the influence of Kathleen Akins and Martin Hahn, and I am grateful for conversations with them about this stuff over the years. Thanks are due as well for conversations on related matters with Richard Brown, Jacob Berger, and David Rosenthal.

References

Akins, K. 2013. Black and White and Color. In *Consciousness Inside and Out: Phenomenology, Neuroscience, and the Nature of Experience*, ed. R. Brown. Studies in Brain and Mind 6, 173–223. Springer Press

Block, N. 2003. Mental paint. In *Reflections and replies: Essays on the philosophy of Tyler Burge*, ed. H. Hahn and B. Ramberg, 165–200. Cambridge, MA: MIT Press.

Dennett, D.C. 1990. Quining qualia. In *Mind and cognition*, ed. W. Lycan, 519–548. Oxford: Blackwell.

Heywood, C.A., A. Cowey, and F. Newcombe. 1994. On the role of parvocellular (P) and magnocellular (M) pathways in cerebral achromatopsia. *Brain* 117(2): 245–254.

Jackson, F. 1982. Epiphenomenal qualia. *The Philosophical Quarterly* 32: 127–136.

Mandik, P. 2012. Color-consciousness conceptualism. *Consciousness and Cognition* 21(2): 617–631. doi:10.1016/j.concog.2010.11.010.

Nordby, K. 1996. Vision in a complete Achromat: A personal account. Retrieved 12 June 2012 from http://consc.net/misc/achromat.html

Nordby, K. 2007. What is this thing you call color: Can a totally color-blind person know about color? In *Phenomenal concepts and phenomenal knowledge*, ed. T. Alter and S. Walter, 77–83. Oxford: Oxford University Press.

Rosenthal, D. 2005. *Consciousness and mind.* Oxford: Clarendon.

Sellars, W. 1956. Empiricism and the philosophy of mind. In *Minnesota studies in the philosophy of science*, 1st ed, 253–329. Minneapolis: University of Minnesota Press.

Part VI
Phenomenal Externalism and the Science of Perception

Chapter 18
The Real Trouble with Phenomenal Externalism: New Empirical Evidence for a Brain-Based Theory of Consciousness

Adam Pautz

> *[We should] reverse the whole programme started by Galileo – we should put these [sensible] qualities back into the physical world again.*
>
> –David Armstrong

The traditional view of the sensible qualities locates them in the head. But within philosophy there has recently been a kind of externalist revolution. While most scientists would still locate the sensible qualities in the head, many philosophers now claim that sensible qualities are really "out there" in the mind-independent physical world and that the function of the brain is just to reveal them to us. In favorable conditions sensory character is determined simply by what mind-independent states you are directly conscious of. The result is a kind of phenomenal externalism. Examples include externalist intentionalism, naïve realism, and active externalism.[1] The stakes are high, because many think that phenomenal externalism represents our best shot at naturalizing consciousness and its intentionality.

[1] For active externalism, see Noë (2004) and O'Regan (2011). Although these authors advertise active externalism as a radical kind of phenomenal externalism (phenomenology fails to supervene on the brain), this is sometimes unclear. For instance, in response to an objection from David Chalmers, Noë (2004, p. 119) changes his view, claiming that his view is that phenomenology is constituted by a slew of sophisticated *beliefs* or *expectations* concerning what the sensory effects of various actions would be. In that case, his view might actually be a version of phenomenal internalism, because (for all he says) the relevant beliefs might be narrow beliefs that supervene on the head. In general, as Block (2012) shows, active externalists do not have any clear view. Partly for this reason, here I will be focusing on other varieties of phenomenal externalism.

A. Pautz (✉)
Department of Philosophy, University of Texas at Austin, 2210 Speedway, Stop C3500, USA
e-mail: apautz@austin.utexas.edu

R. Brown (ed.), *Consciousness Inside and Out: Phenomenology, Neuroscience, and the Nature of Experience*, Studies in Brain and Mind 6, DOI 10.1007/978-94-007-6001-1_18,

My own view is that phenomenal externalism has been a big wrong turn. I favor a kind of internalist counter-revolution. But, for reasons I will explain, I disagree with those who think phenomenal externalism can be refuted very easily on the basis of controversial intuitions about brains in vats (Horgan, Tienson and Graham), inverted spectrum (Shoemaker), actual cases of perceptual variation (Block), and so on.[2] Both sides of the debate have missed the best argument against phenomenal externalism. The real trouble with phenomenal externalism is that it goes against decades of research in psychophysics and neuroscience. The basic point is that, even under ideal conditions (no interfering factors), sensory character is much better correlated with neural patterns in the brain than with anything in the external, physical world. I call this the *problem of correlations* for phenomenal externalism. For this reason most scientists of sensation and perception would probably not take very seriously the kind of phenomenal externalism now being promoted by some philosophers. I have begun developing the argument in previous work. In the present essay, I will go beyond that work, by using impressive new research and by ruling out recent responses.[3]

To make the discussion concrete, I will initially focus on what I call "tracking intentionalism" as a kind of stalking horse. This is a version of externalist intentionalism, which combines the intentionalist thesis that all phenomenal differences among sensory experiences are representational differences with a reductive externalist theory of representation. But while I will focus on tracking intentionalism I would like to stress that I believe my arguments will undermine any reductive variety of externalist intentionalism. Defenders of views in the general vicinity include David Armstrong, Alex Byrne, Fred Dretske, David Hilbert, Christopher Hill, William Lycan, David Papineau, and Michael Tye. I agree with these philosophers that sensory experiences are intentional states that present sensible qualities ostensibly located in the external world or bodily regions; what I will argue against is only their *externalist* variety of the view. Although I will not discuss this here, I believe my empirical arguments also undermine the version of phenomena externalism defended by John Campbell and other "naïve realists".[4]

Many externalists focus narrowly on one sense-modality, without showing how externalism can be developed across the board. To show just how unpromising the externalist picture is, I will consider multiple sense-modalities. In particular, I will target recent externalist views of taste qualities (Smith), smell qualities (Batty), auditory qualities (O'Callaghan), and pain qualities (Tye, Dretske, Hill).

---

[2]See Horgan et al. (2004), Shoemaker (1994) and Block (1999).

[3]See Pautz (2006a, 2010). Hill (2012) provides very interesting externalist responses to the kind research on pain mentioned in Pautz (2006a, pp. 212–213) and elaborated in Pautz (2010). Cutter and Tye (2011) also reply to my empirical objections about pain. I will take their views into account throughout this paper.

[4]For externalist intentionalism, see Armstrong (1999), Byrne and Hilbert (2003), Dretske (1995), Hill (2009), Lycan (2001), Papineau (2012), and Tye (2000). For naïve realism, see Campbell (2002).

My plan is as follows. In Sect. 18.1 I explain tracking intentionalism In Sect. 18.2 I show that, even under ideal conditions, sensory character is much better correlated with neural patterns in the brain than with anything in the external, physical world. In Sects. 18.3 and 18.4 I elaborate in detail two independent empirical arguments against tracking intentionalism based on the correlational data. In Sect. 18.5 I eliminate recent responses which involve defending more elaborate versions of externalist intentionalism. These responses appeal to functional or syntactic features of representations (Lycan, Hill), binding-errors (Hill), valuational or threat-level contents (Cutter and Tye, Hill), complex properties (O'Callaghan), or response-dependent properties (Kriegel). Finally in Sect. 18.6 I will suggest that, although *externalist* intentionalism fails, sensory experiences are indeed nothing but intentional states that present sensible qualities ostensibly located in the external world or bodily regions. However, for empirical reasons, the best way of developing intentionalism is by accepting what David Chalmers called an "Edenic theory" of sensible qualities.

18.1 What Is Tracking Intentionalism?

I start by explaining the basic tracking intentionalist picture that will be my stalking horse. It has three parts.

The *first part* is a reductive and objectivist theory of sensible qualities like colors, smell qualities, taste qualities, audible qualities, and so on. It is *reductionist* in that it holds that sensible qualities are physical properties, in a suitably broad sense of 'physical properties'. It is *response-independent* in that it holds that sensible qualities are not in any way to be defined in terms of effects on perceivers.

So, for instance, maybe colors are reflectance properties, smell qualities and taste qualities are chemical properties of odor clouds and foods, auditory qualities are extremely complex physical properties involving frequency, amplitude, duration, and "critical bands". And maybe perceived shapes are viewpoint-relative but objective properties like *being-elliptical-from-viewpoint-p* and *being-round-from-viewpoint-q* (Hill 2009). Now it is well known that even under optimal conditions multiple physical properties can cause one to be ostensibly conscious of the same sensible quality. In color vision this is known as "metamerism". The same phenomenon occurs in all the sense-modalities. But tracking intentionalists are not perturbed: they simply reduce a sensible quality to the disjunction of all the physical properties that normally give rise to our experience of it.

Tracking intentionalists even say that pain qualities you feel in your body are mind-independent physical properties, for instance, types of bodily damage or "potential" damage. This view faces what I have called the "percipi puzzle" (Pautz 2010). Byrne (2012) has more recently called it the "puzzle of pain" in his discussion of Hill's (2009) somewhat different paradox of pain. On tracking intentionalism, just as colors can exist without experiencers, so can felt *pain qualities!* For pains as well as colors are treated as entirely mind-independent physical properties of external items. For instance, the *very same horrible quality* you feel in your thumb when you hit yourself with a hammer (on, this view, a kind of damage) might occur in an insentient cadaver! However, here I propose to set *a priori* objections to the side. My aim will be to develop new empirical problems.

The *second part* of tracking intentionalism is a broadly *tracking theory* of sensory awareness of *properties*. The rough idea is that you sensorily represent an objective sensible quality (on this view, a physical property), and are thereby aware of it, just in case you undergo an internal state (a "representation") that "registers" or "tracks" the instantiation of that property by external items. By using the term 'tracking' I do not mean to presuppose a simple input-based, causal theory of representation. I use "tracking" in a totally neural way, as a kind of place-holder for a more detailed story.

However, for the purposes of illustration, I will largely focus on views of Michael Tye and Fred Dretske. Tye (2000) reduces the sensory representation relation to the *optimal tracking relation*, that is, the relation: individual *X* is in an internal state that plays functional role *F* and that, under optimal conditions, would be caused the instantiation of property *Y* (or for short, that is *optimally caused by Y*). Dretske (1995) reduces the sensory representation relation to the *indication relation*: individual *X* is in an internal state that plays functional role *F* and that has the function of indicating *Y*. By 'functional role *F*' I mean the special functional role that is supposed to turn unconscious representational states into consciousness ones: maybe some kind of cognitive accessibility. This will not concern us here. While I focus on Tye and Dretske, we will see (in Sect. 18.5) that my arguments also work against more complex versions of externalist intentionalism, including those which appeal to Millikan's (1989) consumer-based approach to representation.

The *third part* of tracking intentionalism is intentionalism about sensory phenomenology. At a minimum, the intentionalist says that if two individuals are ostensibly conscious of, or sensorily represent, the very same sensible qualities (at the same places), then they have phenomenally identical sensory experiences. In cases of illusion and hallucination, the presented sensible qualities do not belong to any external items. Sensory content determines sensory phenomenology. This, or something like it, is extremely plausible. For instance, intuitively, if two individuals are ostensibly conscious of the very same smell or taste qualities, then they must have phenomenally identically smell or taste experiences. If they are ostensibly conscious of the very same auditory properties (from the apparent same direction), they must have phenomenally identical auditory experiences. Intentionalism is just a theoretical gloss on these intuitions.

That, then, is tracking intentionalism. It is undeniably attractive. Indeed, Cutter and Tye (2011, p. 91) have recently said, "tracking [intentionalism] is the most promising view for the philosopher in search a naturalistic account of experience". The reason is simple. Sensory consciousness is *externally-directed*. The sensible qualities certainly *appear* to be out there, in objects, in bodily regions, in foods (or maybe the tongue), and so on. This favors "objectivism" about sensible qualities and sits poorly with a Galilean view that locates the sensible qualities in the head. And if objectivism is true, then tracking intentionalism, or something like it, appears almost inevitable. For, in order to explain in naturalistic terms how the mind can become aware of, or "represent", objective properties out there, objectivists must presumably appeal to *causal* or *indicator* or *teleological* relations between brains and those properties. What other option is there? What else could hook us up to these properties? In that case, what sensible qualities you perceive are fixed by extrinsic factors, namely, your relations to your environment. Since sensory phenomenology is intuitively inseparable from what sensible qualities you perceive, the result is a radically externalist theory of phenomenology. So, while objectivism is a theory of the sensible qualities and tracking intentionalism is a theory of phenomenal character, I think that anyone attracted to objectivism about sensible qualities is under pressure to accept an externalist theory in the vicinity of tracking intentionalism, including for instance Casey O'Callaghan (2002) and Clare Batty (2010).

Of course, the theory I have presented is very simple. Many would like to add some bells and whistles. But I think many are committed to externalist views of sensory character in the general vicinity, including David Armstrong, Alex Byrne, Fred Dretske, David Hilbert, Christopher Hill, William Lycan, David Papineau, and Michael Tye.

Let me mention two caveats. First, Lycan (forthcoming) and Hill (2009) seem to hold that some phenomenal differences among sensory experiences are grounded in functional-syntactic differences, not representational differences. But we will see (in Sect. 18.5) that this cannot save their views from my arguments. Second, Byrne and Hilbert (2003) as well as Hill (2009) express skepticism concerning all existing naturalistic theories of representation. But they still hold that sensory qualities are physical properties of external things, that phenomenal character is (at least largely) determined by the representation of these properties, and that some externalist naturalistic theory of representation is correct (even if we cannot specify it). As we shall see, this is enough to make them vulnerable to my arguments.

So scores of philosophers take the same basic externalist approach. And it is very attractive, because it fits with the externally-directed character of sensory consciousness. The only trouble is that it flies in the face of decades of research in psychophysics and neuroscience. That research shows that consciousness is (in a non-trivial sense) internally-dependent, even if it is also apparently externally-directed.

18.2 It's Only in Your Head: The Neural Basis of Some Phenomenal Facts

Tracking intentionalism is radically externalist. On tracking intentionalism, the character of experiences is not determined by the intrinsic character of their neural correlates. Here is an analogy. The shapes of words do not matter to what they represent. Thus, 'dog' and 'cat' are physically dissimilar, but represent similar animals. In general, there is a sense in which the intrinsic features of the neural content-carriers do not matter to what contents they carry. Likewise, on tracking intentionalism, there is *some* sense (which I will make more precise in Sect. 18.4) in which the intrinsic features of postreceptoral neural processing do not matter to phenomenal character. All that matters to phenomenal character are what physical properties the neural wetware tracks and thereby represents.

But this could not be more wrong. In fact, the whole history of psychophysics and neuroscience shows that *exactly the opposite is true*. In *some* cases, the intrinsic features of the neural wetware does somehow matter, in a way that I will later show to inconsistent with tracking intentionalism (Sect. 18.3). In particular, two facts are relevant.

First, for decades psychophysics has revealed that, even under optimal conditions (no interfering factors), there is *some* sense in which there is an extremely bad correlation between experiences and the external physical properties tracked. What I mean will become clear when consider examples. But let me say at the outset that I do not merely mean that the physical properties tracked are disjunctive or unnatural (because of metamerism). Roughly, what I mean is that, even under optimal conditions, the *structural relations* among experiences (similarity and difference, equal intervals, proportion) are not matched by the *structural relations* among the (disjunctive) external physical properties that those experiences track. True, in some cases, they do match; in other words, there is good external correlation. For instance, under optimal conditions, subjects' reports on when perceived length doubles corresponds to an actual doubling in physical length. But psychophysics has shown that this is the exception rather than the rule. When it comes to taste, smell, pain and sound, there is *bad external correlation*. Here tracking intentionalists have it exactly wrong. The external physical world is just the *wrong place* to look for the basis of qualitative character.

Second, neuroscience has revealed that experiences are much better correlated with neural firing patterns in the brain. What I mean by this, too, will become clear as we go on. But let me say at the outset that I do not merely mean that every distinct experience is co-extensive with a distinct neural correlate in humans, so that every for measurable change in experience there is some measurable change in the nervous system. Some philosophers think that this is enough to refute tracking intentionalism. This is a mistake. It is equally true that going from *thinking about water* to *thinking about aluminum* requires a neural change; but no one would think this undermines externalism about thought about natural kinds. What I mean by good internal correlation is something subtler than the existence of correlations between individual experiences and individual neural states. What I mean is that in *some* cases structural *relations among* experiences (similarity and difference, equal intervals, proportion) are well matched by structural *relations among* their neural correlates. In these cases, while there is bad external correlation, there is *good internal correlation*. In these cases the basis of certain structural facts about phenomenal character are to be found *only in the brain*.

To illustrate these points, I will in the rest of this section provide some empirical background concerning the fascinating science of taste, smell, pain and sound. I will wait until later sections to explain how the science can be definitively shown to be at odds with tracking intentionalism.

18.2.1 Taste

The tracking intentionalist will presumably say that different types of taste qualities are different types of chemical properties that are tracked and thereby represented by our taste experiences. All sensory-phenomenal facts about taste experiences are determined by the chemical types they optimally track and thereby represent.

The trouble is that phenomenal *resemblances and differences* among taste experiences are not well correlated with resemblances and differences among the chemical types they track. Examples of bad external correlation abound. For instance, suppose you taste aspartame and then a stereoisomer of aspartame. The chemical properties that your taste experiences optimally track are *extremely similar*: the compounds only slightly differ in the orientation of two hydrogen atoms. Yet your taste experiences are *extremely different*: the taste of aspartame is sweet while the taste of the stereoisomer is extremely bitter (Walters 1996). Likewise, gentiobiose is bitter, while trehalose has a distinctly sweet taste, even though they are very similar disaccharides composed of two glucose units. Indeed, gentiobiose has an *anomer* (a kind of very similar stereoisomer), namely isomaltose, which tastes sweet (Sakurai et al. 2010). Neohesperidin, which is found in citrus peel, is extremely bitter; removing a single carbon-oxygen bond produces neohesperidin dihydrochalcone, which is extremely sweet.

These are examples of phenomenal difference despite chemical similarity. There are just as many examples of phenomenal similarity despite radical chemical difference. Bitter-tasting compounds form a very heterogeneous lot that includes moderately large organic compounds such as the citrus compound naringin, the large organic acids found in hop oils, small molecules like urea, and even (as we saw above) some sugars. Van der Heijden (1993) listed no fewer than 19 distinct chemical families of bitter substances. Despite being very physically different, these compounds can optimally produce in us very similar bitter experiences. This makes evolutionary sense: they are all bad for us, so the body has no need to distinguish them.

So psychophysics has shown that, in the case of taste, chemical similarities and differences just cannot explain phenomenal similarities and differences. What then explains them? Neuroscience has shown that there is good "internal" correlation: in many cases, the phenomenal similarities and differences in taste experiences are better correlated with similarities and differences among neural states in the taste system than they are with anything in the physical world.

On the tongue there are several types of taste sensitive receptors, each optimally responsive to substances that we regard as having one of the "basic tastes" (sweet, salty, bitter, sour, umami). But at more central locations in the taste system neurons are *broadly turned*, with many neurons responding to more than one taste quality. So when one experiences a particular taste one undergoes a distinctive pattern of neuronal firing across many centrally located neurons. This is called an *ensemble activation*.

The resemblance ordering among tastes is well correlated with the resemblance ordering among such ensemble activations (as determined by multidimensional scaling) in "neural similarity space". There are a number of studies that bear this out. When Schiffman and Erickson (1971) asked humans to make similarity judgments between a number of different solutions, they found that similarities and differences in quality corresponded remarkably well to similarities and differences in ensemble activations in the rat: in general, the greater the phenomenal similarity, the greater the ensemble activation similarity. Likewise, Smith and coworkers (1983) created a multidimensional *neural similarity space* of ensemble activations in the hamster taste system in which distances among points represent degree of similarity. They found that the space is clearly interpretable on the basis of human taste experiences. The ensemble activations of sweet tasting-substances (sucrose, fructose, D-glucose, Na saccharin, and Galanine) are very similar to each other and very different from those of bitter-tasting substance (QHCl and urea). Likewise the ensemble activations of sour-tasting and salt-tasting substance cluster together in different areas.

These studies suggest that *spatial pattern* codes for taste quality: taste is coded by which neurons are activated and to what degree. But more recent studies show that the *temporal pattern* of firing within single neurons also contributes to taste coding. For instance, Di Lorenzo et al. (2009) found that the distinctive temporal patterns in nucleus of the solitary tract (NTS) corresponding to *basic tastes* are very dissimilar, in a way that mirrors those tastes' phenomenal dissimilarity. Further, binary mixtures produce temporal patterns that are *in between* those produced by their respective components. The temporal patterns produced by mixtures were even typically well approximated by a linear superposition of those produced by their components. Indeed, Di Lorenzo et al. report that "the entire [three-dimensional] taste space can be mapped by the temporal characteristics of response in a single cell" (p. 9232).

Of course, taste quality might be coded by both spatial pattern (ensemble activation) and temporal pattern. As Chen et al. (2011) write, "the existence of consistent temporal profiles of response among the responsive neurons for a given taste stimulus enhances the uniqueness of the across-neuron pattern of response by adding a dynamic dimension ... thus the spatial pattern produced by a tastant is sculpted as the response unfolds over time".

So far we have focused on the neural correlates of taste *quality*. There is also a very good correlation between average taste-cell firing rates and taste *intensity*. Due to an anatomical peculiarity, the chorda tympani nerve can be accessed in humans during middle-ear surgery by means of an electrode. In a well-known experiment, Borg et al. (1967) had patients estimate numerically taste magnitudes of certain substances at various concentrations. Then they recorded from taste cells and found *nearly perfect agreement between the neural and phenomenal data* (see also Oakley 1985).

18.2.2 Smell

Now let us turn to smell. As in the case of taste, the tracking intentionalist will presumably say that different types of smell qualities are different types of chemical properties that are tracked and thereby represented by our smell experiences. The phenomenal character of a smell experience is fully determined by the chemical type it optimally tracks.

But, in the case of smell, examples of bad correlation between experiences and the physical properties optimally tracked are even more plentiful than in the case of taste. Cowart and Rawson (2001, p. 568) sum up the situation as follows:

> Available evidence indicates that numerous chemical and molecular features (e.g., molecular weight, molecular mass and shape, polarity, resonance structure, types of bonds and sidegroups) can all influence the odorous characteristics of a chemical. However, no systematic description of how these characteristics relate to particular odor qualities has been developed. In other words, chemicals that bear little resemblance structurally can smell the same, and chemicals that are nearly identical structurally can elicit very different perceptual qualities.

Yet on tracking intentionalism the chemical properties represented by our experiences are supposed to be what fully determines all aspects of the experiences' phenomenal character, including the phenomenal resemblances and differences between them.

Another interesting psychophysical fact is worth mentioning. The tracking intentionalist would presumably say that the *phenomenal intensity* of a smell experience is constituted by the particular *concentration* of the external odorant that it represents. This fits many cases, because in general changes in the level of concentration go with changes in stimulus intensity. But there are counterexamples. In some cases a mere change in the concentration of a chemical can strikingly alter the *quality* and not just *intensity* of olfactory experience: for instance, the smell experience of *thioterpineol* is described as "tropical fruit" at a low concentration, as "grapefruit" at a higher concentration, and as "stench" at a still higher concentration (Malnic et al. 1999).

Here the radically externalist account of phenomenal character promoted by tracking intentionalists is at an explanatory disadvantage. Why should changes in represented concentration sometimes constitute changes in intensity, sometimes changes in quality? In the external world, there is only a difference in degree; but in some cases the quality changes in a categorical way. As we shall see, neuroscience provides the answer. The puzzle is resolved by a more internalist view on which internal neural factors play a role in determining phenomenal character, in a way at odds with tracking intentionalism.

If bad external correlation means that the chemical properties optimally tracked by our gustatory experiences don't explain the phenomenal character of those experiences, then what does explain it? Neuroscience has revealed "good internal correlation", suggesting the explanation is to be found in the brain.

Humans have about 450 *types* of smell receptors on the olfactory epithelium of the nose. (Contrast this with the mere three cone-types in vision or the mere four or five receptor types for taste.) They synapse at the olfactory bulbs, which in turn are connected to the primary olfactory cortex. The primary olfactory cortex is subdivided into several different areas: the anterior olfactory cortex, the olfactory tubercle, the piriform cortex (about which more presently), parts of the amygdala and the entorhinal.

As noted, chemicals that are nearly identical structurally can elicit very different smell experiences. Malnic and coworkers (1999) found that in such cases the very similar chemical produce very different patterns of firing across the smell receptors in mice. So where there is bad external correlation there is good internal correlation.

Interestingly, Malnic et al. also found that at different concentrations the same chemical can sometimes produce radically different *patterns* of activation across the smell receptors. This goes with the fact – just mentioned above as a puzzle on an external view of phenomenal character – that a mere change in the concentration of a chemical can sometimes strikingly alter quality of olfactory experience, not just the intensity.

Indeed, even some enantiomers (chemicals that are mirror image rotations) can smell quite different to us, while others smell the same. For instance, −carvone smells like spearmint while its mirror image +carvone smells like caraway. The best explanation is that the rotated molecules don't fit the same smell receptors (as if you were trying to fit your right hand into your left hand glove). Because of this, they stimulate different receptors. This is because the receptors contain chiral groups, allowing them to respond more strongly to one enantiomer than to the other. Consistently with this, Linster and coworkers (2001) found that enantiomers that smell quite different (as determined by behavioral measures) also produce quite different neural patterns further downstream in the olfactory bulb of *rats*. And those which smell the same produce similar patterns.

There is a striking demonstration of this kind of "good internal correlation" in the case of smell provided by a recent fMRI experiment by Howard and coworkers on *human subjects* (2009). This experiment is an advance in several ways. Most obviously, by contrast to animals, human subjects are able to *report on* phenomenal similarities and difference in their smell experiences. By obtaining their reports, and by performing fMRI scans, they obtained very strong evidence of "good internal correlation" in the case of smell.

In their main experiment, for a reason that will emerge, Howard et al. used chemicals that are physically very different but smell similar (viz. minty, woody or citrus); in other words, they focused on cases where there is "bad external relation" (see Fig. 18.1).

Using these odorants, Howard et al. found "that spatially distributed ensemble activity in human posterior piriform cortex (PPC) coincides with perceptual ratings of odor quality, such that odorants with more (or less) similar fMRI patterns were perceived as more (or less) alike" (2009, p. 932). In particular, Howard and coworkers found that, even though the molecular structures in each of the three families are quite different, they produce very similar ensemble activations in PPC, which are distinct from the activation patterns of the other two categories (Fig. 18.1). They even located ensemble patterns in a *three-dimensional neural similarity space*, and found that neural similarity (represented by distance) coincided very well with phenomenal similarity.

Now I can explain why Howard et al. used structurally very different molecules that smell very similar (minty, woody, or citrus): that is, why they focused on a case of bad external correlation in my sense. The reason is this: because the

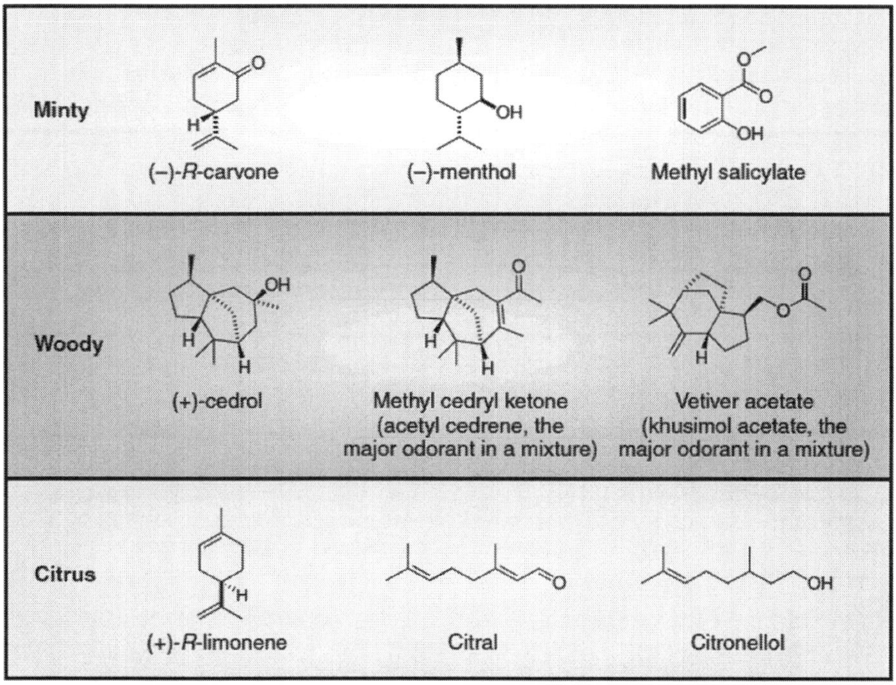

Fig. 18.1 Howard et al. (2008) found that molecular structures that are chemically very different but smell similar (citrus, minty, woody) produced very similar ensemble activation patterns in the PPC. In general, degree of phenomenal similarity or difference coincided with similarity or difference in PPC patterns, not similarity/difference in molecular structure (Reprinted from Margot 2009 with permission)

molecular structures in each of the three families are quite different, we know that the similarities in their ensemble activation representations in the brain are not mere artifacts of similarities in the molecular structures of the odorants used (since they were *not* similar); the only explanation of the observed correlation between ensemble activation similarity and phenomenal similarity is that ensemble activation similarity somehow plays a role in determining phenomenal similarity. As Margot (2009, p. 814) puts it in a discussion of the Howard experiment:

> [Because structurally diverse chemicals were involved] the fMRI effects were not merely reflecting odorant-specific differences [and similarities].... The fMRI effects unequivocally demonstrated that only the PPC ensemble activities are predictive of the category (woody, minty or citrus) of the odor that the subjects smelled. Because the chemical structure of the odors in each odor category are very different, this is strong support for the idea that *the PPC codes [i. e. determines] odor quality rather than structural and chemical similarity [in the odorants tracked]*. (My italics.)

The Howard study on humans is not the only study showing that similarity and difference in olfactory experience to be correlated with similarity and difference in ensemble activations. There are also many studies on animals that show this as well. For instance, Youngentob and coworkers (2006) did a similar study on rats. Of course, by contrast to humans, rats unfortunately cannot report on the degrees of similarities among their smell experiences of chemicals. So to get at the phenomenal structure of their smell experience, a more indirect method is required. Youngentob and coworkers had the rats perform a *confusion matrix task*. The basic idea is that degree of phenomenal similarity corresponds to probability of confusion. Then, using a 2-DG functional mapping technique and multidimensional scaling, they looked at the degrees of neural similarities among the neural ensemble activations set up by the odorants in the rats' olfactory bulb. Here is what they found:

> We found a remarkable predictive relationship between the odorant-specific glomerular activity patterns and the perceptual relationship among the odorants. When the activity pattern for two odorants mapped relatively close to each other in the functional MDS [multidimensional scaling] space, then so did the perceptual data. Alternatively, when the 2-DG activity patterns mapped relatively distant from each other in the MDS space, then so did the behaviorally derived perceptual data.... *Our results support a combinatorial coding model in which the total pattern of bulbar activity is relevant to the production of an odorant's perceptual quality*.... Indeed, our results show neural and perceptual relationships that could not be presumed from any prior notion of molecular similarity among the odorants. There was a greater perceptual and [neural] pattern similarity between pentadecane and santalol, than between either of these odorants and β-pinene, yet both santalol and β-pinene are bridged polycyclic compounds ... (p. 1343; my italics)

In my terms, what they are saying is that they found good internal correlation even when there was bad external correlation. So the results of their experiment on rats are similar to those of the experiment conducted by Howard and coworkers on humans.

18.2.3 Pain

Suppose you have a variety of pains of different intensities in different bodily locations: throbbing pains, prickling pain, stabbing pain, heat-induced pains of various intensities, and so on. On tracking intentionalism, felt pains reduce to types of bodily disturbance, just as colors reduce to reflectance properties; and every phenomenal aspect of the pain reduces to some physical feature of the bodily disturbance represented by the pains. So felt location is just represented location; and differences in *quality* among pains (prickling, stabbing, throbbing, etc.) are constituted by differences in the *types* of bodily disturbance they represent. Now, besides quality and location, you can also focus on the *sensory intensity* of a pain. (This related to but distinct from the *unpleasantness* of the pain, or the *affective dimension* of pain, as we shall see.) On tracking intentionalism, what (possibly complex) aspect or feature of the external stimulus constitutively determines its sensory intensity? As far as I know, tracking intentionalists have simply not addressed this issue. The simplest view would be that the sensory intensity of a pain is fully determined by the *intensity* and *size* of bodily disturbance optimally tracked and so represented.

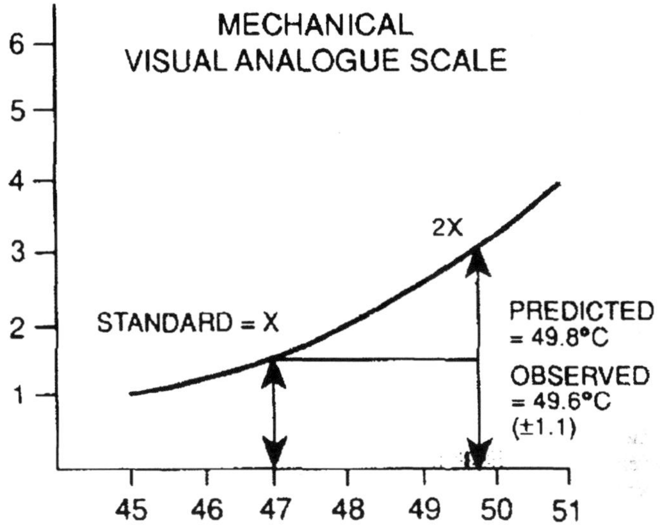

Fig. 18.2 The relationship between noxious temperatures and pain sensation intensity ratings is described by a power function with an exponent of about 3.2 (Reprinted from Price et al. 1994)

But, in fact, even under optimal conditions, pain intensity is very poorly correlated with these factors. For one thing, psychophysics has shown that there is *response expansion*. Even under optimal conditions, the relationship between pain intensity and bodily disturbance is described by a power function with an exponent greater than 1, where the size of the exponent differs for different kinds of stimuli. Stevens et al. (1958) showed this in the case of electric shock. Likewise heat-induced pain intensity is a power function of stimulus-temperature, with an exponent of about 3.2 (see Fig. 18.2). A small increase in temperature can double subjects' ratings of the sensory intensity of pain. Pain intensity is not only a function of stimulus intensity; it is also a function of stimulus area, in a way that cannot be easily summarized (Price 1999).

So psychophysics has revealed that, even under optimal condition, there is a messy relationship between pain intensity and the many aspects of the bodily disturbance tracked. By contrast, many researchers have reported a *perfect correlation* between pain intensity and a single neural parameter, namely firing rates of neurons in the areas of the brain involved in pain. For instance, using noxious temperatures and measuring neural activity with fMRI, Coghill et al. (1999) found *linear* relationships, with different regression coefficients for different areas. They write:

> Many cortical areas exhibit significant, graded changes in activation *linearly related to* pain intensity .... Normalized CBF differences ... confirm that the regression coefficient accurately describes the quantitative relationship between brain activation and perceived pain intensity. For example, the regression coefficient of the medial thalamus was 0.5, and the average psychophysical rating of 50 °C was 15.6. Accordingly, the predicted activation

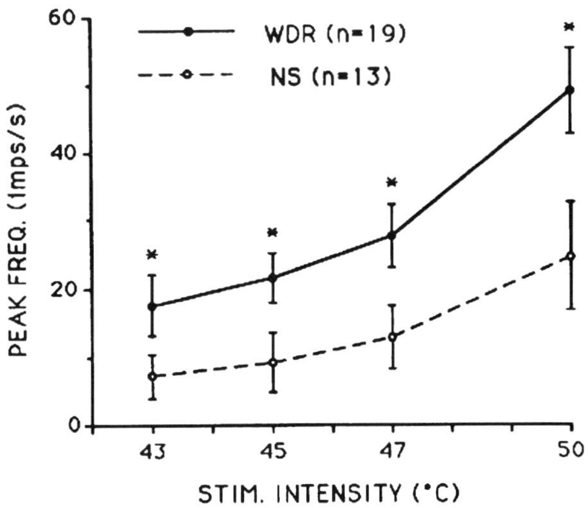

Fig. 18.3 The relationship between temperature and the firing rates wide-dynamic range (WDR) neurons in monkey S1 very closely resembles the psychophysically-derived relationship between temperature and pain sensation intensity in humans (shown in Fig. 18.2) (Reprinted from Kenshalo et al. 2000 with permission)

difference between scans of 50 °C stimulation and rest would approximate 0.5 * 15.6 or 7.8 (in units of normalized CBF). The observed activation difference was 7.07 (in units of normalized CBF). (p. 1936)

Using lasers as their pain stimulus and a different technology – namely MEG – to determine neural response, Timmerman et al. (2001) found a particularly close relationship between human subjects' pain intensity and firing rates of neurons in the *primary somatosensory cortex* (S1) – for short *S1 firing rates*. As they put it, "amplitudes of contralateral S1 activity match precisely the subjects' pain ratings". Kenshalo et al. (2000) also found in a single-unit study that the relationship between temperature and the firing rates of wide-dynamic range (WDR) neurons in monkey S1 very closely resembles the psychophysically-derived relationship between temperature and pain intensity in humans shown in Fig. 18.2. Indeed, just as pain intensity doubles between 47 and 50 °C, so WDR response in monkeys roughly doubles between these temperatures (see Fig. 18.3).

These experiments and a slew of other empirical considerations (Price 2002) suggest that S1 plays a special role in determining pain intensity. For the sake of simplicity, I will sometimes assume this view in what follows; but my arguments would go through on a more distributed view of the neural basis of the sensory intensity of pain as well (e.g. Coghill et al. 1999).

So far I have discussed the sensory dimension of pain. Pain of course also has an affective-motivational dimension. Price (2002, p. 393) describes it as "the moment-by-moment unpleasantness of pain, which consists of emotional feelings that pertain to the present or short-term future, such as annoyance, fear, or distress". Many studies (e.g. Rainville et al. 1997; Hofbauer et al. 2001) actually show that sensory intensity and pain affect can be modulated independently and that, while sensory intensity is based in S1, the affective dimension of pain is based in the anterior cingulate nucleus (ACC).

So, in the case of pain intensity, while there is bad external correlation, there is very good internal correlation. There is messy, uncodifiable relationship between the size and intensity (e.g. temperature) of the various types of bodily disturbances and the S1 firing rates they set up; and in turn there is a linear correlation between these S1 firing rates and pain intensity at the sensory level.

18.2.4 Audition

Finally, there is evidence of "bad external correlation" and "good internal correlation" in the domain of auditory perception.

Here are some examples of bad external correlation in the auditory domain (Moore 2003). There is a relationship between perceived *intensity* of a sound and the amplitude of the sound. But it is one of response *compression*. So, for instance, doubling perceived intensity requires *far* more than doubling amplitude. However, at lower amplitudes, loudness increases more rapidly with increasing amplitude. Perceived loudness is a function not just of amplitude but also of frequency, in a way that resists codification. For complex tones, loudness also depends on bandwidths (as we shall see, this has a cortical explanation). Similarly, there is a relationship between the perceived *pitch* of a sound and the frequency (for complex sounds, fundamental frequency) of the sound. But it too is one of response compression, with the degree of compression depending on frequency level. Perceived pitch also depends on amplitude. In particular, the pitch of tones below 2 kHz increases with increasing amplitude and the pitch of tones above 4 kHz decreases with increasing amplitude. For complex tones, pitch depends on a variety of other factors.

While there is bad external correlation in the auditory domain, there is some evidence of good internal correlation. Let me focus on the case of loudness, because the neural basis of pitch perception remains relatively poorly understood. Relkin and Doucet (1997, p. 2738) write that "the perceived loudness of a pure tone appears to be linked both to the number of spikes fired by single neurons and to spatial spread of excitation in the auditory nerve". Langers et al. (2007) used fMRI to look at neural activity further downstream in the auditory cortex. They found that "cortical activity is more closely related to the perceptual loudness level of sound than to its [external, physical] intensity level" (p. 714) and indeed report "a type of non-linearity ... comparable to that reported in psychophysical studies on loudness perception that employ subjective loudness scaling" (p. 716). On the basis of this study and others, Röhl et al. (2011, p. 1494) conclude that "the most simple interpretation would be, that AC [auditory cortex] is fed by ... the auditory brainstem according to the sound pressure level and the bandwidth of the stimuli, and an additional component is added which is linearly related to the perceived loudness".

So the situation regarding loudness may be similar to the situation regarding pain intensity. In both cases, there is a non-linear relationship between sensory intensity and the physical stimulus. The difference is that, while in the case of pain intensity the relationship is one of *response expansion*, in the case of sound intensity it is one of *response compression*. The explanation is that there is also a compressive relationship between the physical stimulus (amplitude) and average firing rate of auditory cells. And our psychophysical judgments of sound intensity directly correspond to these firing rates.

So much for our brief look at the science of sensation. While many philosophers focus narrowly on one sense-modality, we have considered several. So we can see the "big picture" that emerges. The fact that when it comes to phenomenal character there is "bad external correlation" but "good internal correlation" across the various modalities makes one suspect that there is something very wrong the radically externalist approach promoted by tracking intentionalists, according to which phenomenal character is fully determined by the external physical properties tracked by our experiences. The science is *apparently* at odds with tracking intentionalism. But can this be definitely shown? In the following sections, I will construct arguments that are intended to do exactly that.

18.3 First Argument: The Internal-Dependence Argument

I call my first argument the *internal-dependence argument*. The aim is to *demonstrate* the conflict between tracking intentionalism and science by describing *counterexamples* to tracking intentionalism.

18.3.1 Why Actual Cases Fall Short

Now you might think that *actual cases* involving perceptual variation suffice as counterexamples to tracking intentionalism. Many philosophers have certainly thought so, including Ned Block (1999), Brian McLaughlin (2003), Sydney Shoemaker (2000), and Uriah Kriegel (2009). But their arguments have been ineffective. Before turning to my own counterexamples, it will be helpful to see why. For my counterexamples will be designed to preempt the usual responses.

You are probably familiar with actual cases of perceptual variation. The same bodily disturbance-type can give different individuals slightly different pains. The same substance can taste differently to different individuals. There are, for instance, "supertasters". The same odor cloud can smell differently to different people. The same sound can sound differently. As Block (2010) has noted, in some cases phenomenal variation can even be due to differences in *attention*. All of these cases pose the same problem for tracking intentionalism. In these cases, the individuals involved have *different* experiences of the same stimulus, presumably due to differences in the kind of neural processes discussed in the previous section. But don't we have to say that experiences accurately represent the *same* external physical properties? In that case, phenomenal character is determined by more than the external physical property represented in the world, contrary to tracking intentionalism.

In response to an actual case of variation, tracking intentionalists can always invoke what I shall call the "illusion response" (Tye, Byrne and Hilbert, Batty) or else the "pluralist response" (Kalderon, Smith).[5]

The *illusion response* is especially plausible in cases where there is some kind of interference or abnormality, so that *optimal conditions* do not obtain. The idea is that the individuals' different experiences *do not* represent the same external physical property of the stimulus. One individual represents a physical property that the stimulus does have and the other represents a physical property that it does not have. This representational difference constitutes the phenomenal difference.

The tracking intentionalist might invoke *pluralist response* in cases where optimal conditions obtain, and the individuals involved are normal. The idea is that the individuals are actually *tracking different physical properties* of the stimulus, because of differences between their sensory systems. So the sensible qualities they represent are distinct but equally real properties of the stimulus. They both get it right. This response is "pluralist" because it says that the external world is rich with sensible qualities. This helps the externalist explain perceptual variation without having to posit illusion. So, for instance, a wine might actually have many objective tastes, constituted by overlapping but distinct chemical types. And, due to differences in their taste systems, one individual might perceive one while another individual perceives another (Smith 2007, p. 65). Indeed, to handle more radical cases of *inter-species variation*, the tracking intentionalist will say that foods have various alien tastes that we cannot imagine. The human tastes and the alien tastes are constituted by different chemical properties belonging to the same substances. We track and thereby perceive one range of properties of the substances. Another species might track and thereby perceive a different a totally different range of properties of the substances.

The idea here is that actual differences in neural processing between individuals lead to differences in what external properties they track and thereby represent. In this way, the tracking intentionalist can handle actual cases of variation.[6]

To refute tracking intentionalism, what we need is a case that is invulnerable to both the illusion response and the pluralist response. At this point, some philosophers might be tempted to invoke intuitions about the possibility of far

---

[5]For the illusion response to some cases, see Byrne and Hilbert (2003), Tye (2006), Batty (2010). For the pluralist response to some cases, see Kalderon (2011) and Smith (2007, p. 65).

[6]But I think that some extreme cases of variation, not discussed in the literature, are particularly troublesome for tracking intentionalists and objectivists. As Batty (2010) notes, a large percentage of humans cannot smell *androstenone*. She does not note that, of those who can smell it, half perceive it as having a pleasant sweet floral smell and the other half smell it as having an unpleasant ruinous smell. Here the illusion response would be implausible, given the parity between the groups. And the pluralist response (Kalderon, Smith) is problematic as well. In one version, the pluralist view would have it that the floral smell perceived by the first group is identical with the disjunction of all the molecular types (including androstenone) that are the objective correlate of the perception of that smell among humans; and the ruinous smell perceived by the first group is identical with overlapping but distinct disjunction of all the molecular types (including androstenone) that are the objective correlate of the perception of that smell among humans. On this view, androstenone objectively possesses two *radically different* smells. This is hard to accept.

out hypothetical cases, like brains in vats and spectrum inversion. For instance, since there is an explanatory gap between color experience and the reflectance (light-involving) properties of surfaces, it is quite conceivable that two individuals should accurately track the very same reflectance property but have different color experiences (Shoemaker 1994). But tracking intentionalists just reply that, while this may be conceivable, it is not possible (Tye 2000). Indeed, I would add that such intuitions are just instances of our more general explanatory gap intuition to the effect that experience is modally independent of *all* physical conditions. So physicalists have special reason to be suspicious of them.

My strategy will be quite different. I will describe hypothetical but realistic *coincidental variation cases*. In these cases, there is neural and behavioral variation between the members of different species. Nevertheless, I will simply *stipulate* that, whatever conditions need to be in place in order for two creatures to accurately represent exactly the same properties, those conditions are indeed in place. While there is neural and behavior variation, there is a *complete coincidence* in what objective properties their sensory systems track. Given the vast neural and behavioral differences, I will argue that they would have different experiences. The cases are not just ones of alternative "neural realizations" of the same experience. To establish this verdict I will *not* use dubious intuitions which tracking intentionalists might simply dismiss; I will *argue* for this verdict on the basis of the research in neuroscience and psychophysics discussed previously. But tracking intentionalism (and indeed all versions of externalist intentionalism) delivers the mistaken verdict that the individuals have exactly the same experiences. Given my stipulations, neither the illusion response nor the pluralist response will be available to externalists.

I will consider several cases. By focusing on several cases, we can appreciate the strength of the cumulative case against the externalist program. To answer my argument, externalists would need to develop solutions in every case, instead of narrowly focusing (as they often do) on one sense-modality.

18.3.2 Yuck and Yum

My first case can be introduced *via* an actual case. The berries *actaea pachypoda* (Doll's-eyes) is highly poisonous (and bitter in taste) to humans, but harmless to birds, the plant's primary seed dispensers. They eat it up without problem. It is reasonable to think that while the berries taste horribly bitter to us, they taste different to the birds.

Now this actual case is no problem for tracking intentionalists. They can appeal to the pluralist response. Humans and the birds differ at the receptor level too, so that the brain states that realize their experiences of the berries are caused by different ranges of chemical properties. So the tracking intentionalist can say that the phenomenal difference between the humans and the birds is grounded in their sensorily representing different, but equally real, taste properties of the berries. In short, they can invoke the *pluralist response*.

But with a small twist we do get a counterexample to tracking intentionalism. Just consider a hypothetical *coincidental* variation case in which the brain states of the two individuals involved do optimally track the very same chemical property of the berries. It is still reasonable to think that berries taste differently to them, but tracking intentionalism is inconsistent with this verdict.

In more detail, suppose *Yuck* and *Yum* belong to different species that evolved in separate environments containing some berries. Now you might suppose that Yuck is an actual human – me or you – and Yum is some hypothetical creature. Or you might suppose that Yuck and Yum both belong to hypothetical, human-like species. It does not matter. In any case, the berries are extremely poisonous to Yuck. By contrast, in Yum's environment, the berries are a very important foodsource, since other foodsources are scarce. So Yum's species evolved immunity to the berries. In addition, when Yuck and Yum taste the berries, their taste systems undergo radically different ensemble activation states (spatiotemporal neural patterns discussed in Sect. 18.2). Yuck and Yum also innately disposed respond to their tastes experiences with radically different behaviors. For instance, Yuck vomits and withdraws from it violently, while Yum is drawn to it, rubs his tummy, and so on.

I said that Yuck and Yum undergo different ensemble activations in response to the berries. Let me be more specific. Suppose that the notorious poison dart frog is highly poisonous to both Yuck and Yum. Suppose further that, when Yuck tastes berries, the ensemble activation state he undergoes is quite similar to the one he undergoes when he tastes the dart frog. By contrast, when Yum tastes the berries, the ensemble activation Yum undergoes is radically different from the one is undergoes when he tastes the dart frog, and much more like the one he undergoes when he tastes yummy bananas. In general, the set-up is that the ensemble activation that the berries produce in Yuck is similar to those which he undergoes when he tastes things that presumably taste bad or bitter to him, whereas the ensemble activation that the berries produce in Yum is similar to those which he undergoes when he taste things that presumably taste good (e.g. sweet) to him (see Fig. 18.4). And there are consequent differences in their behavioral responses.

Despite these differences, we can stipulate that Yuck and Yum are similar at the receptoral level. Indeed, we can stipulate that, when they taste the berries, the *postreceptoral* ensemble activation patterns in their taste systems, although different, optimally track the very same complex chemical property of the berries, *C*. This chemical property *C* will likely be a *disjunctive property*, because many different combinations of chemical properties can produce the same response in the taste system. So I am stipulating that their ensemble activation states track the same *disjunction* of chemical properties *C*, the very one with which tracking intentionalists and other objectivists about taste would identify the taste perceived by Yuck and Yum.

That, then, is the case described in non-phenomenal terms. The crucial question is whether Yuck and Yum would have different taste experiences or the same taste experience of the berries.

I think we should say that they would have different experience experiences. We have seen that resemblances and differences in taste quality are much better correlated with resemblances and differences in ensemble activations than with

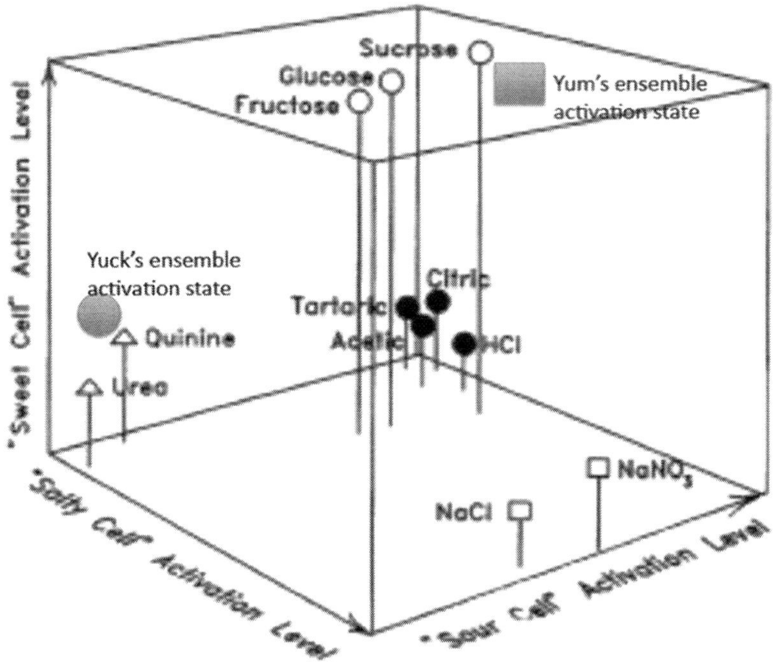

Fig. 18.4 Yuck and Yum's neural representations of the berries occupy different locations in the kind of neural similarity space uncovered by Smith and coworkers (1983) (Adapted from Churchland 1996 with permission)

resemblances and differences in the chemical properties optimally tracked. So the ensemble activation differences between Yuck and Yum *are very good evidence* that they have phenomenally different taste experiences of the doll's eyes berries. By contrast, the fact that they optimally track the same chemical property of the berries is very *poor* evidence of phenomenal sameness.

More specifically, when Yuck tastes the berries his ensemble activation state is very close to those which he undergoes when he tastes the poison dart frog and other characteristically bad-tasting things, whereas when Yum tastes the berries the ensemble activation he undergoes is similar to those which he undergoes when he tastes bananas and other characteristically sweet things. Given that similarities and differences among ensemble activations are the only things in the physical world that correlate well with similarities and differences among taste experiences, Yuck's taste experience of the berries is probably similar to his experience of the poison dart frog and other characteristically bitter-tasting things, whereas Yum's taste experience of the berries is probably similar to his taste experience of bananas and other characteristically sweet-tasting things.

The behavioral differences between Yuck and Yum suggest an independent argument for the same verdict. When Yuck has the berries, he exhibits certain innate responses: he vomits, withdraws, and treats them like other things that presumably taste bad (e.g. bitter) to him. When Yum has the berries, he seeks more, and treats them like other things that presumably taste good (e.g. sweet) to him, like bananas. Even if we are not behaviorists or functionalists, this suggests that they have different experiences of the berries, just as humans and birds to in the actual world.

Note that I only say that, given what we know about the physical basis of taste experience, it is reasonable to suppose that there are *some* phenomenal differences between Yuck and Yum. This is all that is required by the argument. I do not say that we can read off from the physical facts *exactly* what their taste experiences are like. Nor do I say that we can read this off more specifically *just from the character of their neural states*. The functional, sensorimotor differences may play a role too.

So the only reasonable verdict is that Yuck and Yum would have different taste and smell experiences of the berries. But tracking intentionalism delivers the incredible verdict that Yuck and Yum would have phenomenally identical taste experiences of the berries, despite the vast neural and behavioral differences between them when they taste the berries.

By stipulation, Yuck and Yum's ensemble activation states, although different, track exactly the same complex external chemical property, *C*, of the berries. Further, *optimal conditions* obtain. It is not as if one is a genetic freak, or has a malfunctioning taste system. On the contrary, their taste systems, although different, are both working as they were designed by evolution to work. Further, their innate behavioral dispositions in response to the berries, although different, are adaptive, given the difference in the biological significance of the berries to them. On tracking intentionalism, this means that Yuck and Yum both sensorily represent ("perceive") the very same chemical property, *C*, of the berries. In general, their ensemble activation states, although different, represent exactly the same chemical properties.

Consider an analogy. The words 'river' and 'liver' are similar but represent quite different things. Likewise, on tracking intentionalism, even though Yuck and Yum's ensemble activation states occupy different locations in neural similarity space, they represent the very same external chemical properties: on this view the similarity/difference structure of those states in incidental to what they represent.

So far I have just argued that tracking intentionalism implies that Yuck and Yum's ensemble activation states are representationally identical. Now I do not think this is *by itself* bad. (Here I am responding to a query from Fred Dretske.) I agree with tracking intentionalists like Dretske and Tye that ensemble activation states might represent external chemical properties *in some sense*, even if there is "bad external correlation", that is, even if the resemblances and differences among them are not matched by resemblances and differences among the chemical properties. On many views, anything can represent anything: the connection between the intrinsic character of content-vehicles and what they represent is arbitrary.

But recall that on tracking intentionalism phenomenal character is also fully determined by representational content. In particular, tracking intentionalists claim that the *sensory character* of taste experience (sweet, bitter, etc.) is determined by what chemical properties taste experience represents. Once we add this to the tracking intentionalist's commitment to the claim that Yuck and Yum's experiences represent the very same chemical type, we *do* get a bad verdict: that Yuck and Yum have experiences of the berries identical in sensory character (sweet, bitter, etc.), despite the vast neural and behavioral differences between them. For instance, on tracking intentionalism, maybe they both have an *intensely bitter* experience of the berries. *This* verdict is radically implausible. Their ensemble activation states occupy different locations in neural similarity space; and that is the only known predictor of taste quality (sweet, bitter, etc.). And, by stipulation, Yum is drawn to the berries. So by far the most reasonable verdict is that Yuck and Yum would have experiences of the berries (Doll's-eyes) that differ radically in their sensory character (bitter and sweet), just as humans and birds do in the actual world.

Barry Smith (2007) is an objectivist about tastes who handles actual cases of taste variation by accepting the pluralist view that foods have multiple tastes that different individuals can perceive depending on the conditions. But he cannot handle Yuck and Yum similarly. On his view, the taste that Yuck perceives is an enormously complicated property of the berries. (Smith would say that *flavor* is an even more complex property, because unlike taste flavor depends on odor; perhaps flavor is a configuration of sapid, odorous and textural properties tracked through several sense organs. Here I am focusing on taste, but my argument applies *mutatis mutandis* against objectivism about flavor.) Yuck's perceiving the taste presumably supervenes on his bearing some natural relation (perhaps a tracking relation) to it. But I stipulate that Yum bears the same relation to the very same enormously complicated property of the berries. So, even if Smith is right that the berry has multiple objective tastes, he is committed to the claim that Yum perceives very same one as Yuck. The trouble is that this is implausible. It is not the case that they perceive the same taste (as it might be, an extremely bitter one), which Yuck does not like and Yum happens to like. Given that they differ in the kind of ensemble activation states that we know to the best predictors of sensory quality and sensory similarity, and given that they differ in their fine-grained taste-related behavior, the only reasonable verdict is that they ostensibly perceive quite *distinct* tastes (as it might be, a bitter one and a sweet one). The conclusion I draw is that tastes are not complex objective properties of external items.

18.3.3 Sniff and Snort

In the previous section, I discussed an fMRI study by Howard et al. on humans. Let *Sniff* be one of the actual participants of the study. He successively smells the chemical structures represented in Fig. 18.1. Among other things, he reports that his (+R) *limonene* experience resembles his *citral* experience more than his *menthol* experience. In particular, limonene as well as citral presents a citrus smell, whereas menthol presents a mint smell. Why is this?

The explanation is not to be found in the (possibly disjunctive) chemical properties – call them *L*, *C* and *M* – that his smell experiences track, and with which the objectivist would identify the smell qualities perceived by Sniff. There is no evident sense in which limonene is more like citral than menthol. In fact, if anything, limonene is more like *menthol* than citral. This is an actual case of bad external correlation.

Howard et al. claim that the explanation is only to be found in Sniff's brain, in particular, in his posterior piriform cortex (PPC). The explanation for why Sniff's limonene experience resembles his citral experience more than his menthol experience is that Sniff's PPC *neural representation* of limonene is more similar to his *neural representation* of citrus than in his *neural representation* of menthol. As Margot (2009, p. 814) says, "the PPC codes [i. e. determines] odor quality rather than structural and chemical similarity [in the odorants tracked]".

This appears directly at odds with the externalist account of sensory quality promoted by tracking intentionalists. To make the conflict precise, consider a counterfactual situation. In this situation, everything is more or less the same as in the actual world. But, due to some chance differences in our evolutionary history, there are differences in the postreceptoral wiring leading from the receptors in the nose to ensemble activations further downstream in the PPC. In this situation, Sniff's counterpart, *Snort*, participates in the Howard study. Because of the difference in wiring, while Sniff's neural representation of limonene is more similar to his neural representation of citrus than in his neural representation of menthol, Snort's neural representation of limonene is more similar to his neural representation of *menthol* than his neural representation of citral. In short, their PPC representations of limonene occupy quite different positions in the kind of *neural similarity space* studied by Howard et al.

Let us suppose that there are consequent behavior differences. So, while Sniff sorts limonene with citral and not menthol, Snort sorts limonene with menthol and not citral. Since Snort evolved in a different counterfactual situation, he probably does not speak a language that looks like English. But we can suppose that, while Sniff reports that his limonene experience resembles his *citral* experience more than his menthol experience, according to the best translation Snort says "my limonene experience resembles my *menthol* experience much more than my citral experience".

Of course, the most reasonable verdict about this case is that there are phenomenal differences between Sniff and Snort's smell experiences of limonene, citral and menthol. In particular, their smell experiences fall into different phenomenal resemblance-orders. And while limonene presents a citrus smell to Sniff, it presents a minty smell to Snort. For Sniff and Snort's PPC *neural representations* fall into different resemblance orders. And *neural* similarity is the only thing in the physical world known to predict *smell* similarity. In addition, this verdict is the best way of making sense of Sniff and Snort's different sorting and verbal behaviors.

But, if we add some more details to the case, tracking intentionalism delivers the opposite verdict that there are no phenomenal differences among Sniff and

Snort's smell experiences of limonene, citral and menthol, despite the radical neural and behavioral differences between them. For we can stipulate that their neural representations for these chemicals, while different, track under optimal conditions the very same chemical properties of those chemicals, namely them *L*, *C* and *M*. On tracking intentionalism, the smell qualities they perceive are none other than these chemical properties. So, according to tracking intentionalism, Sniff and Snort represent and hence perceive exactly the same smell qualities. In general, on tracking intentionalism, the representational contents of their experiences are *exactly* the same across the actual case and the counterfactual case, if the tracking facts are held constant. There are differences in the "neural content carriers" but no differences whatever in the externally-determined representational contents they carry. Because there are no representational differences between Sniff and Snort, there also can be no phenomenal differences between them, according to intentionalism. *A fortiori*, there can be no differences in the phenomenal *resemblance-orders* of their experiences of limonene, citral and menthol. And, on tracking intentionalism, Snort perceives the same specific objective "citrus" quality in limonene that Sniff perceives. This is so despite the radical neural and behavioral differences between them.

In reply, the tracking theorist might grant that on his theory Sniff and Snort track and thereby represent the odorants as having the same *monadic* olfactory properties, namely *L*, *C* and *M*. But he might insist also that they represent these same properties as standing in *different resemblance-orders*. In particular, Sniff's odor experience represents *L* as more like *C* than *M*, while Snort's odor experience represents *L* as more like *M* than *C*. In this way, the tracking intentionalist can insist that there is a representational difference, and hence a phenomenal difference, between their experiences. Call this the *structure gambit*, because the idea is that their experiences represent different contents about qualitative structure.

But this response fails for a number of reasons. Here I will only mention three. First, in the Howard experiment Sniff and Snort smell the odorants *successively*. So, contrary to the reply, there is no time at which their olfactory experience might have represented them (or the general properties *L*, *C* and *M*) as standing in a certain resemblance-order. Analogy: if you experience three objects *successively*, your experience cannot represent a spatial relation among them. I take this point from Alex Byrne (2003, p. 656). Second, in any case the present reply is inconsistent with tracking intentionalism. I stipulate that Sniff and Snort bear the tracking relation (and other relevant relations) to the *same* conditions. So even if we concede for the sake of argument that Sniff tracks and thereby represents a certain complex structural condition, then Snort represents the *same* structural condition. There is no naturalistic account of how they might represent *different* such conditions. Third, on the "structure gambit", Sniff and Snort perceive exactly the same individual smell qualities, but experience them as standing in different resemblance-orders. But this can be ruled out *a priori*. It is like saying that, while you actually experience blue as more like purple than green, another creature could experience blue as more like *green* then purple. Against this, perceiving individual qualities essentially involves perceiving them to stand in certain resemblance relations.

Clare Batty (2010) is partial to objectivism about smell qualities, even if she does not outright endorse it. She suggests that it can survive arguments based on actual cases of variation. I believe that the hypothetical "coincidental variation" case involving Sniff and Snort poses a difficulty for any version of objectivism that is more serious than the problem she discusses concerning actual cases. Here is the argument. According to the objectivist, when Sniff smells limonene, the smell quality that he perceives is identical with some "objective" property of the odor cloud. Call it *L*. Maybe *L* is a disjunctive chemical property, where the disjuncts are all the combinations of chemical properties that yield that same distinctive smell. The details do not matter. Now, although she does not discuss this issue, Batty (if she outright accepts objectivism) must say that Sniff perceives *L* by virtue of bearing some (perhaps as yet unknown) complex physical relation *R* to *L*. Presumably, *R* would a kind of tracking relation, but my argument is neutral here. Let us stipulate that in the counterfactual situation, on smelling limonene, Snort also bears *R* to the same complex response-independent property *L*, despite his neural and behavioral differences from Sniff. In general, Sniff and Snort both bear to this same property *L* all the naturalistic relations that might ground perceptual representation, despite the neural and behavioral differences between them. Then the objectivist is committed to the claim that on smelling limonene Sniff and Snort perceive the very same smell quality, despite the PPC neural differences and behavioral differences between them. But this is very implausible. The PPC neural differences and behavioral differences between Sniff and Snort make it reasonable to suppose that they ostensibly perceive *distinct* smell qualities. (They don't perceive the same smell quality under different modes of presentation, whatever that might mean; they perceive *numerically distinct* smell qualities.) In particular, Sniff ostensibly perceives a citrus smell quality and Snort perceives a minty one. The conclusion I draw is that smell qualities cannot be objective properties of odor clouds.

18.3.4 Mild and Severe

As I mentioned in Sect. 18.2, the relationship between stimulus intensity and pain intensity is generally one of "response expansion". Pain intensity also depends on stimulus size and duration, in a complex way. In short, there is bad external correlation. For instance, in one experiment, Donald Price (1999) asked subjects to rate their sensory pain intensity on a visual analogue scale (VAS) with a sliding marker in response to noxious temperature. He consistently found that the psychophysical relationship of pain sensation intensity to heat stimuli (45–50 °C, for 5 s) is a positively accelerating power function with an exponents of about 3.0. Of course there might be indeterminacy concerning the precise number. In any case, small changes in temperature yield large changes in perceived pain intensity. This makes evolutionary sense: small changes in stimulus intensity can be very dangerous. For instance, in another experiment, Price found that when subjects perceive a standard stimulus of 47 °C, they determined 49.8 °C (average) to be roughly "twice as intense" as 47 °C, which agrees perfectly with the prediction of his independently-determined stimulus–response curve with a exponent of about 3.0 (see Fig. 18.2 in Sect. 18.2).

Two clarifications. First, we must bear in mind that pain researchers distinguish between the sensory dimension of pain and the affective (unpleasantness) dimension of pain. These can come apart. The same pain can "bother" different people to different degrees. The results I have described concern subjects' ratings of the *sensory* intensity of pain. Price also asked subjects to rate the *unpleasantness* of their pains, and reliably found different results. But this will not concern us at present since I will be concerned with *sensory* intensity (for more on this see Sect. 18.5). Second, you might think the notion of one pain being twice more intense than another makes no sense. But ratio scaling of loudness makes sense, so why not pain intensity? Of course there might be a great deal of indeterminacy. But why can't it true that one pain is *roughly* twice greater than another in intensity? Price provides evidence that rough claims like this are true. However, the argument I am about to present against tracking intentionalism does not strictly speaking require the truth of these particular judgments; even if the judgments are not true, the fact that subjects make them can provide evidence about what their experiences are like.

While pain intensity is related in a non-linear fashion to numerous stimulus features (bad external correlation), it is more proportional to firing rates in the brain (good internal correlation). For instance, as Price (2002, p. 395) reports, numerous studies have found that "stimulus–response functions of WDR [wide dynamic range] neurons to this range of skin temperatures are precise positively accelerating power functions, ones that strongly resemble power functions that are obtained from human subjects who rated these same temperature stimuli" (see for instance Figs. 18.2 and 18.3 in Sect. 18.2). Indeed studies have shown that *sensory intensity* of pain is linearly related to those neural firing rates, as we saw in Sect. 18.2.

All of this goes against tracking intentionalism; indeed it goes against any "externalist" theory of pain. Contrary to this view, the intensity of a painful response to temperature is not merely determined by any features of the stimulus that our pain system tracks. The simplest hypothesis is that it is more directly determined by S1 firing rates. To turn this into an argument into tracking intentionalism, all we have to do is describe a coincidental variation case involving pain.

Let us consider a counterfactual situation in which the psychophysical response curve describing the relationship between noxious temperature and neural response in S1 and other neural regions is *steeper* than it is in the actual world – that is, steeper than those shown in Figs. 18.2 and 18.3 in Sect. 18.2. In other words, the rough exponent is consistently much higher than it is in the actual world (than is, higher than around 3.0).

Let us suppose that this is not because in the counterfactual situation the same noxious temperatures are more of a threat to the organism, or more likely to jeopardize the organism, than they are in the actual world. My idea is that it just happens that in this situation the psychophysical response curve is steeper than in the actual world. While selection pressures might ensure that the response function has an exponent greater than 1, so that we will be sure to avoid increasing noxious temperatures, the precise value (or range of values) of the exponent is evidently a matter of chance. So it can vary across worlds in which noxious temperatures are equally dangerous. (I mention this because it will later on be important to undermining proposals suggested by Cutter and Tye and Hill.)

Now consider a subject of one of Price's experiments in the actual world, where the response function is *less steep* that it is in the counterfactual situation. Call him *Mild*. He experiences temperatures in the noxious range between 45 and 50 °C and undergoes various S1 firing rates. In consequence, he rates his pains using the VAS scale. In addition, he judges 49.8 °C to be roughly twice as intense as 47 °C.

Now consider Mild's counterpart in the counterfactual situation. Call him *Severe*. Like Mild, Severe experiences noxious temperatures between 45 and 50 °C and undergoes increasing S1 firing rates. However, because in this situation humans' psychophysical response function is steeper, Severe's S1 firing rates in response to these same temperatures are much higher than Mild's. There is not just a difference in absolute firing rates; Severe's S1 firing rates increase *more rapidly* than Mild's. So, for instance, while moving from 47 to 49.8 °C roughly doubles Mild's average S1 firing rate, it far more than doubles Severe's S1 firing rate. In consequence of these neural differences, there are also behavioral differences between Mild and Severe. Using the VAS scale, Severe consistently rates his thermal pains as more intense than does Mild. While Mild reports 49.8 °C to be twice as intense as 47 °C, Severe reports 49.8 °C to be "much more than twice as intense" as 47 °C. (Since Mild has a different evolutionary history than Severe, it is unlikely he speaks a language that sounds like English; but suppose that this is the best translation of his report.) Finally, Severe responds to 49.8 °C with much greater urgency than does Mild; his pulse rate is higher; and so on.

Despite these neural and behavioral differences between Mild and Severe, let us stipulate that their pain states track exactly the same properties of the thermal stimuli. In general, whatever kind of relations the tracking intentionalist thinks ground representation (whether they be Tye's simple tracking relations, or Millikan's more complex teleological relations), Mild and Severe bear those relations to exactly the same properties of peripheral stimuli.

So far I have described Severe's situation in neural and behavioral terms. I haven't stipulated anything about his pain experiences. That is the crucial issue.

In my view, Severe would have pains that are more intense than Mild's pains, in response to the same noxious thermal stimuli. There are differences between them at the sensory level, not just the affective level. So for instance, the difference in intensity between Severe's pains at 49.8 and 47 °C is greater than the difference between Mild's pains at those intensities. The case for this is obvious. First, the relationship between noxious temperature and S1 firing rate is steeper in Severe than in Mild, and S1 firing rate is the best-known predictor of the sensory dimension of pain. (Recall that S1 activity codes for the sensory dimension of pain, while the ACC codes for the affective dimension.) Second, the psychophysical and other behavioral differences between Severe and Mild are only explained by the claim that Severe's pains are more intense than Mild's on the sensory dimension.

But, of course, tracking intentionalism delivers the opposite verdict. Indeed, the problem here is not just a problem for self-described "tracking intentionalists" like Tye and Dretske. It is a problem for "externalist intentionalists" in general. So, for instance, Hill does not advocate a specific theory of representation, although he does sometimes appeal to a kind of tracking theory (2009, p. 149, n. 16; pp. 179–80). Like Tye and Dretske, he is somewhat undecided on the issue of exactly what peripheral physical properties various types of pain represent. But, whatever kind of relations the externalist thinks ground representation (whether they be Tye's simple tracking relations, or Millikan's more complex teleological relations), it seems that we can stipulate that Mild and Severe bear those relations to exactly the *same* properties of peripheral stimuli. In that case, the externalist is committed to saying that they have phenomenally identical experiences. Here he cannot appeal to the illusion response or the pluralist response (Sect. 18.3.1). I have simply *stipulated* that, whatever conditions need to be in place in order for two creatures to accurately represent exactly the same properties, those conditions are indeed in place in the case of Mild and Severe.

Cutter and Tye (2011) and Hill (2012) have offered a response to cases like this. Applied to the present case, the idea would be that 49.8 °C is more of a "threat" to Severe than it is to Mild; in particular, it has the property *being bad to degree D to Mild* and the property *being bad to degree D** to Severe. And Severe *somehow* represents the first "valuational" property while Severe represents the second. Because their experiences represent different "valuational properties", the tracking intentionalist can say that they differ at least on the affective dimension of pain, even if he must say that they have exactly the same sensory intensity. But this response fails to apply to my present case. (i) As I have set up the case, 49.8 °C is simply *not* more of a threat to ("more likely to harm") Severe than it is to Mild, as I explained above, so here the valuational response cannot get off of the ground. (ii) In any case, the valuational response does not fit with the science. While S1 activity codes for sensory intensity, ACC activity codes for affect. And Mild and Severe differ in *both* S1 activity and ACC activity. Moreover, their VAS ratings for *sensory* intensity differ. So the only reasonable verdict is that their pains differ on the sensory dimension, not merely the affective dimension. I will have more to say about these issues in Sect. 18.5.

18.3.5 Soft and Loud

My final counterexample to tracking intentionalism concerns the perception of loudness. Tracking intentionalists are objectivists about loudness and other audible qualities, claiming that they objective properties of sound events. My final hypothetical case poses a problem for *any* view that incorporates objectivism about audible qualities, not just tracking intentionalism. Since Casey O'Callaghan (2002) has provided the most sophisticated defense of objectivism about auditory qualities, I will focus on his approach.

To begin with, O'Callaghan holds that particular sounds, the *bearers* of audible qualities, are events of oscillating or interacting bodies disturbing or setting a surrounding medium into wave motion. He holds that auditory qualities are objective, physical properties of these physical events. In the case of loudness, which will be my focus, his basic view is that particular loudness levels are identical with complex properties of these sound events, properties involving amplitude, frequency, "critical bands", and duration. A complex account is needed, because while loudness is mainly related to amplitude it also depends on other parameters. So, for instance, the loudness of a 40 dB pure tone at 500 Hz is the same as that of a 60 dB pure tone at 50 Hz. This is represented in so-called equal loudness curves. It is analogous to metamerism in the domain of color vision: many different combinations of lights can yield the same color perception. So, if we confine ourselves to pure tones, a given loudness level might be identified with the *disjunction* of all the different amplitude-frequency pairs that give rise to a perception of that loudness level. The account would have to be even more complex than this, because loudness also depends on critical bands and duration. In any case, the point is that the objectivist will not identify loudness levels with simple amplitudes or intensities; he will identify them with much more disjunctive, unnatural properties. As O'Callaghan says, these properties will only be of "anthropocentric interest", because they are the objective correlates of human loudness perception.

Now suppose that Soft is an actual human who hears tones of increasing amplitude. As mentioned in Sect. 18.2, there is a non-linear, highly compressive relationship between amplitude and perceived loudness level (holding frequency fixed). Therefore, huge differences in amplitude are needed to generate small differences in perceived loudness. For instance, he judges tone B to be twice louder than tone A when the amplitude of B is ten times louder in intensity (which is related to amplitude). The explanation for why there is a compressive relationship between amplitude and perceived loudness is that there is a matching compressive relationship between amplitude and total neural activity in the auditory channel, as several experiments have shown. As we saw in Sect. 18.2, there is there is evidence that perceived loudness is more proportional to total neural response (even in the cortex) than it is to amplitude and other complex physical parameters. While there is bad "external correlation", there is evidence for better "internal correlation".

Now consider a counterfactual situation in which humans evolved so that the same auditory stimuli normally produce a *greater* total neural response in the auditory channel than they do in the actual situation. In this situation, Soft has a counterpart, Loud. When Loud hears the same tones that Soft hears in the actual situation, the total neural response in his auditory channel increases *more rapidly* than Soft's. (So the case is similar to that of Mild and Severe.) In consequence of these neural differences, there are also behavioral differences. Thus, in this counterfactual situation, humans' subjective estimations of loudness yield stimulus–response functions that are much steeper than those which characterize the loudness perception of actual humans. For instance, Loud and other normal perceivers consistently report that tone B is *much more* than twice as loud as tone A. In addition, Loud and other normal perceivers in this counterfactual situation are more likely than actual humans to *notice* the same auditory events; amplitudes that do not produce discomfort in actual humans, produce discomfort in humans in the counterfactual situation; and so on. Why did humans in this situation evolve auditory systems that differ from our own in amplifying the neural responses responsible for loudness perception? It does not really matter to my argument. Maybe they rely on hearing more than we do; or maybe they evolved in an environment in which they must notice certain sounds more readily.

So Soft and Loud differ in their loudness-related neural and behavioral responses. Yet I also want to stipulate that there is a complete coincidence in the objective auditory properties they track. As Soft hears the tones in the actual world, he perceives increasing loudness levels. As we saw, O'Callaghan identifies these perceived loudness levels with complex, probably disjunctive physical properties of the tones: *D1*, *D2*, *D3*, ... Now, although O'Callaghan does not address this issue, he must hold that Soft perceives, or sensorily represents, *D1*, *D2*, *D3*, ..., because his cortical neural representations (those which realize his auditory experiences) bears *some* naturalistic relation *R* to *D1*, *D2*, *D3*, ... Maybe it is a kind of tracking relation; or maybe it is difficult to specify, because providing a theory of perception, or perceptual representation, is difficult. But there must be some naturalistic facts that determine that Soft perceives *D1*, *D2*, *D3*, ... to the exclusion of all of the other candidates. Now, whatever the naturalistic relation *R* is, I stipulate that Loud and Soft's corresponding cortical neural representations bear to *R* to the same properties, namely *D1*, *D2*, *D3*, ... True, Loud's neural representations involve higher neural firing rates than Soft's, and result in different behavioral responses. But the stipulation here is that they nevertheless track or detect the same objective properties *D1*, *D2*, *D3*, ... of the sound-events. Compare: in different types of mercury thermometers, different mercury heights can track the very same objective temperatures. Therefore, whatever the relation *R* is, the stipulation I am making is apparently possible, and we would need a good reason to think otherwise.

To clarify, my stipulation here is not that Loud and Soft's neural representations bear relation *R* to the same simple *amplitudes* or *physical intensities*. According to O'Callaghan, loudness levels are not mere amplitudes or intensities. Instead, he maintains that they are the more disjunctive, complex properties *D1*, *D2*, *D3*, ... which involve not just intensities but also frequencies and "critical bands". What I am stipulating is that Loud's neural representations, as well as Soft's, bear relation *R* to *these* properties, the very properties with which O'Callaghan identifies the (low) loudness-levels perceived by *Soft* in the actual world.

Now you can see how this creates a problem for O'Callaghan's objectivism about audible qualities as well as tracking intentionalism about the perception of audible qualities. Given the vast neural and behavioral differences between them, together with what we know about the physical basis of loudness perception, the most reasonable view is that Soft and Loud auditory experiences differ phenomenally as regards intensity. Given this, it follows that they ostensibly perceive *different* loudness levels, when presented with the same tones. In particular, when they hear the same sequence of tones, Loud perceives *higher* loudness levels than Soft, which increase more rapidly than those perceived by Soft. It is not the case that they perceive the same loudness levels (properties) via different "modes of presentation" or "mental paint", whatever that means; the correct description of the phenomenal difference is that they ostensibly perceive distinct loudness levels (*pace* Block 2010, p. 25). But, given O'Callaghan's objectivist theory of loudness, together with the natural assumption that perception must be grounded in some naturalistic relation *R*, it follows that Loud perceives *the very same* loudness levels as Soft, which he identifies with *D1*, *D2*, *D3*, ... The case also undermines tracking intentionalism about auditory experience. For, on tracking intentionalism, they have phenomenally identical auditory experiences, despite the vast neural and behavioral differences between them, because their experiences "represent" exactly the same objective properties. I conclude that tracking intentionalism is mistaken. Indeed, we must reject any view on which audible qualities are *objective* properties like *D1*, *D2*, *D3*, ... Maybe they are response-dependent properties. Or maybe they are internal neural properties "projected" onto external sound-events. But they are not *objective* properties like *D1*, *D2*, *D3*, ...

Now you might wonder why *actual* cases of auditory variation are not enough to refute tracking intentionalism and indeed any objectivist theory of audible qualities. The reason is that proponents of such views can appeal to the pluralist response or the illusion response to handle actual cases (Sect. 18.3.1). By contrast, in my hypothetical case of Soft and Loud, I have simply *stipulated* that, whatever conditions need to be in place in order for two creatures to accurately represent exactly the objective audible qualities, those conditions are indeed in place. So, in this case, neither the pluralist response nor the illusion response is available.

O'Callaghan (2009, sect. 3.2.5) considers a partial error theory as a response to "bad external correlation". The idea is that, in the actual world, auditory experiences (accurately) represent individual audible qualities, but also "distort their magnitudes of difference". The proponent of this view might grant that his view entails that Soft and Loud's experiences represent the same individual loudness levels *D1*, *D2*, *D3*, ... on hearing the tones. But he might insist that their experiences (inaccurately) represent those same loudness-levels as standing in different *magnitude relations*, which accounts for the phenomenal difference. This is a version of what I called the "structure gambit" in connection with the case of Sniff and Snort. For a few reasons, it is untenable. (i) There is Alex Byrne's point (2003, p. 656). Since Soft and Loud hear the tones and their apparent loudness-levels successively, and not at the same time, there is simply no time at which their *experiences* might represent all of those loudness-levels as standing in different magnitude relations. (ii) In any case, since Soft and Loud are exactly alike in their relevant naturalistic relations to the environment, there is no obvious naturalistic account of how Soft and Loud's experiences might represent *different* contents involving qualitative structure. (iii) This sort of view can be ruled out *a priori*. For instance, if Soft hears three loudness levels as increasing by equal intervals, then Loud cannot hear the *same* loudness levels as increasing by greater magnitudes. The right description of the case is that Loud is hearing *different, higher* loudness-levels than Soft (which is something that objectivists about loudness cannot accept, as we have seen).

18.3.6 The Official Internal-Dependence Argument

Of course, such cases could be multiplied indefinitely. To refute tracking intentionalism, and the general objectivist treatment of the sensible qualities, only one counterexample is required. So the best way to state the argument is as follows.

1. If tracking intentionalism is true, then in *every* possible coincidental variation case, the right verdict is Same Experiences.
2. But it is much more reasonable to suppose, in at least *some* coincidental variation cases the right verdict is Different Experiences; call this *internal-dependence*
3. So tracking intentionalism is (probably) mistaken.

Let me make two clarifying remarks. First, recently Michael Tye (2009, p. 194) has claimed that no empirical work on the explanatory underpinnings of phenomenology can establish the strong internalist claim that microphysical duplication metaphysically necessitates total phenomenal duplication. But my argument does not depend on the strong internalist claim that microphysical sameness necessarily guarantees phenomenal sameness. It only depends on internal-dependence: internal factors play a role, in the very minimal sense that, in *some* coincidental variation cases, the right verdict is Different Experiences. This weak claim *is* supported by empirical work; and it is enough to refute tracking intentionalism, and indeed (as we shall see in Sect. 18.5) any version of "externalist intentionalism".

Second, the internal-dependence argument also does not depend on any theory of sensory character. For instance, it does not depend on the claim that experience-types are necessarily identical with neural-types in the head, although it might naturally suggest that view. It also does not depend on the somewhat strange view that tastes, smells, sounds and pains are literally in the head. My case for internal-dependence only relies on the empirical findings and is neutral on the philosophical interpretation of those findings. For instance, since the individuals in coincidental variation cases differ in functional and sensorimotor respects, it is also compatible with functionalist and sensorimotor approaches.

Of course, there are potential objections to the internal-dependence argument; but before addressing objections, I would like to put my second empirical argument on the table.

18.4 Second Argument: The Structure Argument

In developing my internal-dependence argument, I used good internal correlation as well as bad external correlation to support internal-dependence, which is a claim about non-actual coincidental variation cases. My structure argument is a totally independent argument. It only depends on bad external correlation, which is well confirmed in psychophysics. And it concerns actual cases. Recall that some of the individuals in my coincidental variation cases were actual individuals. The structure argument is meant to show that, given bad external correlation, their judgments concerning *phenomenal structure* in the actual world come out false. It poses a problem for all objectivists about the sensible qualities. Since tracking intentionalists are committed to objectivism, it undercuts their view.

As with the internal-dependence argument, I will illustrate the structure argument by focusing on a few cases. Given the broad similarities across many sense-modalities, if objectivism is true about one type of sensible quality, then it is true of other types. So the objectivists are committed to a general view, even if they often focus on a single case. By considering a few cases, we will be able to appreciate the cumulative case against their view. I will start with an initial, *prima facie* challenge; the real argument will come afterwards, when we look at potential responses.

18.4.1 Three Illustrations of the Initial Challenge

In the fMRI experiment conducted by Howard (Sect. 18.2), actual subjects made introspective reports on their smell experiences. Suppose that Sniff is one of them and that he makes a report along the following lines:

> *Sniff's report:* The limonene smell quality I experienced is overall more like the citral smell quality than the menthol smell quality.

The structure argument is simple. This introspective report is true. It was just obvious to Sniff. Indeed, to all normal humans, limonene and citral in fact have slightly very similar citrus smells while menthol has a quite different minty smell. But, given bad external correlation, tracking intentionalism would appear to entail that the report is *false*.

Here is why the tracking intentionalist seems forced into accepting an error theory. Sniff makes his report under optimal conditions. So on tracking intentionalism Sniff's smell experiences are veridical: the smell qualities he experiences are identical with the actual objective *chemical characters* of limonene, citral and menthol. (Since many chemical structures might yield the same smell qualities, these might be disjunctive chemical characters.) So Sniff's report is true just in case they satisfy the semantic value of the relational predicate 'x is overall more like y than z' as it occurs in his report. But, because this is an actual case "bad external correlation", they apparently do not. To see this, just look at the representation of these chemical types in Fig. 18.1 in Sect. 18.2. It is very hard to see how the chemical characters of limonene, citral and menthol (in that order) could satisfy the semantic value of 'x is overall more like y than z' in the context of Sniff's report. The chemical character of limonene is overall *quite different* from that of citral. Indeed, if anything, the chemical character of limonene is more like that of *menthol* than that of citral. To answer the challenge here, the tracking intentionalists (who think smell qualities are chemical characters) would have to convince us that, despite appearances, 'x is overall more like y than z' has a semantic value in the context of Sniff's report which is indeed satisfied by the chemical characters of limonene, citral and menthol (in that order). This is what the externalists must do in order to accommodate the truth of that report.[7]

Notice that, because of "good internal correlation", an alternative internalist or response-dependent theory of smell easily accommodates the truth of Sniff's reports. On such a view, the smell qualities are not objective chemical characters. On one version, the relevant smell qualities are literally identical PPC neural types in Sniff's head. On another version, the smell qualities are identical with the dispositions of external odors to produce those neural types. The fMRI study by Howard shows that those neural types *do* stand in the relevant resemblance-order. While there is bad external correlation, there is good internal correlation. So on such theories Sniff's report is straightforwardly true.[8]

Such cases obviously pose a challenge to all objectivists about smell qualities, not just tracking intentionalists. For instance, Batty (2010) defends the view that particular smells are odor clouds and general smell qualities are objective

[7]Many would say that resemblance is always resemblance in respects. They would say that 'x is overall more like y than z' has different semantic values in different contexts, because in different contexts different respects of resemblance can be salient and can be weighted differently (Davies, forthcoming). My initial challenge to objectivists about smell qualities proceeds in full awareness of these points (see also Pautz 2006b, note 4). My challenge to them is to specify the semantic value of 'x is overall more like y than z' *in those contexts in which we make reports of smell similarity*, and also show that the molecular types with they identify the smells really do satisfy this semantic value. The fact that there is "bad external correlation" creates a *prima facie* difficulty here. Incidentally, while I would agree that for *particulars* all resemblance is resemblance in respect of various qualities or properties, I would myself reject this claim when *qualities* themselves are concerned. What are the respects in which color hues, or smell qualities, resemble? We draw a blank. This is because qualities themselves (unlike particulars) have no interesting set of (second-order) properties with respect to which they can be similar or different. So, in some cases, when we say quality *Q1* is more like *Q2* than *Q3*, we arguably use the predicate 'x is more like y than z' to pick out a conceptually primitive comparative resemblance relation, not a relation that we can be unpack by citing some context-specific "respects of resemblance". (Contrary to Byrne (2003), we do not have in mind similarity in "genuine respects", but a basic kind of similarity that is not similarity *in respects* at all.) If this is right, then it makes it even harder for objectivists about the sensible qualities to answer the structure argument. For in that case, in order to show that our similarity judgments about colors, and smells, and so on, are *true* relative to their strict or face-value interpretations, they would have to show that the corresponding reflectance properties, molecular properties, and so on, satisfy the same conceptually primitive comparative resemblance relation.

[8]O'Regan (2011, p. 99) suggests that even such internalist, neural theories of qualitative similarity face a problem. There are different metrics for measuring similarity among neural states. What selects which one is the "right" one? But the proponent of such a theory might claim that our paradigmatic reports about qualitative similarity come out relative to a natural metric for measuring similarity among neural states that can be uncovered by multidimensional scaling (Howard et al. 2009, p. 396). Of course, a totally different approach would be to provide a *functional* account of qualitative similarity (e.g. in terms of similarities in functional role perhaps, or dispositions to form sophisticated similarity beliefs), but no one has developed a plausible account along these lines.

and presumably physical properties instantiated by these odor clouds. Given bad external correlation, how might the objectivist avoid an error theory concerning our resemblance judgments concerning smells?

Sniff's report above concerns resemblances among general qualities or properties as opposed to particular items. But it worth mentioning that we also make judgments about resemblances about particular odors, which we think of as clouds that linger in the air and that we can inhale though our noses. For instance, Sniff will also report that the limonene odor in the air around him is more like the citral odor than the menthol order. This adds to the challenge. On an objectivist account these odors are just collections of limonene, citral and menthol molecules. But, on the face of it, given how different limonene and citral are in their objective chemical characters, there is no obvious sense in which the limonene cloud is more like the citral cloud than the menthol cloud. So on this account it is very hard to see how Sniff's report about the resemblance-order of the odors is true.

There are other impressive cases of bad external correlation involving smell that could be used to illustrate the argument. For instance, −carvone and +carvone are *mirror images*, yet they smell totally different (minty and caraway), because they nevertheless set up totally different ensemble activation states in the brain (Sect. 18.2). So Sniff will report that the smells of −carvone and +carvone are more different (mint and caraway) than the smells of limonene and citral (again, both citrus). But −carvone and +carvone are not *chemically* more different than limonene and citral. Indeed, the opposite is true: limonene and citral are much more chemically more different than −carvone and +carvone. So, on an objectivist theory of smell, it is very hard to see how Sniff's judgments about the resemblances among these odors and their general qualities might be true. Appealing in *some* way to internal factors seems to be the only option (as I will discuss below).

Now let us revisit Soft. Soft is an actual individual in one of the many psychophysical studies on ratio scaling of auditory sensation. He makes the following introspective report:

> *Soft's report:* The apparent loudness of tone B is roughly twice greater than the apparent loudness of tone A.

Now you might be skeptical of such ratio judgments. But such ratio judgments *sometimes* make sense. For instance, we can report on a ratio relationship among the *apparent* lengths of two lines. It turns out that we also quite good at ratio scaling of perceived loudness. Indeed, there is empirical evidence for the validity of ratio scaling of audible intensity. Those working in psychoacoustics generally think that such introspective reports can be true (Gescheider 1997).

But, again, on tracking intentionalism, Soft's report is apparently false. We can suppose that optimal condition obtain. So, on tracking intentionalism, his auditory experiences are veridical. The apparent loudness levels are identical with actual complex, disjunctive physical properties of the tones – call them *D2* and *D1* – involving intensity, frequency, critical bands, and duration. Therefore, on tracking intentionalism, Soft's introspective report is true just in case *D2* is roughly twice greater than *D1*. In other words, Soft's *experience* of B is twice as intense as his *experience* of A, just in case the *physical properties represented* by those experiences stand in this relationship. But there is no obvious sense in which the disjunctive property *D2* is roughly twice greater than the disjunctive property *D1*, in the way that one length property can be twice greater than another. The main issue here is that there "bad external correlation", in particular, response compression. So, in this case, *D2* involves an intensity that is *much more* than twice greater than that involved in *D1*. As a general rule of thumb, for pure tones, doubling loudness requires increasing intensity as a factor of *ten*. So, tracking intentionalists must *apparently* say that Soft's introspective report is false. Indeed, objectivists about loudness such as Casey O'Callaghan (2002) face the same problem. And the problem arises for other audible qualities. For instance, introspective judgments of equal pitch intervals do not correspond to equal frequency intervals. So on objectivism how can these judgments be true?

Those who advocate internalist or response-dependent theories of auditory quality have less of a problem here. On these views, sound qualities are neural responses or else dispositions to produce neural responses, not objective physical properties of sound-events. As noted earlier (Sect. 18.2), many studies show that neural response (especially in the cortex) is more proportional to auditory intensity than anything in the external world. It's still early days, but maybe it will turn out that Soft's total neural response to tone B is actually twice greater than his total neural response to tone A in terms of overall firing rate. Or maybe it is "twice greater" in some functional sense.

Finally, as we saw, in psychophysical experiments on pain, it has been found that on average merely going from 47 to 49.8 °C doubles subjects' perceived pain intensity because it vastly increases the neural response in the pain matrix. This is another actual case of bad external correlation; it is a case of response expansion. Thus, in the actual world, a human subject, Mild, will come out with the following report:

Mild's report: My 49.8 °C pain is roughly twice more intense than my 47 °C pain.

Now if you think ratio scaling of pain makes no sense, and reports like Mild's could not possibly be true, then I could fall back on my previous cases about Soft and Sniff. However in fact that there is evidence that in the right circumstance Mild's report can be true (Price 1999). But, on tracking intentionalism, Mild's report apparently comes out false. On intentionalism, Mild's pains stand in the ratio relation just in case the represented pains do. Since optimal conditions obtain (response expansion is part of the normal function of the pain system), tracking intentionalists must say that the represented pains are the actual peripheral disturbances. Now, in this case, there is no actual tissue damage. So on their view, what are the disturbances in this sort of case? They have not discussed this issue. The physical stimuli, 49.8 and 47 °C? But there is no measure relative to which 49.8 °C is twice greater than 47 °C. And even if there is one, Mild certainly didn't have it in mind when he made his report. So on this option his report is false. Alternatively, the externalist might say that the represented pains are the peripheral neural patterns, *N2* and *N1*. But it is implausible that Mild's experiences represent peripheral neural patterns, because their function is presumably to indicate potential damage or danger instead (more on this in Sect. 18.5). And even if these are the relevant disturbances, Mild's report might come out false. Maybe *N2 less than* twice greater than *N1*, and the response expansion (the magnification of the neural response) occurs further downstream.

Again, an internalist theory of pain has less of a problem accommodating our structure judgments. Indeed, in view of good internal correlation and bad external correlation, an internalist theory seems inevitable. Kenshalo's findings (see Fig. 18.3 in Sect. 18.2) indicate that, just as pain intensity doubles between 47 and 49.8 °C in humans, so average WDR neural response in monkeys S1 roughly doubles between these temperatures. And in a fMRI study directly on humans, Coghill et al. found *linear* relationships between pain intensity and neural response to noxious temperatures in S1 and other brain regions. So, while it's early days, current research suggests that Mild's S1 neural response to 49.8 °C might literally be roughly twice greater than his S1 neural response to 47 °C. So if pains are S1 neural states or states directly supervenient on S1 activity, Mild's report comes out true.

18.4.2 Three Unsatisfactory Responses to the Challenge

Of course, what I have said so far only poses an initial challenge to externalists of various stripes. To complete the structure argument, I would have to eliminate all responses to the initial challenge. I will look at the three most obvious responses (some less obvious ones will be addressed in Sect. 18.5).

Error Theory. One response would be to simply accept my argument that tracking intentionalism and objectivist accounts of sensible qualities in general lead to an error theory concerning structure judgments. So, for instance, O'Callaghan (2009, sect. 3.2.5) briefly considers a *partial* error theory about our judgments about magnitude relations among *audible qualities*. In the case of loudness, the idea is that our beliefs about the individual loudness levels of sounds and their ordinal rankings are generally true, but our beliefs about their ratio relations are false. So, in particular, Soft's report is just false.

But this response is unsatisfactory for three reasons. First, there is empirical evidence in support of the truth of ratio reports concerning apparent loudness, as I already noted. Further, if ratio judgments concerning loudness makes no more sense than ratio judgments concerning level of beauty (say), then it is a wonder that subjects fairly consistently make such judgments at all.

Second, if objectivism is true concerning one range of sensible qualities (e.g. audible qualities), it is presumably true for other ranges (e.g. smell qualities). So, although O'Callaghan does not consider other cases, the objectivist needs a response in every case. Now, even if we accept an error theory about Soft's sophisticated ratio report, it is much harder to accept an error theory about simple judgments of relative resemblance, such as Sniff's report that one smell quality is more like a second than a third. Indeed, such reports about resemblances among general qualities (as opposed to particular items in the environment), such as *blue is more like purple than yellow*, are often thought to be certain *a priori*.

But if this does not convince you, let me make a third point against the error theory. We not only make structure judgments about the sensible qualities apparently possessed by items in the external world; we can also make structure judgments about our own experiences. For instance, consider Sniff's introspective judgment that the citrus smell of limonene *seems* more like the citrus smell of citral than the minty smell of menthol, in other words, that his smell *experience* of limonene is more like his smell *experience* of citral than his smell *experience* of limonene. What is the right account of phenomenal reports like this?

Even though I reject *externalist* intentionalism, I think that the basic intentionalist approach to phenomenology is plausible. On intentionalism, all phenomenal facts derive from facts about the properties presented in experience. (Maybe there are *some* exceptions – for instance, involving visual blur.) So, presumably, Sniff's consecutive smell experiences fall into the relevant phenomenal resemblance-order just in case the successively presented smell qualities do. The phenomenal structure of experiences is inherited from the structure of the qualities successively presented in those experiences. Call this the *inheritance claim*. On intentionalism, there appears to be no other option. The inheritance claim is actually intuitively plausible in general, independently of any theory. As Alex Byrne (2003, p. 645) says: "Why is the experience as of a teal object [phenomenally] similar to the experience as of a turquoise object? Because teal is similar to turquoise".

But, on the error theory of the structure of the sensible qualities, the smell qualities that Sniff is presented with are objective chemical properties that *do not* stand in the relevant resemblance-order. Hence, given the inheritance claim, the error theory *spreads* to Sniff's introspective judgment that the citrus smell of limonene *seems* more like the citrus smell of citral than the minty smell of menthol. The error theory implies that Sniff is even wrong in thinking that that his smell *experience* of limonene is more like his smell experience of citral than his smell experience of limonene. By similar reasoning, the error theory contemplated by O'Callaghan entails that even our introspective judgments about the *apparent* magnitude relations (ratios, equal differences) among sounds are false. But this is hard to accept. It is especially hard to accept an error theory when it comes to Sniff's simple introspective judgment about resemblance-order.

Complex respondent-independent theory. So we need a theory on which our structure judgments come out true. The externalists have two options: a *response-independent theory* of qualitative structure, which does not appeal to our internal neural or behavioral responses to external properties in any way; or a *response-dependent theory* of qualitative structure, which does somehow appeal to our internal neural or behavioral responses.[9]

---

[9]Since tracking intentionalism is meant to be a reductive theory of consciousness, I will be ignoring "primitivism" about the sensible qualities; so the only response-independent theories I will consider will be *reductive* response-independent theories.

Now, given good internal correlation and bad external correlation, our structure judgments "match" our internal neural responses better than the external conditions that prompt them. So a response-dependent theory of qualitative structure seems like an obvious choice; and a response-independent theory seems hopeless. Nevertheless, I will start with the response-independent theory of qualitative structure because such a theory fits best with my targets here, tracking intentionalism and objectivism. The motivation behind such views is that sensory character seems wholly "out there" in the response-independent world. And Michael Tye (2000, p. 163) has actually attempted a response-independent theory of simple color structure judgments like "purple is reddish and bluish". Although he does not generalize to more complex judgments (e.g. judgments of relative resemblance) or other ranges of sensible qualities, his discussion hints at a general strategy for devising such accounts.[10]

The strategy is simple. Structure judgments are correlated not just with internal properties and relations but also with external ones. In some cases, the external correlates are revealed by psychophysics. So corresponding to every response-dependent account there will be a response-independent account. Given bad external correlation and good internal correlation, the external correlates of structure judgments will be more complex and unnatural then the internal correlates. Nevertheless, according to Tye, the subject matter of our structure judgments are the external, response-independent correlates.

So, for instance, psychophysics has revealed that, as a rough rule of thumb, for simple tones with the same frequencies, we judge that loudness has doubled just in case physical intensity (related to amplitude) increases tenfold. So, on a grossly oversimplified response-independent theory, loudness levels are just physical intensities and when we use 'twice greater' than in relation to loudness it takes on a new semantic value: it comes to mean *ten times greater*, although we are *semantically blind* to this! In this way, an objectivist about loudness like O'Callaghan might say that our "doubling" judgments in relation to loudness come out true. Now, in fact, whether one sound appears twice louder than another also depends on a complex variety of other factors, including frequency, duration, and critical bands. Moreover, the relationship between loudness and intensity changes for levels below 40 dB SPL. So, the proponent of a response-independent theory of auditory structure would have to say that loudness levels are extremely complex physical properties and that the semantic value of 'is twice greater than' in auditory contexts is some horrendously complex, disjunctive relation that cannot be easily defined. In order to ascertain this complex relation, we look at our responses. But the relation itself is response-independent.

---

[10]Tye (2000) and also Byrne and Hilbert (2003) defend an interesting *hue-magnitude account* of *color* structure. But even if that account is correct in the case of color (I think there are several problems with it), the same account obviously does not apply to the qualitative structure of smells or audible qualities. So objectivists would need some other accounts here; as we shall, it is very hard to see what accounts they might provide.

Likewise, the proponent of a response-independent theory of qualitative structure might claim that pains are complex bodily disturbances and that in contexts where we use 'twice greater than' in relation to bodily disturbances *D2* and *D1* we are referring to the relation which is the disjunctive psychophysical correlate of our pain-doubling judgments: *D2* and *D1* are noxious temperatures with such-and such spatial extents and *D2* is 33 % greater than *D1*, or *D2* and *D1* are electric shocks applied to the skin with so-and-so spatial extents and *D2* is 25 % greater than *D1*, or . . . .

It would be even more difficult to develop a wholly response-independent theory of qualitative structure in the domain of smell. Such a theory would require that there is a single response-independent relation *R* such that we (or smell experts) judge *smell S1 is overall more like S2 more than S3* when and only when the corresponding molecular types *C1*, *C2* and *C3* stand in relation *R*; and that this relation *R* is the semantic value of '*x* is overall more like *y* than *z*' in our reports of smell similarity of the form *smell S1 is overall more like S2 more than S3*. For, on a purely response-independent theory, only then will such reports come out generally *true*. But the psychophysics of smell is particularly messy. There simply is no single response-independent relation *R* that fills the bill.[11]

Of course, those who favor a wholly response-independent theory of smell similarity might say that, when we use the predicate '*x* is overall more like *y* than *z*' in connection with smells, the predicate comes to express a response-independent relation that only be defined in terms of a "big list": a list of all the chemical structures *C1*, *C2* and *C2* corresponding to the smells *S1*, *S2* and *S3* such that we judge that *S1 is more like S2 more than S3*. This would guarantee that our paradigmatic smell similarity judgments come out true, but it would be it would be totally implausible from a semantic point of view.

[11]Haddad et al. (2008) and Turin (2002) attempt to relate chemical similarity to qualitative similarity. But they are very far from establishing what a response-independent theory of smell similarity would require: that there is a single response-independent relation *R* such that we (or smell experts) judge *smell S1 is overall more like S2 more than S3* when and only when the corresponding molecular types *C1*, *C2* and *C3* stand in relation *R*. First, even when it comes to simple monomolecular odors, their methods are open to counterexamples and do not come close to explaining all of the variance. Second, they do not take into the account the effects of concentration on quality, which can be extreme (Malnic et al. 1999). Third, their methods only apply to simple monomolecular odors. They do not apply to natural odor objects, such as the odors of foods and plants, which are typically mixtures containing tens to hundred of monomolecular components, and whose smells are not at all a function of the smells of their components. Could it be that some future canonical physical description of molecular types will make it evident that there is a relation *R* among molecular types that perfectly tracks our smell similarity judgments? (In an interesting discussion, Davies (forthcoming) brings up a similar question regarding reflectance properties and judgments of color resemblance.) All the evidence suggests that the answer is 'No'. Moreover, we must remember that the proponent of a response-independent theory of qualitative structure would need to show how to accommodate our structure judgments about all types of sensible qualities (loudness levels, pitches, tastes, etc.) in purely response-independent terms. This seems impossible.

Now that we have attempted to elaborate the response-independent theory of qualitative structure, we can see that it is quite hopeless. There are many problems. Let me just mention the simplest one, which also goes deepest. I call it the *metasemantic objection*. The response-independent theory is in part a theory of the content or truth-conditions of our talk about the structural features of sensible qualities. It says that such talk about sensible qualities and their relations is about very complicated response-independent properties and their response-independent relations. But no *metasemantic theory* of how language hooks up to the world would support this view. Any metasemantic theory would instead support a response-dependent theory.

To see this, consider an example. As we saw, on the response-independent theory, when we speak about sounds, the predicate 'x is twice more intense than y' comes to express a horrendously complex response-independent relation *I*. This relation must be extremely complex, because of response compression as well as the dependence of loudness on multiple objective factors. Because the relation *I* is response-independent, on this theory our ratio judgments about sound-intensity have wholly response-independent truth-conditions. By contrast, on a simple response-dependent theory, when we speak about sounds, the predicate 'x is twice more intense than y' expresses a quite different, response-dependent relation *D*. In one version, sounds are external physical events, and *D* is a relation definable along these lines: *x and y normally cause a doubling of total neural response in the relevant auditory channel*. Now, evidently, *D* is much more simple or "natural" than *I*. Further, *D* plays more of a role than *I* in causally explaining our judgments about loudness doubling. Now nearly all physicalist metasemantic theories of content or truth-conditions of our language appeal to causation or naturalness or both. Therefore, given bad external correlation and good internal correlation, any physicalist is more or less *forced* to accept some response-dependent theory of the truth-conditions (contents) of our judgments of loudness scaling. Likewise for judgments about pain intensity and judgments about relative resemblances among smells.

Respondent-dependent theory. So, could tracking intentionalists (Tye, Dretske) and other objectivists about sensible qualities (Hilbert, Byrne, O'Callaghan, Batty) block the structure argument by accepting some kind of response-dependent theory of qualitative structure?

To begin with objectivists certainly cannot accept *some* response-dependent theories. It is part of their view that individual sensible qualities (audible qualities, taste qualities) are *objective, response-independent* properties of external items. So they obviously cannot hold that the sensible qualities are internal neural properties projected onto the external world (projectivism) or dispositions to produce internal neural responses (dispositionalism).

But, on the face of it, tracking intentionalists and other objectivists about sensible qualities apparently could hold that, while sensible qualities are objective physical properties, their *structural relations* are to be explained in terms of the responses they produce in us. On this mixed view, loudness levels are objective physical properties, but when we judge that one is roughly twice greater than another our judgment is true just in case the one normally causes a total neural response roughly twice greater than that normally caused by the other. Likewise, felt pains are objective bodily disturbances, but their sensory intensity is explained in terms of the neural response they typically cause. And smells are objective chemical phenomena, but when we judge that they resemble to a certain degree our judgment is true just in case they normally produce similar ensemble activations in the PPC (the only known physical correlate of smell similarity). If you like, the smells are objective phenomena but they are similar or different *in respect of* their effects on us. One might think that, in this way, even tracking intentionalists and other objectivists about sensible qualities can accommodate the structure judgments of Sniff, Soft and Mild.

However, even this proposal is unsatisfactory. To begin with, my main target here is tracking intentionalism. The response-dependent response to the structure argument is not available to tracking intentionalists for the simple reason that it is *inconsistent with* tracking intentionalism. On the response-dependent theory, the individuals in coincidental variation cases (Yuck and Yum, Sniff and Snort, Mild and Severe, Soft and Loud) have *phenomenally different* experiences of the same objective properties because those properties typically produce different neural responses in them. In particular, their experiences *exhibit different phenomenal structure*. But, as we have seen, tracking intentionalists must apparently say that they have *phenomenally identical* experiences, because the track and hence represent exactly the same external conditions. Only the hopeless *response-independent* theory of qualitative structure appears consistent with the radical externalist theory of phenomenal character promoted by tracking intentionalists.

Now, you might think that, even if tracking intentionalists cannot accept the response-dependent response to the structure argument, others who favor objectivism about sensible qualities (O'Callaghan, Batty) might accept it, if they simply reject tracking intentionalism. But there are serious problems for this response, even if we ignore tracking intentionalism.

First of all, as I have mentioned, even if tracking intentionalism fails, I do believe that there are powerful reasons to accept some version of intentionalism about experience. But the response-dependent theory of qualitative structure is hard to square with *any* version of intentionalism, not only tracking intentionalism. For instance, on the response-dependent theory, Sniff and Snort's smell experiences differ in phenomenal structure, because their neural (PPC) responses differ. But they apparently represent the same individual objective (perhaps disjunctive) chemical types, because they bear the same naturalistic relations to them. According to the objectivist who accepts the response-dependent theory of qualitative structure, what then is the representational difference between their experiences that constitutes the phenomenal difference? It is not enough to say that their different neural responses constitute the phenomenal difference; given intentionalism, this neural difference must be accompanied by a representational difference (for an argument, see the discussion of "quasi-intentionalism" in Sect. 18.5).

Perhaps the objectivist who accepts the response-dependent theory of qualitative structure will reply that Sniff and Snort's experiences do not only represent the objective (chemical) properties of odors; they also represent different contents of the form *the odor clouds have objective properties that cause in me PPC states that resemble to degree D*. Their olfactory systems are "self-centered", because they represent, not only the objective properties of things, but also response-dependent properties concerning the effects of those things on those very systems. It is because Sniff and Snort's experiences differ in these "self-centered" contents that their experiences differ in phenomenal character.

But how could Sniff and Snort's experiences manage to represent such bizarre, complex contents? On the face of it, there is no naturalistic theory of intentionality compatible with this view. (For more on this point, see my response to Kriegel's related view in Sect. 18.5.) Further, in any case, since Sniff and Snort smell the odors consecutively, at no time could their experiences represent such contents involving multiple odors.

There is another decisive problem with the combination of an objectivist theory of individual sensible qualities and a response-dependent theory of their qualitative structure. Evidently, if one loudness level is twice greater than another, then this is an essential feature of the loudness levels. Compare: if one length is twice greater than another, this is an *essential* feature of the lengths. Likewise, intuitively, if two smell qualities resemble to degree *D*, then this is an *essential* feature of them, one they possess in any situation in which they exist. But, on the combination of an objectivist theory of individual sensible qualities and a response-dependent theory of their qualitative structure, these intuitions are false. For, on objectivist theory of individual sensible qualities, these sensible qualities are objective physical properties. Further, those very properties might of course have normally produced quite different neural responses than they in fact do. (This, indeed, is what happens in coincidental variation cases.) On a response-dependent theory of qualitative structure, this means that those very sensible qualities might have had quite different qualitative structure than they in fact do.[12]

This concludes my defense of the structure argument. While others have discussed similar arguments concerning color, I have developed a new version of the structure argument concerning other sensible qualities. And I have shown why the usual responses are unsatisfactory. The conclusion I draw is that tracking intentionalists and objectivists about sensible qualities can provide no satisfactory account of the truth of our ordinary qualitative structure judgments, such as those of Sniff, Soft and Mild. This is perhaps the best way of stating the structure argument against these views.

---

[12]For other problems with the combination of a response-independent (objectivist) theory of sensible qualities and a response-dependent theory of their qualitative structure, see Pautz (2006b).

18.5 No Refuge for Externalist Intentionalists

I have taken tracking intentionalism as my stalking horse. Some phenomenal externalists (Hill, Lycan, Tye, others) have suggested that the kind of tracking intentionalism I have focused on is too simple and that more complex versions of externalist intentionalism avoid my arguments. Therefore I conclude by rebutting their responses to my arguments.

Valuational properties. Recently, Brian Cutter and Michael Tye (2011) have claimed tracking intentionalism can handle cases like that of Mild and Severe. Recall that in response to increasing noxious temperatures Severe's S1 firing rates (known in our own case to be linearly related to sensory intensity) increase much more rapidly than Mild's. He also produces higher pain ratings of sensory intensity the VAS scale. To handle this sort of case, Tye and Cutter retain the claim that it is the representation of objective features of the disturbance (size, intensity) that determines *sensory intensity*, even though studies show that there is at best a complex non-linear relationship here. Since they agree that tracking intentionalism entails that Mild and Severe's experiences of increasing noxious temperatures represent bodily disturbances of exactly the same objective types, they must say that their pain experiences of these noxious temperatures are exactly alike in *sensory* intensity, despite the strong neural and behavioral evidence to the contrary (this was confirmed in correspondence).

However Tye and Cutter would allow that there is *some* phenomenological difference between them. In cases like this, they claim that, while Mild and Severe's experiences represent the same bodily disturbance types, they also represent different *valuational properties*. Their experiences have "layered content". The same disturbance is bad-for-Mild-to-degree-y and bad-for-Severe-to-degree-y, where y is greater than x. They explain badness in terms of aptness for harm but provide no naturalistic explanation of across-species comparisons of *degrees* of badness. This is a serious lacuna in their account. In any case, their view is that Mild's experience represents the disturbance as having the first valuational property while Severe's experience represents the disturbance as having the second one, because their pain systems actually track these different valuational properties. Further, in consequence of this representational difference, their pains differ in *affective phenomenology* or *unpleasantness*, even if they are exactly alike in *sensory intensity*. The (well-documented) difference between *sensory intensity* and *unpleasantness* is subtle. Roughly, Tye and Cutter's idea is that Mild and Severe have pains of exactly the same sensory intensity, but a pain of that intensity *bothers* Severe more than it does Mild. This is supposed to explain the fine-grained behavioral differences between them. This is their answer to my "internal-dependence" argument about pain.

In response to my point that there is "bad external correlation" in the case of pain, Christopher Hill (2012, p. 137) has independently invoked a notion somewhat similar to the notion of badness level. He suggests that it makes sense to claim that the *threat level* of one disturbance is *twice greater than* another. There is a lacuna in his account, because he does not say what exactly what naturalistic facts constitute a doubling of threat. In any case, the idea is that, if Mild (who, recall, is an actual individual) experiences a small increase in temperature from 47 to 50 °C, the threat level might actually double. This might accommodate the truth of his report that his pain intensity doubled, thus answering my "structure argument" about pain. Hill might also say that Mild and Severe represent the same bodily disturbances as having different threat levels. He might say that this means that their pains differ in phenomenology. He might even say that this means that they consequently differ in *sensory intensity*, not merely affective phenomenology. In that case, his view would be somewhat different from that proposed by Cutter and Tye, who claim that there is only an affective difference.

As Tye and Cutter note, *taste experiences* might also represent valuational properties. So the externalist intentionalist might claim that, while the fine-grained sensory character (sweet, salty, bitter, sour, umami) of taste experience is determined by the representation of chemical types (sugar, salt, acid, etc.), the affective character of taste (good or bad) is determined by the representation of valuational properties (good or bad). Now recall the case of Yuck and Yum. The externalist might claim that, while their experiences represent the same chemical property C of the berries, Yuck's experience somehow represents the berries as bad for him (to some degree) and Yum's experience somehow represents them as good or maybe "edible" for him (to some degree). It's not just Yuck and Yum *think* these things; the idea is that their *experiences* comment on the nutritional value of the berries! In consequence, the tracking intentionalism might claim that, while their taste experiences are identical in sensory character, they differ on the affective dimension. For instance, maybe both have the same strongly bitter experience, but strangely Yum really *likes* it.

The view being proposed, then, is that *some* of our experiences represent valuational properties, in addition to objective properties. Call this the *valuational view*. And this is supposed to help answer my arguments.[13]

My Reply. In fact the valuational view does not answer my arguments. First, and most importantly, it does not answer the internal-dependence argument when it comes to *Sniff and Snort* or *Soft and Loud*. Here the external stimuli (odor clouds, sounds) do not differ in "valuational properties". For instance, the same sounds are not "good" for Soft and "bad" for Loud. So here the valuational gambit does not

[13]It might be thought that I should also consider here imperative intentionalism about pain defended by Klein (2007) and Martinez (2011), since externalist intentionalists might naturally use that theory to answer my arguments. But, for two reasons, I will not consider that view separately here. First, the main points I will make about the valuational account apply equally to the imperative account: for instance, it cannot be applied to all of my cases, so it would not afford a general solution to the problems I raise here for externalist intentionalism. Second, elsewhere (2010, note 36) I suggest that it faces especially serious problems. Klein and Martinez (forthcoming) suggest that they can handle *one* of the problems I raise there about degrees of pain [a problem repeated by Cutter and Tye (2011)], but even if they are right I believe that the other problems I list are enough to show that imperative intentionalism about pain is difficult to defend.

get off the ground. It also does not provide an answer to the structure argument. Even if externalists like Hill can use threat levels to accommodate Mild's structure judgments about pain levels, they obviously cannot use them to accommodate Soft's judgment about *loudness* levels, or Sniff's smell resemblance judgment. Since they defend a general theory of phenomenal consciousness, Tye and Cutter and Hill need to say something about these other apparent counterexamples to their theories.

In fact, the valuational view does not help answer my argument about Mild and Severe. Tye and Cutter discuss a version of the Mild and Severe case in which the same disturbance is more of a threat to Severe than to Mild. But, in my present version of the case, I stipulated that this is not true. The noxious stimuli and their aptness to harm are held constant, while there is variation in S1 activity and VAS pain ratings of sensory intensity. Therefore, even if we grant that "degrees of badness" make sense and can be represented by experience (which I question below), the tracking intentionalist is stuck with the mistaken verdict that Mild and Severe's experiences are in every respect exactly alike in phenomenal character.

The proposal of Tye and Cutter also fails when it comes to the kind of case they actually discuss, where the same disturbance is more of a threat to Severe than to Mild. In this kind of case, their proposal at best entails that their Mild and Severe's pains only differ at the affective level, and are exactly the same in sensory intensity. But Tye and Cutter ignore a feature of the case that I have stressed here and elsewhere (Pautz 2010). My argument and the empirical evidence I adduce exclusively concern *sensory* intensity. Pain intensity at the sensory level is linearly related to *S1 firing rates* and is at best related in a complex non-linear fashion to objective features of our bodily disturbances. As for the affective dimension of pain, impressive fMRI studies by Hofbauer and Rainville and others show that it is coded by activity in the anterior cingulate cortex (ACC), as I noted previously (Sect. 18.2). Now I stipulated that Mild and Severe differ in *S1 activity* (not just ACC activity), and their responses on the visual analogue scale *for sensory intensity*. Given these points, the only plausible verdict concerning the case is that their pains differ in *sensory* intensity.

So, both versions of the case of Mild and Severe are in fact counterexamples to tracking intentionalism, even if we take valuational properties into account. The case of Yuck and Yum, too, is a counterexample to tracking intentionalism. As we saw, on the valuational view, their experiences are *identical in sensory character* (because they represent the same chemical type), and only differ in the affective dimension of taste. For instance, maybe both have the same bitter experience, but strangely Yum really *likes* it. This verdict simply does not fit the facts. It is clear that ensemble activation patterns code the sensory character of taste experience (sweet, bitter, etc.), not the affective character. This is shown by the fact that tastes that differ in sensory character but agree in valance (e.g. bitter and sour tastes) are realized by different ensemble activation states the brain. Since Yuck and Yum's ensemble activation states occupy different positions in neural taste space, and since their fine-grained sorting and other behaviors differ, it is totally implausible that they both have (say) a very bitter experience of the berries. Instead, their experiences differ in sensory character. So this case shows that tracking intentionalism fails, even if we take valuational properties into account.[14]

Finally, Hill's interesting threat-level proposal for answering my structure argument about pain faces some problems. For one thing, Hill does not say what naturalistic facts might ground a *doubling* in threat-level. When Mild goes from feeling 47 °C to feeling 50 °C, the probability that he will die does not double. What then could it mean to say that the threat-level doubles? And, again, Hill's proposal evidently does not apply to Soft's judgment about a doubling of perceived loudness. Further, in both of these cases, there is a natural alternative proposal: the appeal to *firing rate*, which is well-defined (unlike "threat-level") and which is well correlated with sensory intensity. This seems to be what is driving our judgments about "intensity". How then could a naturalist avoid the conclusion that our talk of "intensity" in these domains somehow *refers to* overall firing rate or something along these lines?

---

[14]I have another worry about Tye and Cutter's specific version of the valuational gambit. Maybe Millikan's teleological (1989) theory of representation is compatible with their claim that our experiences represent properties like *being bad* or *being poisonous*. (In fact, a problem with her theory might be that it entails that our experiences *only* represent such properties.) By contrast, Tye's own theory of representation appears incompatible with that claim. On his theory, in order for a state to represent a valuational property like *being bad* or *being poisonous,* its tokenings must be *explained by* the instantiation of that property under normal conditions. Despite Cutter and Tye's (2011) interesting efforts to show that this condition is fulfilled, I still have doubts. To see why, notice that, whenever Yuck tastes the berries and his neural representation of the berries is tokened, the following counterfactual is true: if the berries were not poisonous (bad) to Yuck's species (if e.g. we gave them a pill that prevents the action of the poison, or if we somehow removed the poisonous part of the berries), that neural representation still would have been tokened (because the chemical properties of the berries still would have impinged in the same way on the taste system). (See also Pautz 2010, note 15.) The truth of this counterfactual shows that the tokening of the neural representation on particular occasions is not *explained by* the poisonousness, or badness, of the berries; rather, it is only explained by the response-independent chemical property of the berries. A similar counterfactual-based argument would show that the tokenings of pain representations in Mild and Severe is never explained by the "badness" (or badness-to-degree-x) of pain stimuli. So, on Tye's theory of representation, experiences cannot represent these properties. Cutter and Tye (2011, pp. 100–101) argue that the explanatory condition on representation is fulfilled because the badness of the stimuli provides a historical explanation for why *these species were designed by natural selection to token such states that cause withdraw.* This may be true. But, contrary to what they seem to think, it does not follow (and is in fact not true, in view of my counter-factual argument) that *particular tokenings* of these states at the present time are ever explained by the badness of the stimuli, which is what their theory requires. (Compare: a fire alarm might have been designed to ring because fire is dangerous, but particular occurrences of the ring in the present are explained by the presence of smoke, not danger.) Further, even if this problem can be overcome, since Tye and Cutter only attempt to explain how experience represents course-grained evaluative properties like *being bad*, they would still need to explain how experience represents one fine-grained degree (*being-bad-to-degree-x*) rather than another (which would require providing the naturalistic grounds for across-species comparisons of degrees of badness).

Pythagoreanism about sensible magnitudes. Casey O'Callaghan (in discussion) suggested this view to me in the case of loudness, without wholeheartedly endorsing it. (On one way of elaborating the hue-magnitude theory of color defended by Byrne and Hilbert (2003) and Tye (2000), it is also a version of what I will call "Pythagoreanism".) Since the view is somewhat difficult, let's start by seeing how it applies to an extremely simple and fanciful coincidental variation case. Then we will turn to my cases.

Suppose that A and B are two devices. In each device, there is a cylinder of fluid. The devices somehow respond to the lengths of objects in the environment and encode these lengths in terms of the height of the fluid. However, in each case, there is an expansive non-linearity. There is "bad external correlation". In the case of A, a line of n cm results in A's inner fluid rising to n^2 cm. In the case of B, the "response curve" is steeper: a line of n cm results in A's inner fluid rising to n^3 cm. Let us also pretend that A and B have sensations that vary in intensity (rather like auditory experiences or pains), where the intensity is linearly related to inner the fluid level. There is "good internal correlation". (If you like, pretend that they are alien sensations of a sort that we do not have.) Thus, when they both respond to ("track") a line of 2 cm and then a line of 3 cm, A's sensation roughly *doubles* in intensity (because his inner fluid level goes from 4 to 9 cm), while B's sensation *more than doubles* in intensity (because his inner fluid level goes from 8 to 27 cm). In addition, A "judges" that the intensity has doubled, while B "judges" that it has more than doubled.

This provides a schematic illustration of my internal-dependent argument and my structure argument. How could the "tracking intentionalist" or "objectivist" about sensible qualities accommodate the verdict that A and B have sensations of different intensities, and how might he accommodate the truth of their structure judgments? After all, they track the very same objective lengths; and the objective length does not really *double* (or more than double) when it goes from 2 to 3 cm. Given bad external correlation and good internal correlation, how could their sensations be mere representations of external lengths? Isn't it more natural to take an internalist approach, on which their sensations are identical with, or supervene on, inner fluid levels?

The Pythagorean response is meant to save tracking intentionalism and objectivism as follows. Let us say that a line has an *x-value of x* just in case it has a length of n cm and $x = n^2$. And let us say that a line has a y-value of y just in case it has a length of n cm and $x = n^3$. Suppose A and B are presented with a line of 2 cm. The Pythagorean holds that the line instantiates the following three properties: the physical length l (which does not involve number in any way, although we assigns numbers to it), the property *having an x-value of 4* and the property *having y-value of 8*. He accepts an extremely fine-grained view of properties, on which these properties are distinct, even though they are necessarily co-extensive. The idea is that the second two properties are relations to numbers; since they are relations to different numbers, and involve different functions, we should count them as non-identical, even if they are necessarily co-extensive with each other. (Likewise, he would even hold that the property of bearing the length-in-centimeters relation to the number 100 is distinct from, but necessarily co-extensive with, the property of being the length-in-meters relation to the number 1.) Further, according to the Pythagorean, on being presented with a line of 2 cm, A represents ("perceives") the property *having an x-value of 4* and B represents ("perceives") the distinct, but necessarily co-extensive, property *having y-value of 8*. This representational difference constitutes the difference in intensity between their sensations, according to him. This answers the "internal-dependence argument". As for the structure argument, notice that x-values perfectly match A's structure judgments. Thus, when the line goes from 2 to 3 cm, and A judges that the intensity has roughly doubled, the represented x-value roughly doubles (from 4 to 9). So, on this view, his structure judgments come out true. This is not a subjectivist or response-dependent view. A's structure judgments are not about his inner fluid levels. Rather, they are about the represented x-value properties. And these properties are response-independent properties of things (even if they match A's inner fluid levels). For instance, even if A were not around, a line of 2 cm would have an x-value of 4.

Of course, I call this the "Pythagorean view" because it holds that experience represented properties defined in numerical terms. I think that this is an essential feature of the view. Suppose someone just said that, in addition to lengths in cm, the line has two other families of properties, and that, while the lengths do not match A and B's structure judgments, these other properties do. We would be mystified. To make the view comprehensible, he must define x-values and y-values, and explain that he individuates properties extremely finely.

Now that we have a grip on the view, we can see how it might apply to more complex cases, for instance, the case of Soft and Loud. Of course, that case is somewhat analogous to the case of A and B. The main difference is that, while A and B's experiences track non-disjunctive properties (corresponding to individual lengths), Soft and Loud's experiences track extremely complex, disjunctive properties involving amplitude, frequency, critical bands, and so on. The Pythagorean about sensible magnitudes holds that there is a complex function f from these parameters onto numbers that reflects Soft's psychophysical judgments of loudness level. We might call this the S-loudness of a tone. There is a different complex function g from these parameters onto numbers that reflects Loud's (different) psychophysical judgments of loudness level. We might call this the L-loudness of a tone. (Those who work in psychophysics actually have devised scales that reflect our psychophysical judgments, such as the Bark scale and the mel scale.) According to the Pythagorean, the loudness levels that Soft perceives are S-loudness levels, while the loudness levels that Loud perceives are L-loudness levels. So for instance if they hear the same tone, the tone might have both the property *having an S-loudness of 10* and the property *having an L-loudness of 20*. These properties will be necessarily co-extensive, because the very same combinations of physical parameters that yield an S-loudness of 10 yield an L-loudness of 20. But the thought is that they are nevertheless distinct, and that Soft some perceives (or "represents") the first one while Loud perceives the second one. So Pythagoreanism about loudness levels answers my internal-dependence argument. Further, since these properties match their ratio judgments, it also answers my structure argument. Maybe a similar view could be applied to the case of Mild and Severe involving pain intensity (Pautz 2010). This counts as a sophisticated version of tracking intentionalism. The idea is that sensory magnitudes are objective (if complex) properties and sensory intensity is determined by what sensory magnitudes we track.

My reply. The main problem with Pythagoreanism is that it does not provide a general response to my arguments. The mathematical treatment of sensory qualities certainly does not apply to smell qualities or taste qualities, for instance. Therefore the Pythagorean response does not help with my arguments concerning smell and taste. Those arguments are enough to undermine externalist intentionalism and objectivism about sensible qualities more generally.

In fact, for several reasons, Pythagoreanism even fails in the case of loudness. (i) While we can in a rough and ready way *represent* loudness levels in terms of numbers, any *definition* of loudness levels in mathematical terms is totally implausible. There is conventionality and vagueness involved. (ii) To appreciate my next problem, return to the simple case of A and B. The Pythagorean view requires that *having an x-value of 4* is distinct from *having y-value of 8,* even though each is necessarily coextensive with the same length (having a length of 2 cm). Against this, intuitively, what we have here are just two different descriptions of a single length property. (iii) Even if we grant that necessarily coextensive properties can be distinct, the view certainly fails. To appreciate the problem, suppose again that A and B are presented with a 2 cm line. The Pythagorean holds that A's experience represents the property *having an x-value of 4* while B's experience represents the property *having y-value of 8*. But, given that these properties are necessarily co-extensive, what makes it the case that A's experience represents *having an x-value of 4* but not *having a y-value of 8*, while B's experience represents *having a y-value of 8* but not *having an x-value of 4*? The proponent of this view needs a *general theory* of the sensory representation relation, the relation *x sensorily represents property y*, which explains how this might be so. But those theories always appeal to relations like tracking or indication, which cannot distinguish between necessarily co-extensive properties. It would not be enough for the Pythagorean to respond by rejecting this kind of theory. He would need to at least gesture at an alternative general theory of the sensory representation relation that answers the problem. (iv) Supposing we can make sense of Pythagorean properties, they are evidently extremely disjunctive and unnatural. It would seem that firing rates in the auditory channel are much more natural. Since they seem well correlated with auditory intensity, and since they seem to be what causally explains judgments about 'loudness', the naturalistic *must* claim that 'loudness' refers to a property involving firing rates in the auditory channel. By similar reasoning, he *must* say that 'pain intensity' refers to a property involving firing rates in S1 or other regions of the pain-matrix. There is no other reasonable view for the naturalist. This naturally leads to "neural projectivism".

Neural projectivism. This interesting view was suggested to me by Christopher Hill as a possible view of pain intensity. On this view, when you have a throbbing pain in your foot, for instance, there is in your brain an atomic representation *R* of intensity. But this neural representation *R* does *not* represent an actual peripheral state, such as stimulus intensity or size. Bad external correlation means that pain intensity is not directly proportional to any such peripheral stimulus features. Rather, *R* represents the "intensity" of the *internal nociceptive neural signals* set up by the stimulus, which are known to be more proportional to pain intensity ("good internal correlation"). Now, the "intensity" of nociceptive neural signals is just a matter of firing rate. So the idea is that *R* represents a firing rate property of the form *exhibiting firing rate N*. (Of course, even if pain intensity is firing rate, our pain experience doesn't reveal that it is a matter of firing rate, just as a perception of water does not reveal that it is H_2O.) In one version of neural projectivism, *R* represents the firing rates of nociceptive neurons in the spinal cord. In another version, *R* represents the (possibly different) firing rates of nociceptive neurons in the cortex itself (e.g. S1), which are known to be especially well correlated with pain intensity. (I will suggest below that this is the best version.) Thus *R* is a kind of "self-monitoring" representation.

Now for the "projective" element of the view. The neural projectivist further claims that, in the area of the brain responsible for pain, there are other atomic representations, R', R'', ... of other features. By contrast to *R*, these atomic representations *do* represent *peripheral* properties, like *throbbing* and *being located in the foot*. According to the neural projectivist, when you have a pain, the atomic representation *R* of *exhibiting firing rate N* is "combined with" these other representations R', R'', ...The result is a complex representation with the content *there is something with the properties throbbing, being located in the foot, and exhibiting firing rate N*. This content is false: while there is disturbance with the properties *throbbing* and *being located in the foot*, it does not have the property *exhibiting firing rate N*. In that sense, your pain experience projects a property in fact possessed by the central nervous system onto a external bodily region. Maybe this could be thought of as a kind of "binding error", because it involves binding a property that is possessed by one entity (the nervous system) together with properties that are possessed by another entity.

That, then, is neural projectivism. *If* neural projectivism about pain is workable, it answers my arguments about pain. On neural projectivism, Mild and Severe have pains of different intensities in response to the same noxious temperatures, because they "project" different firing rate properties (in fact possessed by their central nervous system) onto the external bodily regions to which the temperatures are applied. This answers my internal-dependence argument. On neural projectivism, when in the actual world Mild judges that his pain has roughly doubled in intensity between 47 and 49.8 °C, his judgment is actually about the neural firing rate projected onto the relevant bodily region. If there really is a linear correlation between neural firing rate and pain intensity, the firing rate literally doubles. So, Mild's ratio judgment is literally true. This answers my structure argument.

Given that there is "good internal correlation" and "bad external correlation" in all sense-modalities, if neural projectivism is right about pains it must be right about other experiences.[15] The idea would be that loudness is neural firing rate in the auditory channel; smell and taste qualities are ensemble activation patterns; color qualities are neural properties of the chromatic channels in the brain; and so on. And all of these sensible qualities are somehow projected onto external events and items, according to the uniform neural projectivist. They are bound together with external locations, shapes, and so on. Such a view might answer my arguments about Yuck and Yum, Sniff and Snort and Soft and Loud. Of course, this would just be a naturalistic version of the traditional Galilean view that locates the sensible qualities in the head. It would be a projectivist version of the "stinking brain" theory (when color is at issue, "colored brain" theory).

Since externalist intentionalists were trying to avoid exactly this view (recall the quote from Armstrong at the start of the paper), one might wonder whether we should count it as a version of externalist intentionalism. I think it does deserve the name. On neural projectivism, many representations really do accurately represent external properties, like location and shape, presumably by way of some kind of tracking or indication relation; and the representation of such properties plays an important role in configuring phenomenology.

My reply. Before I evaluate neural projectivism, let me say that I think the best version of this view holds that the projected neural properties are properties that in fact belong to *cortical* neural assemblies rather than more peripheral neural signals. The reason is that this version of the view is needed to handle the full range of hypothetical "coincidental variation cases". Coghill et al. (2003, p. 8542) report that in some actual cases there is reason to think that "a large portion of the variability of interindividual differences in both the subjective experience of pain and activation of SI and ACC is likely attributable to factors other than differential sensitivity of spinal or peripheral afferent mechanisms [which are often show the *same* levels of activity in such cases]". So, we might add a twist to my case of Mild and Severe: we might stipulate that Mild and Severe's peripheral and spinal nociceptive neurons fire at the same rates in response to the same noxious temperatures, and that there are only differences in S1 and ACC in the cortex. Given that the only differences are in the cortex, in order to explain why Mild and Severe's pains differ in intensity, neural projectivists would have to say that Mild and Severe's pain systems represent different levels of *cortical activity* and project them onto the same bodily regions. And if in this hypothetical case human pain representations only represent cortical neural activity, presumably this is also true in the actual case. The "cortical" version of neural projectivism is also supported by the fact

[15]For instance, our perceptions of the four perceptually prominent elemental or "unique hues" cannot be explained merely in terms of the reflectance properties of external objects. Most researchers assume that it has a cortical basis but for many years that remained elusive. However, a recent breakthrough study by Horwitz and Haas (2012) seems to have gone some way towards uncovering the cortical basis of such perceptions. So there is reason to think that the consistent neural projectivist about pain intensity would have to be a projectivist about the qualitative dimensions of color as well.

that often the best correlations between the phenomenal and the neural are often to be found in the cortex (e.g. smell similarity correlates best with PPC neural similarity).

Now in any version I think neural projectivism is an interesting attempt to come to grips with the kind of empirical problems I have raised. It would explain how sensory consciousness manages to be both externally-directed and internally-dependent. But it faces challenges that seem to me overwhelming. To illustrate, take the case of pain.

To begin with, some background. The *depth problem* or *distance problem* is a well-known problem for naturalistic theories of sensory representation. Consider a cortical representation of size or orientation or some other spatial property. It is not only causally correlated with an external size. It is also correlated with a pattern of firing on the retina, as well as a pattern of firing in the lateral geniculate nucleus. What makes it the case that it represents one of these elements in the causal chain, as opposed to the others?

Typically, theories of sensory representation are designed to solve the depth problem in favor of the distal or "distant" properties. Some theories appeal to the notion of function (Dretske 1995). Intuitively, the function of the cortical representation is to indicate size, not some intermediary retinal state. Other theories appeal to "asymmetrical dependence" (Fodor 1990). The cortical representation tracks neural activity only *because* it tracks shape. So, on these theories, the sensory representation represents the external shape. Indeed, these theories seem to entail that *all* of our cortical sensory representation represent distal properties or conditions. So, for instance, a pain representation would seem to have the function of indicating *noxious temperatures*, which are biologically important because they can harm the organism. The function of the pain representation is not to indicate some parameter in fact instantiated in the brain. In general, the function of sensory representations is to indicate biologically significant distal conditions, like level of sugar, stable reflectance properties, and so on. Hence, on standard theories of sensory representation, all cortical sensory representations represent external properties, not properties in fact possessed by neural assemblies.

By contrast, the neural projectivist takes a non-uniform view concerning what properties are represented by atomic sensory representations. He holds that some atomic sensory representations represent distal properties, like location, orientation or shape. But he also holds that other atomic sensory representations represent properties that are in fact only instantiated by neural assemblies, like nociceptive firing rate as opposed to temperature. (Of course, as a projectivist, he doesn't hold that the sensory system represents these properties *as* instantiated by neural assemblies.) When we have experiences, both sorts of atomic representations are combined into complex representations. The result is a kind of binding error, in which properties that are in fact only instantiated in the head are bound with properties that are instantiated external to the head.

Now I can state my first problem for the view: what is the *general* naturalistic theory of sensory representation, which implies the projectivist's *non-uniform* answer to the depth problem? As I said, standard theories of sensory representation always solve the depth problem in favor of external properties. I cannot think of alternative general theory of sensory representation that in some cases solves the depth problem in favor of external properties and in other cases solves it in favor of "internal" properties. In response, the projectivist would at least have to sketch a general view that might imply his non-uniform view, in order to make his view believable.

In reply, the neural projectivist might claim that it would be *good* if the pain system keeps track of nociceptive firing rates, because nociceptive firing rates correlate better with threat level. By contrast, the physical intensity (e.g. temperature) of the external stimulus correlates poorly with threat level. Since it would be good if the pain system keeps track of nociceptive firing rates, it probably has a representation *R* that represents nociceptive firing rates. (Christopher Hill suggested to something along these lines.)

The trouble is that this does not answer my specific challenge. My challenge was: what is the general theory of representation (sensory representation *X* represents *Y* iff . . .) which entails that *R* represents nociceptive firing rates, and at the same time entails that other atomic sensory representations represent external properties like location and size? At best, the reply only provides a *reason to believe* that the pain system has a representation *R* that represents nociceptive firing rates, in addition to other representations that represent external conditions. It does not explain *what makes this the case*, by showing how it follows from general theory of representation.

(It is also worth mentioning that the reasoning in the reply is somewhat questionable. For instance, it is of course good to keep track of neural computations of size in the sense that it good to engage in behavior in step with those computations; but this does not provide a reason to think that we have sensory representations, or experiences, that *represent* the relevant computations, as opposed to apparent shapes.)

Here is a second problem for neural projectivism. Even if the neural projectivist can come up with a theory of sensory representation on which some of atomic sensory representations represent internal properties while others represent external properties (thus answering my first problem), he needs a theory of how complex representations manage to represent them as *co-instantiated*. For instance, on his view, when you have a pain in your foot, there is in your brain a complex representation that represents the complex condition *there is something with the properties throbbing, being located in the foot, and exhibiting firing rate N*. How is this? On many views, any sensory representation represents a condition by co-varying with it under optimal conditions. But this simple view will not work in this case, because the relevant conditions never obtains and doesn't even obtain in nearby counterfactual situations. Of course, the neural projectivist might claim that the content of this complex representation is determined compositionally (like sentences in a language of thought), thanks to a kind of concatenation relation between atomic neural representations. But just when are two atomic neural representations "concatenated". Is there a functional relation of this relation?

Finally, let me mention a third problem for neural projectivism. Given that there is "good internal correlation" and "bad external correlation" in all sense-modalities, if there is a true general theory of sensory representation that entails neural projectivism in the case of pain (contrary to first worry), then it presumably also entails neural projectivism about other experiences. So neural projectivism about pain stands or falls with neural projectivism about other experiences. But, if the view is odd in the case of pain, it is even odder in the case of other experiences. For instance, the idea would be that loudness level is a matter of firing rates of neurons somewhere in the auditory channel (presumably the cortex), but the auditory system projects internal firing rates onto external events. So the content of an auditory experience might be *there is an event has the property of being at place p and the property of exhibiting firing rate N*. Likewise, brightness and hue are neural properties but the visual system projects them into external objects. So the content of a visual experience might be *there is an object that has the property of being at place p, the property of being round, and the property of undergoing neural activity N*. The problems that I raised above against neural projectivism about pain apply with even more force against neural projectivism about other experiences.

Response-dependent intentionalism. The kind of tracking intentionalism that I have taken as my stalking horse holds that the sensible qualities represented by experience are objective properties of external things. By contrast, *response-dependent intentionalism* holds that the sensible qualities represented by experience are *response-dependent properties*. These response-dependent properties might have the form: *causing, or being disposed to cause, internal neural state N in individual or population I*. The idea is that phenomenal character of experience is partly constituted by the representation of such response-dependent properties. However, the response-dependent intentionalist I am interested in retains the central idea of tracking intentionalism that we represent properties by tracking them. On the basis of actual cases of variation (discussed in Sect. 18.2.1), Uriah Kriegel (2009) has developed this type of view in great detail.

In my view, the best argument for the response-dependent view is that it answers the internal-dependence argument and the structure argument. So, for instance, return to the case of Soft and Loud. On hearing the same tone, they track the same (disjunctive) response-independent physical property of the tone. But on response-dependent intentionalism Soft's auditory experience also tracks and thereby represents the response-dependent property *normally causing firing rate f in the auditory channel of Soft's population*, while Loud's experience tracks and thereby represents the *different* response-dependent property *normally causing firing rate f+ in the auditory channel of Loud's population*. Hence the neural difference between Soft and Loud is associated with a representational difference. This might explain the phenomenal difference between their experiences in intensity. The same kind of account might be applied to the cases of Yuck and Yum, Mild and Severe, and Sniff and Snort. This would answer the internal-dependence argument.

Response-dependent intentionalism might also avoid my structure argument. So, for instance, on this view, smell qualities might be dispositions to produce ensemble activation states in the PPC. Perhaps these dispositions resemble insofar as their neural manifestations resemble. Then, given good internal correlation, our judgments about resemblances among smell qualities will come out true. Indeed, on this view, if two smell qualities resemble to a degree, they do so essentially, because they are essentially dispositions to produce neural states that resemble do that degree. So this view accommodates the intuition (mentioned in Sect. 18.4.2) that the structural features of sensible qualities are essential to them. Likewise for our judgments about the structural features of other ranges of sensible qualities.

In a way, response-dependent intentionalism resembles neural projectivism. Both views hold that the sensible qualities constitutively involve neural responses. The difference is that, while the neural projectivist holds that the sensible qualities are properties of neural responses erroneously projected onto external items, the response-dependent intentionalist holds that they are dispositions to produce neural responses. Since external items really have those dispositions, the response-dependent intentionalist avoids projective error.

My reply. The main problem with response-dependent intentionalism is that there is no good naturalistic theory of how we represent response-dependent properties in experience. Call this the *psychosemantic problem*.

Kriegel (2009) favors Dretske's (1995) theory of sensory representation. But in fact Dretske's theory of sensory representation is incompatible with Kriegel's response-dependent intentionalism. On Dretske view, a brain state *B* belonging to a sensory system represents "the" external property that the brain state has the "function of indicating". Presumably, a brain state *B* does not have the "function of indicating" the biologically unimportant response-dependent property of the form *normally causing brain state B in humans*. (Even more obviously, it does have the function of indicating Kriegel's more complex response-dependent properties: for reasons I will not go into, he claims that they involve dispositions to produce brain states in all actual creatures, including alien creatures, if such there be.) A brain state has the function of indicating a biologically important property: chemical property, bodily disturbance, etc. So on Dretske's theory of representation, we sensorily represent such response-independent properties, not Kriegel's response-dependent properties. That is of course Dretske's view. Many other theories, for instance Tye's (2000) tracking theory, explain representational relations in terms of *causal* or *explanatory* relations. But the dispositional, response-dependent property normally causing brain state *B* isn't causally efficacious in the production of brain state *B*. So such theories, too, are incompatible with response-dependent intentionalism. In general, I see no theory of sensory representation compatible with response-dependent intentionalism.

Kriegel (2012) answers one objection to response-dependent intentionalism, a kind of *circularity objection* he attributes to Robert van Guilick and Joseph Levine. The circularity problem is simply the problem of characterizing the relevant response-dependent properties in non-phenomenal terms, so that they can be appealed to in an intentionalist theory of phenomenal character without circularity. I agree with Kriegel that this problem can be answered: the responses can be characterized in neural terms, for instance. But my psychosemantic my different psychosemantic problem remains. Even if external items possess extremely complex response-dependent properties that can be characterized in non-phenomenal terms, the response-dependent intentionalist still needs an account of what makes it the case that our experiences represent any of them (the standard accounts do not work).[16]

Millikan to the rescue? William Lycan (2006) has suggested that the internal-dependence argument might work only work against versions of externalist intentionalism that incorporate a simple tracking theory of sensory representation. Without developing the details, he suggests that that Millikan's (1989) more complex consumer-based theory of representation might enable the externalist intentionalist to handle the cases I have discussed.

My reply. I think we can see that this response will not save externalist intentionalism even without going into the details of Millikan's sophisticated consumer-based theory of representation. To illustrate, consider Soft and Loud. They have experiences of the same tone that differ in sensory intensity. What, according to the proponent of a consumer-based theory of sensory representation, is represented by their experiences? As Lycan likes to put it, what are the *representata*? There is no good option consistent with externalist intentionalism.

(i) Maybe Millikan's theory implies that Soft and Loud's experiences represent the response-independent physical properties (involving amplitude, frequency, and critical bands) that constitute the loudness, pitch and timbre of the tone, according to the externalist. So, their experiences have the same content. But this option is inconsistent even with the intentionalist thesis that (at least within a sense modality) experiences differ in sensory phenomenology only if they differ in representational content. The externalist intentionalist needs to find *different representata*. (ii) Another option is that Soft and Loud's experiences represent different firing rates in their own auditory systems, and somehow project these onto the same external sound-event. We have seen the problems with this kind of "neural projectivism". (iii) A final option is that their experiences represent different extremely complex Kriegel-style response-dependent properties. But no theory of representation (including Millikan's) is compatible with this option, as we have seen.

---

[16]Suppose Soft and Loud hear the same tone. The tone has a huge set of co-extensive dispositions to cause various neural responses in them and other creatures under various conditions. Even if the response-dependent intentionalist manages to specify a permissive theory of representation on which Soft and Loud sensorily represent such response-dependent properties, he would face a follow-up problem. He does not want to be so permissive as to say that they represent the *same* huge swarm of response-dependent properties, because this would leave him without an account of the phenomenal difference between their experiences. But what could make it the case that Soft sensorily represents one specific response-dependent property within this set and Loud represents a different response-response-dependent property within the set? This might be called the *selection problem* or the *promiscuity problem* (Pautz 2010).

Quasi-intentionalism. Externalist intentionalists are out of options. But maybe they can make a simple retreat. As I defined externalist intentionalism, they are committed to the intentionalist thesis that all phenomenal differences among sensory experiences are constituted by representational differences. Maybe they could reject intentionalism (so understood) and retreat to what I will call *quasi-intentionalism*. On this view, some phenomenal differences among sensory experiences are constitutes by representational differences, while others are constituted by merely *functional* or *neural* differences.

Some philosophers have already defended quasi-intentionalism concerning a certain limited range of cases. Thus Lycan (forthcoming) claims that *subtle affective differences* among experiences are not representational differences; they are mere functional differences concerning effects on desire and behavior. So he would reject the representational account of affective phenomenology defended by Cutter and Tye. Likewise Hill (2012, note 2) has said that "how it seems to one to have an experience [e.g. the apparent simplicity of colors] is determined by two factors – it is determined in part by the representational content of the relevant representation, and in part by the representation's intrinsic [neural] and functional properties". *If* by 'how it seems' he means *phenomenal character* (and not just our inclinations to form sophisticated *beliefs*, e.g. about the simplicity of colors), then Hill endorses quasi-intentionalism about some cases. Some aspects of the phenomenology of experience cannot be explained in terms of representational content, and there are possible cases in which such aspects (apparent phenomenal simplicity perhaps) vary while representational content is held constant. (This *may* not be the correct interpretation because Hill (2009, p. 148) also says that phenomenal character is nothing but the "set of" *represented properties*, suggesting a one-factor view.)

Maybe quasi-intentionalism could be used to explain the more radical forms of phenomenal variation found in my coincidental variation cases. The idea is that, in these cases, the individuals involved (Yuck and Yum, Sniff and Snort, Mild and Severe, Soft and Loud) have experiences that differ radically in their sensory character, but they *have exactly the same representational contents*. What constitutes the phenomenal differences are either merely neural differences or merely functional differences (e.g. tendencies to group stimuli in certain ways). And maybe the quasi-intentionalist could answer the structure argument by explaining facts about qualitative structure in neural or functional terms, rather than in purely response-independent terms.

My reply. Maybe *some* phenomenal differences are not representational, for instance differences in mood or valence. But I think that the suggested response takes quasi-intentionalism too far, and I think that Lycan would agree. Consider Yuck and Yum, Mild and Severe, Sniff and Snort, and Soft and Loud. Contrary to the suggested quasi-intentionalist response, given that their experiences differ in sensory character, they also *differ* in representational content. The phenomenal differences between them cannot be treated as *merely* neural or functional differences.

To see this, let's be fanciful. Suppose you could occupy their points of view on the world, and switch between them. In switching between Soft and Loud the world would *seem* different to you: different loudness-levels would *appear to* attach to the same external sound events. Likewise, the science makes it reasonable to suppose that the experiences of Yuck and Yum, and Mild and Severe, do not just differ in affective valence; they differ in sensory character. So, if you could switch between their points of view, numerically different pain or taste qualities would *appear to be* present in a certain bodily region or in your tongue. So, in some sense of 'representational', their experiences certainly differ in representational content (in what qualities ostensibly present to their subjects), even though they bear the same externally-determined naturalistic relations to the same objective properties.

18.6 An Edenic Theory of Sensory Consciousness?

Now we have puzzle. Sensory consciousness is both "externally-directed" and "internally-dependent". The individuals in my coincidental variation cases are ostensibly conscious of different qualities "out there", owing to the internal neural differences between them, even though they track the same objective properties. How is this? What in the world are these different qualities? We have seen that they are not objective physical properties (chemical properties, types of damage, and so on). In fact, I would argue that they are not properties of extra-cranial items of any sort. The main alternative to this externalist picture is a traditional internalist picture on which phenomenal types are necessarily identical with internal neural types and the sensible qualities are neural properties projected onto items in external space. This peculiar view is sometimes called 'the stinking brain theory' – or, when color is involved, 'the colored brain theory'. But we have seen that this kind of projectivist view too faces enormous problems.

In my view, both of these alternatives share a false presupposition, namely that the sensible qualities must be located *somewhere* in the world. The externalists are wrong to locate the sensible qualities outside the brain. But their opponents are also wrong to kick the sensible qualities upstairs into the brain. Although I cannot argue for this here, I would suggest that the overall best view is that, while our brain portrays the world and our bodies as filled with sensible qualities because that enhances adaptive fitness, they are not real qualities that belong to anything, including the brain itself. They are wholly chimerical. This is what David Chalmers (2006) calls the *Edenic theory*.[17] His arguments for the view are based on *a priori*

[17]Interestingly, Ruth Millikan also defends a kind of error theory of sensory experience. In comments on an earlier version of this paper, she asserts that relations among qualities are "chimerical" and do not obtain among any external items, appealing to the work in neuroscience that I discuss. Millikan (Chap. 2, this volume) makes remarks along similar lines. For questions about her view here and her argument for it, see Pautz (2011).

and phenomenological considerations. By contrast, I think that the best argument requires looking at the kind of research in psychophysics and neuroscience I have discussed here.[18]

References

Armstrong, D. 1999. *The mind-body problem*. Boudler: Westview Press.
Batty, C. 2010. What's that smell. *Southern Journal of Philosophy* 47(4): 321–348.
Block, N. 1999. Sexism, ageism, racism and the nature of consciousness. *Philosophical Topics* 26: 39–70.
Block, N. 2010. Attention and mental paint. *Philosophical Issues* 20: 23–63.
Block, N. 2012. Discussion of J. Kevin O'Regan's Why red doesn't sound like a bell: Understanding the feel of consciousness. *Review of Philosophy and Psychology* 3: 89–108.
Borg, G., H. Diamant, L. Strom, and Y. Zotterman. 1967. The relation between neural and perceptual intensity. *The Journal of Physiology* 192: 13–20.
Byrne, A. 2003. Color and similarity. *Philosophy and Phenomenological Research* 66: 641–665.
Byrne, A. 2012. Hmm... Hill on the paradox of pain. *Philosophical Studies* 161(3): 489–496.
Byrne, A., and D. Hilbert. 2003. Color realism and color science. *The Behavioral and Brain Sciences* 26: 3–21.
Campbell, J. 2002. *Reference and consciousness*. Oxford: Oxford University Press.
Chalmers, D. 2006. Perception and the Fall from Eden. In *Perceptual experience*, ed. T. Szabo Gendler and J. Hawthorne. Oxford: Oxford University Press.
Chen, J.Y., J.D. Victor, and P.M. Di Lorenzo. 2011. Temporal coding of intensity of NaCl and HCl in the nucleus of the solitary tract of the rat. *Journal of Neurophysiology* 105: 697–711.
Churchland, P. 1996. *The engine of reason, the seat of the soul*. Cambridge: MIT Press.
Coghill, R., C. Sang, J. Maisog, and M. Iadarola. 1999. Pain intensity processing within the human brain: A bilateral, distributed mechanism. *Journal of Neurophysiology* 82(4): 1934–1943.
Coghill, R., J. McHaffie, and Y.-F. Yen. 2003. Neural correlates of interindividual differences in the subjective experience of pain. *Proceedings of the National Academy of Science* 100: 8538–8542.
Cowart, B.J., and N.E. Rawson. 2001. Olfaction. In *The Blackwell handbook of perception*, ed. E. Goldstein. Oxford: Blackwell Publishers.
Cutter, B. and M. Tye. 2011. Tracking intentionalism and the painfulness of pain. *Philosophical Issues* 21: 90–109.
Davies, W. Forthcoming. Colour similarities at different levels.
Di Lorenzo, P.M., J.Y. Chen, and J.D. Victor. 2009. Quality time: Representation of a multidimensional sensory domain through temporal coding. *Journal of Neuroscience* 29: 9227–9238.
Dretske, F. 1995. *Naturalizing the mind*. Cambridge: MIT Press.
Fodor, J. 1990. *A theory of content and other essays*. Cambridge: MIT Press.
Gescheider, G. 1997. *Psychophysics: The fundamentals*. Mahwah: Lawrence Erlbaum Associates.
Haddad, R., R. Khan, Y.K. Takahashi, K. Mori, and D. Harel. 2008. A metric for odorant comparison. *Nature Methods* 5: 425–429.

[18]I presented the material in this paper at Rutgers University, Columbia University, and the Australian National University. I would like to thank the audiences on those occasions for very helpful comments. My thanks also to Chris Hill and Casey O'Callaghan for some extremely helpful correspondence on the issues I have discussed. Finally, I would like to thank Richard Brown for his work in organizing the *Consciousness Online* conference in which an earlier version of this paper appeared.

Hill, C. 2009. *Consciousness*. Cambridge: Cambridge University Press.
Hill, C. 2012. Locating Qualia: Do they reside in the brain or in the body and the world? In *New perspectives on type identity*, ed. S. Gozzano and C. Hill. Cambridge: Cambridge University Press.
Hofbauer, R.K., P. Rainville, G.H. Duncan, and M.C. Bushnell. 2001. Cortical representation of the sensory dimension of pain. *Journal of Neurophysiology* 86: 402–411.
Horgan, T., J. Tienson, and G. Graham. 2004. Phenomenal intentionality and the brain in a vat. In *The externalist challenge: New studies on cognition and intentionality*, ed. R. Shantz. Amsterdam: de Gruyter.
Horwitz, G., and C. Haas. 2012. Nonlinear analysis of macaque V1 color tuning reveals cardinal directions for cortical color processing. *Nature Neuroscience* 15: 913–919.
Howard, J.D., J. Plailly, M. Grueschow, J.D. Haynes, and J.A. Gottfried. 2009. Odor quality coding and categorization in human posterior piriform cortex. *Nature Neuroscience* 12: 932–939.
Kalderon, M. 2011. The multiply qualitative. *Mind* 120: 239–262.
Kenshalo, D.R., K. Iwata, M. Sholas, and D.A. Thomas. 2000. Response properties and organization of nociceptive neurons in area 1 of monkey primary somatosensory cortex. *Journal of Neurophysiology* 84: 719–729.
Klein, C. 2007. An imperative theory of pain. *Journal of Philosophy* 104: 517–532.
Klein, C., and M. Martinez. Forthcoming. Naturalism and degrees of pain.
Kriegel, U. 2009. *Subjective consciousness*. Oxford: Oxford University Press.
Kriegel, U. 2012. In defense of self-representationalism: Reply to critics. *Philosophical Studies* 159(3): 475–484.
Langers, D., P. van Dijk, E. Schoenmaker, and W. Backes. 2007. fMRI activation in relation to sound intensity and loudness. *NeuroImage* 35: 709–718.
Linster, C., B. Johnson, E. Yue, A. Morse, Y. Choi, M. Choi, A. Messiha, and M. Leon. 2001. Perceptual correlates of neural representations evoked by odorant enantiomers. *Journal of Neuroscience* 24: 9837–9843.
Lycan, W. 2001. The case for phenomenal externalism. *Philosophical Perspectives* 15: 17–35.
Lycan, W. 2006. *Pautz vs. Byrne & Tye on externalist intentionalism*. https://webspaceutexas.edu/arp424/www/vs.pdf.
Lycan, W. Forthcoming. Block and the representation theory of sensible qualities. In *Themes from block*, ed. A. Pautz and D. Stoljar. Cambridge: MIT Press.
Malnic, B., J. Hirono, T. Sato, and L.B. Buck. 1999. Combinatorial receptor codes for odors. *Cell* 96: 713–723.
Margot, C. 2009. A noseful of objects. *Nature Neuroscience* 12: 813–814.
Martinez, M. 2011. Imperative content and the painfulness of pain. *Phenomenology and the Cognitive Sciences* 10: 67–90.
McLaughlin, B. 2003. Color, consciousness, and color consciousness. In *New essays on consciousness*, ed. Q. Smith. Oxford: Oxford University Press.
Millikan, R. 1989. Biosemantics. *Journal of Philosophy* 86: 281–297.
Moore, B. 2003. *An introduction to the psychology of hearing*. San Diego: Academic.
Noë, A. 2004. *Action in perception*. Cambridge: MIT Press.
Oakley, B. 1985. Taste responses of human chorda tympani nerve. *Chemical Senses* 10: 469–481.
O'Callaghan, C. 2002. *Sounds*. Dissertation, Princeton University, Princeton.
O'Callaghan, C. 2009. Auditory perception. *The Stanford encyclopedia of philosophy*. http://plato.stanford.edu/archives/sum2009/entries/perception-auditory/.
O'Regan, K. 2011. *Why red doesn't sound like a bell: Understanding the feel of consciousness*. Oxford: Oxford University Press.
Papineau, D. 2012. *My philosophical views*. At http://philpapers.org/profile/10091/myview.html.
Pautz, A. 2006a. Sensory awareness is not a wide physical relation. *Noûs* 40: 205–240.
Pautz, A. 2006b. Can the physicalist explain colour structure in terms of colour experience? *Australasian Journal of Philosophy* 83: 535–564.
Pautz, A. 2010. Do theories of consciousness rest on a mistake? *Philosophical Issues* 20: 333–367.
Pautz, A. 2011. *Reply to Ruth Millikan*. https://webspace.utexas.edu/arp424/www/mh.pdf.

Price, D. 1999. *Psychological mechanisms of pain and analgesia*. Seattle: IASP Press.
Price, D. 2002. Central neural mechanisms that interrelate sensory and affective dimensions of pain. *Molecular Interventions* 2: 392–403.
Price, D.D., F.M. Bush, S. Long, and S.W. Harkins. 1994. A comparison of pain measurement characteristics of mechanical visual analogue and simple numerical rating scales. *Pain* 56: 217–226.
Rainville, Pierre, G.H. Duncan, Donald D. Price, M. Carrier, and M.Catherine Bushnell. 1997. Pain affect encoded in anterior cingulate but not somatosensory cortex. *Science* 277: 968–971.
Relkin, E.M., and J.R. Doucet. 1997. Is loudness simply proportional to the auditory nerve spike count? *Journal of the Acoustical Society of America* 101: 2735–2740.
Röhl, M., B. Kollmeier, and S. Uppenkamp. 2011. Spectral loudness summation takes place in the primary auditory cortex. *Human Brain Mapping* 32: 1483–1496.
Sakurai, T., Y. Misaka, M. Ueno, S. Ishiguro, and Y. Ishimaru Matsuo. 2010. The human bitter taste receptor, hTAS2R16, discriminates slight differences in the configuration of disaccharides. *Biochemical and Biophysics Research Communication* 402: 595–601.
Schiffman, S.S., and R.P. Erickson. 1971. A psychological model for gustatory quality. *Physiology and Behavior* 7: 617–633.
Shoemaker, S. 1994. Phenomenal character. *Noûs* 28: 21–38.
Shoemaker, S. 2000. Phenomenal character revisited. *Philosophy and Phenomenological Research* 60: 465–467.
Smith, B. 2007. The objectivity of tastes and tasting. In *Questions of taste*, ed. B. Smith. Oxford: Oxford University Press.
Smith, D., R. Van Buskirk, J. Travers, and S. Bieber. 1983. Coding of taste stimuli by hamster brain stem neurons. *Journal of Neurophysiology* 50: 541–558.
Stevens, S.S., A.S. Carton, and G.M. Schickman. 1958. A scale of apparent intensity. *Journal of Experimental Psychology* 56: 328–334.
Timmermann, L., M. Ploner, K. Haucke, F. Schmitz, R. Baltissen, and A. Schnitzler. 2001. Differential coding of pain intensity in the human primary and secondary somatosensory cortex. *Journal of Neurophysiology* 86: 1499–1503.
Turin, L. 2002. A method for the calculation of odor character from molecular structure. *Journal of Theoretical Biology* 216: 367–385.
Tye, M. 2000. *Consciousness, color and content*. Cambridge: MIT Press.
Tye, M. 2006. The puzzle of true blue. *Analysis* 66: 173–178.
Tye, M. 2009. *Consciousness revisited*. Cambridge: MIT Press.
van der Heijden, A. 1993. Sweet and bitter tastes. In *Flavor science: Sensible principles and techniques*, ed. T.E. Acree and R. Teranishi. Washington, DC: ACS Books.
Walters, D. 1996. How are bitter and sweet taste related? *Trends in Food Science and Technology* 7: 399–403.
Youngentob, S.L., B.A. Johnson, M. Leon, P.R. Sheehe, and P.F. Kent. 2006. Predicting odorant quality perceptions from multidimensional scaling of olfactory bulb glomerular activity patterns. *Behavioral Neuroscience* 120: 1337–1345.

Chapter 19
No Problem

David Hilbert and Colin Klein

Pautz refutes tracking intentionalism. We defend it. In what follows we will attempt to undermine the motivation for some of what he says and discuss one of his arguments in more detail. We don't, however, defend tracking intentionalism because we believe it to be true. There are parts of it we do find plausible (intentionalism with external world content). We defend it, though, because we believe some of Pautz's criticisms raise interesting questions about how to think about perception that are independent of the truth or falsity of tracking intentionalism.

19.1 Perception Is Not Magic

One of the themes of the paper is that there is good evidence for internal states that are well correlated with sensory phenomenology and that, at least for the chemical senses, there are no equally well correlated external states.[1] According to Pautz, this

[1] Although we will only mention one example here, there are a number of difficulties in interpreting the empirical evidence Pautz brings forward concerning what external properties the internal states he discusses might be tracking. Paut's discussion of olfaction relies heavily on the very interesting work done on posterior piriform cortex (PPC) by Gottfried and collaborators (Howard et al. 2009; Gottfried 2010; Zelano et al. 2011). The central conclusion of this work is that PPC contains a distributed representation of *odor objects* (Stevenson and Wilson 2007). Odor objects are learned patterns of more basic odors that correspond to olfactory complexes like the smell of chicken. PPC is thus supposed to function in a way resembling visual object and face recognition. Complaining that there is no simple chemical correlate to PPC activity is thus like complaining that there is no simple physical magnitude corresponding to the activity of visual face cells or to activity in

D. Hilbert (✉) • C. Klein
Department of Philosophy, University of Illinois at Chicago, 1420 University Hall, MC 267 601 S Morgan Street, Chicago, IL 60607, USA
e-mail: hilbert@uic.edu; cvklein@uic.edu

R. Brown (ed.), *Consciousness Inside and Out: Phenomenology, Neuroscience, and the Nature of Experience*, Studies in Brain and Mind 6, DOI 10.1007/978-94-007-6001-1_19,

fact raises a prima facie problem for tracking intentionalism. A typical version of this claim can be found in the conclusion to Sect. 18.2:

> The fact that when it comes to phenomenal character there is "bad external correlation" but "good internal correlation" across the various modalities makes one suspect that there is something very wrong with the radically externalist approach promoted by tracking intentionalists, according to which phenomenal character is fully determined by the external physical properties tracked by our experiences ... and which accords no serious role to internal factors. (p. 12)

Pautz musters detailed empirical evidence in support of this conclusion but the details are unnecessary here. That sensory phenomenology is better correlated with physiology than any external property is a *consequence* of tracking intentionalism in conjunction with very general (and relatively uncontroversial) empirical considerations.

At the heart of tracking intentionalism are two claims. First, that sensory phenomenology is wholly explained by the content of sensory states (intentionalism). Second, that the content of sensory states is to be explained in terms of how they are connected to the external world (tracking). Exactly which internal states track which external states will depend on both the structure of the world and also the structure of the sensory system (including the brain) of the organism. There won't be internal states that track features like acidity without the presence of sensors that respond to the pH of substances in the mouth and without those sensors being connected to neural circuits that process and deliver the information obtained from the sensors. The causal relationships and correlations that underlie tracking depend crucially on internal features of organisms. That the relationship between internal states and phenomenal experience is systematic and not random is also to be expected. Given very general assumptions about physiology and the evolution of nervous systems what is to be expected is that the internal states that track environmental features will have some systematic structure. If we assume that these states are related to perceptual experience and behavior in systematic ways (an unsurprising feature of actual neurophysiology), then good internal correlation falls out directly. Good internal correlation is not only consistent with tracking intentionalism but to be expected.

19.2 Feeling Curved

Good internal correlation thus can't be a threat to tracking intentionalism. And indeed, Pautz's arguments rely heavily on the notion of a "bad external correlation." Now, some degree of independence between the property being tracked and the

---

the ventral stream visual areas involved in object recognition. There are interesting questions for tracking intentionalism here but they are more complicated than the issue of whether there is a simple physical or chemical feature that is correlated with activity in PPC.

internal state doing the tracking is an important part of tracking intentionalism. The particular versions of tracking intentionalism Pautz discusses are very concerned to allow for the possibility of misrepresentation and build their theories accordingly. For example, Tye uses causation under optimal circumstances, rather than plain causation, and Dretske builds representation on indicator functions, rather than plain indication. The inclusion of an optimality requirement and the appeal to function make it possible for sensory states to misrepresent the external world. Since misrepresentation is possible (and actual according to both) external correlation is less than perfect.

Pautz, however, appeals to a more serious mismatch with the external world. In the case of thermal pain, for example, there is an exponential function relating stimulus intensity and judgments of pain intensity. This gives rise to response expansion: a doubling of stimulus intensity gives rise to a more-than-doubling of judged intensity. So according to Pautz, there is a bad correlation between judged intensity and the external stimulus. There is a "perfect correlation" on the other hand, between judgments of pain intensity and internal qualities like S1 firing rates.[2] Problems for tracking intentionalism should follow. We have noted that good internal correlation is compatible with tracking intentionalism. Are "bad external correlations," in Pautz's sense, a threat? You might think so: you might think that bad correlation means poor tracking. This would be a mistake.

First, note that Pautz's notion of "bad external correlation" has nothing to do with the ordinary scientific use of a "bad correlation." In ordinary usage, two variables are correlated if and only if there is an association between them such that information about one *reliably* carries information about the other. A perfect correlation means that the value of one quantity is completely informative about the value of the other. Since 'reliably' is a graded notion, correlation is a graded notion as well. At the lower ends, however, a poor correlation means that the two quantities don't have much to do with each other: knowing the value of one doesn't carry any information about the value of the other.

If internal states and external stimuli were poorly correlated in *this* sense, tracking intentionalism would obviously be in trouble. But that can't be the claim. For all that's been said, a subject's pain intensity judgment lets you predict, perfectly

[2]See p. 10. We grant this latter claim for the sake of argument, but note that it is problematic in a number of respects. Perfect correlations are mathematically improbable in neuroimaging work, even if there is actually a perfect relationship between a variable and neural response (Vul et al. 2009, p. 275). Pautz cites Coghill et al., but they claim only that the correlations are statistically significant, not that they are perfect (1999). S1 is also a problematic place to locate intensity information. Although there are regions of S1 which have activity that is well-correlated with pain intensity judgments (particularly in BA 3a), these regions are not obviously the substrate of pain experience. Large lesions of S1 do not reliably eliminate pain sensation, and stimulation of S1 does not reliably produce pain sensation (Craig 2003, pp. 18–19). The worry is not just the one that Pautz notes, that pain sensation might be more widely distributed. Rather, it is that a linear correlation between intensity ratings and neural responses need not indicate a necessary part of the substrate of pain experience.

accurately, the intensity of the stimulus to which the subject was subjected. So there is a good correlation between intensity judgments and external states, at least in this straightforward sense.

Pautz's claim, so far as we can tell, is something much weaker: that there is no *linear* relationship between judgments and stimuli. Certainly so. But why should the tracking intentionalist care? So long as the judgments reliably carry information about the external state – and on this account, they do – the tracking intentionalists have all they wanted. Linearity is just one of many possible informative relations that can hold between quantities. Any of those are candidates for representation relations; some, for good engineering reasons, might be preferred to others. So where's the problem?

The tracking intentionalist might rest content here. To his credit, however, Pautz gives an argument for why we should prefer linear relationships (though it is not obviously couched as such). Pautz envisions two fluid-filled columns which track lengths, one with a relationship of n^2 to the length and the other with a relation n^3.

> This provides a schematic illustration of my internal-dependent argument and my structure argument. How could the "tracking intentionalist" or "objectivist" about sensible qualities accommodate the verdict that A and B have sensations of different intensities, and how might he accommodate the truth of their structure judgments? After all, they track the very same objective lengths; and the objective length does not really *double* (or more than double) when it goes from 2 cm to 3 cm (p. 37).

Here's a way to reconstruct this argument: there are indefinitely many information-carrying relationships that might hold between a quality and its internal representation. Different relations will give rise to different judgments about the external quality over some range. Yet each of these representations putatively track the *same* property. As they disagree, they can't all be *accurate*. Further, the linear relation most neatly mirrors the world, and so has the best claim to be the accurate one.

Pautz considers one possible response, which is that the different relations track different properties in the world; he ultimately dismisses this as "Pythagoreanism". We'll leave the defense of Pythagoreanism to those who find it attractive. Instead, we suggest that there is another, perfectly reasonable, response available to the tracking intentionalist. Different relations track the same property, but provide *different information* about that property. Properly cashed out, this blunts the force of Pautz's argument.

The brain contains regions with a binary response to stimuli.[3] So consider two possible neural response functions for thermal pain: a continuous one that increases linearly with stimulus intensity, the other a binary function that changes its output over a certain threshold – say 46 °C. Both of these clearly track the same property of the world: the degree of the thermal stimulus. If we imagine these as instantiated in two different organisms, it's also obvious that they would have

[3]In the case of painful thermal stimuli, see Bornhöyd et al. (2002).

different experiences: the continuous neural response would give rise to graduated pain sensations that can't be represented by the binary function. So it is possible to have two representations, each of which track the same property, and yet which give rise to different sensations.

Yet surely, there is no *mystery* here about how that can be the case: the difference in sensation is due to the fact that the two functions carry different information about the world. The amount of information carried by a representation is a function of how many potential states the representational vehicle might be in. A binary representation which can only be in two states can only carry one bit of information: in our example, whether or not the stimulus is painful. The continuous linear stimulus can carry more information: not just whether or not the stimulus is painful, but the intensity of the stimulus. But these are just differences in the amount of information conveyed by the representation, not what it is information *about*: both representations track the same feature of the world.[4]

Indeed, the same point can be made with two *linear* functions. Actual neural response functions are not continuous: there are some discriminations that are too fine for neurons to make. So consider two hypothetical entities *Coarse* and *Fine*. Unlike us, both have linear response functions for thermal pain: a doubling of the stimulus exactly doubles the judged intensity of pain. So there is "good external correlation" in Pautz's sense. Yet *Coarse*'s neural mechanisms can only discriminate with an accuracy of 1 °C, while *Fine* can discriminate with an accuracy of 0.1 °C. So *Coarse* will lump together as similar many states that *Fine* will distinguish. Intuitively, *Coarse* and *Fine* will also have different pain experiences, and their different pain experiences will also be well-correlated with the state of their putative neural mechanisms.

Now, we have a curious case. Both *Coarse* and *Fine's* sensations should have both good internal and external correlation in Pautz's sense. Yet the same stimulus can give rise to different sensations in each. Further, if you're a tracking intentionalist, you have a perfectly good story about how this works: both *Coarse* and *Fine* track thermal stimuli, but the mechanisms by which they track it carry different information. That difference in information is a difference in representation, which gives rise to a difference in phenomenology – exactly as the tracking intentionalist predicts.

What this shows, we think, is that "good correlation" in Pautz's sense is a red herring. What matters for tracking intentionalism is 'correlation' in the old fashioned sense: that is, in the sense of carrying information. There are many ways of carrying information, however, and each gives rise to a different way of tracking the world. Different ways of tracking the world may lump stimuli together as more or less similar: which of these lumpings is preferable is not an *a priori* matter. It is an engineering one: a well-designed system should treat as similar states which need to be treated similarly in output.

---

[4]In the philosophy literature these points were given prominence by Dretske (1981, 1995).

If this is right, then it is an easy step to dispense with Pautz's case of *Mild* and *Severe*. *Mild* and *Severe*, remember, have response functions each with a different steepness. Intuitively, there should be a difference in their response for the very same stimulus. We agree: they are tracking the very same stimulus, but carrying different information about it. Because of this, they respond differently to the same stimulus. Given the granularity of any neural response function, the two functions will carry different information about the very same facet of the world. So it is unsurprising that their phenomenal experience should also differ. But this is a fact that can be fully accounted for by the tracking intentionalist.

19.3 Tracking Systematically

Tracking intentionalism is not threatened either by good internal correlation nor poor external correlation (in Pautz's sense). We believe that Pautz, although he often puts his point in terms of poor external correlation, is actually making an additional, distinct argument. One of the morals that Pautz wishes to extract from his survey of sensory physiology and psychophysics is that, for many senses, there is no external property to be tracked at all. It's not just that the correlation between the property supposedly being tracked and the internal state supposedly doing the tracking is imperfect; it is rather that there are no plausible candidates at all for the properties being tracked. This argument does not apply to all modalities (thermal pain, for example, surely tracks some straightforward external property), and is most plausible in the case of the chemical senses. The form of the argument is familiar from the color literature but Pautz very usefully extends this argument form to a much broader array of sensory properties. We conclude by considering some of these cases.

Pautz offers four examples of cases in which there is supposedly sameness in content combined with difference in experience. These are then generalized to form the internal-dependence argument which, in effect, asserts that some such example is (probably) possible (p. 277). Each case follows the same pattern. There are two individuals who are stipulated to have differing activity in a neural area correlated with perceptual phenomenology while tracking the very same property. Pautz then argues that because of the difference in neural activity it's plausible to suppose that the two are undergoing phenomenally different perceptual experiences. These cases are thus putative counterexamples to tracking intentionalism.

This form of argument immediately runs into a problem. For concreteness take the case of *Yuck* and *Yum*. *Yuck* and *Yum* have very different neural responses to a specific variety of berry. If this is all we are told about the case then we don't really have an argument, just a case that might elicit conflicting intuitions. If *Yuck* and *Yum* were very similar in their neural circuits that process taste information then the empirical evidence Pautz brings forward might support the claim that differing neural activity implies phenomenally different experiences. But one thing we do know about *Yuck* and *Yum* is that they differ in the neural circuits that process taste information since they have significantly different neural activity in response to the same stimulus.

Pautz adds detail to his cases which serves to make plausible the claim that *Yuck* and *Yum* are having phenomenally different experiences. *Yuck* finds the berries similar in taste to poison dart frogs (he shows an amazing willingness to stick things in his mouth) while *Yum* finds the berries similar in taste to bananas and there are corresponding similarities in neural activity for each of them. Their behavioral responses are also very different. Nevertheless when they are tasting the berries they are both representing them to have the very same property. Thus, according to Pautz's intentionalist, there is property that *Yuck* represents and that he represents in a way that is similar to the way he represents a property possessed by poison frogs. That very same property is also represented by *Yum* and his representation of it is similar to his representation of a property of bananas. The added detail, although crucial to establishing that *Yuck* and *Yum* are experientially different, is in some tension with the stipulation that *Yuck* and *Yum* are in states with the same intentional content.

Any version of tracking intentionalism that is able to offer a substantive account of perceptual similarity will be able to offer an account of these types of cases. If berries and poison frogs have similar tastes for *Yuck*, then there must be some similarity in the taste properties they are represented as possessing. It's not enough for similarity that *Yuck* represents the one as having taste A and the other as having taste B. Without some structure to the representations of taste, there will be no basis for judging the tastes to be more or less similar.

So, *Yuck* could represent the tastes by representing (some aspect of) the chemical structure of the objects in which case similarity in taste would track similarity in those aspects of chemical structure. Or *Yuck* could be representing the taste (at least partly) in terms of the effects of the substance on his digestion, in which case similarity in taste would track digestive effects. Given that *Yum* judges different tastes as similar, the one thing we can be sure of is that his representation of taste tracks different aspects of the world from the ones tracked by *Yuck*.

It's only if we think of the content of each taste perception as independent of the contents of the others that it could look at all plausible to make the assignments of content that Pautz stipulates. Perception represents objects in complex and systematic ways that allows comparison across different representations in order to judge of different aspects of perceptual similarity and difference. Note that this follows more or less directly from the considerations about information we advanced in the previous section. Carrying information is not done, as it were, state by state. It works only against the background of an ensemble of different potential states, each of which represents an equivalence class of possible specific states. There may be trouble for some versions of tracking intentionalism here but, without further argument, these kinds of cases pose no problem for tracking intentionalism itself.

19.4 Conclusion

There is much more in the paper than the small set of issues we've commented on, including much more on the issues we raise in our comments. Pautz's paper is rich in philosophical and empirical detail; both are worth engaging with. Nevertheless, we don't think that Pautz has succeeded in refuting tracking intentionalism (although some of the arguments may work against some versions of the theory). In particular, we don't think that Pautz has suceeded in casting doubt on the thesis that the phenomenology of perception can be explained in terms of contents involving external properties.

References

Bornhöyd, K., M. Quante, et al. 2002. Painful stimuli evoke different stimulus–response functions in the amygdala, prefrontal, insula and somatosensory cortex: A single-trial fMRI study. *Brain* 125(6): 1326–1336.

Coghill, R., C. Sang, J. Maisog, and M. Iadarola. 1999. Pain intensity processing within the human brain: A bilateral, distributed mechanism. *Journal of Neurophysiology* 82(4): 1934–1943.

Craig, A. 2003. Pain mechanisms: Labeled lines versus convergence in central processing. *Annual Review of Neuroscience* 26(1): 1–30.

Dretske, F.I. 1981. *Knowledge and the flow of information*. Cambridge, MA: MIT Press.

Dretske, F. 1995. *Naturalizing the mind*. Cambridge: MIT Press.

Gottfried, J.A. 2010. Central mechanisms of odour object perception. *Nature Reviews Neuroscience* 11(9): 628–641.

Howard, J.D., J. Plailly, M. Grueschow, J.D. Haynes, and J.A. Gottfried. 2009. Odor quality coding and categorization in human posterior piriform cortex. *Nature Neuroscience* 12: 932–939.

Stevenson, R.J., and Donald A. Wilson. 2007. Odour perception: An object-recognition approach. *Perception* 36: 1821–1833.

Vul, E., C. Harris, et al. 2009. Puzzlingly high correlations in fMRI studies of emotion, personality, and social cognition. *Perspectives on Psychological Science* 4(3): 274–290.

Zelano, C., A. Mohanty, et al. 2011. Olfactory predictive codes and stimulus templates in piriform cortex. *Neuron* 72(1): 178–187.

Chapter 20
Ignoring the Real Problems for Phenomenal Externalism: A Reply to Hilbert and Klein

Adam Pautz

I am indebted to David Hilbert and Colin Klein for their in-depth response ("No Problem") to my paper "The Real Trouble for Phenomenal Externalists". In Sect. 20.1, I will explain that their main points are actually red herrings directed at arguments I did not make. In Sect. 20.2, I will show that they do not answer the first argument of my paper, the *internal-dependence argument*, since they focus on examples of their own that are quite different from those in my paper and that are indeed no problem for phenomenal externalists. They also never touch at all on my second main argument, the *structure argument*.

20.1 Hilbert and Klein's Red Herring Points

In my paper, my main stalking horse was *tracking intentionalism*, which is in my view the best version of phenomenal externalism. On this view, the sensory dimension of experience is fully determined by the representation of *response-independent* (but possibly viewer-relative) physical properties of external items. Hilbert, along with his coauthor Alex Byrne, has done much to develop and defend exactly this view in the case of color experience.

In my paper, I clearly laid out (in premises-conclusion form) two arguments against tracking intentionalism: the *internal-dependence argument* and the *structure argument*. But Hilbert and Klein's main points do not engage with these arguments; they only count against arguments I did not make and in some cases explicitly disavowed in the paper. The first order of business is to clear this up.

A. Pautz (✉)
Department of Philosophy, University of Texas at Austin, 2210 Speedway, Stop C3500, USA
e-mail: apautz@austin.utexas.edu

R. Brown (ed.), *Consciousness Inside and Out: Phenomenology, Neuroscience, and the Nature of Experience*, Studies in Brain and Mind 6, DOI 10.1007/978-94-007-6001-1_20,

Hilbert and Klein's first red herring point. In my "internal-dependence argument", I appealed in a very indirect way to *good internal correlation*: in *some* cases structural relations among experiences (similarity and difference, equal intervals, proportion) are well matched by structural relations among their neural correlates.

Hilbert and Klein's first red herring point is the topic of their first section, "Perception is Not Magic": "*good internal correlation is not only consistent with tracking intentionalism but to be expected [under tracking intentionalism]*".

This point would only be a criticism of one of my arguments if one of my arguments had had the following extremely simple form: good internal correlation is directly inconsistent with tracking intentionalism (first premise); there is good internal correlation (second premise); so tracking intentionalism is false.

But neither my internal-dependence argument nor my structure argument had this simple form. I did not rely on the premise (which Hilbert and Klein criticize) that good internal correlation alone is directly inconsistent with tracking intentionalism or that tracking intentionalism somehow predicts that good internal-correlation should not obtain. My premises (which I explicitly laid out) were quite different. So, even if it is correct, Hilbert and Klein's consistency point is a red herring.[1]

In fact, as we shall see in Sect. 20.2, Hilbert and Klein actually *accept* the actual premise that I supported using (among other things) "good internal correlation".

Hilbert and Klein's second red herring point. In both of my arguments, I also appealed to "bad external correlation": even under optimal conditions, the structural relations among experiences (similarity and difference, equal intervals, proportion) are *not* matched by the structural relations among the (disjunctive) external physical properties that those experiences track.

[1]It is worth mentioning that Hilbert and Klein's point "*good internal correlation is not only consistent with tracking intentionalism but to be expected [under tracking intentionalism]*" is *not* correct. True, the first part is correct: good internal correlation is *consistent with* tracking intentionalism, in the formal sense of "consistent with". But the second part is incorrect: good internal correlation in my sense is not *to be expected* under tracking intentionalism. In fact, given tracking intentionalism, good internal correlation is *surprising*. So, for instance, on this view, there is no reason to expect that a doubling of sensory intensity involves a doubling of average firing rates. To see this, notice that, on tracking intentionalism, sensation doubles when the representation of external intensity doubles. Further, on tracking intentionalism, anything can represent anything. So, a tripling, or a quadrupling, in internal neural firing rates might represent a doubling in external intensity, provided that it causally-covaries with a doubling in *external* intensity. In fact, a *reduction* in internal firing rates could represent a doubling in external intensity. This is just an instance of the familiar point that there need not be any match between the intrinsic properties of the "content-carriers" and the contents they carry. So, on tracking intentionalism, it would be somewhat of a surprise if a doubling of sensory intensity involves, precisely, a doubling of average firing rates. Likewise, on tracking intentionalism, taste similarity is (presumably) constitutively determined by similarity in the chemical structures represented, as opposed to similarity in internal neural states. It is a radically externalist theory of phenomenology. So, under tracking intentionalism, it should come as a surprise that even under optimal conditions taste similarity is actually *better* correlated with *internal* neural similarity than with *external* chemical similarity. In any case, the issue here is irrelevant, because I did not make the simple "inconsistency" argument that Hilbert and Klein criticize.

Hilbert and Klein's second section, "Feeling Curved", is devoted to their second red herring point:

> Are "bad external correlations," in Pautz's sense, a threat? You might think so: you might think that bad correlation means poor tracking. This would be a mistake.

This point would only be a criticism of one of my arguments if one of my arguments had the simple form: if there is bad external correlation in my sense, our brain states cannot track and thereby represent external properties as tracking intentionalism requires (first premise); there is bad external correlation in my sense (second premise); so our brain states cannot track external properties as tracking intentionalism requires. Call this the *no tracking argument.*

But my actual arguments, the internal-dependence argument and the structure argument, were totally different from this "no tracking" argument. In fact, far from making the simple "no tracking" argument, in my paper I explicitly disavowed it and myself already pointed out that it is fallacious. For instance, I wrote:

> [E]nsemble activation states might [track and thereby] represent external chemical properties... even if there is "bad external correlation" [in *my* special sense], that is, even if the resemblances and differences among them are not matched by resemblances and differences among the chemical properties. (Sect. 18.3.2)

Likewise, in my Mild-Severe case, I *stipulated* that in both individuals under optimal conditions particular S1 firing rates are perfectly correlated with (track) individual noxious temperatures (inter alia), even if the relationship here is complex and non-linear (and hence an example of bad external correlation in my special sense).

Hilbert and Klein's third red herring point. Hilbert and Klein write: "we believe that Pautz is actually making an additional, distinct argument ... that, for many senses, there is no external property to be tracked at all". Call this the *no tracked property argument.* They provide no textual evidence for this belief. There is no place in the paper where I make an argument like this. Indeed in the paper there was plenty of textual evidence that I would *reject* the "no tracked property" argument. (i) Throughout the paper I myself discussed what properties might be tracked in the various senses according to the tracking intentionalist: disjunctive chemical types in the cases of taste and smell, noxious temperatures or types of actual or potential damage in the case of pain, disjunctive properties involving amplitude, frequency and critical bands in the case of sound, and so on. (ii) In developing my "internal-dependence argument", far from suggesting that "for many senses, there is no external property to be tracked at all", I stipulated that the relevant individuals track *the same* external properties. So I did not believe that there are "no external properties to be tracked at all".[2]

---

[2]Hilbert and Klein attribute two additional arguments to me that I did not make. (i) In their footnote 1, they suggest that, on the basis of neuroscientific research, I complain (against tracking intentionalism) that "there is no simple chemical correlate to PPC activity". Here they are

20.2 Do Hilbert and Klein Address the Arguments?

Hilbert and Klein do eventually address *one* of my two actual arguments, my "internal-dependence argument". Unfortunately, in discussing this argument, they again make red herring points: for instance, instead of focusing on my actual Mild-Severe counterexample to tracking intentionalism, they dwell on totally different cases of their own that are not in my paper and that are indeed "no problem" for tracking intentionalists. Moreover, they never address my second major argument, the "structure argument".

Klein and Hilbert's General Response to the Internal-Dependence Argument. My internal-dependence argument concerned hypothetical *coincidental variation cases*, in which two individuals from different species track the same response-independent properties but undergo radically different neural processing and exhibit radically different behavioral dispositions. The argument went like this:

1 *If* tracking intentionalism is true, then in *every* possible coincidental variation case, the right verdict is *Same Experiences:* the individuals involved have experiences that are identical in sensory character, despite their neural and behavioral differences, because they track and thereby represent the same response-independent properties (*Same Content*).
2 But (given the empirical facts) it is much more reasonable to suppose that, in at least *some* coincidental variation cases, the right verdict is Different Experiences; call this *internal-dependence*.
3 So tracking intentionalism is (probably) mistaken.

Now, according to Hilbert and Klein, which premise of this argument should tracking intentionalists reject?

I was pleased to find that Hilbert and Klein accept premise 2 ("internal-dependence"), allowing that in my Mild-Severe and Yuck-Yum cases the right verdict is indeed Different Experiences.

---

attributing to me what might be called the "no *simple* property tracked argument". I did not make this complaint or argument; to the contrary, I stressed that on tracking intentionalism the properties tracked by sensory states at different stages will be enormously *complex*. In my paper, the actual role of the neuroscientific research was to support my premise that in "coincidental variation cases" the right verdict is Different Experiences; and, as we shall see, Hilbert and Klein actually *agree* with this premise. (ii) In their section "Feeling Curved", Hilbert and Klein offer a "reconstruction" of my internal-dependence argument, after quoting from my discussion of a fanciful schematic case. The "reconstructed" argument they attribute to me depends on the claim that a "linear correlation most neatly mirrors the world, and so has the best claim to be the accurate one". I am not sure I understand this argument, so needless to say I did not put it forward in my paper. The premises of my actual "internal-dependence" argument (clearly laid out in my paper and repeated in Sect. 18.2 of the present response) were quite different.

So, even though (as we saw in Sect. 20.1) much of their discussion is devoted to red herrings about 'good internal correlation' and 'bad external correlation', in the end Hilbert and Klein *agree with* the crucial premise that I actually supported on their basis.

Hilbert and Klein recommend that tracking intentionalists instead reject premise 1: the conditional claim that, *if* tracking intentionalism is true, *then* the right verdict in these cases is instead Same Content and hence Same Experience. In other words, they think that, in all these cases, tracking intentionalists can accommodate the (correct) verdict of Different Experiences, contrary to my contention.

This is immediately problematic, because in these cases I simply stipulated that, whatever the tracking intentionalists says "tracking" consists in (whether it is explained using ideas from Fodor, Tye, Dretske, or Millikan), the individuals in coincidental variation cases bear the "tracking relation" to exactly the *same* response-independent properties and states. Given this stipulation, my premise 1 is guaranteed to be true: *if* tracking intentionalism is true, *then* the right verdict should be (implausibly) Same Content and Same Experience. And Hilbert and Klein do not show that this is an impossible stipulation to make.

Hilbert and Klein's Mistreatment of the Mild-Severe case. To illustrate, consider the case of Mild and Severe. Recall that Mild and Severe belong to different human-like species. (Maybe Mild is an actual human and Severe is a member of some human-like species in a different counterfactual situation.) The psychophysical response curve describing the relationship between noxious temperatures and neural response and VAS pain ratings is steeper in Severe than it is in Mild. Nevertheless, I stipulated that their (different) neural responses track the very same response-independent properties of the thermal stimuli. We might call this my *same tracking* stipulation. So if the tracking intentionalist holds that Mild's neural responses track and thereby represent *noxious temperature properties*, as Hilbert and Klein assume (even though it is a controversial issue), then Severe's neural responses track and thereby represent the *very same* noxious temperature properties. In general, my same tracking stipulation entails that Mild and Severe's neural states, although different, can be put into *one-one correspondence*, such that if Mild has a neural state *M1* that tracks noxious thermal information *I*, then Severe has a corresponding neural state *S1* (one involving higher firing rate than *M1*) that tracks the very same noxious thermal information *I*.

Now, since there are obviously some discriminations that are too fine for neurons to make, the relevant thermal information will be less than perfectly precise. In a paper I relied on in my argument, Donald Price (Price 2002) notes that WDR neurons in S1 can differentially respond to a roughly 0.3 °C change in stimulus intensity within a range of painful 45–51 °C skin temperatures, which fits humans' psychophysical performance. Given tracking intentionalism, it follows that a cortical pain state in Mild (for instance) might represent a noxious temperature property like *being roughly between 45.0 and 45.3* °C, as opposed to a perfectly precise temperature-value. Simplifying somewhat, the full information content might be something like: *there is a stimulus in my leg roughly between 45.0 and 45.3* °C.[3] Clearly, my "same tracking" stipulation entails that Severe has a corresponding cortical neural state that tracks *exactly the same* information, at exactly the same level of "grain". These corresponding cortical neural representations in Mild and Severe, although they involve different average firing rates, are both tokened under optimal conditions when and only when (and because) there is something roughly in the *45.0–45.3* °C range touching the leg. In general, Mild and Severe have the same number of possible cortical neural representations of noxious temperatures, and they represent the *same* noxious thermal conditions at exactly the *same* level of grain; it is just that they involve different firing rates. I mention this since, as we shall see, Hilbert and Klein suggest that the informational contents of Mild and Severe's experiences *differ* in "grain". They suggest this only because they ignore my "same tracking" stipulation.

Before I get to that, however, let me clarify why the Mild-Severe case is a counterexample to tracking intentionalism. As we have just seen, given my "same tracking stipulation", if tracking intentionalism is true, then the correct verdict in this case is evidently Same Content (Same Information) and hence Same Experience. Against this, given that pain intensity is in our own case linearly related to neural firing rates throughout the pain matrix, and only related in a complex way to a number of external parameters (temperature, duration, size), clearly the *simplest* and therefore best hypothesis is that pain intensity is directly dependent on firing rates in the pain-matrix. Given this standard view in pain-science, and given that Severe's firing rates in the relevant cortical areas *as well as his* psychophysical responses increase more rapidly than Mild's with increasing temperature, the sensory intensity of his pains increases more rapidly. In short, the correct verdict is Different Experiences, contrary to tracking intentionalism.

In accordance with their general suggested response to my internal-dependence argument, Hilbert and Klein grant my premise that the correct verdict is Different Experience but reject my premise that tracking intentionalism instead implies the (mistaken) verdict of Same Content and Same Experience.[4] Instead, they suggest

---

[3]To say that, on tracking intentionalism, the content of a thermal pain might be something like *there is a stimulus in my leg roughly between 45.0 and 45.3* °C is to simplify in two ways. (i) Since thermal pain depends on stimulus size duration and size as well as temperature, on tracking intentionalism, the real content would be more complex and disjunctive. (ii) Since sensory processes are inherently probabilistic, and the notion of 'optimal conditions' is vague, on tracking intentionalism neural states do not have precise contents.

[4]In their note 2, Hilbert and Klein make a number of helpful empirical points about whether area S1 is the neural locus for pain intensity, some of which I made in my paper. But, as I noted in my paper, this issue, however interesting, is not really relevant to my argument. Hilbert and Klein do not question the key finding of Coghill and coworkers that "many cortical areas [not just S1 but other areas – A.P.] exhibit significant, graded changes in activation linearly related to pain intensity" (1999, 1936). (Incidentally, while Hilbert and Klein question my use of "perfect correlation" to indicate this, I think this is a merely verbal issue.) And – most importantly – they *accept* the premise that I supported (in part) on the basis of empirical research, namely that Mild and Severe (who, recall, differ in their firing rates throughout the pain matrix) have pains of different intensities. The

that, on tracking intentionalism, there is a difference in the "grain" of the contents of Mild and Severe's pain experiences. But, given my "same tracking" stipulation in the Mild-Severe case, how could Hilbert and Klein possibly deny that tracking intentionalism implies Same Content and Same Experience?

What Hilbert and Klein do is to ignore my Mild-Severe case and my "same tracking" stipulation, and instead dwell on totally different cases of their own which are indeed "no problem" for tracking intentionalists because in those cases there are by contrast clear *tracking differences*.

One of their cases is that of Graded and Binary. Graded is like an actual human: he has a mechanism that has many different states *S1*, *S2*, *S3* ... that track relatively fine-grained thermal states. Thus, maybe under optimal conditions *S2* occurs when and only when (and because) there is a stimulus *roughly between 46.0 and 46.3* °C. By contrast, Binary has relatively rudimentary mechanism for detecting temperatures, featuring just two states, *B1* and *B2*. Under optimal conditions, *B1* is tokened just in case (and because) the external temperature is *below 45* °C (below the painful range); while *B2* is tokened just in case (and because) the external temperature is *above 45* °C (within the painful range). So, in the case of Graded and Binary, by contrast to my case of Mild and Severe, there are radical tracking differences between the individuals involved: for instance, unlike Graded, Binary simply has no state that occurs when and only when (and because) there is a stimulus *roughly between 46.0 and 46.3* °C. Another case that Hilbert and Klein discuss is that of Fine and Coarse. Coarse has states that only track thermal states like *there is a stimulus roughly between 46 and 47* °C. By contrast, Fine has many more states than Coarse, which track smaller temperature differences. So, in the case of Fine and Coarse, by contrast to my case of Mild and Severe, there are again clear tracking differences.

(So, while Hilbert and Klein claim that in their own cases "both representations track the same feature of the world", this is misleading. What Hilbert and Klein must have meant is that the individuals' states track (different) features of the same *type*, namely stimulus temperature.)

Hilbert and Klein claim that, if tracking intentionalism is true, then the right verdict in *their own cases* is not Same Content and Same Experience, but Different Content and Different Experience. In particular, if tracking intentionalism is true, then in these cases there is a difference in the *granularity of information*. I agree with this, because, as we have just seen, there are clear *tracking differences* in these cases. (In fact, as I will point out in a moment, in my paper I myself already made the same point about cases like this.)

"neural locus" issue does not matter to my argument, because it does not matter to the plausibility of this premise. It is true that, in the paper, *for simplicity*, I did sometimes assume the standard view that S1 "plays a special role" (perhaps a special causal role). But I noted that this is controversial and not relevant to my argument. (Contrary to Hilbert and Klein, by "S1 plays a special role", I did not have in mind the extremely strong claim that mere S1 activity is alone *necessary and sufficient* for pain, and never suggested that this strong claim is established merely by the finding of linear *correlations* between pain and S1 activity.)

What I do not agree with is their follow-up claim that, if tracking intentionalism is true, then in *my quite different Mild-Severe case* the right verdict is likewise Different Content and Different Experience (as they put it, "they are carrying different information about [the same stimulus]"). (It is revealing that Hilbert and Klein do not elaborate and never say what the difference in information is.) The reason I do not agree with this is simple: whereas in their own cases there are clear *tracking differences*, in my case I made the *same tracking stipulation*, which Hilbert and Klein evidently ignored. Contrary to Hilbert and Klein, given this stipulation, tracking intentionalism implies the mistaken verdict of Same Content and Same Experience in the Mild-Severe case, as we saw above. In particular, given this stipulation, in my actual Mild-Severe case, by contrast to Hilbert and Klein's cases, there is *no difference in granularity*. (Even if there were, how could a mere difference in representational-*granularity* possibly account for the *intensity* differences between Mild and Severe?) Since in this case tracking intentionalism implies the mistaken verdict of Same Content and Same Experience, the case stands as a counterexample.[5]

Indeed, these points were already emphasized in my paper. In Sect. 18.3.1 I noted that tracking intentionalists might provide what I called a "pluralist account" of some cases in which there is perceptual variation between two individuals. On this account, the individuals' experience represent different but compatible information about the same stimulus, because of *subtle tracking differences*. Clearly, the treatment that Hilbert and Klein suggest for their Binary-Graded and Fine-Coarse cases is simply a version of the kind of pluralist account I had already discussed. In my paper, I already emphasized that, since in my own cases (Mild-Severe and the other cases) I made the *same tracking* stipulation, the kind of pluralist gambit Hilbert and Klein have in mind simply does not apply to those cases, contrary what they suggest.

Hilbert and Klein on the Yuck-Yum case. Another coincidental variation case I used to illustrate my internal-dependence argument was that of Yuck and Yum. Yuck and Yum taste some berries that are poisonous to Yuck but an important food-source to Yum. Here again Hilbert and Klein grant that, given the neural and behavioral

[5]To see more clearly that tracking intentionalism implies the verdict of Same Content (and hence the mistaken verdict of Different Experiences) in my Mild-Severe case, consider a fanciful case analogous to my case of Mild and Severe. Suppose there are two devices, Low and High, which indicate (increasing) temperatures by producing sounds of (increasing) pitch. However, suppose that in High the pitches increase more rapidly with increasing temperatures. Thus, when (and only when) the external temperature is roughly between 45.0 and 45.3 °C, both make a distinctive sound, but the pitch of the sound made by High is a bit higher. When the temperature rises to the 45.3–45.6 °C range, they both make sounds of yet higher pitches, only the pitch-increase is higher in the case of High than it is in the case of Low. In this case, the content-vehicles are different (High's pitch-sounds are regularly higher than Low's), but a tracking theory obviously implies that they carry the same bits of thermal information at the same level of grain (e.g. Low's low-pitch noise and High's corresponding high-pitch noise both represent *the temperature is roughly 45.3–45.6 °C*). Likewise, even though Severe's psychophysical response curve is steeper than Mild's, so that Severe's individual S1 states involve higher firing rates than Mild's corresponding S1 states, those different states carry exactly the same thermal information at the same level of grain.

differences, the most reasonable view is that they have different taste experiences, in line with my second premise ("internal-dependence"). But they question my first premise that tracking intentionalism instead implies the (mistaken) verdict of Same Content and Same Experience, suggesting that it might accommodate the correct verdict of Different Contents and Different Experiences.

But again this is immediately problematic because in my Yuck-Yum case, as in my Mild-Severe case, I made the "same tracking" stipulation. However "tracking" is explained, I stipulated that they track the same response-independent properties and conditions of the berries (the same objective information), even though their neural states (the "information-carriers") and behavioral responses are totally different. This just guarantees my premise that that tracking intentionalism implies the (mistaken) verdict of Same Content and Same Experience.

It is no wonder, then, that Hilbert and Klein are unable in their comments to exactly specify how tracking intentionalism implies that there are content-differences between Yuck and Yum, or what those content-differences might be. They make two vague suggestions, but neither is satisfactory. (i) They suggest on behalf of tracking intentionalists a view I had already considered in detail in the paper (Sect. 18.3.3): the *structure gambit*. In the paper I already explained in detail why this sort of view is incompatible with tracking intentionalism (given my *same tracking* stipulation) and generally problematic because of arguments due to Alex Byrne. Hilbert and Klein do not address these objections. (ii) Alternatively, Hilbert and Klein seem to suggest that on tracking intentionalism Yuck and Yum's taste experiences of the berries have *different* contents involving their own digestive systems. They do specify the different contents, so it is difficult to evaluate this suggestion. But, when Yuck and Yum taste the berries prior to digesting them, it is phenomenologically implausible that their *taste experiences* represent conditions involving what happens (or is about to happen) in their stomachs in addition to conditions involving what happens in their mouths. Further, this suggestion is, too, unavailable to tracking intentionalists given my *same tracking* stipulation: the neural states realizing those taste experiences do not track different digestive conditions, but only the same response-independent conditions (information) concerning the berries. How then can *tracking intentionalism* deliver the verdict of Different Content and Different Experience? It cannot.[6]

[6]In some places Hilbert and Klein misrepresent my internal-dependence argument. They claim that "Pautz offers examples in which there is supposedly sameness of content" between two individuals and that I make "the stipulation that Yuck and Yum are in states with the same intentional content". So they attribute the Same Content claim to me. Then they object to the Same Content claim, noting that it is implausible given the neural and behavioral differences between the individuals involved. But my argument does not rely on the Same Content claim. In fact, I *agree with* Hilbert and Klein that it is not plausible; in the paper, I repeatedly endorsed *Different Contents* (because I endorsed Different Experiences and as an intentionalist I hold that Different Experiences entails Different Contents). Instead, my argument relies on a merely *conditional* premise: *if tracking intentionalism* is true, *then* the right verdict should be (implausibly) Same Content (and hence) Same Experience. (In other words, although *I* think the right verdict is Different Contents, I also think *tracking intentionalists* are committed to Same Content.) As we have seen, this *conditional* claim is certainly true, given my "same tracking" stipulation.

Maybe Hilbert and Klein believe that my same tracking stipulation is impossible given what I say about the case. For they write, "Given that Yum judges different tastes as similar [than does Yuck], the one thing we can be sure of is that his representation of taste tracks different aspects of the world from the ones tracked by Yuck." But this is exactly what I argued is false; Hilbert and Klein do not answer the argument. The first step was that there could be two creatures, Yuck and Yum, whose neural states and behavioral responses, in response to some berries, differ radically, but whose neural states track *the same* aspects of the world. The stipulated case is clearly possible, just as it is possible that two thermometers should track the same external thermal conditions, but differ in their internal wiring and "behavioral responses" to those conditions. The second step is that in some cases like this Yuck and Yum have different taste experiences, and (if they are capable of judgment) make different similarity judgments, despite the "same tracking" stipulation. Hence tracking intentionalism is mistaken: sensory experience, and sensory content, is not fully determined by what external-world conditions are tracked.

Hilbert and Klein on the Other Cases? To save tracking intentionalism, one would have to address all the apparent counterexamples I used to illustrate my "internal-dependence argument". But Hilbert and Klein's suggested responses not only fail in my Mild-Severe and Yuck-Yum cases, as we have seen; they also do not show what responses could be developed in the other cases I discuss in the paper, the Soft-Loud and Sniff-Snort cases.

Incidentally, it is worth emphasizing that my internal-dependence argument is directed against not just "tracking intentionalists" but all philosophers who, like Hilbert and Klein, are attracted to (as they put it) "intentionalism and external world content", as I explain in the paper (Sect. 18.5).

Hilbert and Klein Ignore the Structure Argument. In my paper, I developed in detail a *second* major argument, the "structure argument". Like my internal-dependence argument, my structure argument is directed against not just "tracking intentionalists" but all philosophers who, like Hilbert and Klein, are attracted to "intentionalism and external world content". For, on this view, what external-world properties might the huge multitude of sensible qualities (tastes qualities, sound-qualities, etc.) be identified with? The most natural candidates are *response-independent* properties of external things: chemical properties, properties involving wavelength and frequency, and so on. Along with his co-author Alex Byrne, Hilbert holds that colors are response-independent reflectance-types (2003). So, by considerations of parity, he is under pressure to accept a response-independent view in the case of the other senses. But, according to my structure argument, given "bad external correlation" in my sense, proponents of this view cannot adequately accommodate the truth of our ordinary *qualitative structure judgments*.

In the color case, Hilbert (along with Byrne) has defended a "hue-magnitude" of color structure account in response to the *argument from color structure* (2003). My "structure argument" is a novel version of that argument generalized to sensible qualities beyond the colors. However, as I note (footnote 10) in the paper, nothing like the hue-magnitude reply works once we move beyond the color case (in fact in Pautz (2011: footnote 6) I argue that the hue-magnitude reply is problematic and unclear even in the color case). In their comments, Hilbert and Klein do not provide an alternative reply. In fact, they do not discuss my "structure argument" at all.

So, while I am grateful to Hilbert and Klein for their comments, I believe that they do not answer my arguments, the internal-dependence argument and the structure argument. I think it is time we start looking for alternative theories of sensory consciousness.

References

Byrne, A., and D. Hilbert. 2003. Color realism and color science. *The Behavioral and Brain Sciences* 26: 3–21.

Pautz, A. 2011. Can disjunctivists explain our access to the sensible world? *Philosophical Issues* 21(1): 384–433.

Price, D. 2002. Central neural mechanisms that interrelate sensory and affective dimensions of pain. *Molecular Interventions* 2: 392–403.

Part VII
The Ontology of Audition

Chapter 21
What We Hear

Jason Leddington

It is widely assumed (and only rarely argued) that the principal objects of hearing are *sounds*. Thus, Roy Sorensen writes:

> In the course of demarcating the senses, Aristotle defined sound in *De Anima* as the proper object of hearing.... Sound cannot be seen, tasted, smelled, or felt. And nothing other than sound can be directly heard. (Objects are heard indirectly by virtue of the sounds they produce.) All subsequent commentators agree, often characterizing the principle as an analytic truth. For instance Geoffrey Warnock (1983: 36) says 'sound' is the tautological accusative of the verb 'hear'. (2010, 126)[1]

There are two main claims in this passage:

1. Sound is unique in that it cannot be seen, tasted, smelled, or felt, but only heard. That is, sound is the proper object of hearing.
2. Nothing other than sound can be directly heard. That is, if S hears E directly, then E is a sound.[2]

These claims are widely accepted, but I think neither is true. I have only a few remarks to make about (1), which seems to me rather uninteresting, philosophically.

---

[1] Sorensen goes on to argue that "there is a single exception" to this view: "We hear silence, which is the absence of sounds" (126). So, more precisely, according to Sorensen, the immediate objects of hearing are sonic objects (sounds and silences), and when we hear an ordinary object, we do so in virtue of hearing a sonic one.

[2] Note that, even if only sound can be directly heard, it may still be possible to hear sounds indirectly—as in, say, hearing a recording or a radio transmission rather than being present at the live performance.

J. Leddington (✉)
Department of Philosophy, Bucknell University, Lewisburg, PA 17837, USA
e-mail: jl041@bucknell.edu

R. Brown (ed.), *Consciousness Inside and Out: Phenomenology, Neuroscience, and the Nature of Experience*, Studies in Brain and Mind 6, DOI 10.1007/978-94-007-6001-1_21,

But not so (2). It is the principal target of this paper and a point of departure for a great deal of philosophical and empirical work on hearing and sound. Thus, if it falls, it falls with company.[3]

21.1 Seeing Sounds?

According to (1), sounds are unique in that they cannot be seen, tasted, smelled, or felt, but only heard. I think this is easily shown false.

We often perceive things by perceiving their effects. Consider, for instance, the wind: otherwise invisible, we see it in rippling water and swaying branches. And even if we never see the wind directly, "indirect seeing" is still seeing; seeing the wind *on the water*, or *in the trees*, is still seeing the wind. This point is familiar from the philosophy of science: arguably, it is possible to see a subatomic particle by seeing its trail appear in a cloud chamber. Sounds are similarly visible. Get yourself

[3]Four points in passing:

First, it is important that (1) and (2) are mutually independent: that sound can only be heard—(1)—does not impose any restrictions on what else can be heard, whether directly or indirectly; and that only sound can be directly heard—(2)—does not exclude the possibility that sound is perceptually accessible by other means. Therefore, (1) and (2) require independent criticism.

Second, since (1) and (2) are mutually independent, it is unclear exactly what Sorensen has in mind when, in the quoted passage, he refers to "the principle" on which "all subsequent commentators agree."

Third, Sorensen is too hasty in supposing Warnock to agree with either (1) or (2). Since facts about grammar generally don't entail substantive metaphysical or epistemological theses, Warnock's claim that 'sound' is the "tautological accusative" of the verb 'hear' at best *suggests* (2)—that only sound can be directly heard. By contrast, it doesn't even *suggest* (1)—that sound can only be heard—since 'sound' might yet go perfectly well with perception verbs other than 'hear'.

Finally, even if most philosophers accept (2)—that nothing other than sound can be directly heard—a much more extreme view actually seems common among psychologists: not even sounds can be directly heard. Here, for instance, is a passage from a textbook popular in undergraduate psychology courses on "Sensation and Perception":

> Smell and taste are . . . indirect because these experiences occur when chemicals travel through the air to receptor sites in the nose and tongue. Stimulation of these receptor sites causes electrical signals that are processed by the nervous system to create the experiences of smell and taste. Hearing is the same. Air pressure changes transmitted through the air cause vibrations of receptors inside the ear, and these vibrations generate the electrical signals our auditory system uses to create the experience of sound. (Goldstein, 68)

The idea seems to be that, because experiences of smell, taste, and sound lie at the ends of largely intra-cranial causal chains, they cannot be direct experiences of extra-cranial phenomena such as smells, tastes, and sounds. This extreme view of what we directly hear—not sound, but (perhaps?) neural activity—is not only wildly implausible (for one thing, neural activity is, as such, pretty quiet), but it would appear to depend on mistaking the representational content of a neural state for the vehicle of that content. In any case, the arguments of this paper are directed at a less extreme view—claim (2)—that is without question the most common view among philosophers, as the opening quote from Sorensen (2010) attests.

the right speakers and the right music and you'll be able to *see* the sound—in particular, the bass—by seeing the shaking glassware. And the same holds for other sensory modalities. Imagine an earplugged sleeper who wakes to a vibrating bed and unplugs his ears to discover that what he *felt* was (the sound of) the party downstairs (cf. Hamilton 2009, 166).[4] Perhaps these are not cases of direct perception, but they're enough to undermine claim (1): for sense modalities other than hearing, sounds *are* perceivable, even if not *directly*.

In light of these reflections, (1) should be weakened as follows:

> (1′) Sound is unique in that it cannot be *directly* seen, tasted, smelled, or felt, but only heard. That is, sound is the proper object of *direct* hearing.

This is more likely to be true.[5] In any case, so much for claim (1). The rest of this paper targets claim (2).

21.2 Berkeley v. Heidegger

The following exchange occurs early in the first of Berkeley's *Three Dialogues*:

> PHILONOUS. This point then is agreed between us, that *sensible things are those only which are immediately perceived by sense*. You will farther inform me, whether we immediately perceive by sight any thing beside light, and colours, and figures: or by hearing, any thing but sounds: by the palate, any thing beside tastes: by the smell, beside odours: or by the touch, more than tangible qualities.
>
> HYLAS. We do not. (1992, 138)

In short, we immediately or directly perceive only sensible qualities, which, Philonous goes on to argue, exist only insofar as they are perceived. Such a view receives little support from contemporary philosophers. It is widely agreed that what we immediately perceive are not mind-dependent qualities, but mind-independent objects. In particular, what we immediately *see* and *touch* are supposed to be ordinary objects such as horses and tomatoes.[6] This is not to deny that we see

---

[4]Further imagine someone whose visual, tactile, and auditory systems regularly fail. We might devise for her a prosthetic that "translates" an auditory stimulus into a gustatory (and/or olfactory) stimulus in a manner that enables simple communication. She might then literally be said to taste (and/or smell) sounds (but perhaps not directly?). Such "sensory substitution devices" are the subject of extensive and ongoing empirical study. For discussion, see Bach-y-Rita and Kercel (2003).

[5]For a related discussion, see Roxbee Cox (2011, 104–6). Also, note that (1′) remains independent of (2). Claim (1′)—that sound can be directly perceived only by hearing—does not impose any restrictions on what else can be directly heard; and claim (2)—that only sound can be directly heard—does not exclude the possibility that sound is directly perceivable by other means.

[6]This requires qualification. Many contemporary philosophers believe that the immediate objects of visual and tactile perception are not full-blown ordinary objects, but *parts* of them. On this view, what we see or touch, strictly speaking, are not horses, but horse-parts (viz., surfaces). For the

colours and shapes; it is to deny that we see the horse *by* or *in virtue of* seeing its color or shape.[7] Our visual experience of the horse is not "mediated" by the experience of sensible qualities. Similar considerations hold for touch.

Yet the priority accorded to ordinary objects in visual and tactile perception is usually not extended to the other sense-modalities. Focusing on the case of hearing, philosophers typically follow Berkeley in taking the only direct or immediate objects of hearing to be sounds. And even if they reject the Berkeleyan view that sounds are mind-dependent qualities, it still follows that if we hear ordinary objects or events, we do so only in virtue of hearing the sounds that they make. In this respect, contemporary reflection on hearing retains a strong empiricist cast. Here, for instance, is Casey O'Callaghan:

> What do we hear? Sounds are, in the first instance, what we hear. They are the immediate objects of auditory experience in the following sense: whatever else we might hear, such as ordinary objects (bells, trumpets) and events (collisions, typing), we hear it in virtue of hearing a sound. (2009a, 609)[8]

Call this *the Berkeleyan view*. Despite its overwhelming popularity, there are (contra Sorensen) scattered examples of resistance. Consider Heidegger:

> We never really first perceive a throng of sensations, e.g., tones and noises, in the appearance of things . . . ; rather we hear the storm whistling in the chimney, we hear the three-motored plane, we hear the Mercedes in immediate distinction from the Volkswagen. Much closer to us than all sensations are the things themselves. We hear the door shut in the house and never hear acoustical sensations or even mere sounds. In order to hear a bare sound we have to listen away from things, divert our ear from them, i.e., listen abstractly. (1977, 151–2)

Heidegger rejects the idea that the experience of sound mediates between us and the ordinary objects and events that we hear. We do not hear things *in virtue of* hearing the sounds that they make; rather, we hear things *in* hearing their sounds. The experience of the source is immanent in the experience of the sound.[9] In this case, hearing sound is similar to seeing color. We do not see things *in virtue of* seeing their colors; rather, we see them *in* seeing their colors. The experience of the

---

purposes of this essay, I will ignore this complication, and I will write as if what we immediately see or touch are ordinary objects, simpliciter. (For considerations in favor of this commonsense view, see Leddington (2009); for arguments against it, see Bermùdez (2000)).

[7]On this use of the phrase 'in virtue of', see Jackson (1977, 15–20) and Bermùdez (2000, 356–7).

[8]Also see O'Callaghan (2007, 13; 2008b, 318) and Tye (2009, 209 n23).

[9]And it's precisely for this reason that "to hear a bare sound we have to listen away from things, divert our ear from them, i.e., listen abstractly." The idea is that we cannot fail to hear sound sources in hearing sounds, but we nevertheless have the ability to take a distinctively intellectual—and so, not merely experiential—attitude toward the sounds that we hear, regarding them apart from their material sources. As O'Callaghan and Nudds describe it: we have the ability to "attend to sounds *as* independent from their sources" (2009, 15). Arguably, this sort of *listening-as* is necessary for the appreciation of music (cf. Scruton 1997 and 2009). Note, however, that the ability to perceive abstractly is not restricted to audition. We are able to do the same sort of thing in seeing color and shape, and it is arguably integral to the appreciation of much abstract visual art.

tomato is immanent in the experience of its redness. The experience of redness does not mediate your visual experience of the tomato. Similarly, if we hear ordinary objects and events *in* hearing the sounds that they make, then hearing a sound involves an unmediated experience of its source. Call this *the Heideggerian view*.

Against the grain of contemporary thinking, this paper presents considerations in favor of the Heideggerian view. In particular, I argue that the Heideggerian view receives support from reflection on what auditory experience is like. This is important because, as the next section illustrates, arguments for the Berkeleyan view typically appeal to just such phenomenological considerations.[10]

21.3 Phenomenological Independence

While I can simply *see* horses, the Berkeleyan view holds that I can *hear* them only in virtue of hearing the sounds that they make. Why think this? Because it might seem woven into the very fabric of perceptual phenomenology.

Consider seeing color. There is a sense in which colors visually seem to be fused with their bearers. More specifically, colors visually seem to permeate or saturate the things they qualify. In this respect, colors visually seem compresent with their bearers. To say that colors *visually seem* this way is to say that their so seeming is an aspect of visual phenomenology. Reflection on what vision is like therefore suggests that the experience of a color bearer is immanent in every experience of color, and so, that seeing an object is never mediated by seeing its color. Phenomenologically, then, objects themselves appear to be available for direct visual inspection.

By contrast, hearing sound is generally taken to have a very different character. To begin with, sounds are said *not* to be heard as in any way fused with or dependent on the material particulars that make them; in this case, sounds do *not* auditorily seem compresent with their sources. Note that this is a merely negative claim about auditory phenomenology: it tells us only that the sort of phenomenological compresence evident in vision is absent from audition. Call this merely negative claim *Weak Phenomenological Independence*, or (WPI). A related but stronger claim is that sounds are heard as independent of their material sources. This is a positive claim about auditory phenomenology: it tells us that hearing presents sounds in a certain way—namely, as source-independent. Call this positive claim *Strong Phenomenological Independence*, or (SPI). (SPI) is stronger than (WPI) in that (SPI) entails (WPI) but (WPI) does not entail (SPI). If hearing presents sounds as independent of their sources, then sounds do not auditorily seem compresent with

---

[10]A terminological note: I use the verbs 'to hear', 'to auditorily perceive', and 'to auditorily experience' and their cognates interchangeably throughout this paper. Also, I often use 'to perceive' and 'to experience' as short for 'to auditorily perceive' and 'to auditorily experience'. Context should make this clear. (Mutatis mutandis for other sensory modalities.)

their sources. But that sounds do not auditorily seem compresent with their sources does not guarantee that they auditorily seem independent of them.[11]

Both (SPI) and (WPI) find expression in the philosophical literature. For instance, O'Callaghan writes:

> Sounds are unlike ordinary tables and chairs—you cannot grasp or trace a sound—and sounds are not heard to be properties or qualities of tables and chairs, since sounds do not seem bound to ordinary objects in the way that their colors, shapes, and textures do.... Auditory experience presents sounds as independent from ordinary material things, in a way that visual and tactual features are not. (2008a, 804)

In the first sentence, O'Callaghan seems to express a commitment to (WPI) only, while in the second he seems to endorse (SPI).[12] Similarly, here is Matthew Nudds:

> [T]he idea that our experience of sounds is of things which are distinct from the world of material objects can seem compelling. All you have to do to confirm it is close your eyes and reflect on the character of your auditory experience. (2001, 210)

And a few pages later: "[W]hilst sounds appear not to be part of the material world, the same is not true of the objects of sight and touch" (215). Both of these excerpts seem to express a commitment to SPI, not merely to WPI.[13] But perhaps this is unwitting, for Nudds more recently claims that "those writers who have defended phenomenological independence defend the latter claim" (2011, 1). Perhaps so. In any case, that Nudds himself actually had (WPI) in mind in his 2001 paper is at least suggested by one of its central goals—namely, to defend P. F. Strawson's well-known claim that the experience of sound is, as such, non-spatial. According to Strawson, purely auditory experience does not provide any spatial information. Sounds, as we hear them, "have no intrinsic spatial characteristics"; for this reason, a "purely auditory concept of space... is an impossibility" (1959, 65–6). So, Strawson's claim is strictly negative: it is not that sounds auditorily seem

[11]Thanks to Matthew Nudds for drawing my attention to the distinction between stronger and weaker forms of Phenomenological Independence in his commentary during the 3rd Online Consciousness Conference (2011, 1). At the time, I didn't fully appreciate its importance.

[12]Elsewhere, O'Callaghan writes that "a sound seems like such a different sort of thing from a commonplace material object or occurrence" (O'Callaghan 2008b, 319). This superficially resembles (SPI), but it actually doesn't speak to phenomenological independence at all. After all, a color, too, perceptually seems like such a different sort of thing from a commonplace material object or occurrence, but colors don't exhibit phenomenological independence to any degree.

[13]Note that, in saying that our experience of sounds is as of "things which are distinct from the world of material objects," Nudds cannot mean simply that sounds auditorily appear *non-identical* to the world of material objects. After all, *any material particular* is non-identical to the *world* of material objects, and Nudds presumably means to capture the way in which auditory experience suggests that sounds are unusual among the furniture of world. For this reason, he presumably also cannot mean that sounds auditorily appear non-identical to or different from material objects, since this wouldn't distinguish the perceptual appearance of sound from the perceptual appearance of color or any other property-type (cf. the previous note). The only plausible reading seems to be that our experience of sounds is as of things which somehow hang apart from material reality, which is SPI.

independent of spatial—and so, material—reality, which amounts to (SPI), but simply that they *don't* auditorily seem spatial, and so, *can't* auditorily seem compresent with their spatial/material sources. In other words, Strawson is committed to (WPI), but not to (SPI).

Despite their differences, however, both (SPI) and (WPI) encourage the Berkeleyan view that we hear sound sources only in virtue of hearing sounds. Suppose, plausibly, that the following principle is true:

Sonicism:
We hear non-sounds either *in* or *in virtue of* hearing sounds.[14]

Sonicism is based on the idea that auditory experience is through and through a matter of hearing sound. If we hear a non-sound, this is somehow an aspect of hearing a sound. There seem to be two ways in which this might occur: (1) the experience of the non-sound could be immanent in the experience of the sound (the non-sound heard *in* hearing the sound—the Heideggerian view); or (2) the experience of the non-sound could be mediated by the experience of the sound (the non-sound heard only *in virtue of* hearing the sound—the Berkeleyan view). So, if Sonicism is true, the Heideggerian and Berkeleyan views exhaust the range of possibilities for how we hear sound sources. And with Sonicism in the background, it's easy to see how (SPI) and (WPI) encourage the Berkeleyan view.

If the Heideggerian view were true—if the experience of sound sources were immanent in the experience of sounds, and we heard sound sources *in* hearing sounds—then one might reasonably expect that auditory phenomenology would reflect this. In particular, one might reasonably expect that the experience of sound would be such that it seemed to make sound sources available for direct auditory inspection, just as the experience of color is such that it seems to make color-bearers available for direct visual inspection. However, according to (WPI), sound sources do not auditorily seem to be compresent with sounds; therefore, the experience of sound is not such that it seems to make sound sources available for direct auditory inspection. (WPI) thus discourages the Heideggerian view and, against the background of Sonicism, encourages the Berkeleyan view. The same result, of course, holds for (SPI), since it entails (WPI).

Therefore, if either (SPI) or (WPI) is true, then reflection on perceptual phenomenology provides support for the Berkeleyan view of hearing. However, I believe that Phenomenological Independence is false in both of its forms. To demonstrate this it will be sufficient to target (WPI), since (SPI) entails it. My argument is in two stages. First, in Sect. 21.4, I argue for a phenomenological principle that is in deep tension with (WPI). Finally, in Sect. 21.5, I argue against (WPI) directly.

---

[14] I ignore the complication introduced by the possibility of hearing silences, but Sonicism is easily generalized to accommodate it: we hear non-sonic phenomena either *in* or *in virtue of* hearing sonic phenomena (sounds or silences).

21.4 Phenomenological Intimacy

On perceiving something that you do not recognize, it is fitting to ask, "What is that?" Sid hears a sound and asks, "What is that?" Pia answers: "That's my neighbor breaking bottles." Sid and Pia both refer demonstratively to whatever is making the noise; and Pia's claim is true just in case it is, in fact, her neighbor breaking bottles.

Adherents of the Berkeleyan view will typically explain this as follows: Sid and Pia refer to the sound source by "deferred ostension." After all, the only immediate object of perception—and so, possible object of primitive demonstrative reference—is the sound. Michael Martin writes:

> In the case of audition, the primary objects of demonstrative identification are sounds, associated with phrases such as 'that barking' or 'that noise'. One may pick out the source of the sound via picking out the sound itself—we might then understand the demonstrative expression, 'that dog' as involving deferred ostension, perhaps as the descriptive phrase, 'the dog which is actually the source of this sound'. There is a clear contrast between the case of auditory perception of sounds and their sources with the case of colour or shape detection in the case of vision. We do not think of visual demonstrations of objects as proceeding via a demonstration, 'the object which possesses that colour'. (1997, 93)[15]

Martin suggests that auditory experience alone does not enable us to make primitive demonstrative reference to sound sources. Yet his support for this seems to be that we typically *think* of purely auditory demonstrations of sound sources as instances of deferred ostension. But is this true?

Consider a paradigm case of deferred ostension. Jonas points at a cloud of smoke rising over distant treetops and says, "That's a big fire!" and so refers demonstratively to the fire. But note that Jonas isn't pointing at the fire, he's pointing at the smoke. He can't point at the fire, since it's not in view (though he can point toward it). His demonstrative reference to the fire is a case of deferred ostension: it proceeds by means of a descriptive phrase such as 'the fire that is the source of that smoke.' Consequently, his ability to refer to the fire, and our ability to understand him as doing so, is essentially underwritten by knowledge of the causal relationship between fire and smoke. In virtue of this knowledge, we experience the smoke as a sign of the fire. But is this model plausibly applied to auditory experience?

No: we do not typically think of or experience sounds as mere signs of their sources. When Pia enters the house and calls out, "Sid, I'm home!" Sid does not think of or experience this as a *sign* of Pia's presence; rather, it seems to him that he simply *hears Pia call*. In the example above, Jonas doesn't experience the fire as being in view; but Sid *does* experience Pia's calling out as being in auditory view. In particular, it auditorily seems to Sid that he can make primitive demonstrative reference to Pia (and to her calling). Contrary to what Martin says, we do not typically think of purely auditory demonstrations of sound sources as instances

[15] Also see Nudds (2001, 222) and Campbell (1997, 65–6).

of deferred ostension. This is because it auditorily seems that sound sources are available for primitive demonstrative reference, a point recently emphasized by O'Callaghan (2008a, b, 319).

This view of auditory phenomenology can be summed up in the following principle:

Phenomenological Intimacy:
Hearing presents sound sources as available for primitive demonstrative reference.

The question is: is this compatible with Phenomenological Independence, in particular, with (WPI)? Strictly speaking, yes. But there's a catch.

To begin with, let's consider more closely what it means to hear sound sources as available for primitive demonstrative reference. Intuitively, something is available for perceptually-based primitive demonstrative reference only if it is perceptually *given*. This is the point of Russell's notion of *acquaintance* as a demonstrative-thought-enabling relation to an object.[16] So, if Phenomenological Intimacy is correct, then it auditorily seems as though sound sources are things with which we are auditorily acquainted. That is, to use the language of the previous section, it auditorily seems as though sound sources are available for direct auditory inspection. But how could this be compatible with (WPI)?

According to (WPI), sounds never auditorily seem compresent with their sources: it never seems that an experience of a sound source is simply immanent in hearing the sound that it makes. So, (WPI) requires only that, if sound sources *do* seem to be auditorily given (Phenomenological Intimacy), then they *can't* auditorily seem to be given *in* hearing sounds; instead, they must auditorily seem to be given *alongside* sounds. So, we can maintain both (WPI) and Phenomenological Intimacy provided that we're willing to adopt a bipartite view of auditory phenomenology. The problem is that doing so would require that we either reject Sonicism—since it is based on the idea that audition is through and through a matter of hearing sound—or accept that auditory phenomenology is illusory.[17] But not only is Sonicism highly plausible, it gains its plausibility primarily from reflection on auditory phenomenology. Therefore, neither option for reconciling (WPI) with Phenomenological Intimacy seems viable. Arguably, then, retaining a plausible view of auditory phenomenology requires choosing between (WPI) and Phenomenological Intimacy. In my view, this is sufficient to make (WPI) deeply unappealing. Nevertheless, adherents of the Berkeleyan view will choose instead to reject Phenomenological Intimacy. Fortunately, however, there are additional reasons to reject (WPI), and so, Phenomenological Independence *tout court*. To these I now turn.

[16]For Russell's view of acquaintance and its relationship to demonstrative thought, see Russell (1992). For more recent discussion, see, for instance, Campbell (2002). Thanks to Matthew Nudds for encouraging me to introduce the topic of acquaintance into this discussion of Phenomenological Intimacy (2011, 2–3).

[17]Note that Sonicism rules out the possibility of hearing non-sounds *without* hearing any sound as well as the possibility of hearing non-sounds *alongside* sounds.

21.5 Phenomenological Binding

Here's how I see it—or rather, *hear* it. Phenomenological Independence is false in both of its forms. For starters, sounds do not auditorily seem to be independent of ordinary objects and events, as (SPI) requires. Furthermore, contrary to (WPI), sounds auditorily seem compresent with their sources just as much as colors visually seem compresent with their bearers. That is:

> *Phenomenological Binding*:
> Hearing presents sounds as bound to, or fused with, their sources.

If auditory phenomenology is non-illusory, then Phenomenological Binding entails that, in hearing sounds, we hear their sources, which is the Heideggerian view of hearing. So, if Phenomenological Binding is true, it provides strong prima facie support for the Heideggerian view.

Understanding how Phenomenological Binding could be true requires appreciating that, strictly speaking, objects do not make sounds—events do. The static bell is silent; striking it elicits sound. While you may say, "That's the bell," in identifying a sound source, such a claim is implicitly understood as elliptically picking out an event by picking out an object involved in it. The idea that only events can make sounds is part of our ordinary, untutored conception of sound, and ignoring it can lead to confusion. Thinking of the bell as what makes the sound easily leads to thinking of the sound as independent of its source, since, after all, they have very different persistence conditions. On the other hand, if we think of sound sources as events, then we don't have this problem. When the event ends, the sound ceases. Moreover, as the sound changes, the event does, too. And you seem to hear the change in the event, but not merely *in virtue of* hearing a change in the sound; rather, you seem to experience the change in the event *in* experiencing the change in the sound. (This is true even if you don't know in what way the event has changed.)

Suppose that Phenomenological Binding is correct. The question remains: *in what sense exactly* do sounds auditorily seem bound to their sources? There are two main possibilities: (1) like colors, sounds auditorily seem bound to their sources *qualitatively*, as properties; and (2) unlike colors, sounds auditorily seem bound to their sources *mereologically*, as parts to wholes.[18] According to O'Callaghan, sounds are individuals, and they are heard this way, too (2008a). In this case, acknowledging Phenomenological Binding requires adopting (2). However, this seems to me an implausible view of what hearing is like. The sound does not auditorily seem to be *part* of what happens when the hammer strikes the bell; rather, the event auditorily seems to have a certain feature: *it's noisy*. On this view, sounds auditorily seem to permeate or saturate the events that cause them,

[18]Thanks to Casey O'Callaghan for helpfully introducing the distinction between two different ways of hearing sounds as bound to their sources (2011, 2–3). A third but, I think, implausible possibility is: (3) sounds auditorily seem bound to their sources, but the manner of apparent binding is non-specific.

just as colors visually seem to permeate or saturate their bearers. And just as we seem to see *colored objects* rather than objects and their colors, we seem to hear *noisy events* rather than events and their constituent noises. But whatever way we decide this issue, the critical point is that sound sources auditorily seem to be given *in* hearing sounds. Phenomenological Binding is, in any form, incompatible with Phenomenological Independence.

Note, too, that Phenomenological Independence seems to have some seriously undesirable consequences. According to (SPI), hearing presents sounds as independent of their material sources. What exactly would it be to hear sounds in this way? It would be to hear them as only contingently related to their causes—as if they might not have been caused by material events at all. But this is incoherent. Sounds are not merely contingently caused by material events. To be caused by an appropriate sort of material event is part of what it is to be a sound. So, to hear sounds in the way that (SPI) requires would be to hear sounds as if they weren't sounds at all, but something else entirely. Surely this is an undesirable result. In any case, it entails that auditory phenomenology is illusory: sounds are heard as if they were something other than they are.

But perhaps (WPI) can do better. According to (WPI), sounds are not heard as independent of their material sources; instead, auditory experience simply fails to comment on the relationship between sounds and material reality. Sounds are heard neither as connected with nor disconnected from ordinary objects and events. But then where do we so much as get the idea that things make sounds? As Nudds argues, if this Strawsonian view of auditory phenomenology is correct, it can only be in virtue of *multimodal* experience that we experience sounds as related to the ordinary events that we see or feel (Nudds 2001). We see the hands come together, and we hear the clapping sound. At best, then, sounds "appear to be, as Strawson says, correlated with the material world, but they do not appear to be part of it" (Nudds 2001, 215). But correlation is of course a *contingent* relationship. Thus, Nudds continues:

> We can imagine a world of sounds which is dissociated from the world of material objects; we can imagine, too, the sounds we actually hear apart from the things that we see and touch. There appears to be nothing intrinsic to the sounds that we actually hear to connect them with the world of sight and touch. (215)

Consequently, this view faces the same problem as (SPI): in claiming that we perceive sounds as only contingently related to their sources, it commits itself to treating the experience of sound as illusory.

Moreover, (SPI) and (WPI) are incompatible with what seems to be a basic datum of auditory phenomenology: the apparent locatedness of sounds. As O'Callaghan has discussed at length, sounds *auditorily* seem to be located (2007, ch. 3; 2009b, §3). However, this could not be true if, as on (SPI), sounds auditorily seemed to be independent of their material sources (and so, of material things generally), or if, as on (WPI), auditory experience simply failed to comment on the relationship between sounds and their material sources. If sounds are heard as located, then they are heard as *part* of the physical world. (Again, however, the question remains: what sort of part? Are they heard as properties of events? Or as constituents of them? Properties, I think.) In any case, the appropriate response to these difficulties is to reject Phenomenological Independence *tout court*, and to embrace Phenomenological Binding.

If Phenomenological Independence is false, then why have so many philosophers believed it true? I think that they have been misled by the relative epistemological poverty of hearing. It happens that we see things that we do not recognize, but it is far more common that we do not know exactly what we hear. You turn around on hearing a sound behind you to know better what you have heard. The sheer difficulty of identifying many events by their sounds is what leads us to investigate by other means. And it is easy to understand how this might lead us to think of hearing as presenting sounds absent their sources: after all, it seems that I know all about the sound, but I know so little about its source![19] Another way to put the point is that the relative epistemological poverty of hearing encourages us to adopt the sort of intellectualized or "abstract" attitude toward sounds that Heidegger describes as "listen[ing] away from things" (1977, 152). Having taken this attitude, we will hear the sound *as* a purely qualitative something with no apparent connection to material reality. But when we're interested in what hearing is like, we're interested first and foremost in the experience of the "engaged" listener, not in the different attitudes that a "disengaged" listener can take toward what he hears. And I think that the more we reflect on the phenomenology of engaged hearing, the more we come to see that Phenomenological Binding is correct.

Along these lines, Phenomenological Binding is encouraged by a simple imaginative exercise. Imagine striking a bell. Now try to imaginatively subtract or peel away the sound. I think that this is just as difficult as picturing a tomato and trying to imaginatively subtract or peel away its color. Only if we "listen away" from the hammer-strike and the vibrating bell does it seem as though we can perform this imaginative feat. From the engaged perspective, sounds auditorily seem no less bound to the events that cause them than colors visually seem bound to their bearers.[20]

A final, powerful point in favor of Phenomenological Binding is that it is a natural corollary of Phenomenological Intimacy. Indeed, Phenomenological Binding *explains* Phenomenological Intimacy: we hear sound sources as available for primitive demonstrative reference *because* the experience of a sound source seems immanent in the experience of a sound. By contrast, as discussed in the previous section,

[19]Note that the analogy with color experience holds here, too. Imagine trying to identify objects solely on the basis of their colors, without much, if any, information about their shapes or locations. The difficulty is obvious. As in the case of hearing, you would often wish to employ other means, especially touch, and, if this sort of thing occurred often enough, you might be led to think of vision as presenting colors as separate from their bearers; but this, of course, would be a mistake.

[20]This is consistent with the idea that we can think of a sound without thinking of it as having any particular cause. After all, we can think of a color without thinking of it as having any particular bearer. For, just as qualitatively the same color may be borne by very different objects, qualitatively the same sound may be caused by very different events.

the advocate of Phenomenological Independence must reject Phenomenological Intimacy (on pain of rejecting Sonicism). To do so while maintaining that auditory phenomenology is illusory (as argued above) and in the face of a plausible explanation of why we should have been tempted to endorse Phenomenological Independence in the first place—this is surely a pill too bitter. The lesson to draw is that reflection on the phenomenology of auditory experience provides prima facie support for the Heideggerian view of hearing: we hear sound sources directly, *in* hearing the sounds that they make—not, à la Berkeley, merely *in virtue of* hearing those sounds.[21]

References

Bach-y-Rita, P., and S.W. Kercel. 2003. Sensory substitution and the human-machine interface. *Trends in Cognitive Sciences* 7(12): 541–546.

Berkeley, G. 1992. *Three dialogues between Hylas and Philonous, in philosophical works: Including the works on vision*. London: J. M. Dent & Sons Ltd.

Bermúdez, J.L. 2000. Naturalized sense data. *Philosophy and Phenomenological Research* 61(2): 353–374.

Campbell, J. 1997. Sense, reference, and selective attention. *Proceedings of the Aristotelian Society, Supplementary Volumes* 71: 55–74.

Campbell, J. 2002. *Reference and consciousness*. Oxford: Oxford University Press.

Hamilton, A. 2009. The sound of music. In *Sounds and perception: New philosophical essays*, ed. M. Nudds and C. O'Callaghan, 146–182. Oxford: Oxford University Press.

Heidegger, M. 1977. The origin of the work of art. In *Basic writings: From being and time (1927) to the task of thinking*, ed. D.F. Krell, 1964. San Francisco: Harper Collins.

Jackson, F. 1977. *Perception: A representative theory*. Cambridge: Cambridge University Press.

Leddington, J. 2009. Perceptual presence. *Pacific Philosophical Quarterly* 90: 482–502.

Martin, M.G.F. 1997. The shallows of the mind. *Proceedings of the Aristotelian Society, Supplementary Volumes* 71: 75–98.

Nudds, M. 2001. Experiencing the production of sounds. *European Journal of Philosophy* 9(2): 210–229.

Nudds, M. 2011. Comments on Jason Leddington, 'What we hear'. http://consciousnessonline.files.wordpress.com/2011/02/nudds-comments-on-leddintong.pdf

O'Callaghan, C. 2007. *Sounds: A philosophical theory*. Oxford: Oxford University Press.

O'Callaghan, C. 2008a. Object perception: Vision and audition. *Philosophy Compass* 3(4): 803–829.

O'Callaghan, C. 2008b. Seeing what you hear: Cross-modal illusions and perception. *Philosophical Issues* 18: 316–338.

[21]Earlier drafts of this paper were presented at the 2010 Joint Session of the Aristotelian Society and the Mind Association at University College Dublin, the Third Annual Online Consciousness Conference (CO3), and the 2011 meeting of the Central Division of the American Philosophical Association. I am grateful to Matthew Nudds and Casey O'Callaghan for their very helpful and detailed commentaries at CO3, to Sam Wheeler for commenting at the APA session, and to the audiences at all three events, particularly Mark Kalderon and Heather Logue at the Joint Session. Finally, thanks to Matthew Slater and Gary Hardcastle for allowing me to hijack a meeting of our reading group to pick their brains about a much earlier version of this paper.

O'Callaghan, C. 2009a. Sounds. In *The Oxford companion to consciousness*, ed. T. Bayne, A. Cleermans, and P. Wilken, 609–611. Oxford: Oxford University Press.
O'Callaghan, C. 2009b. Sounds and events. In *Sounds and perception: New philosophical essays*, ed. M. Nudds and C. O'Callaghan, 26–49. Oxford: Oxford University Press.
O'Callaghan, C. 2011. Comments on Jason Leddington, 'What we hear',. http://consciousness-online.files.wordpress.com/2011/02/commentary-on-leddington-o_callaghan.pdf
O'Callaghan, C., and M. Nudds. 2009. Introduction: The philosophy of sounds and auditory perception. In *Sounds and perception: New philosophical essays*, ed. M. Nudds and C. O'Callaghan, 1–25. Oxford: Oxford University Press.
Roxbee Cox, J.W. 2011. Distinguishing the senses. In *The senses: Classic and contemporary philosophical perspectives*, ed. F. Macpherson, 101–119. Oxford: Oxford University Press.
Russell, B. 1992. Knowledge by acquaintance and knowledge by description [1911]. In *The collected papers of Bertrand Russell*, vol. 6, ed. J.G. Slater, 147–161. London: Routledge.
Scruton, R. 1997. *The aesthetics of music*. Oxford: Oxford University Press.
Scruton, R. 2009. Sounds as secondary objects and pure events. In *Sounds and perception: New philosophical essays*, ed. M. Nudds and C. O'Callaghan, 50–68. Oxford: Oxford University Press.
Sorensen, R. 2010. Hearing silence: The perception and introspection of absences. In *Sounds and perception: New philosophical essays*, ed. M. Nudds and C. O'Callaghan, 126–145. Oxford: Oxford University Press.
Strawson, P.F. 1959. *Individuals: An essay in descriptive metaphysics*. London: Methuen & Co. Ltd.
Tye, M. 2009. *Consciousness revisited*. Cambridge, MA: MIT Press.

Chapter 22
Audible Independence and Binding

Casey O'Callaghan

In "What We Hear," Jason Leddington (2014) argues against two claims about sounds and hearing. The first is that sounds are proper objects of hearing—that sounds are inaccessible to other senses. The second is that only sounds are heard directly—one hears sound sources only in virtue of hearing sounds. Leddington's main target is the second claim, so it is my focus.

Leddington's case against the second claim turns on arguing against *Phenomenological Independence*, the claim that, as presented in auditory experience, sounds seem independent from ordinary material things and happenings. He claims that auditory experiences present sound *sources* as being available for primitive demonstrative reference (*Phenomenological Intimacy*) and that this tells against the Phenomenological Independence of sounds from sound sources. He also argues that Phenomenological Independence is incompatible with *Phenomenological Binding*, the claim that auditory experiences present sounds *as bound to* their sources.

Since Phenomenological Independence fails, Leddington argues, we do not hear ordinary material things or sound sources indirectly by or in virtue of hearing the sounds they make (*the Berkeleyan view*). Instead, he advocates *the Heideggerian view*, according to which one hears sources *in* hearing their sounds and, therefore, "hearing a sound involves an unmediated experience of its source" (p. 325).

Leddington thus argues from a claim about the apparent relations among objects of auditory awareness to a conclusion about the relations among auditory experiences—from the claim that audible sounds and audible sources seem bound and not independent to the claim that one hears sources in hearing sounds.

My work on sounds and hearing has emphasized the possibility of audition-based demonstrative reference to sound sources, as Leddington mentions (p. 328). I also have argued that locational hearing involves hearing sounds to be located at or near their sources. Notably, audible sounds do not audibly seem to travel

C. O'Callaghan (✉)
Department of Philosophy, Rice University, 6100 Main Street, Houston, TX 77005, USA
e-mail: casey.ocallaghan@rice.edu

R. Brown (ed.), *Consciousness Inside and Out: Phenomenology, Neuroscience, and the Nature of Experience*, Studies in Brain and Mind 6, DOI 10.1007/978-94-007-6001-1_22,

in relation to their sources. I also have argued that sounds and their sources audibly may seem bound or fused. But if I therefore accept what Leddington calls Phenomenological Intimacy and Phenomenological Binding, must I reject Phenomenological Independence?

It is worth pointing out that Phenomenological Intimacy and Phenomenological Independence in fact are consistent. One might hear sounds, hear sound sources, and hear them to be independent; and one might also be auditorily acquainted with and possess the capacity to refer demonstratively to sound sources. With the addition of *Sonicism*, however, Leddington claims that Phenomenological Intimacy cannot be reconciled with Phenomenological Independence. *Sonicism* is the claim that hearing is "through and through" a matter of hearing sounds. In light of this, Leddington claims that the Berkeleyan view and the Heideggerian view exhaust the options—we hear things that are not sounds either *in virtue of* hearing sounds or *in* hearing sounds. Thus, even hearing a sound source directly constitutively involves or depends upon hearing a sound. Leddington thinks Sonicism is not negotiable. Since he holds that the Berkeleyan view requires Phenomenological Independence, Leddington sides with Phenomenological Intimacy and the Heideggerian view. My view is that Sonicism is attractive, but it is not mandatory, so one option is to reject Sonicism. I will return below to this suggestion and to the plausibility of Sonicism.

Even if we assume Sonicism, however, accepting Phenomenological Intimacy does not require rejecting Phenomenological Independence.

Leddington distinguishes *Strong Phenomenological Independence*, the claim that sounds are heard *as* independent from sound sources, from *Weak Phenomenological Independence*, the negative claim that auditory experiences do not present sounds as dependent upon their sources. Since the former implies the latter, Leddington argues only against Weak Phenomenological Independence.

Leddington characterizes Weak Phenomenological Independence as the claim that "sounds do *not* auditorily seem compresent with their sources" (p. 325), where compresence is interpreted as the relation that visible qualities such as colors visibly seem to stand in to the objects that bear them. Thus, Weak Phenomenological Independence holds that "the sort of phenomenological compresence evident in vision is absent from audition" (p. 325). I accept this claim because I hold that sounds are particular audible *individuals* to which audible qualities such as pitch, timbre, and loudness belong, and that audible sounds are not *identical* with ordinary material objects or events. Sounds audibly occur or unfold over time; sounds audibly persist through time and survive changes to their qualities. Since sounds audibly are persisting individuals that bear the familiar audible qualities, sounds themselves do not audibly appear to qualify ordinary material things and happenings. Thus, sounds are not identical with ordinary material objects or happenings, and sounds do not audibly appear to qualify ordinary material things *in the way that* colors visibly appear to qualify ordinary material surfaces and objects or in the way that that textures tactually do. This is the force of the following passage of mine quoted by Leddington (p. 326): "Sounds are unlike ordinary tables and chairs—you cannot grasp or trace a sound—and sounds are not heard to be properties or qualities of tables and chairs, since sounds do not seem bound to ordinary objects in the way that their colors, shapes, and textures do. Auditory experience presents sounds as independent from ordinary material things, in a way that visual and tactual features are not."

Does accepting Leddington's Weak Phenomenological Independence license a view that captures the spirit of Phenomenological Independence? Plausibly, yes. Suppose sounds audibly are *distinct* from sound sources. Distinctness may suggest physical separateness, but it also is fair to say that individual things are distinct if they differ, are not identical, or are distinguishable. Thus, if we can hear individual sounds, if we can hear individual sound sources, and if hearing does not present sounds as identical with sound sources, then this grounds a relatively uncontroversial version of Phenomenological Independence.

One objection is that such apparent distinctness does not suffice for *apparent independence* because non-identical things might nonetheless appear to depend upon each other in some way or another. For instance, one thing can appear to depend causally upon another. One thing might appear to depend for its present existence upon another. And so on. In each case, apparently distinct things do not appear to be wholly independent from each other. So, even though a sound audibly is distinct from its source, if the sound is heard as depending causally upon its source, then it is not phenomenologically independent from the source. By Phenomenological Independence, therefore, one might have in mind something stronger than *apparently distinct individuals*—perhaps the claim that sounds are not ever heard as being dependent for their present existence upon ordinary material things, or that sounds invariably are heard as autonomous from their sources.

Are sounds in any way heard as being dependent upon their sources? Some evidence suggests that sounds are available for attention and demonstrative reference in ways that do not involve attention or demonstrative reference to their sources. Scruton's (1999) discussion of "acousmatic experience" is one example of an attempt to show that this is possible. We can listen or attend to musical sounds in a way that does not obviously involve hearing their sources. In such listening, sounds are not clearly auditorily experienced as bound to their sources or as having source-relative attributes. This is the point of musical listening, according to Scruton. That this is not the normal listening mode does not show that it is impossible. This suggests that sounds are capable of being heard independently from their sources in certain forms of listening; it therefore suggests that we sometimes are capable of hearing sounds in a way that does not present them as being dependent upon their sources.

That this is a possible listening mode does not mean that it is the usual listening mode. It also is plausible that in run-of-the-mill hearing, humans may auditorily experience both sounds and sources, and also may experience sounds as having sources. It is plausible that we do not commonly hear sounds as being wholly distinct from or as completely independent from their sources and, thus, that hearing commonly presents sounds as in some manner dependent upon their sources. Ordinary embedded hearing typically does not involve auditorily experiencing sounds as wholly autonomous with respect to their apparent sources.

Thus, while I accept Leddington's Weak Phenomenological Independence, I prefer to reject his suggestion that its advocates maintain that sounds are "*not* heard as in any way fused with or dependent on the material particulars that make them" (p. 325). Just as there are a number of respects in which we can say that one thing is dependent upon another, there are a number of respects in which we can say that one thing is independent from another. Sounds are not heard as fused with or dependent upon material things and events *in the manner in which visible qualities are seen as fused with or dependent upon visible objects*. The audibly apparent distinctness of individual sounds from individual sources explains what is attractive about Phenomenological Independence without advocating the complete or wholesale phenomenological independence of sounds from sources in each episode of hearing.

Suppose that sounds audibly are distinct from ordinary material things and happenings that are sound sources. And suppose that we accept Weak Phenomenological Independence—the claim that audition does not present sounds as bound with ordinary material things in the manner in which visible colors appear to qualify material surfaces and objects. If we also accept Sonicism, must we therefore reject Phenomenological Intimacy—the claim that hearing presents sounds as available for primitive demonstrative reference rather than mere deferred ostension?

In the case of seeing surfaces and objects, Bermúdez (2000) accepts near visual analogs of Weak Phenomenological Independence, Phenomenological Intimacy, and Sonicism. Bermúdez maintains that one sees three-dimensional objects in a way that is mediated by seeing their facing surfaces, but he nevertheless maintains that vision presents objects as available for demonstrative reference in a manner that is epistemically direct. He therefore accepts a mediated account of seeing ordinary objects but does not reject Phenomenological Intimacy, as Leddington suggests adherents to the Berkeleyan view must (p. 330).

One obstacle to endorsing an auditory account of this type is puzzlement about how awareness as of an individual sound could ground acquaintance with, or epistemically direct awareness as of, a sound source that is distinct from it, so that the sound source is available for demonstrative reference without deferred ostension. This strikes me as Leddington's primary concern. And it leads him to endorse Phenomenological Binding, the claim that we hear sounds "as bound to, or fused with, their sources" (p. 330).

I endorse Phenomenological Binding. Phenomenological Binding does capture the intimacy with which we experience sounds to be related to their sources, and it does help to explain how awareness as of a sound could furnish awareness as of a sound source. It does so because it helps to explain how being aware of a sound could enable one to differentiate a sound source from its surrounding environment, which is a plausible requirement on perceiving a particular. That is why seeing a facing surface may ground acquaintance with and enable demonstrative reference to its object.

To see how Phenomenological Binding in fact is compatible with Weak Phenomenological Independence, it is helpful to distinguish two varieties of perceptually apparent binding. First, properties may be perceptually experienced as belonging to or as bound to their bearers. One sees the redness as qualifying or as spread out across the surface of an object. One feels the texture as being an attribute of the surface. One tastes the flavor as belonging to or as being instantiated by the apricot. But, as discussed above, sounds are not heard as properties or qualities of ordinary material objects or happenings in the way that other straightforwardly sensible qualities are perceptually experienced as belonging to sensible individuals. Instead, sounds are audible individuals to which qualities such as pitch, timbre, and loudness audibly belong.

There is, however, another way in which non-identical things can appear bound or fused. The parts of an object can appear bound or fused to compose a single compound object to which those parts appear to belong. When you see a complex object, such as a table or a chair, its distinct perceptible parts—the legs, the seat, the top . . . —may be visually experienced as being fused or bound together into a single perceptible whole. When you see the facing surface of a table, you may visually experience it to belong to, or to be bound or fused to, a larger object, some of whose parts are hidden from view.

How could this apply to the case of hearing sounds and sources? Sounds are heard as bound to or fused with their sources in the sense that sounds are heard as being mereological parts of complex environmental events that in fact involve sounds. For instance, take the event of an automobile collision. Such an event could occur in a vacuum. When it occurs in a surrounding elastic medium, however, a broader environmental event or happening occurs that includes a sound. The sound is part of an event that involves cars colliding in an elastic medium. One hears the sound, and one hears the broader event that involves the cars and the colliding and the disturbing of the medium. One could not have heard the broader event if not for its sound—had it occurred soundlessly, it would have been inaudible. This is part of the reason some may say one hears the crash in or in virtue of hearing the sound, since hearing the sound enables one to discern the crash from its surroundings. On this account, however, one hears the sound as being a constituent part of the broader collision event. The audible sound is akin to the visible facing surface of the table—the sound determines the audible appearance of the broad environmental event that includes the material objects and happenings that we count among its sources.

This allows that sounds and sound sources audibly are distinct individuals, and it allows that sources are heard in or in virtue of hearing their sounds. It allows that sounds are heard as bound with their sources in the manner of perceptible parts and wholes, but it does not accept that sounds are heard as audible properties or qualities bound to their sources. So, it captures the spirit of Phenomenological Independence while accommodating Phenomenological Binding and Phenomenological Intimacy. And it is compatible with Sonicism.

Should we accept Sonicism? Recall that *Sonicism* is the claim that hearing is "through and through" a matter of hearing sounds and, thus, that hearing a non-sound is an *aspect* of hearing a sound (p. 327). According to Leddington, Sonicism implies that humans hear sound sources only *in* or *in virtue of* hearing sounds, so the Heideggerian view and the Berkeleyan view exhaust the options. Using Leddington's terms, the direct experience of a non-sound is "immanent in" the experience of a sound, or the indirect experience of a non-sound occurs "in virtue of" the experience of a sound. Sonicism, so understood, implies that every episode of hearing a non-sound constitutively involves or depends upon a concurrent episode of hearing a sound (silence may be addressed as a special case).

This is an attractive line of thought. Whenever we hear some ordinary material thing or occurrence, invariably a sound exists to which we are able to direct our auditory attention should we attempt it. This encourages the thought that hearing something (other than silence) always is grounded in hearing a sound and, thus, that each episode of hearing a non-sound constitutively involves or depends for its occurrence upon a concurrent episode of hearing a sound.

But that thought is not mandatory. We need not say that every episode of seeing a material object depends for its occurrence upon seeing its facing surface. Instead, we may simply see an object that possesses a visible facing surface. Similarly, we may simply hear things and happenings that include or possess audible sounds. We need not say that every episode of hearing a non-sound is an aspect of hearing its sound. It may be a necessary condition on hearing an event that it includes an audible sound, or on seeing an object that it possesses a visible surface, but this does not imply that one hears an event in or in virtue of hearing its sound, or that one sees an object in or in virtue of seeing its surface. Sonicism, as Leddington characterizes it, is negotiable.

This raises a deeper concern. As mentioned earlier, Leddington uses claims about phenomenology that concern the apparent relations among objects of auditory awareness to draw a conclusion about the nature of the relationship that holds between auditory experiences of those objects. In particular, the conclusion (the Heideggerian view) is a specific claim about the nature of the dependence that holds between an auditory experience of a sound source and the auditory experience of a sound: one does not experience a non-sound *in virtue of* experiencing a sound; the experience of a non-sound is *immanent in* the experience of a sound.

Whether distinct objects of awareness appear compresent, bound, fused, overlapping, causally related, or otherwise dependent does not, however, have immediate consequences concerning the specific nature of the relationship that holds between the experience of the one object and the experience of the other. In particular, phenomenology that concerns the apparent relations among objects of awareness may be compatible with a range of views about whether and how the perceptual experience of one thing constitutively involves or depends upon the perceptual experience of another. One's account of the perceptually apparent relations among audible sounds and audible sound sources, therefore, lacks immediate or obvious consequences concerning the nature of the relationship that holds between auditory experiences of sounds and auditory experiences of sound sources.

This lesson does not just apply to the decision between the Heideggerian view and the Berkeleyan view—to whether one hears non-sounds *in* hearing sounds or else hears non-sounds *in virtue of* hearing sounds. It extends to the decision about Sonicism—to whether or not one hears non-sounds *in or in virtue of* hearing sounds. That is, to the question whether or not an episode of hearing a non-sound constitutively involves or depends upon a concurrent episode of hearing a sound.

Phenomenological claims concerning the apparent objects of auditory awareness and their audibly apparent relations thus are compatible with an account that is neutral about any relation of priority or dependence that holds between an auditory experience of a sound and an auditory experience of its source.

References

Bermúdez, J.L. 2000. Naturalized sense data. *Philosophy and Phenomenological Research* 61(2): 353–374.

Leddington, Jason. 2014. What we hear. In *Consciousness inside and out: Phenomenology, neuroscience, and the nature of experience*, ed. Brown Richard, 321–334. Dordrecht: Springer.

Scruton, R. 1999. *The aesthetics of music*. Oxford: Oxford University Press.

Chapter 23
Commentary on Leddington

Matt Nudds

Sounds are the objects of auditory experience. They are the individual things that we can attend to in auditory experience. These objects of auditory experience instantiate the acoustic properties of pitch, loudness, and timbre. They appear to be individual things in which these acoustic properties inhere.[1] They do not appear to be properties of material objects in the way that, say, colours appear to be properties of material objects (I say more about this below); nor do they appear to be parts of material objects.[2]

It is of the essence of sounds that they take time, so that the identity of a sound is not fixed by how it is at any time, but depends on the way it unfolds over time. That means that sounds are similar to events and processes, rather than to material objects. So sounds—understood to be the things that we can pick out and can attend to in our auditory experience—appear to be individuals that unfold over time in an event- or process-like way, and that instantiate acoustic properties.

A natural question to ask is where sounds, conceived as individuals, fit into the world of material objects, events, and processes; and in particular how sounds are related to their sources. It doesn't follow from the way sounds appear that they are in fact independent of the things that produce them, but how they stand in relation to those things cannot be answered by simply by reflecting on how they appear.

Because sounds appear to be individuals that instantiate properties and can be individuated as such, I think it's right to say that sounds appear to be independent

---

[1]For an extended defense of this claim, see O'Callaghan (2007, ch. 2).

[2]Of course to say that sounds do not appear to be properties of, or parts of, material objects doesn't mean that they are not; just that they don't appear to be. Similarly to say that the appearance of sounds is determined by acoustic properties doesn't mean that they don't also have non-acoustic properties.

M. Nudds (✉)
Department of Philosophy, University of Warwick, Coventry CV4 7AL, UK
e-mail: matthew.nudds@ed.ac.uk

R. Brown (ed.), *Consciousness Inside and Out: Phenomenology, Neuroscience, and the Nature of Experience*, Studies in Brain and Mind 6, DOI 10.1007/978-94-007-6001-1_23,

of the material objects that produce them—independent, that is, of their sources. Sounds do not appear to be properties instantiated by material objects and so don't appear to depend on them in the way that, say, colours appear to depend on the objects that instantiate them. Sounds appear event- or process-like. Perhaps they are in fact events or processes occurring in material objects. We individuate events occurring in material objects them in terms of the objects in which they occur—events consist in changes occurring in those objects. However, we can individuate a sound without identifying any material object that produces it: sounds "stand alone," they appear to be "pure events" (Scruton 2009, 62–63).

What do these claims about how sounds appear imply about auditory perception, and in particular about the perception of things other than sounds? Hume claimed that, in reflecting on our perceptual experience, we "always suppose the very images presented by the senses, to be the external objects, and never entertain any suspicion, that the one are nothing but representations of the other" (Hume 1751, 118). Hume was making a claim about how our perceptual experience introspectively seems to us. In reflecting on my visual experience of the cup on my desk, it is the cup and its properties (its shape and colour) that seem present in my experience. So reflection on our visual experience reveals it as seeming to present the mind-independent objects and properties that we take ourselves to see. It seems that my visual experience could not be as it actually is were the cup and its properties not present. The same is not true of auditory experience. In reflecting on my auditory experience of a barking dog, it is the *sound* of the barking and its acoustic properties that seem present in my experience, rather than the dog and its properties. It seems that my experience could not be as it actually is were the sound not present, but could be as it actually is were the dog that I take to be making the sound not making it. So reflection on our auditory experience reveals it as seeming to present sounds and the properties of sounds, rather than the mind-independent objects making those sounds that we take ourselves to hear. Because of the way our auditory experience seems, I think it's right to say that whatever we perceive we perceive by perceiving sounds.

In taking sounds and auditory experience to be this way I think I am committed both to what Leddington calls *Weak Phenomenological Independence*, and to *Strong Phenomenological Independence*. But I don't think that either of these independence claims has the consequences that he claims them to have.

If *sounds* have the character that I have described them as having, then they do not appear to be properties of objects (though further argument is required to show that the are not in fact properties of objects); but we often *experience* sounds as produced by events and (arguably) objects of certain kinds—in many cases these are the events and object that did in fact produce them; in such cases sounds are heard as causally dependent on the things that actually produced them. Even in such cases, my experience seems such that it could have been as it actually is even were the things that I experience the sounds as produced by not there. That is, my experience could have been as it actually is were the sounds that I experience not produced by the events and objects that seem to have produced them. For example, I might have an experience of sounds that seem to have been produced by a dog's barking. Either the sounds were in fact produced by a dog's barking, or the sounds were not produced by a dog's barking (but in some other way). In both cases my experience is the same (we can suppose that the two situations are subjectively indistinguishable): I hear the same sounds, and the sounds seem to have been produced by a dog's barking. But only in the first situation is my experience of what produced the sound veridical—only in that situation do I hear the dog.

At one point Leddington rejects Strong Phenomenological Independence on the grounds that it is "incoherent": it would involve hearing sounds "as only contingently related to their causes—as if they might not have been caused by material objects at all" (16). In fact we often hear sounds—produced by wind turbulence and flowing water, for example—that are not caused by material objects. It is also possible to produce sounds that are independent of material objects by using inaudible ultra-sonic energy to directly induce sound waves in air. So the idea of sounds existing independently of material objects is not incoherent. Some sounds are produced by material objects (or events involving material objects); perhaps our ways of individuating sounds are such that we can't make sense of a particular sound that was in fact caused by such an event not having been caused by that event (though that is not obviously so); we can, however, make sense of an indistinguishable sound not having been caused by that event and so of the idea that sounds of that type are only contingently related to that type of cause. So we can conceive of an auditory experience of sounds just like the one we are enjoying that is of sounds not caused by material objects. Leddington's point is perhaps that, even in such cases, the sounds will seem to have been caused by events involving a material objects—that we can't hear the sound as other than seeming to have been caused by such an event. I think that's true of some sounds, but deny that it gives us any reason to reject Strong Phenomenological Independence (or at least the conception of sounds I outlined above).

In a discussion of olfaction, Lycan suggests that olfactory experience has two kinds of content: "that smells represent adaptively significant environmental entities, and they also represent odors. In fact, they represent the environmental entities by representing odors. By smelling a certain familiar odor I also smell – veridically or not – a dog" (Lycan 1996, 148). Whether or not this is true of olfaction, I think we can explain the relation between our experience of sounds and their sources by appealing to this kind of representational structure. It is plausible that our experiences of sounds represent adaptively significant sound sources as well as sounds, and that they represent sound sources by representing the sounds that they produce. If that's right, we should think of auditory experience as having two kinds of content—as representing sounds and as representing the sources of sounds. It follows that hearing a sound normally involves an experience that represents the source of the sound as being some way. When the experience is veridical we hear the source of the sound as well as the sound; but an experience may be *partially* veridical—veridically representing the sound, but misrepresenting the source of the sound. That happens when, for example, we hear a sound that seems to of a dog barking, but is not (because it was produced by something else). In such cases, we hear the sound but not the source of the sound.

Phenomenological Independence is a claim about how *sounds* seem to be. As such it is consistent with a claim about how we experience sounds: that in virtue of our experience representing the events involving material objects that produce them, we experience sounds as seeming to have been produced by those events. If that's right, then it is possible to maintain both that *sounds* appear to be independent of their sources in the way I described above, and that we *experience* sounds as seeming to have been produced by events involving material objects.

Leddington calls the kind of view that I am defending the *Berkeleyan* view. In some ways my view is like Berkeley's view, but in other ways it is not. Berkeley took sounds to be much like I have characterised them to be, but he also thought that the content of auditory experience is exhausted by the sounds phenomenally present in experience. Given that conception of experience (and the explanatory resources available to him), the only way for Berkeley to explain how we hear anything other than sounds is in terms of inference or association between sounds and ideas of the things that produced them; he concluded that "in truth and strictness" nothing other than sounds can be heard. I reject Berkeley's conception of experience and don't think that the content of auditory experience is exhausted by the sounds phenomenally present in experience, so I don't think that the only way to explain how we hear anything other than sounds is in terms of inference or association. Rather, I think that experiences of sounds represent both sounds and the things that produced those sounds.

I think Leddington (11–13) takes the Berkeleyan view to be committed to both Berkeley's view of sounds and his view of experience, and so he contrasts the Berkeleyan view to the *Heideggerian* view. Heidegger says that we "never really first perceive a throng of sensations, e.g., tones and noises, in the appearance of things . . . ; rather we hear the storm whistling in the chimney, we hear the three-motored plane, we hear the Mercedes in immediate distinction from the Volkswagen. Much closer to us than all sensations are the things themselves. We hear the door shut in the house and never hear acoustical sensations or even mere sounds" (Heidegger 1935, 151–152).

We might understand the claim that Heidegger is making here as a claim about auditory attention: that our attention in auditory experience is, for the most part, to the things that make sounds rather than the sounds themselves. He goes on to say that "in order to hear a bare sound we have to listen away from things, divert our ears from them, i.e. listen abstractly" (152). In listening towards or away from something we are directing our attention to or away from it. In fact, it is often very difficult to "listen away from things" in this way: try to describe the sounds made by breaking glass, by a ball bearing rolling over a table, or by screwing up a sheet of newspaper in terms of the bare sounds, that is, in purely acoustic terms that don't mention what it is that makes the sounds. It's almost impossible to do so in anything other than a very general way.

If we reject Berkeley's view of the content of experience then this fact about attention doesn't undermine my claim about how sounds seem. We normally experience sounds as having been produced by certain kinds of events or objects. It may be that we find it very difficult to characterise or describe the sounds we hear other than in terms of the things that seem to have produced them, and that our attention is always to those patterns of similarity amongst sounds that they have in virtue of what seems to have produced them rather than the patterns of similarity that they have in virtue of their acoustic properties. Such patterns of similarity may, in many cases, be far more salient that patterns of acoustic similarity. That might explain why it is difficult to attend to acoustic similarities amongst sounds—to "hear the bare sound". Such an explanation is inconsistent with Berkeley's view, but is consistent with Phenomenological Independence and with my *Berkeleyan* view.

Leddington doesn't take Heidegger to be simply making a claim about auditory attention or auditory experience. As Leddington sees it, Heidegger is committed to the idea that "we hear ordinary objects and events in hearing the sounds that they make" (6), and he takes that to commit him to a claim about the nature of sounds, rather than a claim about our experience of sounds.

Leddington compares hearing sounds to seeing colours: "We do not see things in virtue of seeing their colours; rather, we see them in seeing their colours." I don't think that our experience of objects as coloured helps us understand our experience of the sources of sounds. Leddington is right that we don't see things in virtue of seeing their colours. Our visual experience of objects is neither causally nor phenomenologically dependent on our experience of their colour; in contrast, our auditory experience of material objects is both causally and phenomenological dependent on our experience of the sounds they make.

We have an understanding of what it is to see an object that is independent of our understanding of what it is to see an object as coloured. The same is not true of hearing: we simply have no conception of what it would be to *hear* an object independently of hearing it as making a sound (that is not to say that we have no conception of the object we hear independently of its making a sound). It's right to say that we experience colours as properties of the objects we see only because we have an independent understanding of what it is to see an object and we can visually individuate material objects in ways that are independent of our perceiving their colour. Our seeing an object on any occasion is independent of our experience of its colour on that occasion; we would have seen it on that occasion even if we had experienced it as having a different colour or not experienced it as having any colour at all. That contrasts with hearing. We cannot auditorily individuate material objects independently of perceiving the sounds that they make. Our hearing an object on any occasion is not independent of our hearing the sound it makes on that occasion; we simply would not have experienced it on that occasion had we not experienced the sound that it made on that occasion.

Leddington is right to say that we hear things in hearing sounds. He seems to suggest that this is always the case; I don't think it is always the case, but it *is* often the case. I think he's right, too, to reject Berkeley's conception of auditory experience, and the deferred demonstrative account of hearing the sources of sounds that goes together with Berkeley's view. But I think he's wrong to think the problem with Berkeley's view is his conception of sounds, rather than his conception of experience. A representational view of auditory experience of the kind I sketched above can accommodate both the claim that sounds seem to be independent of the things that produce them and that we can hear the sources of sounds in hearing the sounds they make.

References

Heidegger, Martin. 1935 [1977]. The origin of the work of art. In *Martin Heidegger: Basic writings*, ed. and trans. D. Farrell Krell. New York: Harper and Row.

Hume, David. 1751 [1999]. *An enquiry concerning human understanding*, ed. Tom L. Beauchamp. Oxford: Oxford University Press.

Lycan, William G. 1996. *Consciousness and experience*. Cambridge, MA: MIT Press.

O'Callaghan, Casey. 2007. *Sounds: A philosophical theory*. Oxford: Oxford University Press.

Scruton, Roger. 2009. Sounds as secondary objects and pure events. In *Sounds and perception*, ed. Matthew Nudds and Casey O'Callaghan. Oxford: Oxford University Press.

Part VIII
Multi-Modal Experience

Chapter 24
Making Sense of Multiple Senses

Kevin Connolly

24.1 Introduction

In the *McGurk effect*, a subject views a video of a person saying one set of syllables (e.g. *ga-ga*), while the audio has been redubbed to a second set of syllables (e.g., *ba-ba*). The subject experiences yet a third set of syllables, distinct from the first two sets (e.g., *da-da*) (McGurk and MacDonald 1976, p. 747). The McGurk effect is a crossmodal experience. Crossmodal experiences are a kind of multimodal experience, that is, a kind of experience that involves more than one sense modality. More precisely put, a crossmodal experience is a kind of multimodal experience where an input in one sense modality changes what you experience in another sense modality. In the McGurk effect, for instance, the visual input of seeing the person mouth *ga-ga* changes the auditory input (*ba-ba*) to what you in fact hear (*da-da*).

Tim Bayne (forthcoming) has recently proposed two different interpretations of crossmodal cases such as the McGurk effect. On a strictly causal interpretation, seeing the person mouth *ga-ga* causes you to hear *da-da* instead of *ba-ba*. According to this interpretation, integration occurs between processing in the auditory system and the visual system (more on this process later), but the result of that processing can be fully decomposed into an audio component and a visual component. So, while the processing is multisensory, the content of that processing is not intrinsically multisensory. On a *constitutive* interpretation, on the other hand, the *ga-ga* visual input and *ba-ba* auditory input give you an experience that has constitutively audio and visual content (not just a conjunction of audio and visual

K. Connolly (✉)
Department of Philosophy, University of Toronto, 170 St. George St, Toronto, ON M5R 2M8, Canada
e-mail: kevin.connolly@utoronto.ca

R. Brown (ed.), *Consciousness Inside and Out: Phenomenology, Neuroscience, and the Nature of Experience*, Studies in Brain and Mind 6, DOI 10.1007/978-94-007-6001-1_24,

content). According to this interpretation, the perceptual state that results from the processing cannot be fully decomposed into two unisensory token states, one auditory state and one visual.

Should we hold a constitutive or causal interpretation of crossmodal cases like the McGurk effect? This question can be re-formulated in the following way: in crossmodal cases, are *constitutively multimodal properties* part of your phenomenal content? There are several ways to understand what it means to be a constitutively multimodal property, and later in the paper, I examine some of these options. To start, one option (very roughly) is to hold that a multimodal property is something over and above the properties contributed by each of the sense modalities involved. In this way, a constitutively audio-visual property would be modeled on flavor properties—properties that are arguably not just the conjunction of the properties contributed by each of the sense modalities involved in flavor perception (taste, touch, and retronasal smell). Like flavor properties, multimodal properties might be defined relative to subjects of experience, or they could be defined as objective kinds (see Smith 2013, for a discussion of this issue for flavors).

What does it mean for a multimodal property to be part of your *phenomenal content*. "Phenomenal content," I will hold, is "that component of a state's representational content which supervenes on its phenomenal character" (Bayne 2009, pp. 386–387)? In a McGurk effect case, for instance, the question is whether there is a constitutively audio-visual property in your phenomenal content, or whether it is just an audio property plus a visual property in your phenomenal content.

We can interpret other crossmodal cases constitutively or causally as well. In the motion-bounce illusion, subjects look at a computer display of two disks moving steadily towards each other until they meet. If the subject hears a sound at or around the point of convergence, the disks typically appear to collide and bounce off one another. If the subject does not hear a sound, the disks appear to cross through one another (Sekuler et al. 1997). According to a strictly causal interpretation, the motion-bounce illusion is a case where the sound simply causes you to have a certain visual experience (given the right visual input). According to a constitutive interpretation, on the other hand, it is a case where you have a constitutively audio-visual experience.

Whether we take a constitutive or causal interpretation of crossmodal cases seems to determine, at least at first glance, whether we hold that some of the content of perception is fundamentally multimodal. If we hold a constitutive interpretation of the McGurk effect, for instance, then we hold that at least some of the content of perception is audio-visual. A strictly causal interpretation, on the other hand, does not commit us to that.[1]

In what follows, I argue against various reasons for thinking that content of crossmodal experiences is fundamentally multimodal. In the next three sections,

---

[1] I owe the basic point behind this paragraph to Susanna Siegel, who made the point at *The Unity of Consciousness and Sensory Integration* Conference at Brown University in November of 2011. In the subsequent discussion, Tim Bayne said he held the constitutive interpretation.

I examine three different reasons one might hold that view, and I argue that none of them actually entail fundamentally multimodal content. I close by trying to make sense of crossmodal cases without appealing to fundamentally multimodal content.

24.2 Is Crossmodal Perception Like Flavor Perception?

The constitutive interpretation of crossmodal cases comes in several different varieties. One variety (the weakest, in my view) models the constitutive interpretation after flavor perception, or, at least, one understanding of flavor perception. Such a view is mentioned, although not endorsed, by Fiona Macpherson (2011, p. 449).

Flavor perception is not the product of a single sense. Rather, it arises from the combination of multiple sense modalities, including taste, touch, and retronasal smell (smell directed internally at the food you have just eaten, rather than at external objects) (Smith 2013). For instance, if you plug your nose entirely while eating an orange, you will not be able to detect the flavor of the orange. This is because the sense of smell is necessary for experiencing the flavor. Without it, there is no flavor experience. Flavor experience arises only through the combination of smell, touch, and taste.

On this interpretation of flavor perception, when a particular flavor perception integrates the properties detected by taste, smell, and touch, it creates a new whole: a flavor property. Fiona Macpherson describes what an account of crossmodal cases would sound like if such cases were modeled after flavor perception:

> [W]e can imagine a case where the new information produced was such that it was none of the above—it could not be produced by a single sensory modality, it did not involve crossmodal content of a binding or other kind—it simply consisted of some brand new content. An example of such a case would be one account of flavour experiences. (2011, p. 449)

If the content in crossmodal cases were like the content of flavor perception, then the content would not simply be the sum of the contents of each of the individual sense modalities involved (like the contents of taste, touch, and retronasal smell in flavor perception), but rather something over and above those contents (like flavors in flavor perception). So, on this way of construing the constitutive view, the content of an experience of the McGurk effect is not just an audio content plus a visual content, but a single, new, audio-visual content.

Are such audio-visual properties part of the content of perception? Consider two other properties first: the property of *being a wren* and the property of *being red*. Even for someone with excellent discrimination, there might be fake wrens that are visually identical to real wrens when examined across all the same lighting conditions and angles. Arguably, this suggests that *being a wren* is not a perceptual property at all. The same conclusion does not follow for properties like colors. There is no such thing as a fake red that is visually identical to an authentic red. The idea is that, for red, if you duplicate its appearance properties, you duplicate the property. On the other hand, there can be visually indistinguishable fake wrens or robot wrens.

For a property like *being a wren*, you can duplicate its appearance properties without duplicating the property. Michael Tye registers the same sort of principle for denying that properties are part of the perceptual content:

> It seems plausible to suppose that the property of being a tiger is not itself a feature represented by the outputs of the sensory modules associated with vision. Our sensory states do not track *this* feature. There might conceivably be creatures other than tigers that look to us phenomenally just like tigers. (1995, p. 141)

On Tye's view, the property of being a tiger is not likely to be represented in vision because you could duplicate every single one of its visual features, and still not duplicate the property of *being a tiger*.

Are some of the contents of perception fused multimodal units (fused audio-visual units, for instance)? I think that the answer is no, and one reason why is grounded in the test just described. Call *Q1*, your experience of the familiar ventriloquist and dummy routine, where you hear the sound of the ventriloquists voice as coming from the dummy's mouth, even though it is actually coming from the ventriloquist's lips. Call *Q2*, an experience of a ventriloquism fakery. The ventriloquist, it turns out, is a fraud, and so he has recorded himself and has placed a speaker playing the recording in the dummy's mouth. Now consider the plausible assumption that *Q1* and *Q2* are phenomenally identical experiences: what it's like in *Q1* is exactly what it's like in *Q2*. But quite plausibly *Q*2 represents just a regular auditory property and a visual property, rather than a fused audio-visual property. If that's right, however, we need not hold that the content of *Q1* involves a fused audio-visual property, since we can explain that phenomenal type in terms of an auditory property and a visual property.

We can arrange the same sort of scenario for the McGurk effect. Call *R1* a particular McGurk effect experience: the experience of a subject who views a video of a person saying *ga-ga*, while the audio has been redubbed *ba-ba*, so that the subject experiences *da-da*. Call *R2*, an experience of a fake McGurk effect. *R2* is the experience of a subject who views a video of a person saying *ga-ga*, while the audio has been redubbed to *da-da* (Note that when this scenario was tested in MacDonald and McGurk, 1978, subjects heard *da-da* 100 % of the time). Now consider the plausible claim that *R1* and *R2* are phenomenally identical experiences. Quite plausibly *R*2 just represents an auditory property (of a person saying *da-da*) and a visual property (of a person saying *ga-ga*), rather than a fused audio-visual property. But then we need not hold that the content of *Q1* involves a fused audio-visual property, since we can explain that phenomenal type in terms of an auditory property and a visual property.

Why think that the above cases should be explained as a conjunction of audio content and visual content, rather than as involving fused audio-visual content? One reason is that everyone agrees that audio and visual properties are represented in perception. Unlike fused audio-visual properties, audio and visual properties are uncontroversial candidates for the content of perception. The question is whether fused audio-visual properties are represented in addition to audio and visual properties, not instead of them. If we reject fused audio-visual content, and appeal instead to audio content and visual content, our account of content is also more economical, since we don't need to posit a new kind of property.

Consider another reason for why fundamentally multimodal properties should not be modeled on flavor properties. In the founding study of the McGurk effect, the authors wrote, "A 'fused' response is one where information from the two modalities is transformed into something new with an element not presented in either modality..." (McGurk and MacDonald 1976, p. 747). Note the sense in which the information is transformed into something new. When a subject experiences the McGurk effect and hears *da-da*, this is a new property in the sense that it is neither the input of the auditory system, nor the input of the visual system. But it is not new in another sense: it *can* be the input of the auditory system, and it *can* be the input of the visual system. Those systems can detect that property. On the other hand, the fusion involved in flavor perception is new in a different sense. It cannot be the input of any of the systems involved (taste, touch, or retronasal smell), since those systems cannot detect flavor properties by themselves. In short, the kind of fusion involved in flavor perception does not occur in crossmodal perception.

In the motion-bounce illusion, the crossmodal influence of the sound serves to modulate the particular motion that you see (you see one motion rather than another). But, of course, in a different context you could have seen that motion. It is a new property in the sense that it is not the input of the visual system in the motion-bounce scenario. But it is not new in another sense: it *can* be the input of the visual system. You do not need crossmodal influence to see the motion that you see. In the ventriloquist effect, the sense of vision influences audition. If you are blindfolded as you enter a movie theater, you will hear the sounds of the movie as coming from the sides of the theater. When you are finally unblindfolded, vision influences your audition. Before, you heard the sounds as coming from the sides of the theater. Afterwards, you hear the sounds as coming from the screen. But you could already detect auditory location. The crossmodal influence serves to modulate the auditory location that you experience, as you perceive a new location for the sound. In the McGurk effect, vision influences audition. If you were to cover your ears and then uncover them while watching the video, your visual experience would not change. On the other hand, if you were to cover your eyes and then uncover them, you would hear different syllables in the two experiences. Your auditory perception changes after you see the person's lips move. You see a person saying one set of syllables, while the audio has been changed to a second set of syllables, but you experience yet a third set of syllables. But again, you could already hear syllables and see someone saying them. The crossmodal influence serves to modulate the syllables that you hear (you hear different syllables before and after you uncover your eyes).

24.3 Do We Perceive Audio-Visual Bounces?

In the motion-bounce illusion, audition influences vision. At first, you see the disks passing through one another. Your visual perception of the disk trajectories changes only after the introduction of a sound, and then you see them as colliding with one another. According to a constitutive interpretation of crossmodal cases, it is a case where you have a constitutively audio-visual experience. One variety of such an interpretation is to hold that *being a bounce* is part of the content, where that property is construed as an audio-visual property. What does it mean to be an audio-visual bounce? Matthew Nudds writes, "We often see something happen and hear a sound, and we perceive the sound to have been produced by what we saw happen, we experience the production of the sound" (2001, p. 218). We might construe the "bounce" in the motion-bounce illusion similarly. The idea is that we see the collision and rebound and hear the sound, and we perceive the sound to be produced by the collision, thereby experiencing the production of the sound. The collision causes the sound in an audio-visual bounce.

Nudds defends the view that we experience the production of sound (as in the audio-visual bounce case) by arguing for the more general claim that we can perceive one event causing another. To this end, he claims that we can perceive scrapes, pushes, squashes, and so on (2001; p. 218). Nudds backs up this claim by saying, "For as long as we allow that people possess and use such concepts [like scrapes, pushes, squashes, etc.] and can apply them to things on the basis of perceiving the interactions between, then we should allow that causality, in this sense, can be perceived" (220). Of course, Nudds is right that no one denies that we possess and correctly apply such concepts. But it doesn't follow that those concepts actually pick out scrapes, pushes, squashes, etc. *as perceptible properties*. Plausibly, like many robust concepts, such as the concept EMPTY GAS TANK, we do not apply them based solely on a perception. Rather, we apply them based on a perception and a background belief. If the concepts SCRAPE, PUSH, and SQUASH are like the concept EMPTY GAS TANK in this way, then while we may possess and correctly apply such concepts, it does not follow that scrapes, pushes, and squashes can be perceived.

In the motion-bounce illusion, it might seem at first glance that your perception represents an audio-visual bounce. My claim is that that does not follow, at least from Nudds' considerations. His argument does not actually show that we can perceive one event causing another, so it does not provide a defense of the claim that we experience the production of sound (as in the audio-visual bounce case). Still, there is something right in what Nudds says: we need to think of crossmodal cases like the motion-bounce illusion as events, if we are to understand them. I explore this idea in the next section.

24.4 Do We Need Multimodal Content to Explain Multisensory Integration?

Crossmodal influence modulates properties for a particular purpose, namely, to reconcile them with the properties in another modality (Matthen et al. 2011). That is to say, crossmodal cases involve multisensory integration: "the brain's ability to synthesize the information that it derives from two or more senses" (Stein et al. 2002, p. 227). But why exactly do the inputs in crossmodal experience require integration or reconciliation? Why do the properties represented by one modality have to align with the properties represented by another at all?

As Casey O'Callaghan points out, "[G]iven divergent auditory and visual stimulation, it only makes sense to attempt in a principled manner to reconcile them if they are assumed to share a common source or cause. Otherwise, the notion that there is a conflict that requires resolution is unintelligible" (2008, p. 326). The idea is that in a crossmodal case, the inputs in two different modalities conflict because they are predicated of a common source or cause (whether it be an individual, object, or event). This conflict requires the reconciliation between the inputs, and what we experience is the product of that reconciliation.

I agree with O'Callaghan's claim that in crossmodal cases, the inputs in two different modalities conflict because they are predicated of a common source or cause (whether it be an individual, object, or event). But my claim is that if O'Callaghan's argument is properly understood, it does not entail that those individuals, objects, or events have multimodal content. Roughly and briefly, this is because O'Callaghan's argument is meant only to undermine the view that the content of perception can be exhausted by unimodal content. But such an argument does not compel us to accept multimodal content. This is because the non-unimodal content could be amodal content (that is, modality-independent content—content not shared by the senses, but rather content that outstrips the senses).

Suppose that for the ventriloquist effect, the motion-bounce illusion, and the McGurk effect you did not experience a crossmodal effect. For instance, suppose that you sit down in a movie theater and see people talking on the screen, and cars exploding, but you hear all of the sounds coming from the sides of the movie theater. It is a very unusual experience to see lips moving and hear a sound consistent with those movements, but coming from a different direction. One way to render the data consistent would be to realize the way that a sound system is set up in a movie theater. Instead of this, your sensory system reconciles the auditory and visual inputs for you. You hear the sounds as coming from the screen (although they are coming from the side of the theater).

To take another example, suppose that in the motion-bounce scenario, you simply heard a random sound when the disks intersected, and experienced the disks as crossing through each other rather than bouncing. Once again, that data would require reconciliation. Why was there a random sound? As with the ventriloquist effect, in the motion-bounce illusion, your sensory system reconciles the data. You see the disks as colliding with one another. The sound is heard as the sound of a collision. This makes sense of the random sound.

Suppose that in the McGurk scenario, you saw someone mouthing the syllables *ga-ga*, but heard someone repeating the syllables *ba-ba*. That data would require reconciliation. Typically you hear the syllables that you see a person mouthing, not some other syllables. Seeing someone mouth *ga-ga* while hearing *ba-ba* requires reconciliation. In the McGurk effect, your sensory system performs that task. Importantly, however, even though you are looking at someone mouthing the syllables *ga-ga*, your sensory system does not reconcile that by having you hear the syllables *ga-ga*. Instead, you hear the syllables *da-da*. This might seem to suggest that the auditory and visual inputs are left unreconciled. But McGurk and MacDonald suggest an alternative hypothesis:

> [I]n a ba-voice/ga-lips presentation, there is visual information for [ga] and [da] and auditory information with features common to [da] and [ba]. By responding to the common information in both modalities, a subject would arrive at the unifying percept [da] (1976, p. 747).

When you hear *da-da*, McGurk and MacDonald suggest, this is not a failure to reconcile the ba-voice and the ga-lips. Rather, the ba-voice actually contains some informational features of the sound *da-da*, while the ga-lips contain some informational features of seeing someone say *da-da*. When you hear *da-da*, McGurk and MacDonald claim, you are reconciling auditory and visual data through their common informational features (I explain this further in the next section).

In crossmodal cases, the inputs in two different modalities conflict because they are predicated of a common source or cause (whether it be an individual, object, or event). It might seem at first glance that if we posit individuals, objects, or events as the common source or cause in crossmodal cases, we are positing multimodal content. O'Callaghan, however, is careful not to make that inference. Rather, he says, "[T]here is a dimension or component of perceptual content that must be characterized in multi-modal *or modality-independent terms*. This component either is shared by both vision and audition *or outstrips both the visual and the auditory*" (2008, p. 328, italics added for emphasis; see also pp. 327–332, and O'Callaghan forthcoming, section 5.2). O'Callaghan's point is that we can construe the individuals, objects, or events in two different ways: *either* as both the content of modality one (e.g., audio content) and the content of modality two (e.g., visual content) *or* as neither the content of modality one nor the content of modality two but as content that outstrips them both. If we characterize the individuals, objects, or events in the second way, that is, in modality-independent terms, then we are not positing multi-modal content. We are positing amodal content.

Let's return to the two rival interpretations of crossmodal cases from the introduction of the paper. The idea was that we can take either a constitutive or causal interpretation of crossmodal cases, and that that determines whether we hold that the phenomenal contents of crossmodal experiences are constitutively multi-modal, or whether they are just unimodal. The assumption was that in a McGurk effect case, for instance, there is either a fundamentally multimodal audio-visual property in your phenomenal content or else just an audio property plus a visual property. But suppose that we hold, following O'Callaghan, that crossmodal cases require us to posit individuals, objects, or events so that we can make sense of why reconciliation needs to occur in the first place. Suppose also that we characterize those individuals, objects, and events in modality-independent terms. We then end up with a new position, one where the content of crossmodal cases is neither multimodal, nor simply unimodal, but rather amodal. The lesson is this: O'Callaghan's claim is that we need to posit some sort of common content, shared by different sense modalities, in order to explain why reconciliation needs to occur in crossmodal cases in the first place. But shared content does not entail multi-modal content.

O'Callaghan's main goal in his 2008 article is to argue against the view that unimodal content exhausts perceptual content. As he puts it:

> I wish to argue that understanding cases of cross-modal perception grounds an argument for the claim that there exist consciously accessible aspects of perceptual experience that are not unique or specific to a given experiential modality and that may be shared across modalities. The argument proceeds in two stages. The first aims to show that that there is a dimension or component of perceptual content that must be characterized in multi- modal or modality-independent terms. This component either is shared by both vision and audition or outstrips both the visual and the auditory. (p. 328)

Given that his goal is to argue against the view that unimodal content exhausts perceptual content, O'Callaghan seems satisfied to accept either multimodal or modality-independent (amodal) content, since both are non-unimodal content. At the same time, he clearly does distinguish between the two options. Multimodal content is shared by, say, both vision and audition, while modality-independent (amodal) content outstrips both vision and audition.

24.5 Crossmodal Cases Without Fundamentally Multimodal Content

So far I have argued against various reasons for thinking that crossmodal cases show that at least some of the content of perception is fundamentally multimodal—that is, reasons for thinking that your experience has, say, constitutively audio-visual content (not just a conjunction of an audio content and visual content). I now want to try to make some sense of crossmodal cases without appealing to fundamentally multimodal content.

A 2004 study at Oxford's Crossmodal Research Lab showed that hearing an augmented sound of a crunch makes soft potato chips seem crisper and stale chips seem fresher (Zampini and Spence 2004). In that study, a higher volume of a crunch sound correlated with the chips seeming crisper and fresher, while a lower volume correlated with the chips seeming softer and staler. The study showed that the sensory system is able to reconcile auditory data with gustatory data, in this case by modulating the experience of crispness or freshness.

Take a particular class of perceptible properties (the class of colors, or shapes, or sizes, or locations, or orientations, for instance), and for a substantial portion of its members *x*, *y*, and *z*, *x* is more similar to *y* than it is to *z*. For instance (as a first approximation), for colors, orange is more similar to red than it is to blue. For size, a peanut is more similar to a watermelon than it is to the Empire State Building. A more precise examination of similarity orderings shows that they are often multi-dimensional. Colors, for instance, are comparable along the dimensions of brightness, saturation, and hue (Matthen 2005, p. 111). By utilizing those three dimensions, for a substantial portion of colors *x*, *y*, and *z*, *x* will be more similar to *y* than it is to *z*.

In what follows, I want to show how such a similarity structure might help us to understand crossmodal cases. For the class of crisp things, for instance, we can say of a substantial amount of its members that *x* is more similar in crispness to *y* than it is to *z*. In the Zampini and Spence study, as one's sensory system reconciles a flavor with a sound, the flavor appears more crisp or less crisp, more fresh or less fresh. In everyday situations (outside of the experimental context), when you hear a crunch sound of magnitude *x*, there would be a correlating magnitude of crispness *y*. In the experimental context, when you hear an augmented crunch sound of magnitude *x*, the actual magnitude of the crispness is less than *y*, but you perceive something more similar in magnitude to *y*.

Put another way, modality one detects a property (crunch volume) that can be located on a similarity space. Modality two detects a different property (crispness) that can be located on a similarity space. Certain points on each similarity space correlate with particular points on the other similarity space (crunchiness of magnitude X with crispness of magnitude Y, e.g.). A plausible story is that through learned experience, you build an association between the crunchiness of magnitude X and the crispness of magnitude Y. In crossmodal cases, each modality detects a particular property, one on each of the spaces (the crunchiness space and the crispness space), and these are properties that do not typically correlate. The crossmodal effect is to shift one of the properties, in experience, such that it is closer to its correlating point with the other experienced property. At bottom, this is just a shift along the continuum for a type of property that is already represented in perception. It is not any new kind of property.

My proposal is that hearing an augmented sound of a crunch can make stale potato chips seem crisper because crispness is a kind of property that *can* be reconciled with an aberrant crunch sound of magnitude *x*. Specifically, it can be made more similar to the magnitude of crispness that typically corresponds with the magnitude of that sound. In the Zampini and Spence study, the same holds, *mutatis mutandis*, for the property of freshness.

But now consider our three crossmodal cases as cases that aim at data reconciliation. In the McGurk effect, as your sensory system reconciles a sound with a visual image, it modulates the sound. In most everyday situations (outside of the experimental context), when you see someone mouthing the syllables *ga-ga*, there would be a correlating sound: *ga-ga*. In the experimental context, when you see someone mouthing the syllables *ga-ga*, the actual sound is *ba-ba*, but you hear something more similar to *ga-ga*, namely, *da-da* (I will motivate the claim that these two sounds are more similar shortly).

We know from our own experience that some words sound more similar to each other than others. One piece of evidence for this is that we confuse some words with each other when we hear them, but do not confuse other words with each other. If we break down spoken words into their units, we can tell the same sort of story about these units, or *phonemes*. A phoneme *x* can sound more similar to a phoneme *y* than to another phoneme *z*. Todd M. Bailey and Ulrike Hahn have charted the similarity relations between phonemes in great detail (Bailey and Hahn 2005; Hahn and Bailey 2005). For instance, they argue that "/t/ is more similar to /d/ than to /l/" (where "/t/" represents a phoneme of t) (Bailey and Hahn 2005, p. 339). According to them, this is why "tuck" sounds more similar to "duck" than it does to "luck." Phoneme similarity helps to explain why we sometimes confuse certain words when we hear them, but not others.

We need not commit to a single unified phoneme space, where each phoneme can be ordered in relation to every other phoneme (just as every color can be ordered in relation to every other color). Still, we can say that there are phoneme *spaces*. To use Bailey and Hahn's example, /t/ is more similar to /d/ than to /l/. My claim is that the McGurk effect exploits such spaces. *Da-da* sounds more similar to *ga-ga* than *ba-ba* does. This account dovetails with McGurk and MacDonald's account of the McGurk effect. They speculate that "the acoustic waveform for [ba] contains features in common with that for [da] but not with [ga] . . ." (1976, p. 747). On their view, the similar acoustic waveform is what accounts for the similar sounds of *ba-ba* and *da-da*.

In the McGurk effect, the audio plays one sound (e.g., *ba-ba*), and the visual shows someone mouthing a second sound (e.g., *ga-ga*), but you hear yet a third sound (e.g., *da-da*). My suggestion is that your sensory system reconciles the aberrant sound (*ba-ba*) by making it more similar to the sound that typically would correspond with the image that you see (*ga-ga*). *Da-da* sounds more similar to *ga-ga* than *ba-ba* does.

According to McGurk and MacDonald, the ga-lips also contribute to data reconciliation in the McGurk effect (1976, p. 747). As I mentioned, they claim that the sound *ba-ba* shares some informational features in common with the sound *da-da* (they put this point in terms of a similar acoustic waveform). But they also claim that seeing someone say *ga-ga* shares some informational features with seeing someone say *da-da* (they cite the fact that lip movements for *ga-ga* are frequently misread as lip movements for *da-da*). According to their explanation, hearing *da-da* provides a unique solution to the conflicting visual and auditory data. It reconciles the auditory and visual data through their common informational features.

Typically, when you see someone mouthing "ga-ga," you hear the sound "ga-ga." Notice that in the McGurk effect, the association between seeing someone mouth "ga-ga" and hearing "ga-ga" is not strong enough to make someone hear "ga-ga." Instead you hear "da-da" when the audio is "ba-ba." Still, the weight of the association between seeing someone mouth "ga-ga" and hearing "ga-ga" is strong enough to shift the heard property from ba-ba (which is the input) along a perceptual dimension to da-da (which is what is heard). Why is the auditory pull from ba-ba to da-da, and not all the way to ga-ga? I think the similarity space makes sense of this.

Auditorily, ga-ga is more similar to da-da, than it is to ba-ba. The crossmodal effect is to shift the auditory property, in experience, such that it is closer to its correlating point with the other experienced property, the visual property. What you hear in the McGurk effect is more similar to the auditory correlate of what you see. Again, this is just a shift along the continuum for a type of property that is already represented in perception, rather than a new kind of property.

In the ventriloquist effect, as your sensory system reconciles an auditory location with what you see, it modulates the auditory location. Typically, when you see lips moving and hear a sound consistent with the lip movements, the location of that sound is the moving lips. In the ventriloquist effect, when you see the lip movements, the actual auditory location is from elsewhere, but you experience the location as from the moving lips. The ventriloquist effect operates on auditory location. In the ventriloquist effect, your sensory system reconciles an aberrant auditory location (e.g., the location of the sides of a movie theater) by making it more similar to the auditory location that typically would correspond with the image that you see (e.g., the movie screen).

Both the McGurk effect and the ventriloquist effect are cases where auditory and visual data conflict, and in both cases, vision is dominant. That is, in both cases, the auditory data reconciles with the visual data. Vision is not always dominant, however. In the motion-bounce illusion, for instance, as your sensory systems reconcile a visual image with what you hear, it modulates the visual image. Typically, when you see two objects coincide and hear a sound when they do, you see the motion we call "bouncing." But in the motion-bounce illusion, when you see the two objects coincide, you hear a random sound, but you experience the "bouncing" visual motion. In the motion-bounce illusion, your sensory system reconciles an aberrant sound by making the image that you see more similar to the visual motion that would typically correspond with that sound (a "bouncing" motion).

24.6 Conclusion

I have argued against various reasons for thinking that crossmodal cases show that at least some of the content of perception is fundamentally multimodal—that is, reasons for thinking that your experience has, say, constitutively audio-visual content (not just a conjunction of an audio content and visual content). In Sects. 24.2, 24.3, and 24.4, I presented three different reasons for thinking that content of crossmodal experiences is fundamentally multimodal. My claim was that none of these reasons actually entail the conclusion that crossmodal experiences involve fundamentally multimodal content. These reasons do not show that cases like the ventriloquist effect, the McGurk effect, and the motion-bounce illusion must involve fundamentally multimodal content. In Sect. 24.5, I then tried to make some sense of crossmodal cases without making reference such content. This is just a start, but it yields a general two-pronged approach. The first prong is to evoke unimodal features (such as crunchiness and crispness in the Zampini and Spence case). But a unimodal approach is not in itself sufficient. For as O'Callaghan points out, in crossmodal cases, the inputs in two different modalities conflict because they are predicated of a common source or cause (whether it be an individual, object, or event). The second prong is to posit individuals, objects, and events, conceived of in amodal terms (that is, as modality-independent content—content not shared by the senses, but rather content that outstrips the senses). Such an account steers clear of what I was trying to avoid. Making sense of crossmodal cases does not require us to posit multimodal content.[2]

References

Bailey, T.M., and U. Hahn. 2005. Phoneme similarity and confusability. *Journal of Memory and Language* 52(3): 339–362.

Bayne, Tim. 2009. Perception and the reach of phenomenal content. *The Philosophical Quarterly* 59(236): 385–404.

Bayne, Tim. Forthcoming. Building block or unified fields: How should we model the unity of consciousness? In *Sensory integration and the unity of consciousness*, eds. D. Bennett and C. Hill. MIT Press.

Hahn, U., and T.M. Bailey. 2005. What makes words sound similar? *Cognition* 97(3): 227–267.

Macpherson, Fiona. 2011. Cross-modal experiences. *Proceedings of the Aristotelian Society* 111(3): 429–468.

Matthen, Mohan. 2005. *Seeing, doing, knowing: A philosophical theory of sense perception*. Oxford: Clarendon.

Matthen, Mohan, Alex Byrne, Fiona Macpherson, Susanna Siegel, and Barry Smith. (2011). *Description of activities for a partnership development grant from the social sciences and humanities research council of Canada*. Unpublished Grant Proposal.

McGurk, H., and J. MacDonald. 1976. Hearing lips and seeing voices. *Nature* 264: 746–748.

Nudds, Matthew. 2001. Experiencing the production of sounds. *European Journal of Philosophy* 9(2): 210–229.

O'Callaghan, C. 2008. Seeing what you hear: Crossmodal illusions and perception. *Philosophical Issues* 18: 316–338.

O'Callaghan, C. (forthcoming). Perception and multimodality. In *Oxford handbook to philosophy and cognitive science*, eds. Eric Margolis, Richard Samuels, and Stephen Stich. New York: Oxford University Press.

Sekuler, R., A.B. Sekuler, and R. Lau. 1997. Sound alters visual motion perception. *Nature* 385: 308.

Smith, Barry C. 2013. Taste, philosophical perspectives. In *Encyclopedia of the mind*, ed. Harold E. Pashler. Thousand Oaks: SAGE Publications, Inc.

[2]This paper benefited due to comments from Matthew Fulkerson, Bernard Katz, Eric Liu, Mohan Matthen, Barry C. Smith, and Charles Spence.

Stein, Barry E., Paul J. Laurienti, Mark T. Wallace, and Terrence R. Stanford. 2002. Multisensory integration. In *Encyclopedia of the human brain*, ed. V. S. Ramachandran, 227–241. New York: Academic Press.

Tye, Michael. 1995. *Ten problems of consciousness*. Cambridge: MIT Press.

Zampini, M., and C. Spence. 2004. The role of auditory cues in modulating the perceived crispness and staleness of potato chips. *Journal of Sensory Studies* 19: 347–363.

Chapter 25
Explaining Multisensory Experience

Comment on Kevin Connolly's "Making Sense of Multiple Senses"

Matthew Fulkerson

25.1 Introduction

Our experience of the world involves a number of senses, including (but perhaps not limited to) sight, hearing, touch, taste, and smell. These senses are not isolated from one another. They work together, providing a robust and coherent awareness of our environment. Consider entering a good restaurant: one sees the décor and the other patrons, smells the pleasing odors wafting from the kitchen, hears the pleasant music and sound of conversation, feels the comfort of the seating, and, finally, savors the taste of the food. It seems obvious that, in some sense at least, our perceptual awareness of the restaurant is multisensory. Saying exactly what it is for perceptual awareness to be multisensory is more challenging than it appears, however.

One might suppose, for instance, that there is no single "experience of the restaurant." To say that our awareness of the restaurant is multisensory is just shorthand for saying that it involved many distinct perceptual experiences contributed by different senses. This notion of what it means for an experience to be multisensory is not especially robust or interesting: we have one experience and then another, or perhaps we have several different experiences at the same time.[1] In some cases, this probably is what we mean by multisensory.[2] But this can't be the whole story. Consider what happens when the food is tasted: at this moment the aroma, taste, feel, and temperature seem to blend into a novel whole. Anyone who has eaten a favorite

[1]As we'll see, this way of carving up perception into separate 'experiences' generates some problems in our understanding of multisensory interaction. See Byrne (2009) for a pointed criticism of the philosophical notion of 'experience.'

[2]As when a restaurant critic describes the overal meal as a delightful, 'multisensory' experience.

M. Fulkerson (✉)
Department of Philosophy, UCSD, 9500 Gilman Drive #0119, La Jolla, CA 92093, USA
e-mail: mtfulkerson@gmail.com

R. Brown (ed.), *Consciousness Inside and Out: Phenomenology, Neuroscience, and the Nature of Experience*, Studies in Brain and Mind 6, DOI 10.1007/978-94-007-6001-1_25,

meal with a bad cold, or when it is at the wrong temperature, or after it's been ground up in a food processor, can attest to the influence of many senses on our experience of food. In these cases, it does not seem as though there are several separate experiences going on at the same time, but rather that there is one, unified experience of the flavor that results from the coordinated operation of more than one sense (Auvray and Spence 2008). Call these types of multisensory experience *multimodal*. Such experiences are not at all explained by the mere conjunction of distinct sensory experiences.

In addition to multimodal experiences, there are also cases of *crossmodal* experience, where the operations of one sensory modality influence or make a difference in the operations of another. Empirical studies reveal that the senses have strong influences on one another (see, e.g., Calvert and Thesen 2004; Spence and Driver 2000; Ernst et al. 2007). In crossmodal cases too, something more than mere conjunction of distinct perceptual experiences seems required. But how exactly ought we distinguish those cases where the senses are merely co-occurrent from those where they somehow blend into one another, from those that have strong influences on each other?[3] And what to make of the many other forms of interaction that are also more than mere conjunctions, but that do not fully blend into single experiences or involve direct influence on another modality?

The recent realization that perceptual modalities are often deeply intertwined might lead some to call for abandoning the very notion of a unisensory experience (see e.g., Shimojo and Shams 2001; also Spence and Driver 2000). Multisensory experience, on this perspective, requires radically jettisoning our standard conceptions of perceptual experience, at least for a range of paradigm multisensory interactions. For convenience, call those who want to resist such radical moves "sensory conservatives" and the view they defend "sensory conservatism." (A defender of sensory conservatism for vision is Pylyshyn 2006). In this volume, Kevin Connolly defends a limited sensory conservatism for crossmodal (but not multimodal) experiences. Now, to be clear, sensory conservatism does not deny that many of our perceptual experiences are multisensory. It simply takes such experiences to be nothing more than conjunctions of unisensory experiences (or, more accurately, to be a complex formed *somehow* by a combination of nothing but unisensory components). The sensory conservative holds the intuitive and highly plausible view that a perceptual experience is multisensory if it involves more than one sense.

As a sensory moderate, I welcome the parsimony and intuitive appeal of the conservative viewpoint, but believe it ultimately fails to do justice to the complexity and messiness of actual sensory interactions. On the other hand, I don't think we ought to abandon the very concept of a sensory modality as the radicals suggest. Making space in this middle ground is not easy, since,

[3] I attempt in Fulkerson (2011) to describe and motivate one way of distinguishing these distinct forms of multisensory interaction.

unlike the conservative, we can't simply take for granted that there are perfectly individuated, informationally-encapsulated sensory modalities. For this reason, we cannot account for multisensory interactions as mere conjunctions of such constituents without saying quite a bit more about how we are carving up the individual modalities and experiences. After all, whether there are such constituents is one of the main ongoing theoretical questions.

At any rate, I will set these worries aside for now, and focus my efforts elsewhere. In particular, I distinguish two versions of sensory conservatism, and use these more precise formulations to put some pressure on the conservative position. As we'll see, Connolly denies the first version, and only accepts the second for a limited range of cases. This makes him a rather lukewarm defender of the conservative line. Indeed, there is an inherent tension in both trying to defend the conservative position because of its relative simplicity and explanatory parsimony, while ackowledging that the view is true only for a limited range of cases.

25.2 Clarifying the Target: Sensory Conservatism

First, make the simplifying assumption that there are only five distinct sensory modalities: audition, vision, touch, olfaction, and gustation. Second, let each modality have a set of sensible features, so that $\{a_1 \ldots a_n\}$ is the set of sensible features available to audition; $\{v_1 \ldots v_2\}$ is the set of features available to vision, and so on. For convenience, label these sets of sensible features A, V, T, O, and G. Finally, a perceptual experience E has content E(F), where F is the set $\{f_1 \ldots f_n\}$ of sensible features represented by E (for my purposes, this is equivalent to the claim that F is the *content* of E).[4]

Using this terminology, we can say that for a sensory conservative an experience E(F) is multisensory if F contains sensible features from more than one sensory modality. So an experience that represents a blue dot and a C# is multisensory, since it represents features available to two distinct sensory modalities (I discuss this issue in more detail in my 2011). This account is intuitive inasmuch as it assumes that we already have a good grasp of how to individuate sensory modalities and the sensible features that belong to them.

When Connolly suggests that crossmodal cases can be entirely explained by appeal to unimodal features, he is defending perceptual conservatism with respect to

[4]Of course, these are gross simplifications. In addition to represented features, perceptual content will also have a spatial distribution and (perhaps many) internal relations (like binding). And there are likely many more than five modalities, often with obscure or entangled contents. I focus in what follows on this simplified account, since I believe it helps clarify the position Connolly defends. But as we'll see, these simplifications can make the conservative viewpoint seem more plausible than it actually is.

certain crossmodal cases. In order to properly assess this position, we must clarify the issues and present as precise a formulation of the options as possible.[5]

We must distinguish two versions of sensory conservatism. The first is defined by the following thesis:

The Proprietary Content Thesis (PC): No sensory content is shared among the senses.

The paradigm for this kind of view is Fodor's (1981) classic modular view of the senses as (among other things) hard-wired, informationally-encapsulated, domain specific input systems. Still, we can give a more precise formulation. Let P be the collection of modality-specific sensory feature sets (so, e.g., P_V is the set of visual features; P_A is the set of auditory features, etc.). Now, PC is the claim that the members of P are pairwise disjoint: $P_i \cap P_j = \oslash$. That is, there are no sensory features found in more than one modality, and therefore no sensory content (representing sensory features) is to be found in more than one sensory modality.

PC is one way to resist the idea that perceptual experiences are inherently or radically multisensory. If an experience contains content from more than one sensory modality, then, according to PC, we can always *decompose* this content into its consituent modalities. This accords well with the intuitive idea that a multisensory experience is nothing more than a combination of unisensory components. Note that PC can be violated, and the conservative position undermined, even if we restrict the represented properties to the so-called 'basic' sensibles (if, for example, two or more senses represent such basic features as *number*, *location*, *size*, or *shape*, then PC would be violated). According to PC, even in cases where the senses seem to represent the same sensory features, they do not literally *share* content. Instead, each sensory modality represents that feature in its own proprietary format (this story can be told in several different ways).[6]

The second version of sensory conservatism is defined by the following thesis:

The Exhaustive Content Thesis (EC): The content of perceptual experience consists only in the sensible features found in the individual modalities.

More formally, for any perceptual experience E(F), for all $f \in F$, $f \in P$. Note that the truth of EC is independent of the truth of PC. First, EC can be false when PC is true. This would be the case, for instance, if there were novel features not available in any of the individual modalities, but which occur in (multisensory) perceptual experiences. Call this the possibility of *multimodally emergent* content. Conversely, there could be contents shared among the individual senses (violating PC), yet

[5]What follows is my attempt to give a more precise and careful account of the view Connolly defends. It is hoped that it avoids some of the many difficulties that arise trying to discuss relations between experiences, where many levels of explanation (including qualitative, informational, and functional) interact.

[6]The main argument of O'Callaghan (2008) seems to be directed at PC. He argues after all that there must be some "shared content" between distinct sensory modalities. Such contents are, in a real sense, inherently multisensory.

these modality-specific features may nevertheless exhaust the contents of perception (ensuring the truth of EC). In all likelyhood, the most interesting and cohesive versions of sensory conservatism will endorse *both* PC and EC.[7]

So where does Connolly stand on these theses? It seems he believes that both PC and EC are false for some range of perceptual experiences. He agrees with Nudds (2001), O'Callaghan (2008), and others that PC is (perhaps often) violated. Indeed, it's *very* difficult to see how one could defend PC without subscribing to either a extremely naive view of the senses or some version of Fodorian modularism (e.g., Pylyshyn 2006).[8] But Connolly also thinks EC is false, at least for our experiences of flavor. Flavor experiences, he allows, seem to have emergent content that is not found in P. Since he denies both theses associated with sensory conservatism, it's difficult to see any general theoretical motivation behind Connolly's position.

Instead, Connolly's focus is on particular cases: whereas some (notably for him, Bayne) seem to hold that certain paradigm crossmodal experiences violate EC, Connolly does not believe they do. The paradigm cases he discusses are the McGurk Effect, The Motion-Bounce Illusion, and the Ventriliquism Effect.[9] It's not clear if these cases are thought to form a natural kind, allowing us to generalize to the falsity of EC for all crossmodal cases, or if his claims are just restricted to these three cases. At any rate, we can ask what hangs on whether these crossmodal cases involve emergent content or not? Connolly seemingly asks this question: *can* we explain crossmodal cases as representing only features in P? He then proceeds to give a consistent account of just such an explanation. Yet it's not clear what hangs on this mere possibility, especially if we allow that the members of P may be shared among multiple sensory modalities. Without maintaining PC, several modalities could share complex sensory features like *cause* or *speech* or even *bounce*. Indeed, such shared contents offer the best explanation of crossmodal experiences, since differences in the shared or overlapping contents would explain the need for such contents to be reconciled. There are many known mechanisms for such reconciliation, usually falling under the rubric of *multisensory integration*. Such mechanisms include sensory suppression, dominance, and facilitation (see Calvert et al. 2004). This is an interesting and rich notion of multisensory interaction, one that would pose many problems for a typical sensory conservative. But Connolly does not really take issue with this possibility. In fact, he seems to think it is right.[10]

---

[7] I believe these theses are not always clearly distinguished in Connolly's discussion.

[8] To appreciate how naive the view would have to be, consider that even Aristotle's account of the senses (which individuates them according to their unique or 'proper' sensibles) allowed that the senses shared sensory features, which he called 'common' sensibles.

[9] As he does a fine job describing these effects, I will not redescribe them here.

[10] I base this claim on the following (representative) passage: "The idea is that in a crossmodal case, the inputs in two different modalities conflict because they are predicated of a common source or cause (whether it be an individual, object, or event). This conflict requires the reconciliation between the inputs, and what we experience is the product of that reconciliation" (p. 357). Now, it's possible that the senses could predicate features of the same objects and events without sharing content, and perhaps this is Connolly's view (and so maybe he wants to also deny PC). But such a

His main target is the idea (contra EC) that there are emergent sensible features (e.g., *sound-sources, bounces*) that are represented only when more than one sensory modality is combined. He thinks this idea is false for the paradigm crossmodal illusions (though it may well be true for other multimodal experiences). Connolly's argument, essentially, is that we can explain crossmodal cases by appeal only to unisensory contents. Whenever there is a purported content not in P, he suggests we can find some conjunction of elements in P that would do as well explaining the target experience. There is no need, he suggests, to posit such emergent content. In order to properly assess this claim, we need to discharge some of our simplifications.

If a unisensory experience just is one representing only members of V or T or G, then *what*, if EC is false, is being combined to generate the emergent content? Experiences qua experiences are not at the right level of explanation, and neither is the level of content. But Connolly seems concerned about accounting for "a new kind of property," as though what is being considered is the possibility that two distinct sensory features can be combined in experience and thereby generate a novel, emergent sensible feature. This kind of metaphysical emergence *would* be quite mysterious. But multimodally emergent content is not metaphysically emergent at all. In fact, such emergent content is entirely metaphysically innocent. We are talking about perception as an information processing/representational system, and there is nothing ontologically problematic about a sensory process that combines distinct sensory inputs and produces simplified, coherent representations of more complex features. This confusion is apparent in the way Connolly frames the debate in terms of a causal-constitutive distinction (a distinction at the level of processes) but then worries about novel properties (at the level of ontology). The causal-constitutive distinction does not make any sense at the level of features or experience, only at some lower level of functional or physiological sensory processing, where there are no worries at all about ontological parsimony or economy. Consider an analogous case of testimony. I hear from one source that my friend Jill is in the next room. I hear from another source that Jill is expecting a child. From this I come to the belief that *a pregnant woman is in the next room.* The content of this belief is not provided by either source alone, nor is it merely a conjunction of the two source reports. But there is nothing ontologically mysterious about where the belief gets its content![11]

Setting aside the misplaced metaphysical worries about ontological austerity, it must be that, for some range of typical crossmodal cases, Connolly denies that there are any processes or subsystems that serve to represent content over and above the

---

view is only plausible if we assume that sensible features alone enter into perceptual content. But as noted above, this is a gross simplification. Indeed, to even make sense of co-predication, some shared spatial or temporal contents will be required. For this reason, I believe this quote commits Connolly to the falsity of PC.

[11]The analogy with testimony can be fruitfully extended to cover crossmodal cases: such cases arise when the testimony of multiple senses come into conflict. When one sense/source is more trusted, cases of sensory dominance arise, generating typical crossmodal illusions. Notice that such an account only makes sense if the two sources "share" content.

content available in the individual modalities.[12] So, he is denying that there are any modality-independent "causal detectors" or similar higher-level systems that take as input perceptual information from multiple sources and have as output novel contents. For example, a system taking as input an auditory speech sound and a visual mouth movement and (using temporal and spatial proximity among other factors) generating as output the novel multimodal content *sound-source*.

It's not clear that we are given much reason for rejecting such modality-independent systems (indeed, as noted above, Connolly seems to let them in by accident). The main motivation seems to be that it is *possible* that there are no such systems, and as this would be a more elegant or parsimonious account, we ought to accept it. As he writes, "If we reject fused audio-visual content, and appeal instead to audio content and visual content, our account of content is also more economical, since we don't need to posit a new kind of property" (p. 355). But ultimately this is an empirical question (though one desperately in need of conceptual clarification). And there is little reason to think our perceptual systems are wholly subject to constraints of parsimony (if they were, wouldn't we have only a single sensory modality?). And there seems to be strong evidence that there are systems above the level of the individual modalities that both make a difference to and contribute to perceptual contents. Once the door is opened to such systems (for example, for flavor perception, causal awareness, etc.) it's very difficult to use parsimony as a reason for closing the door in other cases. And as noted, there are few general theoretical motivations for the sensory conservative view. Ultimately, I don't see us adjudicating these matters through philosophical argument alone. While much work is needed to clarify our target and to distinguish the various forms of sensory interaction, real progress will be made only when we applying these more precise formulations to the actual data. We need to look more closely at the empirical evidence—the constituent sensory systems and their interactions—in order to tease apart the contributions of the individual modalities from the higher-level systems that integrate and coordinate the input from those modalities.

25.3 Conclusion

We find ourselves in the following position: our experience of the world is largely multisensory. We experience the world with all of our senses, and these senses interact at many levels of processing, and in many different ways. The empirical research literature, and philosophy of mind along with it, is beginning to recognize

[12]Actually, Connolly *seems* even to allow such contents that 'outstrip' the sensory modalities, without realizing that this would be a violatin of EC. He writes, "If we characterize the individuals, objects, or events in the second way, that is, in modality-independent terms, then we are not positing multimodal content. We are positing amodal content" (13). But amodal content of this kind is clearly emergent content. If this is right, then it's simply not clear what position Connolly intends to defend.

the importance of these interactions. Instead of focusing on the individual senses in unrealistic isolation (along the lines popularized by Fodor's modular account), philosophers and psychologists are starting to take seriously the idea that the individual senses are deeply intertwined, and that our perceptual experience is inherently multisensory. This move towards a more interactive perspective might make it seem as though there were a clearly defined notion of multisensory interaction, and that it can be easily contrasted with our intuitive notion of unisensory experience. While it may be true, in some narrow sense, that all of our perceptual experiences are multisensory, in reality our perceptual experiences are subserved by a wide range of distinct sensory interactions. Sometimes, for some purposes, we focus on mere conjunctions, other times we focus on the functional interactions between the systems that generate our perceptual experiences, and yet other times we focus on the contents of our experiential states. These different purposes yield distinct and often incompatible notions of multisensory interaction.

The embrace of a multisensory perspective of perceptual experience is a good thing. The idea that the senses are wholly separate and disconnected forms of experience is surely wrong. But we should not suppose on this basis that there are no sensory modalities. The senses are a messy, heterogeneous, complex jumble of distinct interactions at many levels of explanation, but depending on our purposes we can isolate patterns of unity and coherence characteristic of our intuitive notion of the senses. Understanding sensory experience requires that we look more closely at the specific (and pervasive) interactions between the senses, but also that we keep in mind those elements that make the individual senses significant and interesting. That is, we ought to preserve what is important about the individual senses while acknowledging the many varied interactions between them. This is sensory moderation. It is, of course, a difficult task, but these are still early days in our thinking about sensory interaction.

References

Auvray, M., and C. Spence. 2008. The multisensory perception of flavor. *Consciousness and Cognition* 17(3): 1016–1031.

Byrne, A. 2009. Experience and content. *The Philosophical Quarterly* 59(236): 429–451.

Calvert, G., and T. Thesen. 2004. Multisensory integration: Methodological approaches and emerging principles in the human brain. *Journal of Physiology-Paris* 98: 191–205.

Calvert, G., C. Spence, and B.E. Stein. 2004. *The handbook of multisensory processes*. Cambridge, MA: MIT Press.

Ernst, M.O., C. Lange, and F.N. Newell. 2007. Multisensory recognition of actively explored objects. *Canadian Journal of Experimental Psychology* 61(3): 242–253.

Fodor, J. 1981. *The modularity of mind*. Cambridge, MA: MIT Press.

Fulkerson, M. 2011. The unity of haptic touch. *Philosophical Psychology* 24(4): 493–516.

Nudds, M. 2001. Experiencing the production of sounds. *European Journal of Philosophy* 9(2): 210–229.

O'Callaghan, C. 2008. Seeing what you hear: Crossmodal illusions and perception. *Philosophical Issues* 18: 316–338.

Pylyshyn, Z. 2006. *Seeing and visualizing: It's not what you think*. Cambridge, MA: MIT Press.
Shimojo, S., and L. Shams. 2001. Sensory modalities are not separate modalities: Plasticity and interactions. *Current Opinion in Neurobiology* 11(4): 505–509.
Spence, C., and J. Driver. 2000. Attracting attention to the illusory location of a sound: Reflexive crossmodal orienting and ventriloquism. *Neuroreport* 11: 2057–2061.

Part IX
Synesthesia

Chapter 26
Seeing as a Non-Experiental Mental State: The Case from Synesthesia and Visual Imagery

Berit Brogaard

26.1 Seeing: The Traditional View

According to a traditional analysis of the verb 'to see', 'seeing' denotes a kind of veridical, experiental mental state.[1] 'John saw that Mary cried' entails both that John had a visual experience representing Mary crying and that there is a non-deviant causal route from Mary's crying to John's visual experience (Lewis 1988).[2] 'John saw Mary' entails both that John had a visual experience representing Mary and that there is a non-deviant causal route from Mary to John's visual experience, and 'John saw Mary cry' entails both that John had a visual experience representing Mary crying and that there was a non-deviant causal route from a crying event with Mary as the agent to John's visual experience.

The last construction is one in which 'see' combines with a so-called unsupported clause. As James Higginbotham points out, unsupported clauses are clauses that exhibit 'none of the internal inflectional structure of a full sentence or a clausal complement: neither tense, nor infinitival *to*, nor progressive *–ing*.' (1983: 102). Consider:

(1)

(a) John saw Mary cry
(b) We like carrots raw
(c) I consider John smart

[1] 'Seeing' can perhaps also denote unconscious states. I shall set aside this potential use of 'seeing' here. For some considerations against this use, see Siegel (2006).

[2] I shall here set aside the purely epistemic use of 'seeing', as it occurs in 'I see what you are saying'. Used this way, 'see' is roughly synonymous with 'understand' or 'empathize with'.

B. Brogaard (✉)
Departments of Philosophy, Center for Neurodynamics & Brogaard Lab for Multisensory Research, University of Missouri, St. Louis, MO, USA
e-mail: brogaardb@gmail.com

R. Brown (ed.), *Consciousness Inside and Out: Phenomenology, Neuroscience, and the Nature of Experience*, Studies in Brain and Mind 6, DOI 10.1007/978-94-007-6001-1_26,

'Mary cry', 'carrots raw' and 'John smart' are unsupported clauses. Higginbotham argues that 1(a)–(c) cannot be paraphrased using 'that'-clauses, as in:

(2)

(a) John saw that Mary was crying
(b) We like (it) that carrots are raw
(c) I consider that John is smart

2(a)–(c) appear to mean something quite different from 1(a)–(c). For 2(a) to be true, John need not have observed the crying event but merely needs to have seen signs indicating that Mary had just been crying. So, unlike 1(a), 2(a) could be true if John saw Mary after she had been crying. Unlike 1(b), 2(b) entails that carrots (in general) are raw. Finally, unlike 1(c), 2(c) entails that John is smart.

Higginbotham suggests that 'seeing' constructions with unsupported clauses should be analyzed as follows (using Barwise's situation-semantics and 1(a) as an example):

There is an s, and John saw s, and s $\in$ $[[\text{Mary cry}]]^M$

This is to be read as follows: There is an s such that John saw s and s is in the extension of event type: Mary cries. One reason given in favor of the event analysis is that (i) 'seeing' constructions with unsupported clauses are referentially transparent, that is, they do not admit of opaque readings, and (ii) the event analysis predicts that this is so.

Object-seeing is subject to more constraints that it may initially seem. Siegel (2006) argues convincingly that phenomenology constrains object-seeing. In one of her examples a subject S is looking through a window in a skyscraper. Franco, an individual who is the exactly color of the sky, is suspended from invisible fibers in S's line of sight. Though S is looking directly at Franco, she does not see Franco in any intuitive sense of 'see'. The reason S fails to see Franco, despite looking at him, is that visual experience cannot represent objects that do not stand out from the background. S's visual experience fails to represent Franco as being within her line of sight.

'Seeing-as' is a special construction that allows us to use 'seeing' to report illusions or aesthetic interpretations (Church 2000). 'John saw the sign as red' entails that John saw a sign but it leaves it open whether the sign was red or not. Likewise, 'Mary saw the painting as inspired by Monet' entails that Mary saw the painting but it leaves it open whether the painting in fact was inspired by Monet.

In all of these cases, it is assumed that seeing involves a veridical visual experience. I am going to use subscript 1 to mark this particular mental state. The links between visual experience and seeing_1 can be articulated as follows.[3]

[3]I have included the veridicality requirement, although one could argue that it is captured by the causal constraint. For discussion of the causal constraint on seeing, see Lewis (1988) and Kvart (1993).

Experience-Seeing Bridge Laws

S $sees_1$ that p iff S has a veridical visual experience that represents p, and there is a non-deviant causal route from p to S's visual experience.

S $sees_1$ E iff S has a veridical visual experience that represents E, and there is a non-deviant causal route from E to S's visual experience.

S $sees_1$ A iff S has a veridical visual experience that represents A as being within S's line of sight, and there is a non-deviant causal route from A to S's visual experience.

S $sees_1$ A as F iff S $sees_1$ A, and S has a visual experience that represents A as F.

While I do think 'seeing' can be used to denote $seeing_1$ states, I will argue that this sense of 'seeing' is marginal and that 'seeing' more commonly denotes two different types of (conscious) seeing, viz. $seeing_2$ and $seeing_3$, which are neither strictly veridical nor experiental. $Seeing_2$ is non-experiental and veridical, whereas $seeing_3$ is neither experiental nor veridical. Throughout this paper I shall take 'experiental mental state' to refer to low-level perceptual states, i.e., states that have neural correlates in visual cortex (in the case of vision).

My argument for there being a kind of seeing that is non-experiental and veridical rests on considerations of synesthesia, a relatively rare neurological condition in which stimulation in one sensory or cognitive stream involuntarily leads to associated experiences in a second unstimulated stream (Cytowic 1989). I argue that not all cases of visual synesthesia are genuine experiental phenomena, contrary to influential claims made by Ramachandran and Hubbard (2003). I use this observation to show that there is a kind of seeing that is non-experiental but that this type of mental state can still be assessed for veridicality. I then argue that the verb 'to see' more commonly denotes this type of mental state rather than the experiental type.

My argument for there being a kind of seeing that is neither experiental nor veridical rests on considerations of visual imagery. I show that the nature of visual imagery and introspection creates a need for a kind of seeing that does not require veridicality.

26.2 Grapheme-Color Synesthesia

One of the most common forms of synesthesia is grapheme-color synesthesia, in which numbers or letters are seen as colored. But lots of other forms of synesthesia have been identified. Here I shall focus exclusively on types of synesthesia that involve visual images, or what I will just call 'visual synesthesia'.

One mark of visual synesthesia is that images are either seen as projected out onto the world or in the mind's eye (Dixon et al. 2004). Another mark is that it exhibits test-retest reliability (Baron-Cohen et al. 1987; Eagleman et al. 2007): Colors, shapes or other attributes identified by the subject as representative of her synesthetic experiences in the initial testing phase are nearly identical to colors, shapes or other attributes identified by the subject as representative of her synesthetic experiences in a retesting phase at a later time (see Fig. 26.1).

Age/graph	0	1	2	3	4	5	6	7	8	9
3	/	B	Y	G	P	R	Bl	W	Br	R
4	/	B	Y	G	P	R	Bl	W	Br	R
5	Go	B	Y	G	P	R	DBr	W	Br	R
6	Go	B	Y	G	P	R	DBr	W	Br	R
7	B	B	Y	G	P	R	Br	W	Br	R
8	B	B	Y	G	P	R	Bl	W	Br	R

Fig. 26.1 Example of test-retest reliability of synesthetic experience in one of our associator grapheme-color synesthetes from age 3 to 8 (*Go* gold, *B* blue, *Y* yellow, *G* green, *P* purple, *R* red, *Bl* black, *DBr* dark brown, *Br* brown, *W* white)

An open question about visual synesthesia is whether it is more like visual experience or more like visual imagery or imagination. According to Ramachandran and Hubbard (2003), synesthesia is a genuine experiental, or "sensory," phenomenon. As they put it:

> Work in our laboratory has shown that synaesthesia is a genuine sensory phenomenon ... The subject is not just 'imagining the colour', nor is the effect simply a memory association (e.g. from having played with coloured refrigerator magnets in childhood). (2003: 51)

Some of the evidence listed in favor of treating synesthesia as a genuine experiental phenomenon is that it is automatic and sometimes projected out into the world. Synesthetes often describe the phenomenology of their color experiences as experiental. One of our subjects FS (at age 54), for example, describes his synesthetic experiences as follows:

> The colors are not out there. [...] I think it's related to imagery. It feels like imagining that something has a color. But I am not just imagining it. I think it's perceptual. The phenomenology is sensory. [...] I had no idea that this was unusual until I noticed that other people don't have it.

Another of our subject RS (at age 5) offers a similar description in an interview with an experimenter:

> RS: Sometimes I see it. Sometimes it's out in front of me.
> E: Is it like seeing something or thinking that something is the case?
> RS: It's like seeing something, and my brain is telling me.
> E: Is it exactly like seeing something?
> RS: Both, you know
> E: Can everyone see the same colors as you can when they think about numbers?
> RS: Not everyone, because not everyone thinks very well about numbers.
> E: Are the colors in your head?
> RS: Yes, and I am seeing them too
> E: Are numbers printed in colors?
> RS: No, they are printed in black but that's because they don't know what colors they are

The argument for treating synesthetic experience as experiental is not just based on self-reports. Some synesthetes have been said to experience a pop-out effect in visual search paradigms in which some characters elicit synesthetic experience. For

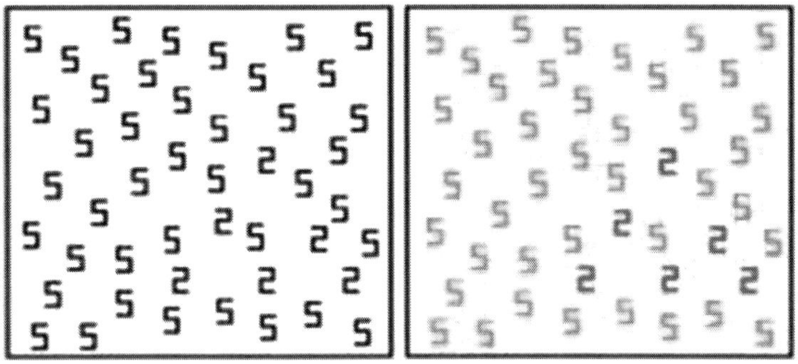

Fig. 26.2 When normal subjects are presented with the figure on the *left*, it takes them several seconds to identify the hidden shape. Some grapheme–color synesthetes instantly see the triangular shape because they experience the 2s and the 5s as having different colors

example, if a cluster of 2s is embedded in an array of randomly placed 5s, normal subjects take several seconds to find the shape formed by the 2s, whereas grapheme–colour synesthetes who experience a pop-out effect instantly see the shape (see Fig. 26.2; Ramachandran and Hubbard 2001; Smilek et al. 2001).

Visual search paradigms are supposed to be indicators of whether synesthetic experience requires focal attention. If synesthetic experience does not require focal attention, then the idea is that digits with unique synesthetic colors should capture attention, which would lead to highly efficient identification of digits. If, on the other hand, synesthetic experience requires focal attention, then synesthetic colors do not capture attention and the identification process should be inefficient (Edquist et al. 2006). Perceptual features must be processed early enough in the visual system for them to attract attention and lead to segregation (Beck 1966; Treisman 1982). So the appearance that synesthetic experience can lead to pop-out and segregation indicates that synesthesia is an experiental phenomenon (Ramachandran and Hubbard 2001, 2003).

However, while a significant number of grapheme-color synesthetes are more efficient in visual search paradigms than controls, this does not clearly show that attention is not required for synesthetic experience. In one subject PM, it was shown that quick identification of graphemes occurred only when the graphemes that elicit synesthetic experience were close to the initial focus of attention (Laeng et al. 2004). Smilek et al. (2003) used a variation on the standard visual search paradigm to test subject J's search efficiency. J was shown an array of black graphemes on a colored background, some of which induced synesthetic experience. The colored background was either congruent or incongruent with the synesthetic color of the target. The researchers found that J was more efficient in her search when the background was incongruent than when it was congruent. This indicates that the synethetic colors attracted attention when they were clearly distinct from the background. Edquist et al. (2006) carried out a group study involving 14 grapheme-color synesthetes and 14 controls. Each subject performed a visual search task in

Fig. 26.3 Synesthetes interpret the middle letter as an A when it occurs in 'cat' and as an H when it occurs in 'the'. The color of their synesthetic experience will depend on which word the grapheme is considered

which a target digit differed from the distractor digits in terms of its synesthetic color or its display color. Both synesthetes and controls identified the target digit efficiently when the target had a unique display color but the two groups were equally inefficient when the target had a unique synesthetic color. The researchers concluded that for most grapheme-color synesthetes, graphemes elicit synesthetic color only once the subject attends to them. This indicates that synesthetic colors cannot themselves attract attention because they are not processed early enough in the visual system.

Another reason to think that not all cases of color experience in grapheme-color synesthesia are genuinely experiental is that their appearance seems to depend on interpretation of visual experience. In Fig. 26.3, for instance, synesthetes assign different colors to the middle letter depending on whether they interpret the string of letters as spelling the word 'cat' or the word 'the'. For example, one of our child subjects, a 7-year old female, experiences the middle letter as red when she reads the word 'cat' and the middle letter as brown when she reads the word 'the'. This suggests that it is not the shape of the letter that gives rise to the color experience but the category or concept associated with the letter (Cytowic and Eagleman 2009: 75).

The fact that the very same grapheme can trigger different color experiences in synesthetes depending on the context in which it occurs suggests that synesthetes need to interpret what they visually experience before they experience synesthetic colors.

Though Ramachandran and Hubbard (2003) argue that grapheme-color synesthesia is experiental, they admit that the linguistic context can affect synesthetic experience. They presented the sentence 'Finished files are the result of years of scientific study combined with the experienced number of years' to a subject and asked her to count the number of 'f's' in it. Most normal subjects count only three 'f's' because they disregard the high-frequency word 'of'. Though the synesthete eventually spotted six 'f's' she initially responded the way normal subjects do. Ramachandran and Hubbard suggest that these contextual effects can be explained by top-down factors. Below I will argue that visual experience processed in early visual areas probably is not affected by top-down factors. If that is right, then top-down influences cannot explain the contextual effects. A better explanation is that interpretation of experiental information is required for synesthetic experience.

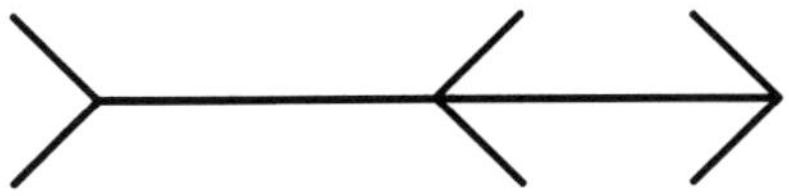

Fig. 26.4 The Müller-Lyer Illusion. The *line* segments between the *arrows* have the same length but they seem to have different lengths

A first step in interpreting what is experienced visually involves a move from visual experience to a visual seeming. Roderick Chisholm (1957) distinguished between three uses of 'appear' words: Phenomenal, comparative and epistemic. Visual seemings are the mental states denoted by phenomenal uses of 'seem'. 'This seems like yesterday' is an example of a comparative use of 'seem'. The 'seem' that occurs in these constructions denote visual seemings or epistemic seemings. Unlike visual seemings, which probably have a neural correlate in the visual system (the ventral stream), epistemic seemings are kinds of belief states that are inferred from other belief states. Suppose I hear on the radio that a tsunami is going to cause flooding in my area and say to my housemates 'It seems like a good idea to evacuate'. In this case, Chisholm would say that the seeming is epistemic, because it is not grounded in the phenomenology of my visual experience. I propose that a definitive mark of epistemic seemings is that they recede in the presence of a defeater if the subject is rational (Brogaard 2012; Brogaard 2013). It may seem like a good idea to evacuate my house because the radio host announced that there will be flooding in my area but if the radio host comes back on the radio and says that the earlier warning was a hoax, then it will no longer seem like a good idea to evacuate. Non-epistemic seemings, on the other hand, persist in the presence of a defeater. If the roads look wet, then they will continue to look wet even if you tell me that the city painted them as a step in their "drive safe" campaign.

Whereas epistemic seemings are belief states, visual seemings clearly are not belief states. You can believe that p even if it visually seems to you that not-p. For example, I can believe that the horizontal lines in the Müller-Lyer optical illusion are the same length even though it seems as if they have different lengths (see Fig. 26.4).

I have argued elsewhere that we need to distinguish between visual experience and visual seemings (Brogaard 2010). I could have a visual experience that represents a rock visually located in front of me. But I may fail to notice the rock, in which case it would not visually seem to me as if there is a rock in front of me.

Presumably the notion of a visual seeming is the folk-theoretical equivalent of what cognitive scientists call 'high-level perception' (Chalmers et al. 1992). This kind of perception, or visual seeming, can be influenced by such things as beliefs, goals and external context. In Fig. 26.3 your visual experience of the middle letter in 'cat' and 'the' represents the same grapheme when you read 'cat' and 'the' but they seem different to you on the two occasions. When you read 'cat', it visually seems to you that the letter is an A. When you read 'the', it visually seems to you that the letter is an H. For grapheme-color synesthetes the color of the visually experienced grapheme depends on which word they focus on.

In grapheme-color synesthesia, then, it is not the visually experienced grapheme that is experienced as colored but the grapheme represented by a visual seeming. The visual experience represents three lines that do not add up to an H or an A. The visual seeming state, on the other hand, represents three lines that do add up to either an A or an H. It is most likely at this stage that synesthetic color is attributed.

26.3 Mathematical Synesthesia

There is further reason to think that at least some forms of synesthesia are not experiental. Though hyperactivity has been measured in visual cortex in some synesthetes in functional magnetic resonance imaging paradigms (Aleman et al. 2001; Nunn et al. 2002; Sperling et al. 2006), other synesthetes experience visual synesthesia without any hyperactivation in visual cortex. One such case is that of JP. In 2002, at the age of 32, JP was a victim of assault and was subsequently diagnosed with a bleeding kidney and an unspecified head injury. After the incident JP began to see complex geometrical figures when looking at moving objects and mathematical formulas. He describes static objects as not having smooth boundaries, and he says that he sees motion in "picture frames." He hand-draws what he sees and does so with great precision (see Fig. 26.5).

After testing JP for synesthesia using the standard test-retest reliability assessment (Baron-Cohen et al. 1987; Eagleman et al. 2007), we carried out an fMRI study to compare brain activation during exposure to image-generating

Fig. 26.5 Image hand-drawn by subject JP

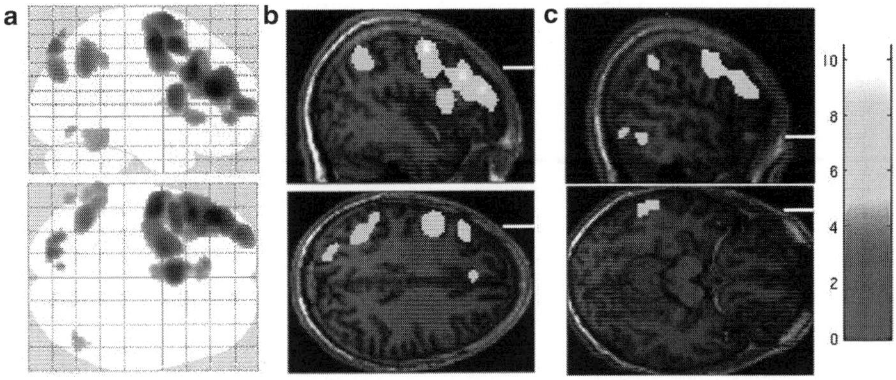

Fig. 26.6 The image shows the areas with increased activity in response to image-inducing formulas compared to non-inducing formulas (the control state) (Brogaard et al. 2012)

mathematical formulas and non-inducing formulas (Brogaard et al. 2012). The study demonstrated that JP processes the image-generating formulas and the non-inducing formulas differently. Image-generating formulas triggered increased activity in left-hemisphere parietal and frontal areas as well as the inferior temporal gyrus compared to non-inducing formulas (the control state) (Fig. 26.6). The two most effective sites were IPS and lateral precentral gyrus. Both IPS and lateral precentral gyrus have been implicated in numerosity estimation and counting (Dehaene et al. 2004; Piazza et al. 2006). During exposure to image-generating formulas there was no right-hemisphere activation compared to the control state.

As noted above, previous studies of visual synesthesia have reported increased activation in the visual cortical areas (striate cortex and V4/V8) (Aleman et al. 2001; Nunn et al. 2002; Sperling et al. 2006). However, we did not find any increased activity in the visual cortical areas in response to the image-generating formulas compared to the control state. The only region of the ventral cortical visual system in which we found increased activation compared to baseline was inferior temporal gyrus, which has been implicated in the representation of global shape (Denys et al. 2004).

These findings, however, turn out to be consistent with the findings of a previous study of subject DT, who has synesthesia and savant syndrome (Bor et al. 2007). Despite the fact that DT reports intensely colored synesthetic images associated with numbers, the study did not find increased activation in DT's visual cortical areas compared to controls. Bor et al. (2007) suggest that DT may have a different type of synesthesia than the experiental form that is more commonly studied.

In Brogaard et al. (2012) we suggest that JP, like DT, has synesthetic experiences that are more conceptual in nature than experiental and that his visualization gives rise to his exceptional drawing abilities. Our study does not directly show whether the increased activation in the left hemisphere comes from seeing the complex geometrical images or from processing meaningful mathematical formulas. But there was no increased activity compared to the control state in the visual cortical areas during processing of the meaningful formulas. So if the meaningful formulas did indeed give rise to vivid visual images, as reported by JP, then the visual images must have been generated by these activation sites. Recent studies also suggest that parietal areas, including some of the areas activated in JP during processing of image-generating formulas, are activated during episodic memory retrieval (Wagner et al. 2005). However, in the case of episodic memory retrieval original sites of activation are normally reactivated (Rissman and Wagner 2012; Danker and Anderson 2010; Kahn et al. 2004). So if the areas with increased activity in JP during processing of image-generating formulas are engaged in memory, these areas presumably are not just responsible for the retrieval of information but also for the generation of visual information.

Despite the fact that JP and DT's synesthetic experiences are more conceptual in nature than experiental, numbers and formulas nonetheless visually seem to them to have colors, shapes or geometrical structure. In the case of JP, the visual seeming states are grounded in activity primarily in the parietal and frontal lobes. In the case of DT, the visual seeming states are grounded in activity primarily in the temporal and frontal lobes. So despite the visual intensity of their synesthetic appearances, the studies indicate that no areas in visual cortex are involved in producing their visual seeming states.

26.4 Seemings and Seeings

As visual seeming states involve a layer of interpretation and needn't involve visual cortical activity, they are not truly experiental. They are kinds of cognitive states that are penetrable by higher-level brain activity.

It may be argued that there is no difference in this regard between visual seeming states and visual experience. Macpherson (2012), for example, argues that an older study carried out by Delk and Fillenbaum (1965) indicates that our beliefs about the characteristic color of an object can affect the color we experience objects as having. In the study, the experimenters cut out shapes of objects from a uniformly colored piece of paper. Some shapes represented objects that are characteristically red (for example, an apple, a heart, a pair of lips). Some shapes represented objects that are not characteristically red (for example, a circle, a square, an oval, a bell, a mushroom). Each of the cutout shapes were placed in front of a colored background that could be changed from yellow through orange to red. Subjects were asked to tell the experimenters to adjust the background until the color was the same as the shape in front. The researchers found that when the object represented by a shape had a characteristically red color, the subjects selected a background color that was redder than the color they selected when the shape was of an object that was not characteristically red. On the basis of these observations Macpherson argues that the subjects' beliefs about the colors of the objects represented by the cutout shapes penetrate their perceptual experiences of the cutout shapes.

I think, however, that there is good reason to believe that the study does not show that people's beliefs affect the phenomenology and content of their visual experiences. We cannot see an object as having a characteristic color unless the object visually seems to be a certain kind of object. As seeming states and visual experiences are different mental states, seeing an object as having a characteristic color requires forming visual seeming states about the objects seen. What the studies show, then, is not that people's visual experiences are penetrable by their beliefs about the characteristic colors of objects but rather that their states of seeming are penetrable in this way.

The same sorts of considerations that provide evidence against taking Delk and Fillenbaum's (1965) results as suggestive of cognitively penetrable visual experiences also provide reason against thinking that visual experience and visual seeming states both represent high-level properties. Siegel (2005, 2011) offers the following argument for thinking that visual experience contains high-level properties.

Let E1 be a visual experience of someone who has the ability to recognize elm trees (expert) and who is looking at an elm tree, and let E2 be the visual experience of someone who does not have the ability to recognize elm trees (novice) and who is looking at the same tree in the same viewing conditions. The expert finds the tree familiar, the novice does not. So there is a difference in the overall phenomenal character of their experiences. The argument can be articulated as follows:

The Argument from Phenomenal Contrast

(1) The overall phenomenology of which the phenomenology of E1 is a part differs from the overall phenomenology of which the phenomenology of E2 is a part (*familiarity effects*).
(2) If the overall phenomenology of which the phenomenology of E1 is a part differs from the overall phenomenology of which the phenomenology of E2 is a part, then there is a phenomenological difference between E1 and E2 (*cognitive penetration*).
(3) If there is a phenomenological difference between E1 and E2, then E1 and E2 differ in content (*representationalism*).
(4) If there is a difference in content between E1 and E2, it is a difference with respect to K-properties represented in E1 and E2.

The conclusion is that the difference in overall phenomenology between the novice and the expert is grounded in a difference between what the novice and the expert's visual experiences represent.

The problem with this argument is that it does not get off the ground without assuming that the Elm tree seems different to the expert and the novice (see Brogaard 2010). But if this is so, then the overall difference in phenomenology between the expert and the novice stems from a difference in the phenomenology of the two distinct states of seeming. To the expert, the tree seems to be an Elm, it seems familiar and it seems different from the nearly identical neighboring tree which is not an Elm. To the novice, the tree just seems to be a tree, it doesn't seem familiar and it seems exactly similar to the neighboring tree.

But, now, visual seeming states and visual experiences are distinct mental states; so premise (2) is false. How things visually seem needn't affect how visual experience represents the world. For example, even if my visual experience represents exactly 21 students, it may visually seem to me that there are at most 20 students in my class room because of an unusual configuration of the class room. The content of my visual seeming does not penetrate my visual experience.

Despite not being truly experiental, visual seeming states form the ground on the basis of which we normally make judgments concerning what we see or don't see. Let us call visual seeming states that are appropriately connected to the states of affair that they represent *seings*$_2$. I propose the following bridge laws between visual seemings and seeings$_2$:

Seeming-Seeing Bridge Laws

S sees$_2$ that p iff it visually seems to S that p, the seeming is veridical, and there is a nondeviant causal route from p to S's visual seeming.
S sees$_2$ E iff it visually seems to S that E takes place, the seeming is veridical, and there is a nondeviant causal route from E to S's visual seeming.
S sees$_2$ A iff it visually seems to S that A is within his line of sight, the seeming is veridical, and there is a non-deviant causal route from A to S's visual seeming.
S sees$_2$ A as F if S sees$_2$ A, and it visually seems to S that A is F.

Grapheme-color synesthetes don't see$_1$ graphemes as colored; they see$_2$ them as colored. When they are presented with the words 'cat' and 'the', they first interpret what standardized grapheme they see$_1$. Once it seems to them that a particular standardized grapheme is present, they see$_2$ that grapheme as colored. This state of seeing$_2$ is a non-experiental state, viz. a visual seeming that satisfies certain constraints.

Similar remarks apply to the cases of the mathematical synesthetes JP and DT. Even though JP and DT's synesthetic experiences are not grounded in activity in the visual cortex, both subjects see$_2$ numbers and formulas as shapes or geometrical patterns. As in the case of standard grapheme-color synesthetes, the seeing here is non-experiental despite being highly visual.

As seeing$_2$ is grounded in a non-experiental visual seeming state, it itself is non-experiental. Though no one has previously made this clear, the English word 'see' is most commonly used to denote this type of non-experiental mental state. The following exchange illustrates this.

Prosecutor: Are the gang members you saw present here today?
Witness [looking around]: Yes.
Prosecutor: So you saw the 18 gang members on April 17?
Witness: Yes.
Prosecutor: If you saw the 18 gang members on April 17, then why did you tell the police that you didn't know how many there were?
Witness: I couldn't see that there were 18 gang members.

This exchange indicates that 'seeing' here denotes seeings$_2$. To say that the English word 'see' most commonly denotes seeing$_2$ states is not to say that 'see' cannot be used to denote seeing$_1$ states. One strategy of skeptics about seeing is to trigger a switch from seeing$_2$ to seeing$_1$. The following exchange demonstrates this sort of skeptical move.

A: I saw_2 Alex in his office yesterday.
B: Did you really see_1 Alex? Or did you merely see_1 the part of him that wasn't hidden behind his desk?
A: Well, if you really must be that anal about it, then, yes, I only saw_1 the part of him that wasn't hidden behind the desk.

B's skeptical question draws A into considering what she saw_1. B could follow up with further skeptical inquiries:

B: And are you sure you really saw_1 Alex? Or did merely see_1 a person that looked like Alex? Speaking of which, are you even sure you saw_1 a person?

Shifting from $seeing_2$ to $seeing_1$ is one way we question what people think they know on the basis of their visual experiences. $Seeing_2$ entails having low-grade knowledge, whereas $seeing_1$ entails having high-grade knowledge. We will return to the relation between seeing and knowledge below.

26.5 Visual Imagery

Bourget (2010) argues that some uses of 'see' introduce a hyperintensional context just like 'belief'.[4] Suppose the heartbroken Lois Lane takes a strong hallucinatory drug and then utters the following (these are modifications of Bourget's examples):

(3)

(a) Wow, I see Superman on my left. That's a really strong drug.
(b) I see Superman spinning in front of me, even though I know he isn't even here.
(c) I see Superman all over the place. Maybe I should stop taking the drug.

The sentences in (3) seem intuitively true. But the result of substituting 'Clark Kent' for 'Superman' is false. So 'see' is hyperintensional on some of its uses.

It may be objected that when 'see' is used in the above ways, the locutions are idioms just like 'The sun is rising'. We know that the latter location is literally false. But it is nonetheless prevalent among English speakers. Despite being false, it conveys something true.

However, I think there are several good reasons to believe that 'see' does not merely appear to have a third sense because it is used idiomatically in these constructions. One is that this use of 'see' is far too widespread to be idiomatic. 'The sun is rising' is an idiom and we might say that 'to rise' is used idiomatically here. But this is the only type of construction in which it is so used. 'To see' is systematically used as a hyperintensional verb.

A second reason is that even if we were to grant that 'to see' is a term of art, an expression to be assigned a meaning by philosophers, the third sense of 'see' may well remain in place. It is far from settled what introspection is. Philosophers

[4]Bourget merely argues that 'see' has intensional uses. However, I think his argument shows that it has hyperintensional uses as well.

who defend a perceptual (or inner sense) theory of introspection, for example, might allow for non-veridical states of seeing. Suppose we ask Elizabeth who is introspecting her visual image, what she sees. When she replies 'I see a man walk down the stairs' or 'I see that a man is walking down some stairs', she might well be saying something true if the perceptual theory of introspection is correct. On the view of introspection I am inclined to think is correct, our beliefs about our mental states and about the past are caused and justified by introspective (phenomenal) seemings, which in turn are grounded in mental images or other mental states. On this view, it is quite natural to regard introspective seemings that satisfy certain constraints as kinds of seeings. One model that lends some support to this view of introspection is the reactivation model, the most up-to-date model of visual memory (Rissman and Wagner 2012; Danker and Anderson 2010; Kahn et al. 2004; Kosslyn 2005). According to this model, memory storage consists in a strengthening of the perceptual and cognitive pathways that originally processed the information. The hippocampus does not store information for later retrieval (Squire et al. 2004: 296; Squire 1987; Mishkin 1982). Instead it plays a role akin to that played by attention in maintaining working memory (Serences et al. 2009).[5] It encodes associations among the components of individual events and thus functions as a control center in the strengthening process (Eichenbaum 2004). Over time the hippocampus no longer is needed to maintain the neural networks. How does this model lend support to a perceptual theory of introspection? Well, some of our beliefs about the present are justified by visual seemings, which are grounded in visual experience. If the reactivation model of memory is correct, then visual memory retrieval consists in an activation of the visual pathways involved in the original visual experiences. This makes it plausible that some of our beliefs about the past are justified by introspective seemings, which are grounded in memory images. This argument, of course, does not extend to beliefs about our mental states but it is plausible that introspective seemings justify both beliefs about the past and beliefs about our mental states, in which case the point generalizes.

A third reason to think that 'see' does not merely appear to have a third sense because it is used idiomatically is we can see how the third sense of 'to see' may have evolved from the second sense. A seeing_2 state is a visual seeming state that stands in a non-deviant causal relation to the state of affairs it represents. Likewise, a seing_3 state is an introspective seeming state that stands in a non-deviant causal relation to a visual image (a hallucination, a memory or an imagination) that it represents. Suppose you ask Krista about her recent involuntary memory recalls:

You: What do you see?
Krista: I see a little girl sitting on the floor alone
You: What else do you see? Is she crying?
Krista: She has tears in her eyes
You: What else do you see? Are you there?
Krista: No, I am not there. The girl is not me.

---

[5] Serences et al. (2009) call their model of working memory 'the sensory recruitment model'.

This exchange is a powerful illustration of $seeing_3$ as an introspective seeming that is causally connected to a visual image (the hallucination, memory or imagination) in a non-deviant way.

A skeptic about $seeing_3$ can, of course, easily push the standards for seeing in the direction of $seeing_2$. A skeptic might ask the subject whose brain he is poking. 'Are you *really* $seeing_2$ stars everywhere?'. The subject might then reply: 'No, I am not really $seeing_2$ stars. They just appear to be there in front of my eyes'. And, as we have already seen, a skeptic about $seeing_2$ can likewise push the standards for seeing in the direction of $seeing_1$, witness 'Did you really see Alex, or did you just see the front of his upper body?'

What the skeptics about seeing are doing when they question seeing claims is quite similar to what skeptics about knowledge are doing when they question knowledge claims. Skeptics about knowledge may use an analogous type of discourse to raise the standards for knowledge, witness 'Do you really know that this is Providence Airport? Are you completely sure?'

On one increasingly popular view of knowledge, seeing is a determinate of the determinable knowledge (Williamson 2000; Brogaard 2011). On this view, memory states, belief states, seeming states and perceptual states can all count as knowledge states, provided that they satisfy certain further constraints (e.g., veridicality and safety). On this view, $seeing_3$, like the other kinds of seeing, is a kind of knowledge, viz. introspective knowledge.

We can articulate the links between visual images (hallucinations, memories, imaginations) and $seeings_3$ as follows.

The Image-Seeing Bridge Laws

S $sees_3$ that p iff it introspectively seems to S that p, and there is a non-deviant causal route from the image-based proposition p to S's introspective seeming.
S $sees_3$ E iff it introspectively seems to S that E is occurring, and there is a non-deviant causal route from the image-based appearance of E's occurrence to S's introspective seeming.
S $sees_3$ A iff it introspective seems to S that A is within his line of sight, and there is a non-deviant causal route from the image-based appearance of A to S's introspective seeming.
S $sees_3$ A as F if S $sees_3$ A, and it introspectively seems to S that A is F.

The overlap among the three analyses of the concept of seeing suggests that there really are three distinct types of mental states that all count as states of seeing. 'To see' is likely a polysemous word denoting three different, but related, mental states.

26.6 Conclusion

On a traditional view of seeing, seeing is a visual experience that stands in a nondeviant causal relation to the state of affairs represented by the experience. Here I have argued that this kind of seeing is not the most common one denoted by the

English verb 'to see'. I have argued that there are three distinct, but analytically similar, states of seeing.

Considerations of color-grapheme synesthesia, mathematical synesthesia and visual imagery lend evidence to this hypothesis. Synesthesia is a neurological condition in which stimulation in one sensory or cognitive stream involuntarily leads to associated experiences in a second unstimulated stream (Cytowic 1989). In grapheme-color synesthesia, a letter or number triggers an experience of color. Grapheme-color synesthesia is normally considered a kind of illusory visual experience in which numbers or letters are experienced as colored. However, cases in which one and the same grapheme gives rise to different colors in grapheme-color synesthetes depending on the linguistic environment in which the grapheme occurs suggest that not all synesthetic experience is truly experiental. Synesthetes cognitively process the grapheme presented to them before they experience color. This suggests that synesthetic experience is a kind of visual seeming. Visual seeming states differ from visual experience in terms of their representational richness and their neural correlates. Some mathematical synesthetes have rich visual representations of numbers and formulas without any hyperactivity in virtual cortical areas. Their neural activity appears to be limited to parietal or temporal areas as well as frontal areas.

The English verb 'to see' normally denote visual seeming states that stand in a non-deviant causal relation to the states of affair they represent. For example, it is perfectly acceptable to say that we saw Alex, even if we only saw the front side of his upper body.

In the final section of the paper, I argued that the English verb 'to see' can also function as a hyperintensional verb that denotes a kind of non-veridical mental state. The way we talk about inner images provides reason for thinking that there are states of seeing of this kind. Reporting on a visual image, it is perfectly acceptable to say things like 'I see a girl walk into the living room' or 'Stop poking my brain, I see stars everywhere'. This third sense of 'seeing', I argued, is not idiomatic, as locutions containing them are too abundant to be idioms. Furthermore, it is not hard to see how 'seeing' may have evolved from denoting only veridical mental states to denoting also non-veridical ones. Seeing_2 is a visual seeming that stands in a non-deviant causal relation to a state of affair represented by the seeming. Likewise, seeing_3 is a visual image that stands in a non-deviant causal relation to an introspective seeming state whose content overlaps that of the visual image.

Seeing_2 is probably the most common type of mental state denoted by the verb 'to see'. You may have a visual experience of a tiny bird in the horizon. It doesn't follow from this that it seems to you that there is a tiny bird in the horizon. In the envisaged scenario, it is true that you see_1 a tiny bird in the horizon. But if I asked you what you saw, your answer wouldn't be that you saw a tiny bird in the horizon. In one sense of 'seeing', you were not in a position to see that there was a tiny bird in the horizon. The state of seeing involved here is seeing_2. 'Seeing' thus appears to denote three related kinds of seeing, which is to say that the English verb 'to see' is polysemous.

Acknowledgements I am grateful to David Bourget, Alex Byrne, David J. Chalmers, Ophelia Deroy, David Eagleman, Kristian Marlow, Sydney Shoemaker and Juha Silvanto for discussion of these and related issues. Special thanks to Ophelia Deroy for written comments on the paper.

References

Aleman, A.A., G.-J.M. Rutten, M.M. Sitskoorn, G. Dautzenberg, and N.F. Ramsey. 2001. Activation of striate cortex in the absence of visual stimulation: An fMRI study of synesthesia. *Neuroreport* 12: 2827–2830.

Baron-Cohen, S., M. Wyke, and C. Binnie. 1987. Hearing words and seeing colors: An experimental investigation of synesthesia. *Perception* 16: 761–767.

Beck, J. 1966. Effect of orientation and of shape similarity on perceptual grouping. *Perception & Psychophysics* 1: 300–302.

Bor, D., J. Billington, and S. Baron-Cohen. 2007. Savant memory for digits in a case of synaesthesia and Asperger syndrome is related to hyperactivity in the lateral prefrontal cortex. *Neurocase* 13: 311–319.

Bourget, D. 2010. Intensional and phenomenal uses of perceptual verbs, Chapter in ANU Dissertation.

Brogaard, B. 2010. Do we perceive natural kind properties?, *Philosophical Studies* 162(1) (2013): 35–42.

Brogaard, B. 2011. Primitive knowledge disjunctivism. *Philosophical Issues* 21: 45–73.

Brogaard, B. 2012. Perceptual reports. In *Oxford handbook for the philosophy of perception*, ed. M. Matthen. New York: Oxford University Press.

Brogaard, B. 2013. Phenomenal seemings and sensible dogmatism, forthcoming. In *Seemings and justification: New essays on dogmatism and phenomenal conservatism*, ed. C. Tucker. Oxford: Oxford University Press.

Brogaard, B., S. Vanni, and J. Silvanto. 2012. Seeing mathematics: perceptual experience and brain activity in acquired synesthesia. Neurocase (in press)

Chalmers, D.J., R.M. French, and D.R. Hofstadter. 1992. High-level perception, representation, and analogy: A critique of artificial intelligence methodology. *Journal of Experimental and Theoretical Artificial Intelligence* 4: 185–211.

Chisholm, R.M. 1957. *Perceiving: A philosophical study*. Ithaca: Cornell University Press.

Church, J. 2000. 'Seeing As' and the double bind of consciousness. *Journal of Consciousness Studies* 7: 99–111.

Cytowic, R.E. 1989. *Synesthesia: A union of the senses*. New York: Springer.

Cytowic, R.E., and D.M. Eagleman. 2009. *Wednesday is indigo blue*. Cambridge: MIT Press.

Danker, J.F., and J.R. Anderson. 2010. The ghosts of brain states past: Remembering reactivates the brain regions engaged during encoding. *Psychological Bulletin* 136: 87–102.

Dehaene, S., N. Molko, L. Cohen, and A. Wilson. 2004. Arithmetic and the brain. *Current Opinion in Neurobiology* 14: 218–224.

Delk, J.L., and S. Fillenbaum. 1965. Differences in perceived colour as a function of characteristic color. *The American Journal of Psychology* 78(2): 290–293.

Denys, K., W. Vanduffel, D. Fize, K. Nelissen, H. Peuskens, D. Van Essen, and G.A. Orban. 2004. The processing of visual shape in the cerebral cortex of human and nonhuman primate: A functional magnetic resonance imaging study. *Journal of Neuroscience* 24: 2551–2565.

Dixon, M.J., D. Smilek, and P.M. Merikle. 2004. Not all synaesthetes are created equal: Projector versus associator synaesthetes. *Cognitive, Affective, & Behavioral Neuroscience* 4: 335–343.

Eagleman, D.M., A.D. Kagan, S.S. Nelson, D. Sagaram, and A.K. Sarma. 2007. A standardized test battery for the study of synesthesia. *Journal of Neuroscience Methods* 159: 139–145.

Edquist, J., A.N. Rich, C. Brinkman, and J.B. Mattingly. 2006. Do synaesthetic colours act as unique features in a visual search? *Cortex* 42: 222–231.

Eichenbaum, H. 2004. Hippocampus: Cognitive processes and neural representations that underlie declarative memory. *Neuron* 44: 109–120.
Higginbotham, J.T. 1983. The logic of perceptual reports: An extensional alternative to situation semantics. *Journal of Philosophy* 80: 100–127.
Kahn, I., L. Davachi, and A.D. Wagner. 2004. Functional-neuroanatomic correlates of recollection: Implications for models of recognition memory. *Journal of Neuroscience* 24: 4172–4180.
Kosslyn, S.M. 2005. Mental images and the brain. *Cognitive Neuropsychology* 22: 333–347.
Kvart, I. 1993. Seeing that and seeing as. *Noûs* 27: 279–302.
Laeng, B., F. Svartdal, and H. Oelmann. 2004. Does color synesthesia pose a paradox for early-selection theories of attention? *Psychological Science* 15: 277–281.
Lewis, D. 1988. Veridical hallucinations and prosthetic vision. In *Perceptual knowledge*, ed. J. Dancy. Oxford: Oxford University Press.
Macpherson, F. 2012. Cognitive penetration of colour experience: Rethinking the issue in light of an indirect mechanism. *Philosophy and Phenomenological Research* 84: 24–62.
Mishkin, M. 1982. A memory system in the monkey. *Philosophical Royal Society of London Biology* 298: 85–92.
Nunn, J.A., L.J. Gregory, M. Brammer, S.C. Williams, D.M. Parslow, M.J. Morgan, R.G. Morris, E.T. Bullmore, S. Baron-Cohen, and J.A. Gray. 2002. Functional magnetic resonance imaging of synesthesia: Activation of V4/V8 by spoken words. *Nature Neuroscience* 5: 371–375.
Piazza, M., A. Mechelli, C.J. Price, and B. Butterworth. 2006. Exact and approximate judgements of visual and auditorynumerosity: An fMRI study. *Brain Research* 1106: 177–188.
Ramachandran, V.S., and E.M. Hubbard. 2001. Synaesthesia: A window into perception, thought and language. *Journal of Consciousness Studies* 8: 3–34.
Ramachandran, V.S., and E.M. Hubbard. 2003. The phenomenology of synaesthesia. *Journal of Consciousness Studies* 10: 49–57.
Rissman, J., and A.D. Wagner. 2012. Distributed representations in memory: Insights from functional brain imaging. *Annual Review of Psychology* 63: 101–128.
Serences, J.T., E.F. Ester, E.K. Vogel, and E. Awh. 2009. Stimulus-specific delay activity in human primary visual cortex. *Psychological Science* 20: 207–214.
Siegel, S. 2005. Which properties are represented in perception? In *Perceptual experience*, ed. T. Szabo Gendler and J. Hawthorne, 481–503. Oxford: Oxford University Press.
Siegel, S. 2006. How does visual phenomenology constrain object-seeing. *Australasian Journal of Philosophy* 84: 429–441.
Siegel, S. 2011. Cognitive penetrability and perceptual justification. *Nous* 46(2): 201–222. doi:10.1111/j.1468-0068.2010.00786.x.
Smilek, D., M.J. Dixon, C. Cudahy, and P.M. Merikle. 2001. Synaesthetic photisms influence visual perception. *Journal of Cognitive Neuroscience* 13: 930–936.
Smilek, D., M.J. Dixon, and P.M. Merikle. 2003. Synaesthetic photisms guide attention. *Brain and Cognition* 53: 364–367.
Sperling, J.M., D. Prvulovic, D.E.J. Linden, W. Singer, and A. Stirn. 2006. Neuronal correlates of colour-graphemic synesthesia: A fMRI study. *Cortex* 42: 295–303.
Squire, L.R. 1987. *Memory and brain*. New York: Oxford University Press.
Squire, L.R., C.E. Stark, and R.E. Clark. 2004. The medial temporal lobe. *Annual Review of Neuroscience* 27: 279–306.
Treisman, A. 1982. Perceptual grouping and attention in visual search for features and for objects. *Journal of Experimental Psychology. Human Perception and Performance* 8(2): 194–214.
Wagner, A.D., B.J. Shannon, I. Kahn, and R.L. Buckner. 2005. Parietal lobe contributions to episodic memory retrieval. *Trends in Cognitive Science* 9: 445–453.
Williamson, T. 2000. *Knowledge and its limits*. Oxford: Oxford University Press.

Chapter 27
Synesthesia: An Experience of the Third Kind?

Ophelia Deroy

Over a good part of my scientific career, I've spent much time and energy chasing down an elusive creature known as synesthesia. Early in this quest, I thought I'd caught up with it: I was poised, ready to snare it – only to watch it get away. Apparently, my first synesthesia-catcher was too small, and insufficiently flexible, to capture a critter at once so large and agile. (Marks 2011, p. 47)

27.1 Introduction

What is it like to have a synesthetic experience? Most synesthetes have stressed "having trouble putting into words some of the things (they) experience" as if they had to explain "red to a blind person or middle-C to a deaf person".[1] The current definition of synesthesia as a condition in which "stimulation in one sensory or cognitive stream leads to associated experiences in a second, unstimulated stream"[2] leaves the question open: What do these 'associated experiences' consist in?

Brogard's paper, I contend, provides a thought-provoking answer to this question. True, the main point in her paper consists in distinguishing various kinds of visual conscious states, or 'seeings', but synesthesia comes to play a key role in her argument. She argues that there is room, conceptually, to think about a kind of visual conscious state, reported as 'seeing', which differs both from visual perception

---

[1]Cytowic (1989).

[2]Hubbard (2007), p. 193.

O. Deroy (✉)
Centre for the Study of the Senses, Institute of Philosophy, School of Advanced Study, University of London, Senate House, Malet Street, London WC1E 7HU, UK
e-mail: ophelia.deroy@gmail.com

R. Brown (ed.), *Consciousness Inside and Out: Phenomenology, Neuroscience, and the Nature of Experience*, Studies in Brain and Mind 6, DOI 10.1007/978-94-007-6001-1_27,

(which, in her terms, is experiential, richly representational and veridical) and from visual imagery (or/and what she considers to be our introspective take on visual imagery, which turns out to be non experiential, but representational and veridical relative to visual imagery).[3] Let's grant that being representational, veridical, or experiential are non-overlapping, exhaustive and appropriate conditions on which to evaluate conscious mental states; let's also put aside the arguments regarding mental imagery and our introspective take on it. There is then no reason to object that there can be a third kind of state, which could be defined as non experiential yet representational (although less richly, and as just seen, more weakly) and non-veridical. A second step in Brogaard's paper is to show that the extension of this 'third kind' of state is not empty. Her argument here consists in arguing that synesthesia, at least the varieties that come with a conscious visual concurrent, fall in this third category. This last step, I contend, is more disputable.

My starting point is, of course, a little different. I am more interested in the problem of finding the right category for synesthetic experiences, as we know them, than in kinds of seeings. Little clarity is to be expected here from the analysis of verbal reports. When asked to describe what she 'sees', a color grapheme synesthete is likely to respond, as quoted in Broggard's paper (p. 408) "The colors are not out there. [. . .] I think it's related to imagery. It feels like imagining that something has a color. But I am not just imagining it. I think it's perceptual. The phenomenology is sensory". So does this synesthete see colors? Yes and no. Perhaps she thinks about 'seeing' in one sense, and then in another. The main lesson of verbal reports is that synesthetic experiences are not subjectively clearly like perceptual experiences or imagery.

No blame should be put here on the synesthete's confusion. The same confusing description is to be found in the scientific and philosophical literature on synesthesia.[4] Across the years and papers, synesthetic experiences have been characterized as sensory (Ramachandran and Hubbard 2003), perceptual (e.g. Palmeri et al. 2002; Rich and Mattingley 2002; Segal 1997), conceptual (Simner 2007) or related to mental imagery (Galton 1880; Rader and Tellegen 1987; see Craver-Lemley and Reeves in press for discussion), illusions, hallucinations or after images (Lycan 2006; Sagiv et al. 2011). That they could be 'non experiential visual seemings' (Brogaard, present volume) is an interesting proposal to consider.

Now obviously, one reason for which synesthetic experiences have received so many different labels comes from the fact that people do not all agree on how many kinds of conscious visual states there are. It is however relatively uncontroversial, from both a common sense and empirical perspective, that there are at least two different kinds of conscious visual experiences: One which corresponds to

[3]Brogaard introduces the term of 'introspective visual seemings' and relates them to visual imagery. I will not discuss this proposal and keep the more common and general category of visual mental images.

[4]See for instance Marks (2011) for a review.

perception and the other to imagining or mental imagery.[5] But how should these be defined? Moreover, do perceptual and mental imagery exhaust the domain of conscious visual experiences? Is there for instance some third (or fourth, or fifth) kind of state which is phenomenologically similar, or close, is reported as 'seeing' and yet presents distinct characteristics?

The idea that we have to make room for some kind of conscious visual states that is distinct from perceptual and imagistic states falls in what I call a "claim of the third kind". Claims of the third kind are not unusual, at least in philosophy. For one thing, visual illusions can count as a distinct kind of state that is (or at least could be) phenomenologically or *subjectively* indiscriminable from perceptual states, but fails to veridically represent – or relate to – an independent object.[6] As rightly argued by Brogaard here and discussed by others before (Auvray and Deroy 2013; Segal 1997; see Macpherson 2007 for an extended discussion), synesthetic experiences do not straightforwardly fail to be representational or even veridical, at least in a weak sense. In other terms, it is possible to re-describe the systematic association between inducers and concurrents in such a way that the presence of a certain concurrent will represent the presence of a certain inducer (or set of inducers). Going one step further, recent research suggest that some features or dimensions of the concurrent[7] co-vary with differences in the object or property – like differences in shape. For instance, for each synesthete, the occurrence of certain conscious visual states (e.g. of certain shades of green, blue, yellow, etc.) correlates with the presence of certain letters in the outer world (e.g. B for green, G for blue, C for yellow, etc.). Some difficulties can arise because of cases where thinking about the letter B is sufficient to elicit the synesthetic color, in which case the conscious state does not indicate the presence of an independent state of affair. Such cases are nonetheless rare (see Dixon et al. 2000; Spiller and Jansari 2008). There might be other difficulties if the very same visual experience can occur when the synesthete faces a letter B surrounded by a colored halo of that very green – in which case the very same conscious state can correlate with two distinct states of affairs. Let's put these (interesting) difficulties aside. Synaesthetic colors are caused by the presence of an object of a certain kind and reliably co-varies to this kind of object. In this sense, they can serve a representational function – and be reliable indicators.

This leaves open the question: Are synesthetic experiences really perceptual? At this point, the definitional problems turn into methodological problems. Even if people agree on what perception and imagery are, they might disagree on how to practically determine that a specific state is perceptual or imagistic. Which tests should be conducted, and do they all bring equal evidence?

[5]Many people who are sceptical of the pictorial nature of visual imagery would disagree with this statement.

[6]In this case, conscious states are not just individuated on internalist grounds.

[7]The content of color-grapheme experiences co-varies with external objects present the environment – and several candidates to explain the variations more finely have been advanced (such as shape, Hubbard et al. 2005; Brang et al. 2011; or phonemial similarities between the letters, Witthoft and Winawer 2006).

In the case of synesthesia, an agreement has emerged on what is sufficient to qualify as a synesthete.[8] If one consistently pairs the same type of color (or concurrent) to the same type of letter (or inducer) across distant or large number of trials and if this pairing is fast and involuntary. then one counts as a synesthete.[9] As these tests cannot be performed out of normal memory, one seems to have to see letters in certain shades in order to pass them. It remains unclear though what these tests say on what sort of conscious state it is[10] and whether this state is perceptual.

Certain cases of mental imagery can satisfy the conditions of being consistent and involuntary, as well as conscious. For instance, seeing the lip movements of a famous singer performing on TV with the sound off is likely to give rise to an auditory image of what he is singing; if one's repertoire of songs is large enough, one will be able to pair lip movements to auditory images for many songs – like synesthetes do for many colors and letters. The induction of crossmodal imagery (Spence and Deroy 2012) can also be involuntary, as suggested by the fact that seeing silent lip movements can induce a significant increase in auditory cortex activation, even in the absence of any auditory speech sounds (e.g., Calvert et al. 1997; Hertrich et al. 2011; Pekkola et al. 2005) or that the tactile exploration of an object in the dark activates visual areas (Lacey et al. 2010).

While everyone agrees that there is a difference between perception (caused by the outer world and co-varying with it) and imagery (possibly independent of the outer-world, and not necessarily co-varying with it), cases of spontaneous, automatic, imagery raise problems as they seem to be diagnosed as perceptual.

Can't we think then of other differences between imagery and perception that could help? Two criteria are usually applied: First, the mental image of a certain object is supposed to be less vivid than the visual experience of the very same object which obtains in perception ('perceptual experience' for short). Second, a mental image is supposed to be under a form of control, whereas a perceptual experience is not. Unfortunately, these two criteria do not deliver a simple answer when they are applied to synesthesia. As individuals differ in terms of the vividness of their mental imagery (e.g. Cui et al. 2007), synesthetes reporting vivid colors might just be vivid visual imagers (Rader and Tellegen 1987). The lack of control over the content of the concurrent experience is a very generally acknowledged (and reported) aspect of synesthesia, and seems to do better. This said, several cases of partial control over the content of the concurrent are also documented. In Rich et al.'s (2005) study, 15 % of the synesthetes quizzed reported complete voluntary control over their synesthetic concurrents. Nearly half of them (46 %) also reported being able to increase the vividness of their synesthetic experience as a result of attention (see also Rich and Mattingley 2003; Sagiv and Robertson 2005).

[8]Whether these conditions are necessary is debated, see Simner (2012).

[9]Eagleman et al. (2007).

[10]Auvray and Deroy (2013), Deroy and Spence (submitted).

Table 27.1 Comparisons between the key characteristics of perceptual experiences, visual imagery and synesthesia

Visual experience	Caused by an external object	Co-varying with external features	Vivid	Control over content	Giving rise to beliefs
Perception	+	+	+	–	+
Visual imagery	+/–	+/–	+/–	+/–	–
Synesthesia	+	+	+/–	+/–	+/–

Another key difference can come from the spatial character of the experience. In the case of perception, one sees the color 'out there', while in mental imagery, one's visual experience seems to be in 'the mind's eye' (i.e. not 'out there'). But once more, this difference does not deliver a clear-cut verdict when applied to synesthetic colors. Some synesthetes – or rather, some synesthetes in some circumstances (Hupe et al. 2011) – experience the synesthetic concurrent 'out there' and not just in their mind (Dixon et al. 2004 for the distinction).[11]

Another aspect, which is more a matter of anecdotal reports than systematic study, shows that experience of synesthetic colors can give rise to beliefs, like perceptual experiences and unlike visual imagery. Individual synaesthetes confess having for a long time believed for instance that Rs were blue, or P orange (Duffy 2001 for an example). This mostly confirms what was said earlier: Synesthetic colors are subjectively similar to perceptual experiences.

Whether synesthetic colors should be classified as a kind of perceiving, a kind of imagining or something else is then no easy matter, as summarized in Table 27.1.

27.2 An Experience of the Third Kind: The Neurological Criterion

Brogaard suggests that a difference can be found in terms of what she calls the 'experiential' character of visual conscious states, by which she means in terms of neural correlates of the conscious state. She takes " 'experiential mental state' to refer to low-level perceptual states, i.e., states that have neural correlates in visual cortex (in the case of vision)."

I have no disagreement with this physicalist addition to the discussion. The fact that kinds of mental states, although primarily individuated in terms of psychological function, can also be individuated in terms of certain neurological

[11]As Hupe et al. (2011) remark: "*Both data from phenomenology and psychophysics now clearly indicate that the experience of synesthetic colors is far from being equivalent to the experience of real colors (for most, if not all, synesthetes), contrary to early enthusiastic claims based on surprising observations obtained but of single individuals, sometimes with poor methodological controls. Nonetheless, the experience of synesthetic colors must bear some connection to the experience of real colors*" (Hupe et al. 2011, p. 7).

properties is at the very heart of the scientific investigation of the mind. What might be questionable is that the neurological properties underlying kinds of mental states will always receive a structural or neuroanatomical characterization, noticeably in terms of localization. Sometime the neurological difference will also be functional.

Coming back to Brogaard's strategy, the problem does not come from her resorting to neurological differences to argue for synesthetic experiences being 'a third kind' of state. The problem is rather about the precise evidence that is offered. The non-experiential aspect of the claim is built here on a single case study, which does not look representative of synesthesia, and dominant cases like color-grapheme. First, it is not clear that the mathematical synesthesia of JP, described by Brogaard here, requires the presence of an external inducer. As said earlier, only very few synesthetes seem to have visual experiences when only imagining or thinking about a certain inducer (Dixon et al. 2004 for the only case). If JP has automatic visual images of diagrams when seeing or thinking about mathematical formulas, or if thinking about the formula is a sufficient inducer of visual conscious states, the case is fascinating but different from other cases of synesthesia where the presence and characteristics of the inducer matter. Indeed, it is also not clear that JP's experiences co-vary with features of the external objects – as synesthetic color-grapheme do. At the end, I do not see why visual experiences of diagrams induced by mathematical formulas cannot count as (non intentional, not controlled, complex) visual imagery.

It is fair, I think, to try and stick to the most documented cases of synesthesia when it comes to discussing the neural correlates of synesthetic colors. Now, intra-modal cases such as color-grapheme synesthesia come with the activation of primary visual areas – V1 and most often V4, crucially. As noted in Rouw et al. (2011) in their review of the brain-imagery data collected about synesthesia, "*synesthetic color activation is found related to visual cortex, but this is not restricted to V4. As is reported in 'real' color processing (e.g., Beauchamp et al. 1999; Schluppeck and Engel 2002), synesthetic color has been found to activate a broader range of areas in ventral occipito-temporal cortex.*" (Rouw et al. p. 218). There is then no straightforward neurological difference in terms of localization between perceptual experience of colors and synesthetic ones, that could help one count the later as 'non experiential' across synesthetes (see Table 27.2).

Turning to cross-modal cases of synesthetic colors, like color-hearing, the same is true – except that in this case, V4 and the auditory cortex will be activated (Nunn et al. 2002). More generally, if a state is to count as non-experiential in Brogaard's sense by not correlating with activation in primary sensory areas, then every case of synesthesia is experiential, because synesthesia is defined and studied through the actual presentation of a sensory stimulus.

Pushing the non-experiential argument further requires making the neurological criterion more specific. What is needed for a state to count as non-experiential is not just that it does not activate primary sensory areas at all: it should not activate the primary sensory areas that one would expect given the experience reported. This could work quite well for crossmodal cases, like colored-hearing for instance, where

Table 27.2 Comparison of neuroimaging studies of synesthetic color experiences

First author, year	Number of synesthetes	V1 activation during synesthetic color experience	V4 activation during synesthetic color experience	V4 activation to real color in synesthetes
Hubbard et al. (2005)	6	No	Yes	N.A.
Nunn et al. (2002)	13	No	Yes	Yes
Sperling et al. (2006)	4	No	Yes	Yes
Van Leeuwen et al. (2010)	21	N.A.	Yes	Yes
Gray et al. (2006)	15	N.A.	No	Yes
Paulesu et al. (1995)	6	No	No	N.A.
Rich et al. (2006)	7	N.A.	No	Yes
Rouw (2007)	18	No	No	N.A.
Weiss et al. (2005)	9	N.A.	No	No

See Rouw et al. (2011) for a more detailed review
The table presents the activation documented in V1 and V4 during synesthetic color experiences, and activation in V4 in synesthetic participants during perceptual experiences of colors. Nearly half of the studies report activation of V4 during synesthetic color experiences and, as noted by Rouw et al. (2011), other areas of the visual cortex are active during synesthetic experiences (but insufficiently documented to be part of a systematic review). What's more, the apparently weak evidence concerning the increased activation in V1 due to synesthesia (noted only in 2 studies) must be taken carefully as increase in V1 might have been obscured in other studies by activation during both baseline and experimental conditions

V1 is not activated although subjects report visual experiences (this bracketing the fact that V4 receives projections from other areas). This works less well for color grapheme, as in this case what seems to occur is that there is more or a different pattern of activation in the primary sensory areas that one would expect based on a comparison with neuro-typical, non-synesthetic individuals. The neuroanatomical criterion needs then to be re-formulated in the following way:

Experiential/perceptual criterion: States of kind K are not experiential (neither perceptual nor imagistic) if they do not activate the primary sensory areas usually associated (in neuro-typical subjects) to the kind of experience one reports or show anomalous activation in the primary sensory areas that are associated to this kind of experience.

Is this revision sufficient? Formulated in this way, the neuroanatomical criterion is not yet enough to determine that synesthetic color experiences are non-experiential. Without entering into details, it seems unlikely that looking at sensory areas only to characterize synesthesia is not going to succeed, as synesthesia involves a network of areas (Rouw 2011 for a recent review). The most important question here is to know whether synesthesia is grounded in structural (e.g. Hyperconnectivity, Hubbard et al. 2011; Rouw and Scholte 2007) or functional (e.g. dishinibition, Cohen Kadosh et al. 2007; Grossenbacher and Lovelace 2001) differences. Alternatively, one might need to 'zoom in', as specific neurons located in color-sensitive areas have been found to be active in synesthetic color experiences and to behave differently from the neurons involved in real color perception. Finding the neural correlates of synesthetic colors, that will eventually distinguish them eventually from the kind of color experiences involved in real perception, requires looking both beyond and deeper within the primary sensory areas.

27.3 Synesthesia and Incorporation: Do We Need to Posit a Third Kind of Visual Experience?

I want to consider an option not envisaged above. Its advantage is to be more compatible with the actual state of neurological evidence, as well as with the kind of mixed evidence and subjective reports that synesthesia gives rise to. This proposal goes against the "third kind claims". It only recognizes two kinds of conscious visual states, one involved in perception and one involved in mental imagery. What synesthesia shows is not the existence of a third kind of conscious visual states, but the incorporation of mental imagery within perception.

The easiest way to define incorporation comes from Perky cases (Perky 1910). Back in the early twentieth century, Cheves Perky showed that the content of people's visual imagery could be influenced by perception. Participants were asked for instance to imagine a banana while facing and fixing a white wall. At the beginning of the task, the state they were in was a canonical state of visual imagery. As the experiment went on, a picture of a banana would appear on the wall, causing a change in the content of visual imagery: Participants reported a change in the orientation of the banana they were visualizing, consistent with the orientation of the banana projected on the wall.

Various follow-ups and interpretations of Perky cases have been given, none of which have proved robust (Segal and Gordon 1969). According to one interpretation of the initial results, being involved in an imagery task raises the threshold for visually detecting images, so that people do not detect or have a conscious experience of the banana on the wall. In order to explain the change in the imagined content, the imagery state must have been influenced by the concurrent unconscious perceptual processing. The resulting Perky state is still a (transformed) state of imagery. According to a second interpretation though, participants are detecting and consciously perceiving the banana, and stop imagining it at some point. What explains the results is that they are subjectively mistaken and take a perceptual state for an imagined one. One difficulty with this interpretation is that it does not account for other Perky cases where for instance participants were asked to image the New York skyline, while a tomato was projected on the wall. In this case, the content did not switch totally to become identical with the content they should have perceived. Participants reported imagining New York skyline under sunset. This leads to a third and last interpretation: The participants had a conscious, albeit faint, perception of the pictures on the wall as they were engaged in the imagery task, without probably recognizing what these pictures were of. The imagery state then incorporated this conscious perceptual state. In both the first and the third interpretation, the resulting experience is an imagery state with a mixed content: The content comes both from what was voluntarily imagined and from what as involuntarily consciously or unconsciously perceived. The content is not experienced as a mixture: It is experienced as the unified, cohesive content of the imagined state, albeit one with surprising features. Participants are also supposed to report having at some point felt that their imagery was not in their control and been surprised to experience for instance the banana in a different orientation.

Here, though, it is important to notice that thinking about incorporation is detachable from the questions concerning the robustness or various interpretations of Perky effects. Even if there is no Perky-incorporation, or if it is less frequent than Perky said (Perky 1910), this does not mean that incorporation does not exist. Perky cases provide an illustration of what it means for one conscious state to incorporate another state, as well as the way incorporation occurs. The content of a conscious imagery state can be involuntarily changed or enriched by the progressive incorporation of the content of a perceptual experience of the same modality without the initial state stopping to be felt as imagery, but still loosing some of its key features, like control over content.

Synesthesia, I reckon, can be thought of as a kind of incorporation, with the following differences: First, incorporation occurs in the reverse direction as a perceptual state (e.g. of a letter) is changed or enriched by the incorporation of a conscious mental image (e.g. a color); Second, this incorporation is not involuntary, but also fast, as the mental image is not progressively generated but appears rapidly (after the stimulus has been perceived or attended to, as demonstrated by the now converging evidence that synesthesia does not give rise to pop out effects, see Mattingley et al. 2006; Rich and Mattingley 2003, 2010; Sagiv et al. 2006). Like in the Perky cases, the initial state determines the subjective character of the experience for the subject: The incorporated state is felt as imagery in the Perky cases, and as perceptual for synesthesia. The mixed origin of the content gets un-noticed, and subjects experience a single (incorporated) content in a unified perceptual way. A parallel here can be done with states whose content results from multiple sensory sources, like speech, and are experienced in a single way, for instance as purely auditory (see Spence and Bayne, in press, for discussion). Finally, like in Perky cases, the resulting state can loose some key features: In the synesthetic case, the overall perceptual state can be under a limited form of control, or feel as if it is (as indicated by some reports).

The incorporation account has several benefits. It explains why the overall synesthetic experience is vivid and 'feels' perceptual, although part of its content is not really perceived. It explains why synesthetic experiences depend, like perception, on the presence of an external object/ stimulus: One needs a perceptual state for the imagery content to be generated and incorporated into. The account can also be reconciled with the current state of neurological investigation regarding hyperconnectivity or lack of inhibition, at least in the sense that it takes no commitment on what grounds the incorporation relation at the neural level. Of course, one question which might be raised is whether the same neural mechanisms that are (or could be) at stake in one direction of incorporation (i.e. Perky one, where mental imagery incorporates perception) are also at stake in the other direction (i.e. the synesthetic one, where perception incorporates mental imagery). If that was the case, one might want to predict a higher propensity to Perky phenomena among synesthetes. The claim has not been tested yet, to my knowledge. This said, people being more prone to 'synesthetic induction' under hypnosis (Cohen Kadosh et al. 2009; Terhune et al. 2011) are the most suggestible, and suggestibility is measured in terms of susceptibility to absorption – of which incorporation is a kind (Tellegen and Atkinson 1974). According to the incorporation account, it could that synesthetes were not necessarily just higher imagers in the concurrent of the modality (a claim that is for the moment only backed up by subjective reports, see Barnett and Newell 2008) but also and mainly people who are more suggestible and subject to incorporation between imagery and perception.

Another interesting aspect of this account is that it explains why intramodal cases would be so dominant in synesthesia. Although crossmodal Perky effects are being documented, the most discussed and reported cases occur between visual imagery and visual perception, between closely related areas. If synesthesia is seen as a case of incorporation, along the lines of Perky-effect, then the reverse incorporation is also much more likely to occur within vision, that is by the incorporation of visual imagery in visual perception.

27.4 Conclusions

How similar or different synesthetic experiences are from perceptual experiences and imaginings is an important question, which Brogaard's paper is right to raise. Contrary to its proposal, though, I do not think that synesthesia has yet given us enough reasons to posit an experience of the third kind – that is, for instance, non-experiential visual seeming. What synesthesia shows is probably a form of fast, involuntary incorporation of mental imagery into the content of the perceptual experience of the inducer, which results in state which still feels like perception. One important theoretical and practical conclusion of this account is that in synesthesia, we should not separate the concurrent from the experience of the inducer (Auvray and Deroy 2013; Deroy in press). Most philosophers (Gray 2001a, b; Wager 1999, 2001, but see Macpherson 2007 for an exception) have been talking about synesthetic colors forgetting they are experienced not just in close connection with letters, but as *intrinsic parts* of the letter – that is of what it looks (and the same is true for sounds, tastes, etc.). The idea of the incorporation of color into a perceptual state is here more appropriate and gives another way to understand the etymology of 'synesthesia': not as different sensory objects being perceived together and bound in perception, but as a sensory and an imagined object being consciously experienced together with the second being absorbed into the initial perceptual experience.

Acknowledgments This research has been conducted thanks to the European Commission FP7 programme (IEF, 4CB). Thanks to Richard Brown and Fiona Macpherson for comments and suggestions on this material.

References

Auvray, M., and O. Deroy. 2013. How do synaesthetes experience the world? In *Oxford handbook of philosophy of perception*, ed. M. Matthen. Oxford: Oxford University Press.

Barnett, K.J., and F.N. Newell. 2008. Synaesthesia is associated with enhanced, self-rated visual imagery. *Consciousness and Cognition* 17: 1032–1039.

Beauchamp, M.S., J.V. Haxby, J.E. Jennings, and E.A. DeYoe. 1999. An fMRI version of the Farnsworth-Munsell 100-hue test reveals multiple color-selective areas in human ventral occipitotemporal cortex. *Cerebral Cortex* 9(3):257–263.

Brang, D., S. Coulson, and V.S. Ramachandran. 2011. Similarly shaped letters evoke similar colors in grapheme-color synesthesia. *Neuropsychologia* 49: 1355–1358.

Calvert, G.A., E.T. Bullmore, M.J. Brammer, R. Campbell, S.C. Williams, P.K. McGuire, P.W. Woodruff, S.D. Iversen, and A.S. David. 1997. Activation of auditory cortex during silent lipreading. *Science* 276: 593–596.

Cohen Kadosh, R., K. Cohen Kadosh, and A. Henik. 2007. The neuronal correlate of bidirectional synesthesia: A combined event-related potential and functional magnetic resonance imaging study. *Journal of Cognitive Neuroscience* 19: 2050–2059.

Cohen Kadosh, R., A. Henik, A. Catena, V. Walsh, and L.J. Fuentes. 2009. Induced cross-modal synaesthetic experience without abnormal neural connections. *Psychological Science* 20: 335–346.

Craver-Lemley, C., and A. Reeves. In press. Is synesthesia a form of mental imagery? In *Multisensory imagery: Theory and applications*, eds. S. Lacey and R. Lawson. New York: Springer.

Cui, X., C.B. Jeter, D. Yang, P.R. Montague, and D.M. Eagleman. 2007. Vividness of mental imagery: Individual variability can be measured objectively. *Vision Research* 47: 474–478.

Cytowic, R.E. 1989. *Synesthesia: A union of the senses*. New York: Springer.

Deroy, O. In press. Synaesthesia and parasitic qualia. In *Phenomenal qualities*, ed. P. Coates. Oxford: Oxford University Press.

Deroy, O., and C. Spence. Submitted. Why we are not all synaesthetes (not even weakly so).

Dixon, M.J., D. Smilek, C. Cudahy, and P.M. Merikle. 2000. Five plus two equals yellow. *Nature* 406: 365.

Dixon, M.J., D. Smilek, and P.M. Merikle. 2004. Not all synaesthetes are created equal: Projector versus associator synaesthetes. *Cognitive, Affective, & Behavioral Neuroscience* 4: 335–343.

Duffy, P.L. 2001. *Blue cats and chartreuse kitten: How synesthetes color their worlds*. New York: Henry Holt & Company.

Eagleman, D.M., A.D. Kagan, S.S. Nelson, D. Sagaram, and A.K. Sarma. 2007. A standardized test battery for the study of synesthesia. *Journal of Neuroscience Methods* 159: 139–145.

Galton, F. 1880. Visualised numerals. *Nature* 21: 252–256.

Gray, R. 2001a. Synaesthesia and misrepresentation: A reply to Wager. *Philosophical Psychology* 14: 339–346.

Gray, R. 2001b. Cognitive modules, synaesthesia and the constitution of psychological natural kinds. *Philosophical Psychology* 14: 65–82.

Gray, J.A., D.M. Parslow, M.J. Brammer, S. Chopping, G.N. Vythelingum, and D.H. Fytche. 2006. Evidence against functionalism from neuroimaging of the alien colour effect in synaesthesia. *Cortex* 42: 309–318.

Grossenbacher, P.G., and C.T. Lovelace. 2001. Mechanisms of synesthesia: Cognitive and physiological constraints. *Trends in Cognitive Science* 5: 36–41.

Hertrich, I., S. Dietrich, and H. Ackermann. 2011. Cross-modal interactions during perception of audiovisual speech and nonspeech signals: An fMRI study. *Journal of Cognitive Neuroscience* 23: 221–237.
Hubbard, E.M. 2007. Neurophysiology of synesthesia. *Current Psychiatry Reports* 9: 193–199.
Hubbard, E.M., A.C. Arman, V.S. Ramachandran, and G.M. Boynton. 2005. Individual differences among grapheme-color synesthetes: Brain-behavior correlations. *Neuron* 45: 975–985.
Hubbard, E.M., D. Brang, and V.S. Ramachandran. 2011. The cross-activation theory at 10. *Journal of Neuropsychology* 5: 152–177.
Hupe J-M, Bordier C, Dojat M (2011) The neural bases of grapheme–color synesthesia are not localized in real color-sensitive areas. Cerebral Cortex 1687. doi:10.1093/cercor/bhr236
Lacey, S., P. Flueckiger, R. Stilla, M. Lava, and K. Sathian. 2010. Object familiarity modulates the relationship between visual object imagery and haptic shape perception. *NeuroImage* 49: 1977–1990.
Lycan, W. (2006). Representational theories of consciousness. In *The Stanford encyclopedia of philosophy*, cd. E.N. Zalta ⟨http://plato.stanford.edu/archives/win2006/entries/consciousness-representational/⟩
Macpherson, F. 2007. Synaesthesia, functionalism and phenomenology. In *Cartographies of the mind: Philosophy and psychology in intersection series: Studies in brain and mind*, ed. M. de Caro, F. Ferretti, and M. Marraffa, 65–80. Amsterdam: Springer.
Marks, L. 2011. Synaesthesia, now and then. *Intellectica* 55: 47–80.
Mattingley, J.B., J.M. Payne, and A.N. Rich. 2006. Attentional load attenuates synaesthetic priming effects in grapheme-colour synaesthesia. *Cortex* 42: 213–221.
Nunn, J. A., Gregory, L. J., Brammer, M., Williams, S. C. R., Parslow, D. M., Morgan, M. J., ... Gray, J. A. (2002). Functional magnetic resonance imaging of synesthesia: Activation of V4/V8 by spoken words. Nature Neuroscience, 5(4), 371–375.
Palmeri, T.J., R. Blake, R. Marois, M.A. Flanery, and W. Whetsell. 2002. The perceptual reality of synesthetic colors. *Proceedings of the National Academy of Science of the United States of America* 99: 4127–4131.
Paulesu, E., J. Harrison, S. Baron-Cohen, J.D.G. Watson, L. Goldstein, J. Heather, and C.D. Frith. 1995. The physiology of coloured hearing A PET activation study of colour-word synaesthesia. *Brain* 118: 661–676.
Pekkola, J., V. Ojanen, T. Autti, J.P. Jääskeläinen, R. Möttönen, A. Tarkiainen, and M. Sams. 2005. Primary auditory cortex activation by visual speech: An fMRI study at 3 T. *Neuroreport* 16: 125–128.
Perky, C.W. 1910. An experimental study of imagination. *The American Journal of Psychology* 21: 422–452.
Rader, C.M., and A. Tellegen. 1987. An investigation of synesthesia. *Journal of Personality and Social Psychology* 52: 981–987.
Ramachandran, V.S., and E.M. Hubbard. 2003. The phenomenology of synaesthesia. *Journal of Consciousness Studies* 10: 49–57.
Rich, A.N., and J.B. Mattingley. 2002. Anomalous perception in synaesthesia: A cognitive neuroscience perspective. *Nature Reviews Neuroscience* 3: 43–52.
Rich, A.N., and J.B. Mattingley. 2003. The effects of stimulus competition and voluntary attention on colour-graphemic synaesthesia. *NeuroReport* 14: 1793–1798.
Rich, A.N., and J.B. Mattingley. 2010. Out of sight, out of mind: Suppression of synaesthetic colours during the attentional blink. *Cognition* 114: 320–328.
Rich, A.N., J.L. Bradshaw, and J.B. Mattingley. 2005. A systematic, large-scale study of synaesthesia: Implications for the role of early experience in lexical-colour associations. *Cognition* 98: 53–84.
Rich, A.N., M.A. Williams, A. Puce, A. Syngeniotis, M.A. Howard, F. McGlone, and J.B. Mattingley. 2006. Neural correlates of imagined and synaesthetic colours. *Neuropsychologia* 44: 2918–2925.
Rouw, R. 2011. 'Special Cases': Neural mechanisms and individual differences in synaesthesia. *Journal of Neuropsychology* 5: 145–151.

Rouw, R., H.S. Scholte, and O. Colizoli. 2011. Brain areas involved in synaesthesia: A review. *Journal of Neuropsychology* 5(2):214–242.

Rouw, R., and H.S. Scholte. 2007. Increased structural connectivity in grapheme-color synesthesia. *Nature Neuroscience* 10: 792–797.

Sagiv, N., and L.C. Robertson. 2005. Synesthesia and the binding problem. In *Synesthesia: Perspectives from cognitive neuroscience*, ed. L.C. Robertson and N. Sagiv, 90–107. New York: Oxford University Press.

Sagiv, N., J. Heer, and L. Robertson. 2006. Does binding of synesthetic color to the evoking grapheme require attention? *Cortex* 42: 232–242.

Sagiv, N., I. Ilbeigi, and O. Ben-Tal. 2011. Reflections on synaesthesia, perception, and cognition. *Intellectica* 55: 81–94.

Schluppeck, D., and S.A. Engel. 2002. Color opponent neurons in V1: A review and model reconciling results from imaging and single-unit recording. *Journal of Vision* 2(6):480–492.

Segal, G. 1997. Synaesthesia: Implications for the modularity of mind. In *Synaesthesia: Classic and contemporary reading*, ed. S. Baron-Cohen and J.E. Harrison, 211–224. Malden: Blackwell.

Segal, S.J., and P.E. Gordon. 1969. The Perky effect revisited: Blocking of visual signals by imagery. *Perceptual and Motor Skills* 28: 791–797.

Simner, J. 2007. Beyond perception: Synaesthesia as a psycholinguistic phenomenon. *Trends in Cognitive Sciences* 11: 23–29.

Simner, J. 2012. Defining synaesthesia. *British Journal of Psychology* 103: 1–15.

Spence, C., and O. Deroy. 2012. Crossmodal mental imagery. In *Multisensory imagery: Theory and applications*, eds. S. Lacey and R. Lawson. New York: Springer.

Spence, C., and T. Bayne. in press. Is consciousness multisensory? In: *Perception and its modalities*, ed. M. Matthen and D. Stokes. Oxford: Oxford University Press.

Sperling, J.M., D. Prvulovic, D.E.J. Linden, W. Singer, and A. Stirn. 2006. Neuronal correlates of colour-graphemic synesthesia: A fMRI study. *Cortex* 42: 295–303.

Spiller, M.J., and A.S. Jansaria. 2008. Mental imagery and synaesthesia: Is synaesthesia from internally-generated stimuli possible? *Cognition* 109: 143–151.

Tellegen, A., and G. Atkinson. 1974. Openness to absorbing and self-altering experiences ("absorption"), a trait related to hypnotic susceptibility. *Journal of Abnormal Psychology* 83: 268–277.

Terhune, D.B., S. Tai, A. Cowey, T. Popescu, and R. Cohen Kadosh. 2011. Enhanced cortical excitability in grapheme-color synesthesia and its modulation. *Current Biology* 21(23):2006–2009.

Van Leeuwen, T.M., K.M. Petersson, and P. Hagoort. 2010. Synaesthetic colour in the brain: Beyond colour areas. A functional magnetic resonance imaging study of synaesthetes and matched controls. *PloS One* 5: e12074.

Wager, A. 1999. The extra qualia problem: Synaesthesia and representationalism. *Philosophical Psychology* 12: 263–281.

Wager, A. 2001. Synaesthesia misrepresented. *Philosophical Psychology* 14: 347–351.

Weiss, P.H., K. Zilles, and G.R. Fink. 2005. When visual perception causes feeling: Enhanced cross-modal processing in grapheme-color synesthesia. *NeuroImage* 28: 859–868.

Witthoft, N., and J. Winawer. 2006. Synesthetic colors determined by having colored refrigerator magnets in childhood. *Cortex* 42: 175–183.

Chapter 28
Varieties of Synesthetic Experience

Berit Brogaard

In her response to my "Seeing as a Non-Experiental Mental State: The Case from Synesthesia and Visual Imagery" Ophelia Deroy presents an argument for an interesting new account of synesthesia. On this account, synesthesia can be thought of as "a perceptual state (e.g. of a letter)" that is "changed or enriched by the incorporation of a conscious mental image (e.g. a color)." Deroy argues convincingly that Perky cases, in which the content of visual imagery is partially constituted by the content of perceptual experience, possibly are best understood as incorporated, or mixed, mental states (Deroy, Chap. 27, this volume; Perky 1910). But even if Perky cases are not truly mixed conscious states, Deroy argues, it is quite plausible that some of our mental states have perceptual as well as imagistic elements. Cases of synesthesia are good candidates to be exactly these kinds of mixed states.

Deroy's account of synesthesia no doubt provides a correct and fruitful description of many cases of synesthesia, particularly cases of the more common kinds, such as week-color synesthesia and grapheme-color synesthesia. Synethetes with these forms of synesthesia often describe their experiences exactly in this kind of mixed way. As I mentioned in the paper Deroy addresses, one of our subjects FS provides just this kind of mixed description of his synesthesia:

> The colors are not out there. [...] I think it's related to imagery. It feels like imagining that something has a color. But I am not just imagining it. I think it's perceptual.

One important lesson to draw from this account, if correct, Deroy argues, is that "in synaesthesia, we should not separate the concurrent from the experience of the inducer" (Deroy, Chap. 27, this volume; Auvray and Deroy 2013). In other words, we tend to focus on the effects incurred by the inducer, for example, the color experience that occurs in grapheme-color synesthesia as a result of exposure to

B. Brogaard (✉)
Departments of Philosophy, Center for Neurodynamics & Brogaard Lab for Multisensory Research, University of Missouri, St. Louis, MO, USA
e-mail: brogaardb@gmail.com

R. Brown (ed.), *Consciousness Inside and Out: Phenomenology, Neuroscience, and the Nature of Experience*, Studies in Brain and Mind 6, DOI 10.1007/978-94-007-6001-1_28,

graphemes. But, as Deroy correctly points out, many synesthetes do not experience the concurrent in isolation from the inducer. Grapheme-color synesthetes typically experience numbers *as* colored. They may even believe that numbers are colored, until someone or something makes them question their belief. Synesthete Patricia Lynne Duffy, for example, tells us that she thought everyone experienced graphemes in the same colors as she did. The turning point came when she was 16:

> I was sixteen when I found out. The year was 1968. My father and I were in the kitchen, he, in his usual talk-spot by the pantry door, my sixteen year-old self in a chair by the window. The two of us were reminiscing about the time I was a little girl, learning to write the letters of the alphabet. We remembered that, under his guidance, I'd learned to write all of the letters very quickly except for the letter 'R'.
>
> "Until one day," I said to my father, "I realized that to make an 'R' all I had to do was first write a 'P' and then draw a line down from its loop. And I was so surprised that I could turn a yellow letter into an orange letter just by adding a line."
>
> "Yellow letter? Orange Letter?" my father said. "What do you mean?"
>
> "Well, you know," I said. "'P' is a yellow letter, but 'R' is an orange letter. You know – the colors of the letters."
>
> 'The colors of the letters?' my father said.
>
> It had never come up in any conversation before. I had never thought to mention it to anyone. For as long as I could remember, each letter of the alphabet had a different color. Each word had a different color too (generally, the same color as the first letter) and so did each number. The colors of letters, words and numbers were as intrinsic a part of them as their shapes, and like the shapes, the colors never changed. They appeared automatically whenever I saw or thought about letters or words, and I couldn't alter them.
>
> I had taken it for granted that the whole world shared these perceptions with me, so my father's perplexed reaction was totally unexpected. From my point of view, I felt as if I'd made a statement as ordinary as "apples are red" and "leaves are green" and had elicited a thoroughly bewildered response. I didn't know then that seeing such things as yellow P's and orange R's, or green B's, purple 5's, brown Mondays and turquoise Thursdays was unique to the one in two thousand persons like myself who were hosts to a quirky neurological phenomenon called synesthesia. Later in my life, I would read about neuroscientists at NIH and Yale University working to understand the phenomenon... But that day in the kitchen, my father and I, never having heard of synesthesia, both felt bewildered. (Excerpt from *Blue Cats and Chartreuse Kittens*)

This sort of case is very common. Duffy's experiences are not simply experiences of colors in response to graphemes but experiences of colored graphemes.

I agree with Deroy that focusing on the concurrent and ignoring the inducer when describing synesthetic experience could lead to misleading accounts of what is actually going on in many cases of synesthesia, I am, however, skeptical of there being one correct account of all forms of synesthesia. In the case study I report on as well as in a case study reported on by Bor et al. it appears that the concurrent is experienced as separate from the inducer (Brogaard et al. 2012; Bor et al. 2007). Subject JP, who acquired synesthesia following a brutal assault, describes the complex geometrical images he experiences in response to mathematical formulas as an interpretation or "understanding" of the formulas. DT, a high-functioning autistic savant and synesthete, characterizes the colored shapes he experiences in response to operations on numbers as sometimes emerging before he becomes conscious of the numbers. For example, the result of multiplying two numbers is the colored shape that fits between the shapes representing the multiplied numbers.

He first experiences the colored shape, the multiplication product, and then notices the number it represents. A similar phenomenon occurs when he recites the decimal points of irrational numbers like Pi. He describes his recitation experience as walking through a Pi landscape. As he walks through this landscape of colors and shapes, the actual digits occur to him one by one.

Another synesthete in our lab LS, a vision-to-sound synesthete who was born profoundly deaf in both ears, also exhibits a clear dissociation between inducer and concurrent in some of his synesthetic experiences. When people familiar to him enter the periphery of his visual field, it gives rise to a "ping" sound experience. Strangers that he has never encountered or has paid no attention to in the past do not. LS, who also suffers from face blindness, relies on his "pings" to recognize faces. The pings come from wherever the face is located in space, despite LS not being consciously aware of having seen anyone. At one time he was in the lobby of the Museum of Natural History in London. The lobby was full of people. LS scanned the lobby and suddenly he heard a "ping." He walked through the crowd to the other side of the room. There he found a fellow synesthete and friend sitting with his head pointed downward reading the museum guide. LS had not actively been looking for anyone. He had just been looking around to decide in which direction he wanted to go. In LS's case the inducer (i.e., a familiar face) is not experienced at all while the concurrent (i.e., the "ping" sound) is. LS's "ping" experiences thus are evidently not of the mixed variety. They are purely auditory.

One lesson to draw from this is that different kinds of synesthetic experience may compose different kinds of mental state. This is to be expected, however, as different kinds of synesthesia likely proceed via different mechanisms. Some kinds of grapheme-color synesthesia likely occur via cross-activation between color areas in the visual cortex and the adjacent visual word form area, as proposed by Hubbard and Ramachandran (Hubbard et al. 2006; Ramachandran and Hubbard 2001a, b). These types of grapheme-color synesthesia are good candidates to be of the mixed kind suggested by Deroy.

But other kinds of synesthesia clearly do not proceed via this mechanism. LS's vision-sound synesthesia likely is due to a reorganization of regions of auditory cortex during the first 5 years of his life. JP and DT's mathematically induced color-shape experiences could be due to memory associations that have become automatic over time.

Other forms of synesthesia may arise as a result of disinhibited feedback from an area of the brain that binds information from different senses (Armel and Ramachandran 1999; Grossenbacher 1997; Grossenbacher and Lovelace 2001). Armel and Ramachandran (1999), for example, report on a case of a patient PH, who was seeing visual movement in response to tactile stimuli following acquired blindness (Armel and Ramachandran 1999). As PH was blind, he could not have received the information via standard visual pathways. It is plausible that the misperception was a result of disinhibited feedback from brain regions that receives information from other senses.

The little-discussed cases of drug-induced synesthesia, which occur as a result of hallucinogens such as mescaline, psilocybin and LSD, may also turn out to be due to disinhibited feedback (Sinke et al. 2012). It is doubtful, however, that drug-induced synesthesia and congenital synesthesia have exactly the same underlying mechanism, as the former differs from the latter in nearly every respect. Even the very experience of drug-induced synesthesia at the time at which it occurs appears notably different from most cases of congenital synesthesia. The psychological effect of hallucinogens is typically described as a dream-like state accompanied by a diminished sense of self, a decrease of self-control, a change in time perception and vivid visual illusions and hallucinations and sometimes synesthetic experience. Though music and sounds are the most frequent inducers of synesthesia during drug intoxication, all sorts of sensory input, including olfactory, gustatory, haptic, pain and emotional stimuli, can induce synesthetic experience. Drug-induced synesthesia also may fail to exhibit the test-retest reliability that is characteristic of other forms of synesthesia, though the judge is still out.

Perhaps ‘synesthesia’ is best construed as an umbrella term covering a variety of interesting forms of cross-modal perception. Deroy discusses a case of seeing the lip movements of a famous performer on television with the sound off and involuntarily experiencing an auditory image of the song. With a large enough repertoire of songs, the phenomenon might satisfy the standard synesthesia conditions as well as grapheme-color synesthesia does. While Deroy is not particularly sympathetic to such lip-movements/sound phenomena being rendered a type of synesthesia, my intuitive response is “why not?”

References

Armel, K.C., and V.S. Ramachandran. 1999. Acquired synesthesia in retinitis pigmentosa. *Neurocase* 5: 293–296.

Auvray, M., and O. Deroy. 2013. How do synaesthetes experience the world? In *Oxford handbook of philosophy of perception*, ed. M. Matthen. Oxford: Oxford University Press.

Bor, D., J. Billington, and S. Baron-Cohen. 2007. Savant memory for digits in a case of synaesthesia and Asperger syndrome is related to hyperactivity in the lateral prefrontal cortex. *Neurocase* 13: 311–319.

Brogaard, B., S. Vanni, and J. Silvanto. 2012. Seeing mathematics: perceptual experience and brain activity in acquired synesthesia. Neurocase (in press)

Grossenbacher, P.G. 1997. Perception and sensory information in synaesthetic experience. In *Synaesthesia: Classic and contemporary readings*, ed. S. Baron-Cohen and J.E. Harrison, 148–172. Malden: Blackwell Publishers, Inc.

Grossenbacher, P.G., and C.T. Lovelace. 2001. Mechanisms of synesthesia: Cognitive and physiological constraints. *Trends in Cognitive Science* 5: 36–41.

Hubbard, E.M., S. Manohar, and V.S. Ramachandran. 2006. Contrast affects the strength of synesthetic colors. *Cortex* 42: 184–194.

Perky, C.W. 1910. An experimental study of imagination. *The American Journal of Psychology* 21: 422–452.

Ramachandran, V.S., and E.M. Hubbard. 2001a. Psychophysical investigations into the neural basis of synaesthesia. *Proceedings of the Royal Society London B Biological Sciences* 268: 979–983.

Ramachandran, V.S., and E.M. Hubbard. 2001b. Synaesthesia: A window into perception, thought and language. *Journal of Consciousness Studies* 8: 3–34.

Sinke, C., J.H. Halpern, M. Zedler, J. Neufeld, H.M. Emrich, and T. Passie. 2012. Genuine and drug-induced synesthesia: A comparison. *Consciousness and Cognition* 21(3): 1419–1434.

Part X
Higher-Order Thought Theories of Consciousness and the Prefrontal Cortex

Chapter 29
Not a HOT Dream

Miguel Ángel Sebastián

29.1 Introduction

In 'On a confusion about the function of consciousness', Ned Block (1995–2002) famously maintained that our folk psychological term 'consciousness' equivocates between two concepts: 'access-consciousness' and 'phenomenal consciousness'. The first one has to do with the processing of information. When I look at the cup of coffee in front of me I take in plenty of information: the cup is located in front of me, to the left of my computer, it has cylindrical shape and red color and it is filled with a black liquid. When I consciously see the cup, my brain processes all this information and this information is typically available for further reasoning (deciding to drink the coffee), motor control (moving my hand toward the cup), etc. Understanding the mechanisms that underlie these processes constitutes what Chalmers (1996) calls 'the easy problem of consciousness'. It is, no doubt, a very complicated issue given the complexity of our brains, but the research in neurosciences has made huge amounts of progress in recent years and it is, from a philosophical perspective, relatively unproblematic.

Nevertheless, there is more to consciousness than this information processing. When I see my cup, there is *something it is like for me* to see it; a reddish way, among others, *it is like for me* to have this experience. This is phenomenal consciousness and explaining it is what constitutes *the hard problem of consciousness* (Chalmers 1996).

The relation between access and phenomenal consciousness is an important issue that cannot be settled without a further clarification of the notions involved. Even so, some form of access seems to be essential to phenomenal consciousness, for

M.Á. Sebastián (✉)
UNAM/LOGOS, Tumaco 17, 28027 Madrid, Spain
e-mail: msebastian@gmail.com

R. Brown (ed.), *Consciousness Inside and Out: Phenomenology, Neuroscience, and the Nature of Experience*, Studies in Brain and Mind 6, DOI 10.1007/978-94-007-6001-1_29,

it is platitudinous that when one has a phenomenally conscious experience, one is in some way aware of it. Let me call this kind of access '*Awareness*' following Block (2007).

Higher-Order Representational (HOR) theories of consciousness maintain that *Awareness* is a form of representation. That is to say, phenomenally conscious states are states that are the object of some sort of higher-order representation. The kind of representation that is required by the theory makes a basic difference among HOR theorists.[1] Nonetheless, I want to draw an orthogonal distinction to make the target of the argument I am about to present clear. My target in this paper will be theories that maintain that *Awareness* is a form of cognitive access, the same cognitive access that underlies the ability to report – more precisely, higher-order theories that maintain that the cognitive ability that makes it possible to report the content of a mental state is essential to phenomenally conscious mental states. My opponent holds a higher-order cognitive position characterized by the following three claims:

Higher-Order Cognitive

1. Consciousness requires *Awareness*.
2. *Awareness* requires the right kind of Higher-Order Representation.
3. The right kind of Higher-Order Representation depends on the cognitive accessibility that underlies reporting.[2]

This position has been paradigmatically held by Higher-Order Thought (HOT) theorists.[3] According to HOT theories, a mental state M is conscious if and only if

[1]The main concern is whether higher-order states are belief-like or perception-like. The former are called Higher-Order Thought (HOT) theories (Gennaro 1996; Rosenthal 1997, 2005) and the latter Higher-Order Perception (HOP) or 'inner-sense' theories (Amstrong 1968; Carruthers 2000; Lycan 1996). According to the former theories, when I have a phenomenally conscious experience as of red I am in a mental state with certain content, call this content 'RED'. For this mental state to be phenomenally conscious, there has to be, additionally, a higher-order thought targeting it, whose content is something like 'I AM SEEING RED'. On the other hand, HOP theories maintain that what is required is a (quasi-) perceptual state directed on to the first-order one. A second point of disagreement is whether a given state is conscious in virtue of its disposition to raise a higher-order representation (Carruthers 2000) or by being actually the target of a higher-order representation (Rosenthal 1997, 2005); this is the difference between dispositional and actualist HOR theories. According to dispositional HOR theories, the higher-order representation that renders the *Awareness* of the first-order one doesn't have to be actual; i.e., there is no need for the higher-order representation to happen actually, what is needed for a mental state to be conscious is a disposition to be the object of such a higher-order representation.

[2]Note that organisms lacking our ability to report being in a particular mental state might still have the same kind of cognitive accessibility that we have. Hence, lacking the ability to report does not prevent that one can have higher-order representations.

[3]Not all Higher-Order theories are committed to these three claims. Consider, for instance, Carruthers (2000)'s dispositionalist view. According to Carruthers, phenomenally conscious states are, roughly speaking, states that are recognized as representations by a Theory of Mind. Each experience would, at the same time, be a representation of some feature of the world (for example, a representation of red) and a representation of the fact that we are undergoing such an experience (a representation of seems red), through the consumer system that is the Theory of Mind.

there is another belief-like mental state (a Higher-Order Thought) to the effect that one is in M. Being conscious requires being *Aware* of oneself as being in a certain mental state and this *Awareness* is explained as being the target of the appropriate HOT (e.g. a HOT that is non-inferentially caused). The greatest exponent of this theory, David Rosenthal, explicitly endorses the correspondence between HOTs, and hence conscious mental states, and the ability to report being in a particular mental state. In 'Thinking that one thinks' Rosenthal (2005, chapter 2) writes:

> [G]iven that a creature has suitable communicative ability, it will be able to report being in a particular mental state just in case that state is, intuitively, a conscious mental state. If the state is not a conscious state, it will be unavailable to one as the topic of a sincere report about the current content of one's mind. And if the mental state is conscious one will be aware of it and hence able to report that one is in it. *The ability to report being in a particular mental state therefore corresponds to what we intuitively think of as that state's being in our stream of consciousness.* (Op. cit., p.55, my emphasis)

I will focus on Rosenthal's HOT theory in my criticism for I consider it to be the quintessence of theories that hold a higher-order cognitive position. The position that I will be defending, call it *non-cognitive position*, maintains that *Awareness* does not depend on the cognitive accessibility that underlies reporting. Therefore, it maintains, *pace* HOT theories, that there can be cases of phenomenal consciousness on which subjects might not be able to report due to a failure in the cognitive access.

In the next two sections, I will provide empirical evidence in favor of the premises of my argument. Section 29.4 presents my argument against HOT theories and in Sect. 29.5 I consider some possible objections and offer a rejoinder.

29.2 The Neural Correlate of Cognitive Accessibility for Visual Experiences: dlPFC

The evidence for the neural correlate of the cognitive accessibility, in the case of visual experiences, is provided by an experiment performed by Lau and Passingham (2006). This experiment suggests that such cognitive accessibility depends on the dorsolateral prefrontal cortex (dlPFC).

The experiment is based on a visual discrimination task with metacontrast masking. Metacontrast masking takes place when a target stimulus is followed, after a short period of time called '*Stimulus Onset Asynchrony*' (SOA), by a mask that shares a contour with it, leading to a reduction in perceived brightness and to degraded perception of the spatial shape of the target (Haynes and Rees 2003).

If these mindreading capacities do not depend on the cognitive accessibility that underlies reporting, as it plausibly doesn't, then Carruther's theory illustrates an example of a Higher-Order theory that is not jeopardized by the success of my argument.

In Sect. 29.4 I will present a hypothetical cognitive HOR theory that might be immune to my argument.

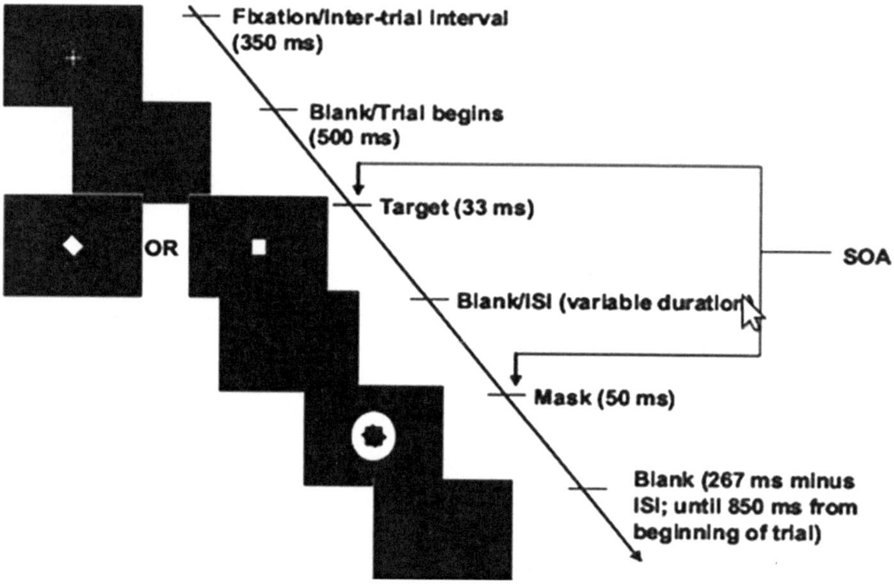

Fig. 29.1 Lau and Passingham's experimental set up (Lau and Passingham 2006)

Subjects in the experiment are asked to fixate their gaze and they are presented with one of two possible stimuli, either a square or a diamond on a black background. After a short variable period of time, the SOA, a mask is presented. The mask overlaps with part of the contour of both possible stimuli but it does not overlap with any of them spatially (See Fig. 29.1).

Subjects in the experiment have to perform two tasks after the presentation of the target and the mask:

1. Decide whether the target stimulus was a diamond or a square.
2. Indicate whether they actually saw the target or were simply guessing in the previous task.

The first question is intended to measure the objective *performance capacity* of the subjects: how good they are at identifying the target stimulus. The second question is intended to measure the *perceptual certainty* of the subjects: how confident they are on having seen the stimulus. This subjective report, according to the authors and to HOT theories, is an indication of phenomenal consciousness.

Figure 29.2 shows the result as a function of the SOA, the interval between the presentation of the target stimulus and the mask. The presence of the mask has nearly no influence on the performance capacity (represented by a continuous line) nor on the perceptual certainty (represented by the dotted line) when presented before or close to the stimulus. As the SOA increases, the mask interferes with the perception of the target stimulus and both, the performance capacity and the perceptual certainty decrease until a certain point where the influence of

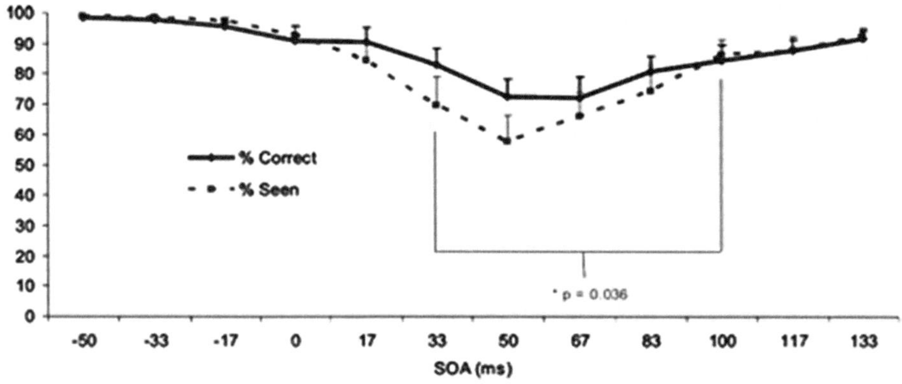

Fig. 29.2 Performance capacity (% correct) vs. Perceptual certainty (% seen) (Lau and Passingham 2006)

the mask starts to diminish, having no effect at all when it is presented much later than the stimulus. The resulting curves have a U-shape, where two points, corresponding to different SOAs, with the same performance capacity and two points, also corresponding to different SOAs, with the same perceptual certainty can be identified.

The interesting finding is that we can detect two conditions under which the performance capacity of the subjects is the same but such that they differ in their perceptual certainty. Whereas in one (short SOA), subjects tend to report having guessed when they were asked about the identity of the stimulus; in the other (long SOA), subjects are fairly confident of having seen it. For HOT theories, the subject is phenomenally conscious only in the second case where she reports having seen the stimulus.

Lau and Passingham performed an fMRI study on the subjects of the experiment. Their study revealed that the long SOA condition was associated with a significant increase in activity in the left mid-dorsolateral prefrontal cortex (mid-dlPFC, Brodmann's area 46).

My opponent maintains that *Awareness* depends on the cognitive accessibility that underlies reporting. In the Lau and Passingham experiment, subjects report having seen the stimulus in the long SOA condition but not in the short one. Hence, we may assume that they are phenomenally conscious of the stimulus only in the long SOA condition. Since HOTs are associated with reporting abilities, Lau and Passingham have found the "neural residence" of HOTs, at least for visual higher-order thoughts (thoughts of the form 'I SEE A SQUARE').[4] Rosenthal explicitly

[4]Lau and Passingham maintain that consciousness should be associated with perceptual certainty. Lau (2008) explicitly endorses this view. He maintains that consciousness depends on Bayesian decisions on the presence of the stimuli relying upon a learning process and on the firing pattern of the first-order representations. Lau named his view 'Higher-Order Bayesian Decision Theory'.

accepts the evidence from this experiment as showing that the neural correlate of HOTs is in the dlPFC:

> There is, however, some evidence that states are conscious when, and only when, a distinct neural state occurs in mid-dorsolateral prefrontal cortex (area 46) (Lau and Passingham 2006), and it is reasonable to explore identifying these neural occurrences with the posited HOTs. Rosenthal (2008, p. 835).

On the other hand, the defender of the non-cognitive position would maintain that the curve corresponding to phenomenology could be somewhere in between the two curves in Fig. 29.2 (% correct and % seeing) and is not impressed by the fMRI data. The reason is that she would have predicted exactly this result: the judgment of having seeing the stimulus, which corresponds to a HOT, is reflected in the prefrontal cortex.[5]

So, does the Lau and Passingham's experiment bring some light to the debate between higher-order cognitive and non-cognitive approaches? I think it does, but precisely in the opposite direction from which the authors intended. If HOTs live (or at least a significant part of their neural correlate is) in the dlPFC, as the experiment suggests, and there were a case of phenomenology without activation of dlPFC, HOT theories would be in trouble. It's time for dreaming.

29.3 Dreams and dlPFC

Revonsuo (2000) defines dreams as "a subjective experience during sleep, consisting of complex and organized images that show temporal progression". Dreams are phenomenally conscious experiences, experiences that are similar in many respects to the ones we have during wakefulness. Our dreams are highly visual, with rich colors, shapes and movements, and include sounds, smells, tastes, tactile sensations, and emotions, as well as pain and pleasure (Hobson et al. 2000).

Dreams can be so similar to our waking experiences that the dreamer may be uncertain whether he is awake or asleep. This platitude has been taken for granted by most philosophers. It has, for instance, led philosophers to wonder whether we can distinguish the two states or even whether one could actually be dreaming constantly. This has been considered by Plato, Aristotle and most famously in Descartes' skeptical argument in the First Meditation. The view that dreams are conscious experiences has been explicitly endorsed, in the philosophical field, by, among others, Kant, Russell, Moore, and Freud (Malcolm 1959, p.4). Most contemporary philosophers working on dreams also hold this view (see for instance Ichikawa 2009; Ichikawa and Sosa 2009; Metzinger 2003, 2009; Revonsuo 2006; Sosa 2005).

It is unclear to me why, a proposal along these lines should be considered a Higher-Order Representationalist one. See fn. 12.

[5]This possibility has been suggested by Ned Block in the Second Consciousness Online Conference (http://consciousnessonline.wordpress.com)

I do not intend to argue that dream experiences are exactly like waking experiences. According to Tononi (2009, p.100), dreaming experiences in comparison to waking experiences are characterized by disconnection from the environment, internal generation of a world-analogue, reduction of voluntary control and reflective thought, amnesia and a high emotional involvement. Furthermore, dream reports may include phenomena that resemble neuropsychiatric conditions such as distortion of time perception, perceived distortion of body parts, bizarre illogical situations, prominence of negative emotions, anxiety and fear, and misidentification syndromes like erroneously recognizing a familiar person despite the lack of any obvious physical resemblance (Karim 2010). The only point that is relevant for the purpose of this paper is that we have dreams and that dreams include phenomenally conscious visual experiences.[6]

Sleep is traditionally divided into two phases: non-rapid eye movement (NREM) sleep and rapid eye movement (REM) sleep.[7] The succession of this two phases is called a sleep cycle, and, in humans, it lasts for approximately 90–110 min. There are 4–5 cycles per night. It has been established that dreams occur during (though probably not exclusively) REM phase of sleep.

Although there is some controversy as to whether or not there are dreams that occur during NREM, there is no doubt that we dream during REM phase. If subjects were awakened from that stage of sleep and asked whether they had dreamed, they would say yes at least 80 % of the time. What happens in the brain during this period?

29.3.1 Neurophysiology of Sleep

During sleep there is a global reduction in metabolic activity and blood flow in the brain. Compared to resting wakefulness, the decrease during NREM phase can reach a 40 % as shown by positron emission tomography (PET) studies (Braun et al. 1997). At the cortical level, activation is reduced in the orbitofrontal and anterior cingulate and dorsolateral prefrontal cortex – Broadmann area 46 (See Braun et al. 1997, table 1 p.1177).

During REM sleep some areas are even more active than in wakefulness, especially the limbic areas. In the cortex, the areas receiving strong inputs from the amygdala, like the anterior cingulate and the parietal lobe, are also activated (Maquet et al. 1996, table 1 p.164). On the other hand, the rest of the parietal cortex, the precuneus and the posterior cingulate are relatively inactive (Braun et al. 1997, table 2 p.1178).

---

[6]Some philosophers have tried to resist this claim. I will present their views and offer a rejoinder in Sect. 29.5.2.

[7]A more fine-grained categorization of the NREM phase can be done based on EEG, EOG, and EMG patterns. For details see Tononi (2009).

What is relevant for this discussion is that there is a selective deactivation (compared with wakefulness) of the dlPFC (Braun et al. 1997; Maquet et al. 1996, 2005; Muzur et al. 2002) during REM phase.[8] Specifically, Maquet et al. showed a very significant reduction in the activity of the area identified by Lau and Passingham (left dorsolateral prefrontal cortex).

All of these regional activations and inactivations are consistent with the differences in mental states between sleep and wakefulness (see Schwarz and Maquet 2002; Tononi 2009). In particular, the deactivation of the dlPFC, which is associated with executive abilities such as expectancy, volitional control and working memory in wakefulness (Fuster 2008), fits in well with the common loss of self-reflective awareness and rational control in dreams (Kahn 2007).

According to Lau and Passingham's experiment, the neural correlate of HOTs lies in the dlPFC; there is an increase in its activity when subjects report having seen the stimulus in comparison with the situation in which they report not having seen it and having guessed – despite the lack of difference in their performance in both situations. If HOTs were constitutive of phenomenal consciousness we would expect its neural correlate to be active during dreams. However, empirical evidence suggests the opposite. Given these elements the reader can easily anticipate my argument against HOT theories.

29.4 The Argument

In this section I present the argument against HOT theories in more detail.

Let me start with a simple argument against *cognitive theories of consciousness* in general. I call 'cognitive theories of consciousness' those theories that maintain that the cognitive accessibility that underlies reporting is constitutive of phenomenal consciousness. One example of such cognitive theories is, as we have seen, Rosenthal's HOT theory. Another example is Michael Tye's PANIC theory (Tye 1997, 2002). According to Tye, phenomenally conscious mental states are states whose content is Poised, in the sense that it is available to first-order belief-forming and behavior-guiding systems; Abstract, meaning that the intentional content is not individuated by the particular things represented; and Non-conceptual in the sense that it is not structured into concepts. Contrary to HOT, PANIC is a first-order theory. It does, however, endorse the claim that phenomenal consciousness depends on the cognitive accessibility underlying our ability to report – on the plausible assumption that it is the same one as the one that underlies belief-forming and behavior-guiding.

The argument against *cognitive theories of consciousness* has the form of a *reductio ab adsurdum*:

---

[8] In the Maquet et al. study, subjects were controlled for dreaming (the subject maintained steady REM sleep during scanning and recalled dreams upon awakening). This control is missing in the Braun et al. study.

(Anti-Cognitive)

1. Phenomenal consciousness depends on the cognitive accessibility that underlies reporting.
2. The cognitive accessibility that underlies reporting, in the case of visual experiences, depends on the left dorsolateral prefrontal cortex (dlPFC).
3. dlPFC is necessary[9] for phenomenally conscious visual experiences (From 1 and 2).
4. We have phenomenally conscious visual experiences during the REM phase of sleep.
5. dlPFC is deactivated during the REM phase of sleep.
6. dlPFC is not necessary for phenomenally conscious visual experiences (From 4 and 5).
7. Phenomenal consciousness does not depend on the cognitive accessibility that underlies reporting (From 1 to 6).

Premise 1 is the common claim of what I have called *cognitive theories of consciousness* and the assumption of the argument. Premise 2 is supported by Lau and Passingham's experiment. As I have presented it, the neural correlate of the difference between subjects reporting seeing the target stimuli and not seeing it is in the left dorsolateral prefrontal cortex. (3) follows from these two premises.

It is hard to deny that we have conscious experiences during sleep and that those experiences include conscious visual experiences. These experiences typically happen during the REM phase of sleep (4). However, as we have seen, there is empirical evidence showing a selective deactivation of the dlPFC during the REM phase (5). (4) and (5) together suggest that the activation of the dlPFC is not required for having a phenomenally conscious experience and lead us to the claim that the dlPFC is not necessary for consciousness (6). (3) and (6) are contradictory claims, what lead us to reject premise 1, *QED*.

This argument might, however, be invalid. The reason is that one can deny that (6) follows from the conjunction of (4) and (5). This possibility is explored by Lau himself. According to Lau's theory (Lau 2008), the role of dlPFC is to work as a Bayesian decision system that tries to make "accurate judgments" about the inputs of the sensory cortex. The increase in the noise signals in the sensory cortex during REM phase in comparison to NREM, accompanied by a deactivation of the dlPFC, explains dreams as a malfunction of the decision system.

> By this definition, one hallucinates while dreaming; in dreams we consciously perceive stimuli that are not really there . . . Dreams are more likely to be reported during a stage of sleep that is characterized by rapid eye movement (REM), and brain activity of relatively high frequency and intensity. Let us assume that the overall signal during REM-sleep is higher. If the brain maintains the same criterion for detection over alternations of REM and non-REM sleep, it would be predicted that false positives are a lot more likely during REM-sleep, because of the higher signal intensity. (op.cit., p.41)

[9]Modal claims in this argument are obviously to be read as restricted to *beings like us in worlds with the same laws as the actual one*.

Dreams are for Lau similar to hallucinations. According to Lau, during sleep the dlPFC is deactivated and, therefore, malfunctioning, making the wrong *judgments*.[10] Lau can, hence, accept (4) and (5) while resisting (6): the dlPFC is malfunctioning due to its deactivation, but its *judgments*, right or wrong, are still required for phenomenal consciousness.

In order to properly evaluate Lau's claim, further details about how the decision mechanism is supposed to work and how the decrease of activity in the dlPFC is related to this mechanism need to be added. We need an explanation of how the decrease in the activity of the dlPFC during REM is related to the failure "to set an appropriately high criterion during REM sleep" so that "one mis-classifies noise as stimuli." (op.cit, p.41). Such an explanation has to be compatible with the fact that the perceptual certainty, which according to Lau corresponds to phenomenal consciousness, is accompanied with an increase in the activity of the dlPFC in the original experiment. It is an open question whether a satisfactory answer can be provided and an empirical issue whether the dlPFC works in this way. If Lau were right then (Anti-Cognitive) would be an invalid argument.

This line of reasoning can be endorsed by defenders of first-order cognitive theories like Tye's PANIC. It seems reasonable to think of the dlPFC as a filter. A state would be available for reporting – and hence poised – if the dlPFC let its content go through; in other words, if the dlPFC decides that the signal arriving corresponds to sensory input and not to noise. A similar reply could be provided by a particular kind of Higher-Order Theory, call it Indexical Higher-Order Representational Theory (IHOR). According to IHOR, in the case of visual conscious experiences, the first-order state with the content *SQUARE* is accompanied by a higher-order indexical thought, encoded in the dlPFC, with the content 'I SEE THIS' pointing to the first-order one.[11,12]

---

[10]Lau has maintained, in private conversation, that, contrary to HOT, the under-activation of the dlPFC during REM phase is favorable to his theory because in dreams perceptual judgments are wrong.

[11]If one is interested in this strategy, one would have to elaborate on the mechanisms on which such a demonstration would rely.

[12]Those willing to endorse Lau's model of cognitive accessibility will maintain that there are two states involved. The relation between these two states distinguishes higher-order and first-order theories. Lau and Passingham (2006) seem to be silent among the two kinds of theories.

On the one hand, a first-order theory maintains that there is a merely causal relation between the two states, which we can call ANIC and PANIC taking Tye's theory as a model, and that both states have the same intentional object, say the square.

On the other hand, IHOR maintains that the relation between a first-order and the higher-order one is not only causal but intentional. Whereas the first-order state has the square as its object, the higher-order one has the first-order one as its intentional object. IHOR has to make room for cases in which there is no first-order state, cases of misrepresentation. It is unclear to me what would be the phenomenology of cases in which the demonstration fails and there is no first-order state the higher-order one is pointing to. For a discussion on related issues derived of such an intentional relation see Block (2011), Rosenthal (2011) and Weisberg (2011).

This strategy is, however, not available for HOT theories. According to HOT theories, the higher-order state is not indexical as in IHOR, but something like 'I SEE A SQUARE' in the previous example. If dlPFC encodes HOTs, we would expect an increase in its activity as the content of conscious phenomenology increases, because we would expect more frequent updates in the corresponding HOTs. HOT theory seems to be committed to the claim that there is a monotonic relation between the content of conscious experiences and the activity of the neural correlate of HOTs. It is, therefore, unable to accommodate the data about the brain activity during dreams as we have just seen, blocking thereby the inference from (4) and (5) to (6) in the argument.

In the next section I will discuss possible replies that the defender of HOT theories can endorse against the argument and offer a rejoinder.

29.5 Replies

29.5.1 HOTs Have a Different Neural Correlate During Dreams

One possible way to resist the argument would be to maintain that HOTs have two different neural correlates. During wakefulness, dlPFC is the neural correlate for visual HOTs, whereas during sleep HOTs have a different neural correlate. This way, one blocks step 3 in (Anti-Cognitive), because, in spite of the fact that the cognitive accessibility that underlies reporting in the case of visual experiences depends on the dlPFC, it only does so during wakefulness and, therefore, it is not true that the activity of the dlPFC is necessary for conscious visual experiences (3).

That kind of dissociation seems, however, implausible. Having another area responsible for HOTs during dreams would require a functional duplication and mutual exclusion. Imagine that we have another area that is the neural correlate of dreams during sleep,[13] let me refer to this area as 'the sleep neural correlate of HOT '(SNCHOT). When we have a visual experience during wakefulness, the neural correlate of the corresponding HOT is in the dlPFC, and not SNCHOT, which is not differentially activated as the fMRI in the Lau and Passingham's experiment shows. On the other hand, during dream experiences, dlPFC is deactivated and the neural correlate of the HOT would be SNCHOT. The question is: why do we need SNCHOT?

[13]A plausible candidate could be the anterior cingulate. As we have seen this area is strongly activated during the REM phase. Furthermore, the anterior cingulate communicates to the relevant sensory and limbic areas.

REM sleep seems to be exclusive to marsupial and placental mammals (Winson 1993). It is, therefore, reasonable to assume that the only organisms capable of dreams are those at the top of the pyramid of evolution. The plausibility of SNCHOT depends on the function of dreams during sleep; a function that should require HOTs. If dreams have no function, it seems unreasonable to assume that changes in brain activity during REM phase appear to give rise to HOTs in other areas that were not present during wakefulness, and the only area they are present during wakefulness seems to be the dlPFC.

Most of the theories of dreaming yield dreams as epiphenomenal.[14] This has been explicitly claimed by Flanagan:

> [Dreams are] a likely candidate for being given epiphenomenalist status from an evolutionary point of view. P-dreaming [phenomenal experiences during sleep] is an interesting side effect of what the brain is doing, the function(s) it is performing during sleep. To put it in slightly different terms: p-dreams, despite being experiences, have no interesting biological function. I mean in the first instance that p-dreaming was probably not selected for, that p-dreaming is neither functional nor dysfunctional in and of itself (Flanagan 1995, p.9).

Sometimes it is held that dreams are the result of noise activity or a by-product of the changes in brain activity during sleep. This option is considered by the Activation-Synthesis theory (Hobson and McCarley 1977), where dreams are the result of the forebrain responding to random activity initiated at the brainstem; the improved AIM (Activation, Input-ouput gating, Modulation) model (Muzur et al. 2002) or by Lau (2008), as we have just seen.

Solms (1997) has recently defended the Freudian view that the function of dreams is to protect sleep. However, Solms does not attribute any functions to the content of dreams, and therefore HOTs, and he also regards dreams as hallucinations that the weakened frontal reflective systems mistake for real perception.

Other theories maintain that dreams have a function in memory processing (Crick and Mitchison 1983; Foulkes 1985; Hobson et al. 1994), in which case there is no function for HOTs and dreams merely reflect the corresponding memory processing – processes that do not require any HOT.

One exception is Revonsuo (2000).[15] According to him, the function of dreams is "to simulate threatening events and to rehearse threat perception and threat avoidance". But this function can also be performed during wakefulness, so the same structures that we use while we are awake could be used during sleep.

As long as one cannot make the case for the function of HOTs in dreams, and I seriously doubt that it can be made, we have no additional reason for defending the possibility of having an additional neural structure, SNCHOT, which differs from dlPFC. There seems to be no reason for a duplication of the HOT machinery.

---

[14]In the intended sense here, something is epiphenomenal if and only if it lacks biological function. This sense should be contrasted with the sense in which something is epiphenomenal if and only if it lacks causal impact whatsoever.

For a review of these epiphenomenal theories see Revonsuo (2000).

[15]See also Franklin and Zyphur (2005) for an extension of Revonsuo's proposal.

If this is right, and dlPFC is the neural correlate of HOTs responsible for visual experiences, then we have good reasons for believing that there are no visual HOTs during dreams and therefore a good support for (3).

An alternative objection would deny that we have phenomenally conscious experiences during sleep. This is the next objection I am going to consider.

29.5.2 We Do Not Have Conscious Experience During Dreams

A different possibility to block the argument is to reject premise (4). The common sense position maintains that dreams are conscious experiences; a position that has been maintained by philosophers, psychologists and neuroscientists, but not without exception.

The common sense position has been famously rejected by Malcolm (1959) who asserts that it leads to conceptual incoherency "... the notion of a dream as an occurrence that is logically independent of the sleeper's waking impression has no clear sense." (op.cit., p. 70). Malcolm maintains that we have no reason to believe the reports given by awakened subjects, for there is no way to verify them: they could be cases of "false memory".[16] It could be that processes during REM phase are all non-conscious and that on awakening there is a HOT targeting the content of memory and thereby making it conscious.

Whereas Malcolm denies that there are dreams, Dennett (1976) has defended a skeptical position. Dennett presents an alternative account in which dreams could be unconscious memory loading processes.[17] According to Dennett, before establishing whether dreams are conscious we need an empirical theory of dreams and that it is an "open, and theoretical question whether dreams fall inside or outside the boundary of experience" (op.cit., pp.170–171). Dennett goes a step further, claiming that we have some empirical evidence indicating that dreams are not conscious experiences, for they fail to satisfy well confirmed conditions for conscious experience like the activation of the reticular formation (op.cit., p.163).

This position has been challenged by Revonsuo (1995) who provides empirical evidence to the effect that there is in fact activity of the reticular formation and important neurophysiological similarity between dreaming and wakefulness.

> From the standpoint of the thalamocortical system, the overall functional states present during paradoxical sleep and wakefulness are fundamentally equivalent, although the handling of sensory information and cortical inhibition is different in the two states... That is, paradoxical sleep and wakefulness are seen as almost identical intrinsic functional states in which subjective awareness is generated. (Llinas and Pare 1991, p.522, quoted in Revonsuo 1995)

---

[16]Rosenthal, in conversation, points in this direction.

[17]It is not worth discussing the value of the proposal itself, for it is only intended to present a skeptical argument showing that there can be alternative explanations to dreamer's reports when awakened.

Unfortunately that would not impress my opponent. According to HOT theory, consciousness necessitates the presence of a HOT; HOTs are absent during dreams, so dreams are unconscious experiences.

Skepticism about dreams being phenomenally conscious experiences is based on the fact that the access to dreams is retrospective: we recall the dream when we are awakened and we have no reason for trusting these reports. There are cases, however, in which some people are aware of being dreaming. This is the case of lucid dreams. In lucid dreams, the dreamer is able to remember the circumstances of normal life and to act deliberately upon reflection.

Although lucid dreams have been reported since Aristotle, many have had their doubts about the reality of these episodes. Dennett endorses this skepticism; he considers that the report of lucid dreams is consistent with the hypothesis that dreams are unconscious episodes and that the subject is dreaming that she is aware of being dreaming. The empirical evidence suggests, nonetheless, that Dennett's hypothesis is wrong.

During REM sleep all skeletal muscle groups except those that govern eye movements and breathing are profoundly inhibited (LaBerge 2000); this fact makes it very difficult to collect evidence in favor of lucid dreams beyond subjects' reports upon awaking. Nevertheless, Rowarg et al. (1962) showed that some of the eye movements of REM sleep correspond to the reported direction of the dreamer's gaze. Based on this evidence, LaBerge et al. (1981) could provide evidence in favor of lucid dreams. They trained subjects and asked them to make distinctive patterns of voluntary eye movements when they realized they were dreaming. These prearranged eye movement signals were recorded by the polygraph records during REM, proving that subjects had indeed been lucid during uninterrupted REM sleep. Furthermore, LaBerge and Dement (1982) recorded lucid dreamers who were asked to either hold their breath or breath rapidly (in their lucid dreams), marking the interval of altered respiration with eye movement signals. The subjects reported having accomplished the agreed-upon tasks a total of nine times, and in every case, a judge was able to correctly predict, on the basis of the polygraph recordings, which of the two patterns had been executed. These results have been replicated by other laboratories (For a review see LaBerge 1988).

The experiments on lucid dreams provide evidence that we have conscious experiences during sleep, and give us the opportunity to record reports to that effect. The main reason for skepticism is dissolved: there are conscious dreams. In lucid dreams, subjects can report having an experience. One might be willing to concede that, independently of the preferred theory of consciousness, when subjects report having an experience they are entertaining a HOT. If dlPFC is the neural correlate of HOTs we should expect an increase in its activity in these cases.

Some authors have hypothesized that the deactivation of the dlPFC observed during REM sleep does not occur during lucid dreams. Dreams are conscious experiences characterized, among other things, by reduced voluntary control and reflective thought. These characteristics fit well, as we have seen (Fuster 2008), with the independent hypothesis that the dlPFC is involved in volitional control and self-monitoring. For this reason, a reactivation of the dlPFC is expected during lucid dreams (Hobson et al. 2000; Kahn and Hobson 2005; Tononi 2009). Preliminary empirical evidence for this hypothesis has been obtained from a recent study by Voss et al. (2009). This study shows that lucid dreaming in trained participants is associated with increasing electroencephalography (EEG) power, especially in the 40-Hz range, over frontal regions during REM sleep. Furthermore, Wehrle et al. (2005, 2007) use fMRI to study brain regional activation during lucid dreams and show that in lucid dreams not only frontal but also temporal and occipital regions are highly activated in comparison to non-lucid dreams. Hobson (2009) also refers to preliminary fMRI data gathered by M. Czisch, R. Wehrle and M. Dresler showing that dream lucidity is correlated with increased activation of the cortical areas including the dlPFC.[18]

My opponent can still try to resist the argument by maintaining that we have conscious experiences during lucid dreams but not during ordinary dreams, for only during lucid dreams can the subject report on them (according to her, reporting is inextricably linked to HOTs). However, distinguishing lucid dreams from other dreams in such a way that there is phenomenology associated to the former but not to the latter seems to be something of a reach.

29.6 Conclusions

Some philosophers have argued that phenomenal consciousness requires a certain form of *Awareness*, and that this *Awareness* depends on the cognitive accessibility that underlies reporting. Higher-Order Thought theories of consciousness are an example of this position.

Lau and Passingham's experiment provides good evidence for believing that the neural correlate of the reporting access to our visual conscious experiences depends on the dorsolateral prefrontal cortex (dlPFC). This would be, accordingly, the most plausible candidate to be the neural correlate of visual HOTs. The evidence seems to suggest that visual HOTs are not necessary for consciousness, because their neural correlate is highly deactivated during the phenomenally conscious experiences we have when we sleep: dreams.

I have argued that we have no reason to believe that visual HOTs are implemented by another area during sleep. The defender of HOT theory can embrace a skeptical position as to whether we have conscious dreams. This position, which runs against common sense, has been refuted by empirical evidence (lucid dreams).

---

[18]I am not sure about how to make this reactivation of the dlPFC compatible with Lau's hypothesis about the role of the dlPFC in dreams. Recall that this hypothesis might be endorsed by other cognitive theories, such as PANIC, to block my argument.

The position remaining for HOT theory is a not very plausible one, according to which, there would be an ontological dichotomy with regard to dreams (some dreams are phenomenologically conscious and others are not).[19]

References

Amstrong, D. 1968. *A materialist theory of the mind*. London: Routledge.

Block, N. 1995–2002. On a confusion about the function of consciousness. In *Consciousness, function, and representation: Collected papers*, vol. 1, ed. N. Block. Bradford: Bradford Books.

Block, N. 2007. Consciousness, accessibility, and the mesh between psychology and neuroscience. *The Behavioral and Brain Sciences* 30: 481–548.

Block, N. 2011. The higher order approach to consciousness is defunct. *Analysis* 71(3): 419–431.

Braun, A., T.J. Balkin, N.J. Wesenten, R. Carson, M. Varga, P. Baldwin, S. Selbie, G. Belenky, and P. Herscovitch. 1997. Regional cerebral blood flow throughout the sleep wake cycle. An H2(15)O pet study. *Brain* 120: 1173–1197.

Carruthers, P. 2000. *Phenomenal consciousness: A naturalistic theory*. Cambridge/New York: Cambridge University Press.

Chalmers, D.J. 1996. *The conscious mind: In search of a fundamental theory*, 1st ed. New York: Oxford University Press.

Crick, F., and G. Mitchison. 1983. The function of dream sleep. *Nature* 304: 111–114.

Dennett, D.C. 1976. Are dreams experiences? *Philosophical Review* 73: 151–171.

Flanagan, O. 1995. Deconstructing dreams: The spandrels of sleep. *The Journal of Philosophy* 92(1): 527.

Foulkes, D. 1985. *Dreaming: A cognitive-psychological analysis*. Hillsdale: Erlbaum.

Franklin, M., and M. Zyphur. 2005. The role of dreams in the evolution of the human mind. *Evolutionary Psychology* 3: 59–78.

Fuster, J. 2008. *The prefrontal cortex*, 4th ed. London: Academic.

Gennaro, R.J. 1996. *Consciousness and self-consciousness: A defense of the higher-order thought theory of consciousness*. Amsterdam: John Benjamins.

Haynes, L., and G. Rees. 2003. What defines a contour in metacontrast masking? *Perception* 32: 48.

Hobson, A. 2009. REM sleep and dreaming: Towards a theory of protoconsciousness. *Nature Reviews Neuroscience* 10: 803–813.

Hobson, J.A., and R.W. McCarley. 1977. The brain as a dream state generator: An activation-synthesis hypothesis of the dream process. *The American Journal of Psychiatry* 134: 1335–1348.

[19] I am very grateful to David Pineda and Rubén Sebastián for comments on a previous draft.

A previous version of this paper was presented on the 3rd Consciousness Online Conference and the LOGOS's GRG. An earlier ancestor was presented in the Cognitive Science talks at CUNY Graduate Center in summer 2010. I am very grateful to Marc Artiga, Richard Brown, Jake Berger, Michal Klincewicz, Stevan Harnad, Marta Jorba, Hakwan Lau, Dan Lopez de Sa, Pete Mandik, Manolo Martínez, Myrto Mylopoulos, David Rosenthal, and very especially to Josh Weisberg, Matthew Ivanowich and two anonymous referees for their comments.

Financial support for this work was provided by the Committee for the University and research of the department of Innovation, Universities and Company of the Catalonia government and the European Social Fund; by the DGI, Spanish Government, research project FFI2009-11347, the Consolider-Ingenio project CSD2009-00056 and by the AGAUR of the Generalitat de Catalunya (2009SGR-1077).

Hobson, J., E. Pace Schott, and R. Stickgold. 1994. *The chemistry of conscious states*. Boston: Little, Brown.
Hobson, J., E. Pace-Schott, and R. Stickgold. 2000. Toward a cognitive neuroscience of conscious states. *Behavioral and Brain Science* 23: 793–842.
Ichikawa, J. 2009. Dreaming and imagination. *Mind and Language* 24(1): 103–121.
Ichikawa, J., and E. Sosa. 2009. Dreaming, philosophical issues. In *The Oxford companion to consciousness*, ed. A.C. Tim Bayne and P. Wilken. New York: Oxford University Press.
Kahn, D. 2007. Metacognition, recognition and reflection while dreaming. In *The new science of dreaming*, ed. D. Barrett and P. McNamara. Westport: Praeger.
Kahn, D., and J.A. Hobson. 2005. A comparison of waking and dreaming thought. *Consciousness and Cognition* 14: 429–438.
Karim, A.A. 2010. Transcranial cortex stimulation as a novel approach for probing the neurobiology of dreams: Clinical and neuroethical implications. *International Journal of Dream Research* 3: 17–20.
LaBerge, S. 1988. Lucid dreaming in western literature. In *Conscious mind, sleeping brain. Perspectives on lucid dreaming*. New York: Plenum.
LaBerge, S. 2000. Lucid dreaming: Evidence and methodology. *The Behavioral and Brain Sciences* 23(6): 962–963.
LaBerge, S., and W. Dement. 1982. Voluntary control of respiration during REM sleep. *Sleep Research* 11: 107.
LaBerge, S.P., L.E. Nagel, W.C. Dement, and V.P. Zarcone. 1981. Lucid dreaming verified by volitional communication during REM sleep. *Perceptual and Motor Skills* 52: 727–732.
Lau, H. 2008. A higher-order Bayesian decision theory of perceptual consciousness. *Progress in Brain Research* 168: 35–48.
Lau, H., and R. Passingham. 2006. Relative blindsight in normal observers and the neural correlate of visual consciousness. *Proceedings of the National Academy of Sciences of the United States of America* 103: 18763–18768.
Llinas, R., and D. Pare. 1991. Of dreaming and wakefulness. *Neuroscience* 44: 521–535.
Lycan, W.G. 1996. *Consciousness and experience*. Cambridge: The MIT Press.
Malcolm, N. 1959. *Dreaming*. London: Routledge and Kegan Paul.
Maquet, P., J. Peters, J. Aerts, G. Delore, C. Degueldre, A. Luxen, and G. Franck. 1996. Functional neuroanatomy of human rapid-eye-movement sleep and dreaming. *Nature* 383: 163–166.
Maquet, P., P. Ruby, A. Maudoux, G. Albouy, V. Sterpenich, T. Dang-Vu, and S. Laureys. 2005. Human cognition during rem sleep and the activity prole within frontal and parietal cortices: a reappraisal of functional neuroimaging data. *Progress in Brain Research* 150: 219–227.
Metzinger, T. 2003. Being no one: The self-model theory of subjectivity, illustrated edition edn. Cambridge, MA: The MIT Press.
Metzinger, T. 2009. *The ego tunnel. The science of the mind and the myth of the self*. New York: Basic Books.
Muzur, A., E.F. Pace-Schott, and J.A. Hobson. 2002. The prefrontal cortex in sleep. *Trends in Cognitive Sciences* 6: 475–481.
Revonsuo, A. 1995. Consciousness, dreams, and virtual realities. *Philosophical Psychology* 8: 35–58.
Revonsuo, A. 2000. The reinterpretation of dreams: An evolutionary hypothesis of the function of dreaming. *The Behavioral and Brain Sciences* 23(6): 877–901.
Revonsuo, A. 2006. *Inner presence. Consciousness as a biological phenomenon*. Cambridge, MA: MIT Press.
Rosenthal, D.M. 1997. A theory of consciousness. In *The nature of consciousness*, ed. N. Block, O.J. Flanagan, and G. Guzeldere. Cambridge, MA: MIT Press.
Rosenthal, D.M. 2005. *Consciousness and mind*. Oxford, New York: Oxford University Press.
Rosenthal, D.M. 2008. Consciousness and its function. *Neuropsychologia* 46(3): 829- 840.
Rosenthal, D.M. 2011. Exaggerated reports. Reply to block. *Analysis* 71: 431–437.
Rowarg, H.P., W.C. Dement, J.N.J.N. Muzio, and C. Fisher. 1962. Dream imagery: Relationship to rapid eye movements of sleep. *Archives of General Psychiatry* 7: 235–258.

Schwarz, S., and P. Maquet. 2002. Sleep imaging and neuropsychological assessment of dreams. *Trends in Cognitive Sciences* 6: 23–30.
Solms, M. 1997. *The neuropsychology of dreams: A clinico-anatomical study*. Mahwah: Erlbaum.
Sosa, E. 2005. Dreams and philosophy. *Proceedings and Addresses of the American Philosophical Association* 79: 718.
Tononi, G. 2009. Sleep and dreaming. In *The neurology of consciousness: Cognitive neuroscience and neuropathology*, ed. S. Laurey and G. Tononi. Amsterdam: Elsevier.
Tye, M. 1997. *Ten problems of consciousness: A representational theory of the phenomenal mind*. Cambridge, MA: The MIT Press.
Tye, M. 2002. *Consciousness, color, and content*. Cambridge, MA: The MIT Press.
Voss, U., R. Holzmann, I. Tuin, and J.A. Hobson. 2009. Lucid dreaming: A state of consciousness with features of both waking and non-lucid dreaming. *Sleep* 32: 1191–1200.
Wehrle, R., M. Czisch, C. Kaufmann, T.C. Wetter, F. Holsboer, D.P. Auer, and T. Pollmaecher. 2005. Rapid eye movement-related brain activation in human sleep: A functional magnetic resonance imaging study. *Neuroreport* 16: 853–857.
Wehrle, R., C. Kaufmann, T.C. Wetter, F. Holsboer, D. Auer, T. Pollmaecher, and M. Czisch. 2007. Functional microstates within human REM sleep: First evidence from fMRI of a thalamocortical network specific for phasic REM periods. *European Journal of Neurosciences* 25: 863–871.
Weisberg, J. 2011. Abusing the notion of what-it's-like-ness: A response to Block. *Analysis* 71: 438–443.
Winson, J. 1993. The biology and function of rapid eye movement sleep. *Current Opinions in Neurobiology* 3: 243–248.

Chapter 30
Sweet Dreams Are Made of This? A HOT Response to Sebastián

Josh Weisberg

In his paper "Not a HOT Dream," Miguel Ángel Sebastián (2014) argues that a certain species of higher-order representational theory is empirically undermined by data from the dreaming brain. Some higher-order (HO) theories, notably David Rosenthal's higher-order thought (HOT) theory, seem committed to the claim that a crucial brain area implicated in phenomenal consciousness is the dorsolateral prefrontal cortex (dlPFC). However, this region shows reduced activation in REM sleep when phenomenally conscious dreams occur, threating the HOT view. In this commentary, I will present Sebastián's argument (modified a bit, as I'll explain below) and then offer a response on behalf of the HOT approach. I contend that Sebastián's attack falls short—there are a number of plausible rejoinders open to a defender of the HOT view. But I wish to stress at the outset that I think Sebastián presents a strong challenge to the theory and what's more, a challenge rooted in empirical data. This is the proper way to approach consciousness, rather than taking endless detours to zombie worlds and the color-deprived prisons of super scientists.

30.1 The Argument

First, a stipulation. I will assume that we are all using the terms 'consciousness' and 'phenomenal consciousness' in the same way: to mean experience where there is "something it's like for the subject." And this does not entail that phenomenal consciousness is irreducible or "intrinsic" in a special way. It's clear that Sebastián is

J. Weisberg (✉)
Department of Philosophy, University of Houston, 513 Agnes Arnold Hall,
Houston, TX 77204, USA
e-mail: jweisberg@uh.edu

R. Brown (ed.), *Consciousness Inside and Out: Phenomenology, Neuroscience, and the Nature of Experience*, Studies in Brain and Mind 6, DOI 10.1007/978-94-007-6001-1_30,

not arguing on conceptual grounds that the HOT view can't explain consciousness; rather, he is arguing that the view is refuted by the empirical evidence. So there is no charge of conceptual confusion about the target explanandum here.

Sebastián's argument against the HOT view (which he labels "Anti-Cognitive") is as follows:

Anti-Cognitive:

1. Phenomenal consciousness depends on the cognitive accessibility that underlies reporting.
2. The cognitive accessibility that underlies reporting, in the case of visual experiences, depends on the left dorsolateral prefrontal cortex (dlPFC).
3. dlPFC is necessary for phenomenally conscious visual experiences (From 1 and 2).
4. We have phenomenally conscious visual experiences during the REM phase of sleep.
5. dlPFC is deactivated during the REM phase of sleep.
6. dlPFC is not necessary for phenomenally conscious visual experiences (From 4 and 5).
7. Phenomenal consciousness does not depend on the cognitive accessibility that underlies reporting (From 1 to 6). (2014)

It is a reductio, using the contradiction between 3 and 6 to derive its conclusion. But note that Sebastián does not directly mention HOT theory. Instead, he attacks the idea that the "cognitive accessibility that underlies reporting" is necessary for phenomenal consciousness and argues separately the HOT theory is committed to the claim that phenomenal consciousness is explained by (or reduces to) that access. In my reconstruction of his argument, I will drop this formulation in favor of a direct argument against HOT theory. I think it confuses matters to bring in issues of "cognitive access" and reporting here. While there is a close connection between higher-order awareness, cognitive access to conscious mental states, and our ability to report those states, higher-order theorists do not present their views in this way and it raises suspicion that Sebastián may be creating a straw man by foisting an "access consciousness" view on them. Rosenthal, for example, rejects Block's access/phenomenal distinction, and Block himself deploys a different term ("reflexive consciousness") to pick out higher-order views.[1] What's more, as Sebastián notes (2014), Rosenthal denies that reporting ability is necessary for higher-order thought. Given these complications, I think it better to frame the argument as directly attacking a proposal for the *realization* of higher-order awareness, one that is at least tentatively endorsed by some higher-order theorists.[2]

[1]The initial phenomenal/access distinction is in Block (1995). For Rosenthal on Block's distinction, see Rosenthal (2002). For Block's reflexive consciousness, see Block (2001).

[2]I do reconsider Sebastián's cognitive access claim in Sect. 30.4, where I present alternative realizations for HOT.

The reconstructed argument (we can call it "Anti-HOT") runs as follows:

Anti-HOT:

1. REM dreams are conscious
2. dlPFC is not active in REM dreams
3. HOT is realized by dlPFC activity
4. There is consciousness without dlPFC activity (from 1 and 2)
5. Therefore, there is consciousness without HOT (from 3 and 4)

Here, the key premise is that HOT is realized by dlPFC activity—there is no detour through access and reporting. I believe this captures the real thrust of Sebastián's argument and nothing essential is lost by the simplification.[3] Further, it's clear that Rosenthal at least has tentatively endorsed something in the direction of 3, so there's no need, for the sake of this debate, to saddle the HOT view with the constitutive connection to access and reporting. Premises 1 and 2 are defended by Sebastián in his paper and I'll consider them below. 3, as Sebastián rightly points out, gains its support from the work of Lau and Passingham (2006), where they found that dlPFC activity seemed to correlate with phenomenal consciousness, as measured by confidence judgments of subjects in a metacontrast masking task. I will consider below whether this is the only reading of that data, and I'll also consider if there are any plausible alternative realization stories available for the HOT view. 4 and 5 follow from the first three premises, under the assumption that dlPFC activity is necessary for HOT. I'll consider the necessity (and sufficiency) claims about dlPFC and HOT as well.

30.2 Are REM Dreams Conscious?

Premise 1 holds that dreams occurring during REM sleep are phenomenally conscious (I'm calling them "REM dreams"). As Sebastián notes, this is indeed the commonsense position and it is a view held by many philosophers and scientists as well (2014, p. 427). But he acknowledges that the idea is not universally endorsed; famously, Malcolm rejected it for verificationist reasons and Dennett has raised methodological worries about it as well.[4] And while I believe that Malcolm's verificationism and Dennett's "first-person operationalism" are unnecessarily restrictive principles, I think it's worth noting the empirical difficulties involved in the neuroscientific study of dreams.

Subjects must be woken up and asked whether they were dreaming. This introduces a potential memory confound, as Dennett rightly notes: perhaps my dreams were not conscious, but my current memory of them is. And how could

---

[3]This does not, however, provide an argument that also challenges Michael Tye's "PANIC" theory (1995). But I am skeptical of Sebastián's attempt to assimilate Tye's and Rosenthal's views in this way. So, following the lead of Sebastián's title, I will focus solely on the HOT view.

[4]Malcolm (1959) and Dennett (1976).

one tell the difference between this possibility and the presence of accurately recalled conscious dreams? Sebastián is aware of this worry and responds by citing work on "lucid" dreams, where subjects seem to voluntarily report (nonverbally, of course) that they are *currently* dreaming—this apparently sidesteps the memory concern. But it turns out that lucid dreams are correlated with activation of dlPFC, so Sebastián's main point is lost. We don't have a confirmed case of conscious dreaming without dlPFC. But he responds that

> The position remaining ... is a not very plausible one, according to which there would be an ontological dichotomy with regard to dreams (some dreams are phenomenologically conscious and others are not) (2014).

I am not sure why this is such a worry, especially given the rarity of lucid dreaming. The HOT theorist (in a "Dennettian" spirit) asks for evidence of conscious dreams. Sebastián offers lucid dreams as an example. But that is a case where dlPFC is active. We are left with the claim that non-lucid dreams are sufficiently like lucid dreams to qualify as conscious. But why should we think this, given the worry of the memory confound? Indeed, there is a crucial anatomical difference between the two—dlPFC activation—suggesting that they are not alike. Perhaps the "lucidness" of lucid dreams just is the presence of phenomenal consciousness.

But even if this skeptical concern is waived (and I fully agree that commonsense holds that dreams are conscious), there is still a more moderate kind of worry we can press about premise 1, one that may have impact on the challenge to premise 2. It may be that the phenomenology of dreams is much less rich than the phenomenology of conscious sensory experience. We may believe that dream experiences are as rich as waking experiences, but that could be an illusion or a product of reconstructive memory. This is not to say that dreams are not conscious; rather, it's like the illusion of clear phenomenology all the way out to the periphery of our visual field.[5] There might be less there than meets the "mind's eye." It's not implausible to think that we don't represent *every* sensory detail of our dream worlds. And if there's less present in dream experiences, there's less for the dlPFC to do in dreams, even if the HOT theory is correct. My own dream experiences are not particularly vivid, at least in the sensory domain. Instead, it is the emotional content that stands out. I seem to have the gist of where I am and what's going on, but it's not often the case that I recall vivid sensory experience. And even if I do, that tends to be in a single modality—an intense sound or sight.

The question of the *richness*, rather than the existence, of conscious dreams helps bring out the great methodological difficulties of empirical research on dreams. How might we establish whether dreams are "thick" or "thin" in this sense?[6] Even the dedicated phenomenally-conscious-dream realist must acknowledge this point. And if things are thin, then reduced activation of dlPFC is not a problem. Indeed, it might be expected on a HOT view. I'll now consider reasons of this sort for rejecting premise 2 of Anti-HOT.

[5] See Dennett (1991). See also Schwitzgebel (2011).

[6] Cf. Hurlburt and Schwitzgebel (2007).

30.3 Is There dlPFC Activity in REM Sleep?

Premise 2 holds that dlPFC is not active in REM dreams. But it is more accurate to say that dlPFC activity decreases during REM dreams when compared to waking experience. This is taken by Sebastián to imply that the activity needed for the instantiation of HOT is missing. But this implication can be challenged.[7] The PET studies used to determine brain activity note which regions have the most change from waking to REM sleep. While they do find that frontal activity drops, they do not establish that all activity ceases there. Indeed, that is not the case—some metabolic activity plausibly continues in dlPFC in REM sleep.[8] But this means that it's not accurate to say that dlPFC is "deactivated"; rather, it is less active than it is during waking. If there is still some residual activity occurring, then it may be that this is enough to realize the HOT in question. This would be implausible, perhaps, if the content of REM dreams were as *rich* as that of waking experience. But as noted above, there is good reason to doubt this claim. The content of most dreams is intuitively sparser than the content of waking experience. Further, our intuition of richness in dreams, such as it is, may be unreliable. We may confabulate the richness of dream phenomenology or we may enrich our conscious memory of dreams beyond what was present in the actual event. Either way, there is less work required of HOT and so less work required for dlPFC. The reduction of activity therefore does not entail the absence of HOT.

In the discussion of a forerunner of Sebastián's paper during the "Consciousness Online 3" conference, Sebastián responded in some detail to this line of argument.[9] He contended that while it *may* be possible for the HOT theorist to explain the reduction of activity in dlPFC in this manner, there is a further fact which undermines the HOT position. There is evidence that dlPFC is also more active during *non-REM* (NREM) sleep than it is in REM sleep. But it is agreed by all parties that no conscious dreams occur in NREM, so the activity present in dlPFC can't be associated with phenomenal-consciousness imparting HOT.

Here, the HOT theorist can respond that dlPFC activity may not be *sufficient* for HOT, though it is (perhaps) necessary. If so, in NREM sleep, the activity in dlPFC may indicate something other than HOT. And there may be reasonable evidence that this is the case. Studies by Tononi and colleagues (i.e. Massimini et al. 2004) show that during NREM, long-range slow "delta" waves propagate throughout the cortex.[10] The function of these waves may be to control "spike timing-dependent

[7] David Rosenthal endorsed this sort of response in conversation. Thanks to both David Rosenthal and Hakwan Lau for helpful comments on the CO3 talk this paper is based upon.

[8] Muzur, Pace-Schott, and Hobson write, "We are aware that increased delta activity does not always mean (complete) inactivity... Rather than evaluating the absolute metabolism of the prefrontal cortex, we consider 'deactivation' of the prefrontal cortex in terms of relative activity" (2002, 476). Thanks to Richard Brown for noting this point in discussion during CO3.

[9] See http://consciousnessonline.com/2011/02/18/not-a-hot-dream/

[10] See also Ioannides et al. (2009) and Tian et al. (2006).

synaptic plasticity" leading to "synaptic consolidation … or downscaling" (6869). It turns out that a locus of the activity generating these waves is a region including dlPFC. So it may be that during NREM sleep, that region of brain is involved in a different task, explaining the increase of activity. When REM sleep occurs, the dlPFC is not involved in generating slow-wave synaptic consolidation and its activity drops accordingly. But there is no reason to think it drops below the level required for the sparsely-contented HOTs needed for REM dreams.

Sebastián then argued that it is implausible to think that one brain region might instantiate different functions at different times. But given the known plasticity of the brain and its massively complex interconnected circuitry, I am not sure we should expect a simple one region-one function mapping. Indeed, the "distributed networks" approach to neuroscientific modeling (e.g. Sporns 2010) holds that any anatomical region might be implicated in a range of psychological processes.[11] So the mere presence of activity in dlPFC need not indicate HOT in NREM sleep. Note also I am not embracing the claim attacked by Sebastián that there might be one realization for HOT in waking experience and another in dreams ("SNCHOT"). While I am not sure this is as implausible as Sebastián thinks, I am not endorsing it here. Rather, the claim is that when there is HOT, dlPFC realizes it. But dlPFC can do other things as well. This effectively captures the data at issue.

So it seems to me that Sebastián has failed to establish premise 2, undermining his attack. He has not ruled out the presence of all activity in dlPFC and so there may yet be enough to realize HOT. But it also strikes me as a rather restricted reading of the HOT theory to tie it to dlPFC in such a tight way. It may be that dlPFC realizes a particular kind of HOT content, present in certain sorts of conscious experiences (including, perhaps, lucid dreams). But dlPFC activity may not be even a necessary component of HOT in general. I will turn to this question now as I challenge premise 3 of Anti-HOT.

30.4 Is HOT Realized by dlPFC Activity?

Premise 3 holds that HOT is realized by dlPFC activity. As noted, the main support for this claim comes from Lau and Passingham (2006) and it is tentatively endorsed by one of the main HO theorists, David Rosenthal (2008; Lau and Rosenthal 2011). But it is not obvious to me that this is the only realization story the HOT theorist can embrace or even if it is the best one. Other theorists defending versions of HO theory suggest alternative realization bases for their views. It is worth looking at the evidence for these positions and then considering if premise 3 can be rejected.

At the outset, though, a concern must be addressed. Sebastián argues that his only target is "higher-order cognitive" theories, theories involving the cognitive access

[11] See also Van Orden et al. (2001) and Anderson (2007, 2008). Thanks to Cameron Buckner for alerting me to this point and for the references.

involved in reporting. And what's more, this access requires dlPFC. This fact, he contends, undermines HOT theory, but leaves other nearby HO views unscathed.[12] So he may respond at this point that giving up premise 3 removes the view he is concerned with—it is not a minor alteration. But this is an overly-narrow reading of the HOT theory. HOT theory holds that conscious states are states we are suitably aware of being in, and this awareness, in turn, is explained by the presence of a thought-like metarepresentation. It is certainly not entailed by the view that HOT be realized by dlPFC activity. Sebastián contends, however, that dlPFC activity is implicated in the "cognitive access that underlies reporting" and so, given his formulation, provides the tight tie with HOT theory.

But here the HOT theorist can plausibly reject this overly-close tie. HOT theory rejects the claim that reporting is necessary for consciousness. All that's needed for consciousness, on the view, is the awareness provided by HOT, whether or not we can report what we're aware of. And while it's true (as Sebastián notes in quoting Rosenthal (2014)) that it is intuitive that if we are *not* aware of a state we can't report it, this only means that HOT (and so consciousness) is needed for reporting, not that reporting ability is constitutive of HOT. The question then becomes, are there other ways to explain this "cognitive access" that do not require dlPFC? I see no reason not to think so beyond the suggestive evidence in Lau and Passingham. But this is an *empirical claim* about realization. And thus it is open to the HOT theorist to seek out other realization bases if the claim does not pan out.

And in any event, it is not clear to me that dlPFC activity tracked in Lau and Passingham (2006) corresponds to HOT. Lau and Passingham (L&P) asked subjects to make confidence judgments: i.e., "how confident are you that you saw the square (or diamond)?" As confidence rose, there was a corresponding increase in dlPFC activity. L&P argue that confidence judgments track phenomenal consciousness—confidence goes up as subjects become more conscious of the target stimulus. But another possibility is that dlPFC is involved in parsing out signal from noise in conditions where that's not clear from lower-level processing alone (cf. Lau 2008). As such, dlPFC may only be involved in HOT in certain conditions: conditions where it's hard to see the stimulus. At others times, HOT may not need to rely on the parsing of dlPFC. Indeed, that may occur in the L&P trials where no diamond or square is seen. Subjects are still conscious of the background, the monitor screen, their proprioceptive sensations, the sound of the lights and the AC, etc.[13] So it's not at all clear that L&P license the claim that dlPFC is necessary for HOT, even though it is suggestively implicated in these experiments and is just the sort of "frontal" process that HO theorists expect to find correlated with experience.

But what are the alternatives open to HOT theory? To the extent that independent cases can be made for these claims, we can reasonably reject premise 3. I will briefly sketch three proposals. The first, inspired by Peter Carruthers view, connects HOT with "theory of mind" and the *medial* prefrontal cortex (mPFC). The second, taken

[12] Carruthers (2000) and Lau (2008), for example.

[13] Cf. Ivanowich, this volume.

from the work of Antonio Damasio, involves higher-order mappings and the anterior cingulate cortex (ACC). The third is more in line with the "distributed networks" approach mentioned above: Hans Flohr's proposal for distributed HOT instantiated by large-scale neuronal cell assemblies involving NMDA receptors. All three are independently plausible live options for the HOT theorist, in my opinion.

To begin, one credible idea is that HOTs are products of a "theory of mind" (ToM) system, one that automatically employs "theoretical" representations of mental states, both of ourselves and others.[14] ToM posits mental states in order to predict and explain complex patterns of behavior. The theory can then be targeted back at ourselves, delivering a form of higher-order thought. Peter Carruthers has defended a version of higher-order theory which explicitly appeals to ToM, albeit a "dispositional" version of the view. But it is also open to an "actualist" HOT view to appeal to ToM—there is nothing about the ToM aspect of Carruthers' view necessitating a move to dispositionalism. On such a view, we are phenomenally conscious when we are occurently aware of our own states via application of ToM. This, in turn, gives us an alternative target for a realization base.

There is a considerable amount of literature on ToM and the brain and unsurprisingly there is a fair bit of controversy. But one leading theory is that ToM is at least partially realized by the *medial prefrontal cortex* (mPFC) (Saxe 2009). This prompts us to consider what happens in mPFC during REM dreams. And we find that mPFC is active.[15] What's more, J. Alan Hobson, whom Sebastián cites as providing key evidence about dlPFC activity, concludes that ToM is indeed active in REM dreams (Kahn and Hobson 2005). This hypothesis provides a "frontal" locale for HOT, but avoids the problems dlPFC seems to have with REM dreams.

A second proposal comes from the work of Antonio Damasio. His theory of core consciousness, particularly as laid out in his (1999) book *The Feeling of What Happens*, provides another possible realization base for HOT. Damasio explicitly notes his view's affinity with HO views, including Rosenthal's. Damasio holds that consciousness occurs when "higher-order maps" actively track both the "proto-self" and sensory cortices. These HO maps in essence represent the self and its current state. This is very much in the spirit of the HOT theory, which holds that mental states are conscious when we are conscious of ourselves as being in them (Rosenthal 2005). Damasio offers a detailed neurological sketch of his theory. He holds that activity in the anterior cingulate cortex (ACC) is crucial to realizing the posited higher-order maps: ACC seems to possess the right connective and functional profile. This provides us with our second alternative realization base. And it turns out that ACC is highly active in REM dreams. Indeed, Damasio cites Hobson's work establishing the high activation of ACC in REM dreams as evidence for his view. If ACC activity realizes HOT, then REM dreams are no threat to HOT theory.

[14]See Nichols (forthcoming) for an overview.

[15]Braun et al. (1997) and Nofzinger et al. (1997); see Nir and Tononi (2010) for overview.

In the course of laying out his view, Damasio notes that we should resist the temptation of neural "phrenology"—the idea that punctate brain regions will instantiate particular psychological functions. This parallels my comments above on the distributed networks approach. With this in mind, we can consider the third alternative realization for HOT, Hans Flohr's proposal of distributed neural assemblies involving NMDA-sensitive synapses.[16] Flohr explicitly endorses a higher-order approach (also citing Rosenthal and other HO theorists), holding that conscious states are states we are aware of ourselves as being in. He then argues that the higher-order awareness constituting consciousness is realized by a special type of large-scale neuronal cell assembly. These assemblies are marked by the presence of synapses sensitive to the neurotransmitter NMDA. NMDA synapses implement the binding mechanisms which produce the distributed assemblies realizing HOT. Flohr cites research showing the role of NMDA receptors in anesthesia: an important class of anesthetic drugs disrupts these synapses and renders subjects unconscious. Interestingly, in smaller doses, these sorts of drugs—for example, ketamine—have ego-bending, consciousness altering effects. Flohr takes this to show that NMDA synapses underwrite both the presence of consciousness and its particular character.

Flohr's view provides yet another alternative to the claim that dlPFC activity realizes HOT. And it has the virtue of avoiding the neural "phrenology" warned of by Damasio. And when it comes to REM dreams, there is evidence that NMDA synapse activity is implicated in the process of "long-term potentiation" crucial to memory formation (Winson 1990). Thus, we find an active role for NMDA synapses during REM sleep and we gain an explanation, perhaps, of why dreams are conscious. Dreams are conscious because the NMDA synapses instantiating HOT are active during REM sleep in order to consolidate long-term memories. Therefore, Flohr's proposal avoids the REM dream worry as well.

Obviously, these are speculative proposals, but I think they all have plausibility and are not just ad hoc moves to save the view in the face of Sebastián's challenge. Nor do I think that embracing one of these proposals amounts to abandoning the HOT theory, even if its construed as more tightly tied to the "cognitive accessibility that underlies reporting" than I've allowed. All the alternative proposals I've described can reasonably explain the connection between HOT and reporting: the neural mechanisms discussed realize HOT, making subjects appropriately aware of their mental states. And this awareness seems necessary for reporting on our mental states. Unless it can be argued that HOT *couldn't* be realized by these neural bases, I do not see a worry about the access needed for reporting.

This concludes my challenge to premise 3. I think this is the best place to question Sebastián's argument, not because I think the other premises can't be convincingly challenged, but because I think there is still considerable distance between the HOT theory as developed by Rosenthal and others and our theoretical knowledge of the brain. While I agree that the Lau and Passingham result is important and highly suggestive for the HOT view, it would surprise me if that we're the end of the story,

---

[16]Flohr (1995, 1999).

rather than the beginning. With that in mind, I wish to close by praising Miguel Ángel Sebastián's fine contribution to the *empirical* debate over HO theory. While I disagree with his conclusion, I am in full support of his methodology here: try to figure out the empirical commitments of a "philosophical" theory of consciousness and then go get your hands dirty with the messy data of science. This is the only way to reach the sweet dream of a satisfying theory of consciousness. On this point at least, who am I to disagree?[17]

References

Anderson, M. 2007. Evolution of cognitive function via redeployment of brain areas. *The Neuroscientist* 13: 13–21.

Anderson, M. 2008. Circuit sharing and the implementation of intelligent systems. *Connection Science* 20(4): 239–251.

Block, N. 1995. On a confusion about a function of consciousness. *The Behavioral and Brain Sciences* 18(2): 227–287.

Block, N. 2001. Paradox and cross purposes in recent work on consciousness. *Cognition* 79(1–2): 197–219.

Braun, A.R., et al. 1997. Regional cerebral blood flow throughout the sleep-wake cycle. An H2(15)O PET study. *Brain* 120: 1173–1197.

Carruthers, P. 2000. *Phenomenal consciousness: A naturalistic theory*. Cambridge: Cambridge University Press.

Damasio, A. 1999. *The feeling of what happens*. New York: Houghton Mifflin Harcourt.

Dennett, D.C. 1976. Are dreams experiences? *Philosophical Review* 73: 151–171.

Dennett, D.C. 1991. *Consciousness explained*. Boston: Little Brown.

Flohr, H. 1995. Sensations and brain processes. *Behavioral Brain Research* 71: 157–161.

Flohr, H. 1999. NMDA-receptor-mediated computational processes and phenomenal consciousness. In *Neural correlates of consciousness*, ed. T. Metzinger, 245–258. Cambridge, MA: MIT Press.

Hurlburt, R., and E. Schwitzgebel. 2007. *Describing inner experience: Proponent meets skeptic*. Cambridge, MA: MIT Press.

Ioannides, A.A., et al. 2009. MEG identifies dorsal medial brain activations during sleep. *NeuroImage* 44: 455–468.

Kahn, D., and A. Hobson. 2005. Theory of mind in dreaming: Awareness of feelings and thoughts of others in dreams. *Dreaming* 15(1): 48–57.

Lau, H. 2008. A higher-order Bayesian decision theory of perceptual consciousness. *Progress in Brain Research* 168: 35–48.

Lau, H., and R. Passingham. 2006. Relative blindsight and the neural correlates of visual consciousness. *Proceedings of the National Academy of Science* 103: 18763–18769.

Lau, H., and D.M. Rosenthal. 2011. Empirical support for higher-order theories of conscious awareness. *Trends in Cognitive Sciences* 15(11): 508–509.

Malcolm, N. 1959. *Dreaming*. London: Routledge and Kegan Paul.

Massimini, M., R. Huber, F. Ferrarelli, S. Hill, and G. Tononi. 2004. The sleep slow oscillation as a traveling wave. *Journal of Neuroscience* 24: 6862–6870.

[17]Thanks to Richard Brown, Jake Berger, Matthew Ivanowich, Michal Klincewicz, Hakwan Lau, Myrto Mylopoulos, David Rosenthal, and Miguel Sebastián for helpful discussion.

Muzur, A., E.F. Pace-Schott, and J.A. Hobson. 2002. The prefrontal cortex in sleep. *Trends in Cognitive Sciences* 6(11): 475–481.

Nichols, S. Forthcoming. Mindreading and the philosophy of mind. In *The Oxford handbook on philosophy of psychology*, ed. J. Prinz. New York: Oxford University Press.

Nir, Y., and G. Tononi. 2010. Dreaming and the brain: From phenomenology to neurophysiology. *Trends in Cognitive Sciences* 14(2): 88–100.

Nofzinger, E.A., et al. 1997. Forebrain activation in REM sleep: An FDG PET study. *Brain Research* 770: 192–201.

Rosenthal, D.M. 2002. How many kinds of consciousness. *Consciousness and Cognition* 11(4): 653–665.

Rosenthal, D.M. 2005. *Consciousness and mind*. Oxford: Clarendon.

Rosenthal, D.M. 2008. Consciousness and its function. *Neuropsychologia* 46(3): 829–840.

Saxe, R. 2009. Theory of mind: Neural basis. In *Encyclopedia of consciousness*, ed. W. Banks. Oxford: Academic.

Schwitzgebel, E. 2011. *Perplexities of consciousness*. Cambridge, MA: MIT Press.

Sebastián, M.A. 2014. Not a HOT dream. In *Consciousness inside and out: Phenomenology, neuroscience, and the nature of experience*, 415–432. Dordrecht: Springer.

Sporns, O. 2010. *Networks of the brain*. Cambridge: MIT Press.

Tian, S., et al. 2006. Assessing functioning of the prefrontal cortical subregions with auditory evoked potentials in sleep–wake cycle. *Neuroscience Letters* 393: 7–11.

Tye, M. 1995. *Ten problems of consciousness: A representational theory of the phenomenal mind*. Cambridge, MA: MIT Press.

Van Orden, G., B. Pennington, and G. Stone. 2001. What do double dissociations really prove? *Cognitive Science* 25: 111–172.

Winson, J. 1990. The meaning of dreams. *Scientific American* 263(5): 86–96.

Chapter 31
The dlPFC is not a NCHOT: A Reply to Sebastián

Matthew Ivanowich

31.1 Introduction

In his article, Sebastián attempts to present an empirically-based argument against HOT theory. Specifically, he argues that while empirical studies by Lau and Passingham (2006) demonstrate that activation of the dorsolateral prefrontal cortex (hereafter, "dlPFC") is necessary for the kind of cognitive accessibility that underlies the ability to report visual experiences, it cannot be necessary for phenomenal consciousness of visual experiences since we are phenomenally conscious when we dream even though the dlPFC is inactive. This is a problem for HOT theory, which holds that a state isn't phenomenally conscious unless it is reportable.

In this commentary I will briefly describe what I take to be a promising response on behalf of the HOT theorist: namely, that the dlPFC isn't a neural correlate of the sort of HOT that is required to make an experience phenomenally conscious.

In other words, Sebastián's argument targets only a restricted version of HOT theory—a version which, in addition to the claim that we have conscious visual experiences when we have a higher-order thought to the effect that we're in a certain state, makes the further empirical claim that these sorts of HOTs are realized by the dlPFC. (Call this version of the theory "d-HOT".[1])

[1] Thanks to Josh Weisburg for this terminology.

M. Ivanowich (✉)
Department of Philosophy, The University of Western Ontario, 403-680 Wonderland Rd. N., London, ON N6H 4T6, Canada
e-mail: mivanowi@uwo.ca

R. Brown (ed.), *Consciousness Inside and Out: Phenomenology, Neuroscience, and the Nature of Experience*, Studies in Brain and Mind 6, DOI 10.1007/978-94-007-6001-1_31,

In Sebastián's argument, Lau and Passingham's (hereafter, "L&P") experiments are supposed to provide support for the claim that HOT theory should be committed to d-HOT theory.[2] However, I argue that an alternative interpretation of the function of the dlPFC allows the defender of HOT theory to resist this move.

31.2 The Function of the dlPFC

For example, in experiments by Heekeren et al. (2004), subjects undergoing an fMRI were asked to decide whether an image presented on a screen was a face or a house. Based on their findings, they concluded that the dlPFC appears to compare the outputs from lower-level sensory regions and use a subtraction operation to compute perceptual judgments about the identity of the stimulus. Furthermore, they found that dlPFC activation levels were correlated with the level of certainty or confidence about the judgment.[3]

In other words, Heekeren et al. claim that the function of the dlPFC is to decide what the subject is seeing on the basis of the strength of sensory information. *If* it makes this decision, it activates (and its level of activation represents the certainty of the decision). However, when the input is too noisy or brief to permit identification, it doesn't activate.

Thus, it's possible that activation of the dlPFC doesn't correspond to the sort of HOT that is responsible for phenomenal visual awareness (a HOT to the effect that one is having a visual experience of a certain sort). More likely, dlPFC activation reflects a confidence judgment about the categorical identity of stimulus, which is then itself the target of a HOT.

Moreover, this view about the function of the dlPFC fits with a large body of neuropsychological data: for example, Damasio (1994, 1999) has argued both that the dlPFC is involved in categorization and that damage to it does not result in deficits of consciousness. Moreover, Pollen (2008) reviews a variety of evidence which seems to suggest that dlPFC damage doesn't impair visual awareness.

---

[2]Specifically, L&P's experiments involved forced-choice judgments as to whether a visual stimulus was a square or a diamond, followed by a second forced-choice judgment about "whether they actually saw the identity of the target or simply guessed what it was." (L&P, p.18763) Both Sebastián and L&P interpret the subject's responses to the "Seen or Guessed?" question as reflecting the presence or absence of phenomenally conscious visual experiences. Moreover, since the only brain region which showed differential activation relative to the two conditions (seen or guessed) was the dlPFC, it's presumed the function of the dlPFC to realize the sort of HOTs that make an experience phenomenally conscious.

[3]Specifically, they found that the level of dlPFC activation was both (i) proportional to the difference in output between face- and house- brain regions and (ii) correlated with how difficult the decision was (such that its activation was highest when the evidence is strongest, and noisy/brief signals show lower levels of activity).

31.3 HOTs, Categorization, and Indeterminate Content

Of course, HOT theory does require that one apply certain concepts in order to have the appropriate HOT, and this requires categorization mechanisms. However, that mechanism need not be the dlPFC, since it's at least conceivable that the HOT to the effect that one is having a visual experience of a certain sort need not deploy concepts like 'SQUARE' or 'DIAMOND'. (Similarly, HOT theory is not committed to the claim that one needs concepts like 'ELECTRON MICROSCOPE' to have a phenomenal visual experience of one).

On this view, when the dlPFC doesn't activate (for example, in the short SOA condition of the L&P experiments) subjects could nevertheless have a phenomenally conscious visual experience of an indeterminate shape that is neither a square nor a diamond. This possibility is fully compatible with HOT theory—for example, Rosenthal (2009) claims that, e.g., one can be "aware of one's perception of an 'A' as a perception of some alphanumeric character or other, but not as a perception of an 'A'".

31.4 Two Potential Problems

However, although the above view seems to offer the defender of HOT theory a promising way of resisting Sebastián's argument, there are in fact two potentially serious problems lurking just around the corner: First, the above interpretation doesn't explain why categorization performance was matched in the long and short SOA conditions of the L&P experiments. Second, if the HOTs which make an experience phenomenally conscious are realized elsewhere, why was the dlPFC the only region where differential activation could be identified in the long and short SOA conditions?

In response to the first problem, recall that Heekeren et al. not only found that dlPFC activation was correlated with categorization abilities, but moreover, dlPFC activation *levels* correlate to increased confidence in judgements about the certainty or reliability of one's visual experience. This fact is particularly noteworthy here because, as mentioned above, Sebastián and Lau & Passingham interpret the subject's responses to the "Seen or Guessed?" question as reflecting the presence or absence of phenomenally conscious visual experiences. However, subjects' responses to this question are more naturally interpreted as reflecting judgments of *perceptual certainty*—how confident subjects are about the contents of their visual experiences.

So it's not surprising that we find dlPFC activity when subjects are confident that they actually saw the stimulus rather than merely guessed it. (Likewise, it's not surprising that subjects report guessing in cases when the dlPFC is not confident enough about the contents of visual experiences to reach the threshold necessary to activate.) However, this is nevertheless compatible with the possibility that low-level sensory information about the stimulus can affect behavior in forced-choice situations (e.g., blindsight) in the absence of dlPFC activation. In other words, although dlPFC activity may be *sufficient* to report the identity of a stimulus, it perhaps isn't *necessary* for such reporting.[4]

In response to the second problem, note that even in the short SOA condition of L&P's experiments subjects can not only have phenomenally conscious experiences of indeterminate shapes that are neither squares nor diamonds, but moreover they consciously see the computer screen, the room that they're in, and presumably many other things. Thus, one wouldn't necessarily expect to find difference in activation levels in brain regions that code for the HOTs responsible for phenomenal visual experience in the long and short SOA conditions.

What's more, it's compatible with HOT theory to hold that it's unlikely that there is a single, isolated region of the brain that codes for HOTs. Rather, HOTs may be much more widely distributed across neural architecture. Indeed, it's even possible that the dlPFC may occasionally *play a role* in HOTs despite not being necessary for them to generate phenomenally conscious experiences. (In other words, perhaps the rest of the neural machinery of HOT is sufficient to have a phenomenally conscious visual experience, though perhaps one with indeterminate content.)

References

Damasio, A. 1994. *Descartes' error: Emotion, reason, and the human brain*. New York: Putnam.

Damasio, A. 1999. *The feeling of what happens: Body and emotion in the making of consciousness*. New York: Harcourt.

Heekeren, H.R., S. Marrett, P.A. Bandettini, and L.G. Ungerleider. 2004. A general mechanism for perceptual decision-making in the human brain. *Nature* 431: 859–862.

Lau, H.C. 2008. A higher-order Bayesian decision theory of consciousness. *Progress in Brain Research* 168: 35–48.

Lau, H.C. 2010. Theoretical motivations for investigating the neural correlates of consciousness. *WIREs Cognitive Science* 2: 1.

Lau, H.C., and R.E. Passingham. 2006. Relative blindsight in normal observers and the neural correlate of visual consciousness. *Proceedings of the National Academy of Sciences of the United States of America* 103: 18763–18768.

[4]It should be noted that Lau (2008, 2010) explicitly endorses the view that phenomenal consciousness requires perceptual certainty. As he puts it, "Awareness occurs when a percept is correctly judged to be sufficiently reliable." (2010, p.4) However, there doesn't seem to be any empirical evidence to support this interpretation. Lau cites studies on a blindsight patient ("GY") by Sahraie et al. (1997) which showed that increased levels of dlPFC was correlated with an awareness (increased certainty) about whether or not there was a movement in the blind area of GY's visual field. However, this increased activity was not correlated with a phenomenal visual experience – rather, the patient merely "reports an 'awareness,' a 'knowing,' or a 'feeling' that something has moved, although he denies any experience of "seeing" as such." (Sahraie et al. 1997, p.9406)

Pollen, D. 2008. Fundamental requirements for primary visual perception. *Cerebral Cortex* 18: 1991–1998.

Rosenthal, D. 2009. Perceptual certainty and higher order awareness: Comments on Hakwan Lau. *NYU/CNRS workshop: Perception, action, and the self.* https://wfs.gc.cuny.edu/DRosenthal/www/DR-Lau-NYU.pdf

Sahraie, A., L. Weiskrantz, J.L. Barbur, A. Simmons, S.C. Williams, and M.J. Brammer. 1997. Pattern of neuronal activity associated with conscious and unconscious processing of visual signals. *Proceedings of the National Academy of Sciences of the United States of America* 94: 9406–9411.

Chapter 32
I Cannot Tell You (Everything) About My Dreams: Reply to Ivanowich and Weisberg

Miguel Ángel Sebastián

One of the main problems for the scientific study of consciousness is methodological. At least *prima facie*, the kind of knowledge we have of our own experiences is direct and not mediated by an inference process. This kind of knowledge contrasts with the kind of knowledge we have of others' experiences, which relies on the observation of their behavior and their reports.

Collecting data for the scientific study of consciousness requires scientists to go beyond their own personal experiences and study others' states. This, in turn, requires that subjects report or act a certain way depending on their experiences. Although we can, at least typically, report on our own experiences, there are two important methodological worries if:

1. There are experiences on whose content we cannot report.
2. There are experiences in circumstances in which we cannot report.

Theories that I have called *Higher-Order Cognitive*, like Rosenthal's HOT theory, maintain that the mechanisms that render a state phenomenally conscious depend on the kind of cognitive access that underlies our ability to report the content of the state. These theories deny that – in beings like us, with our reporting abilities unimpaired – (1) is possible. The question at this point is how we can empirically falsify this kind of theory. In order to do so, we would need a case of an experience on which the subject cannot report; but, if this were the case, how can we know that the subject is undergoing an experience? The paper I have presented offers a possible reply to this question.

The results of Lau and Passingham's experiment suggest that the neural correlate of the cognitive access that underlies our ability to report lies in the dlPFC, making this area the most plausible candidate to implement the required kind of HOTs.

M.Á. Sebastián (✉)
Programa de Maestría y Doctorado en Filosofía, Instituto de Investigaciones Filosóficas, UNAM, Circuito Mtro Mario de la Cueva, Ciudad Universitaria, Del. Coyoacán, México D.F. 04510
e-mail: msebastian@gmail.com

R. Brown (ed.), *Consciousness Inside and Out: Phenomenology, Neuroscience, and the Nature of Experience*, Studies in Brain and Mind 6, DOI 10.1007/978-94-007-6001-1_32,

Rosenthal endorses this later idea. Such a commitment is an empirical one and can be empirically falsified. We shouldn't – by no means – think of this as a weakness of the theory; quite the opposite, because the connection between reportability and consciousness is, in any case, a posteriori. Now, if there are circumstances in which a subject undergoes experiences without the corresponding activity in the dlPFC, then the kind of cognitive theories under consideration would be jeopardized. Dreams seem to present such a case.

The problem is that dreams are instances of (2) and someone might raise doubts on whether dreams are conscious experiences. In "Not a HOT Dream" (Sebastián 2014) I presented the case of lucid dreams in favor of the reality of dreams as conscious experiences, given that subjects are able to make some simple reports during these episodes. Surely, as Weisberg (2014) notes and I make clear in the paper, it is an open possibility for my opponent to accept that lucid dreams are conscious but not so ordinary dreams. First of all, I guess that most would not find this possibility really plausible and I think that this is a desperate move. But, more importantly, there is some empirical evidence suggesting that ordinary dreams are accompanied by mental imagery. In these experiments (I mention Roffwarg et al. (1962)'s one in the paper), subjects (whose eyes movements are monitored during sleep) are awaken during REM sleep; they report their dreams, and scenes requiring a determining control of gaze are selected. It has been observed a correlation between the movement of the eyes and the movements required to motorize these scenes. For example, in an experiment by Dement and Kleitman (1957), a sleeper looked up and down during REM sleep followed by his report that he dreamed of climbing up a series of ladders looking up and down as he climbed. Similar results have been found in studies with REM sleep behavior disorder. This condition is characterized by a loss of muscle atonia (paralysis) during REM phase. Leclair-Visonneau et al. (2010) showed that when rapid eye movements accompanied goal-oriented motor behavior during REM sleep behavior disorder (e.g. grabbing a fictive object, hand greetings, climbing a ladder) the great majority were directed towards the action of the patient (same plane and direction) and they suggest that, when present, rapid eye movements imitate the scanning of the dream scene.

I find the second of Weisberg's proposals to block the argument more appealing. He acknowledges the low level of activity in dlPFC, but he rightly stresses that this doesn't mean that there is no activity at all. It might be the case that the remaining activity corresponds to a few HOTs which would account for dreams. Weisberg's interesting suggestion here is that it may be the case that the phenomenology of dreams is much less rich than the phenomenology of waking experience. In favor of this proposal, Weisberg appeals to his own dreams and hold that they are not especially vivid, at least in the sensory domain. It would be of no help to contrast the content of my dreams with Weisberg's ones, for they might easily differ. However, it is possible to explain Weisberg's claim that his dreams' content is sparse rather than rich and that "the content of most dreams is intuitively sparser than the content of waking experience" (2014) as a problem of memory. It is a well known fact that we tend to quickly forget the content of our dreams (some people even think of themselves as not having dreams at all), something that scientists know and try to avoid controlling the waking up conditions in the lab and recording reports directly upon awakening in the REM phase.

Weisberg also suggests the possibility that we may confabulate the phenomenological richness of our dreams. It might be the case that our dream experiences are sparse and that we enrich our conscious memory of dreams beyond what was present in the actual event. I think that Weisberg is right and this is a serious possibility, but a possibility for any kind of post-presentational report, not only in reports about the content of our dreams. In any case, given the low level of activity in dlPFC during REM sleep, the content of our dreams would have to be dramatically sparser than the content of our awaken experience. This kind of speculative reply is especially problematic for the kind of theories we are considering to a point where it is doesn't seem plausible. The reason is that HOT theories already claim that awaken phenomenology is not as richer as it might seem to be. Let me elaborate.

Based on Sperling (1960)'s experiment and some more recent results (Landman et al. 2003; Sligte et al. 2008), Ned Block (2007, see also Block 2011) argues that phenomenology overflows cognitive access. Roughly the insight of Block's *mesh argument* is the following:

When presented with a 3×4 array of letters quickly flashed on a computer screen, subjects in Sperling's experiments report having seen a bunch of letters arranged in a block but they are unable to report the identity of most of them. The reason for this result is the limited capacity of the working memory, the memory buffer that encodes the information we can report on. The interesting case comes from a second condition where a tone is played after the array ceases to be visually present. This tone cues subjects to report one single row. In this case, subjects are able to report the identity of all the letters in the cued row. Block concludes that the best explanation for this result is that the content of experience overflows what we have cognitive access to, because subjects report having seen all the letters and they were able to report the letters when they were cued, in spite of the fact that the letters were not visually present.

In reply to this argument defenders of some form or other of HOT theory (Rosenthal 2007; Brown 2012; Brown and Lau forthcoming) have maintained that the content of phenomenology might not be as rich as some might have thought. In the Sperling's case presented above, our experience would represent an array of alphanumeric characters without thereby representing any determinate character. Furthermore, it has been theorized that something similar usually happens in our everyday experience. For instance, Lau and Brown (forthcoming) suggest that despite our thinking that we see color in the periphery of our visual field we might not experience any determinate color in this area. Independently on whether we can make sense of a color experience which is not an experience of any particular color or of an experience that represents alphanumeric characters without thereby experientially representing any particular alphanumeric character, this line of reply maintains that the content of our experiences lacks all the details that it, at least *prima facie*, might seem to have. Now, in reply to my argument, defenders of HOT might claim that the phenomenology of dreams is "thin" rather that "thick"; the problem is that according to their theories the content of awaken experiences is, arguably, already "thin".

Ivanowich (2014) takes a different route. He argues that what I call Higher-Order Cognitive position can be consistent with the lack of expected activity in dlPFC during dream because one can resist, Ivanowich argues, the idea that required HOTs are realized in dlPFC. Ivanowich claims that it is possible that dlPFC activation reflects a confidence judgment about the categorical identity of stimulus, which is then itself the target of a HOT. In Ivanowich's interpretation of Lau and Passingham's experiment, subject's reply to the question on whether they had seen the target or they were just guessing their reply reflects a judgment about their experiences. This kind of interpretation would be committed to the idea that in order to reply to a question about our perception some kind of additional judgment is required, but it seems to me that we reply to these questions solely in virtue of our experience, without the need of any further judgment. Imagine you are lying in a beach with a friend. He suddenly asks you: "have you seen that plane?", referring to a plane that just crossed over your heads. In order to reply this question there is no need to make any judgment about the categorical identity of the stimulus, in case there was one, and you can reply to this question solely in virtue of the experience you have undergone. Be that as it may, Ivanowich interpretation is, I think, untenable precisely because of the problems that he foresees. Let me comment on them.

The first one is that performance capacity is matched between the long and the short SOA condition in the experiment. Ivanowich mentions a study by Heekeren et al. (2004) in favor of his interpretation, where it is suggested that the function of the dlPFC is to decide what the subject is seeing on the basis of the strength of the responses of sensory information. In particular, as Ivanowich puts down, they noted that dlPFC activity correlated with the difficulty in the decision task. The problem is that, in the Lau and Passingham's experiment, in both – the short and the long SOA – conditions the performance capacity is the same. This suggests that the "strength of the responses of sensory information" is the same – for otherwise we would expect a variation in the performance capacity as it happens when we modify the SOA – and, therefore, that the activity of the dlPFC seems not to correspond to a "more difficult" decision judgment as Ivanowich following Heekeren would predict.

The second problem is also pressing. Ivanowich seems to concede that there is a phenomenological difference in the experiences of the subjects during the short and the long SOA conditions. However, the only region that shows a difference in activity in the fMRI study that Lau and Passingham performed is dlPFC. Both Ivanowich and Weisberg stress that there might be a whole bunch of other experiences that the subjects undergo while performing the task: subjects are still conscious of the background, the monitor screen, their proprioceptive sensations, the sound of the lights and the AC, etc. If this is the case, one might suggest, adding a visual experience as of a square or as of a diamond would not make much of a difference in the overall experience; we would not expect much of a change in the brain activity and it might be the case that fMRI technology is not fine-grained enough to find further differences in areas that implement HOTs. There are two important considerations that should be remarked in reply at this point:

The first one is that we should assess empirical theories in the light of our current scientific research; the claim that dlPFC encodes HOTs fits the data whereas the claim that there might be other areas encoding them and that fMRI measurements are not fine-grained enough to capture the expected changes remains in the speculative domain.

The second one is that subjects are focusing their attention in a certain point in the screen where the stimulus will appear. It is well known that attended objects are more phenomenologically salient that unattended ones (just move your attention away from this paper to the proprioception of your toes). The stimulus is neither like an element in the periphery nor like an unattended stimulus, which might present *defused phenomenology*. Even if elements like proprioception, the light noise or the monitor screen are part of the content of the subjects' phenomenology (a not very plausible assumption, according to the theories we are dealing with, given the capacity restrictions of the kind of memory that underlies our ability to report), the square or the diamond would be the most salient ones, because they appear in the position the subject is gazing at and they occupy the locus of attention. I do not find it very plausible the claim that we cannot find any brain difference that matches these differences in phenomenology. On the contrary, we would expect to see differences in the brain areas responsible for making some information and not other available to the working memory (and therefore to report) and, according to the theories under consideration, making the content conscious.

Finally, in the last section of his paper, Weisberg rightly notes that my argument targets only HOT theories that rely on the cognitive access that underlies our ability to report and that the insight of HOT theories can still be kept while giving up on cognitive access. I agree with him; my only aim in this paper was to undermine the idea that cognitive access is required for having an experience, a thesis that is clearly endorsed in Rosenthal's HOT theory. Weisberg mentions two alternatives: one that relates consciousness and a theory of mind (Carruthers 2000), according to which higher-order representations would be realized in the medial prefrontal cortex and Damasio (2000)'s proposal which links activity in sensory cortex with representations of the current states of the organism. Weisberg notes that both are "in the spirit of the HOT theory, which holds that mental states are conscious when we are conscious of ourselves as being in them" (2014). Although I agree with this, it is doubtful, however, that one needs to appeal to higher-order representations to account for this idea.[1]

[1] See Sebastian (forthcoming) for an account of this transitivity principle unpacked as self-ascription of properties in same-order terms. Such a self-ascription makes use of Damasio's proto-self but without any need to postulate higher-order representations; in other words, the relation between, say, ACC activity and activity in the sensory cortex, is causal but not representational. It links and modulates the connection between the proto-self and the sensory cortex.

32.1 Conclusions

Higher-Order Cognitive theories, like HOT, maintain that phenomenal consciousness depends on the cognitive access that underlies our ability to report. Lau and Passingham's experiment suggests that such an access depends on the dlPFC. Against this conclusion Ivanowich offers an alternative interpretation of the results – in keeping with Heekeren et al. theory about the role of dlPFC – but this interpretation leaves the match in the performance capacity of subject in the short and long SOA unexplained.

The dlPFC is highly deactivated during dreams. This fact jeopardizes HOT theories on the assumption that dreams are phenomenally conscious experiences. Empirical evidence in favor of the reality of this later fact comes from lucid dreams. One can theorize, as Weisberg does, that it might be the case that ordinary dreams radically differ from lucid ones (the former but not the later be phenomenally conscious experiences), but common sense and empirical evidence do not recommend this alternative. Weisberg also notes that, even if conscious, the content of our dreams might be sparser than what we thought, so that the remaining activity in dlPFC account for these experiences. However, in the light of our current knowledge, this doesn't seem to be a satisfactory reply at all given the low level of activity in the dlPFC during REM phase and the commitments of Higher-Order Cognitive theories.

The argument I have presented advocates that Higher-Order Cognitive theories like HOT are wrong. As Weisberg notes, there are other Higher-Order theories in the spirit of Rosenthal's HOT theory that remain untouched. This is true insofar as they are not committed to the idea that phenomenal consciousness depends on the cognitive access that underlies our ability to report.

References

Block, N. 2007. Consciousness, accessibility, and the mesh between psychology and neuroscience. *The Behavioral and Brain Sciences* 30: 481548.

Block, N. 2011. Perceptual consciousness overflows cognitive access. *Trends in Cognitive Sciences* 12: 567575.

Brown, R. 2012. The myth of phenomenological overflow. *Consciousness and Cognition* 21: 599–604.

Carruthers, P. 2000. *Phenomenal consciousness: A naturalistic theory*. Cambridge/New York: Cambridge University Press.

Damasio, A. 2000. *The feeling of what happens: Body and emotion in the making of consciousness*. Harvest Books.

Dement, W., and W. Kleitman. 1957. The relation of eye movements during sleep to dream activity: An objective method for the study of dreaming. *Journal of Experimental Psychology* 53: 339–346.

Heekeren, H.R., S. Marrett, P.A. Bandettini, and L.G. Ungerleider. 2004. General mechanism for perceptual decision-making in the human brain. *Nature* 431: 859–862.

Ivanowich, M. 2014. Commentary on "Not a HOT Dream". In *Consciousness inside and out: Phenomenology, neuroscience, and the nature of experience*, ed. Brown Richard, 1–10. Dordrecht: Springer.

Landman, R., Spekreijse, H., and Lamme, V. 2003. Large capacity storage of integrated objects before change blindness. *Vision Research* 43(2): 149–164.

Lau, H., and R. Brown. Forthcoming. The emperor's new phenomenology? The empirical case for conscious experience without first-order representations. In *Festschrift for ned block*, ed. Adam Pautz and Daniel Stoljar. Cambridge, MA: MIT Press.

Leclair-Visonneau, L., D. Oudiette, B. Gaymard, S. Leu-Semenescu, and I. Arnulf. 2010. Do the eyes scan dream images during rapid eye movement sleep? Evidence from the rapid eye movement sleep behavior disorder model. *Brain: A Journal of Neurology* 133: 1737–1746.

Roffwarg, H.P., W.C. Dement, J.N. Muzio, and C. Fisher. 1962. Dream imagery: Relationship to rapid eye movements of sleep. *Archives of General Psychiatry* 7: 235–258.

Rosenthal, D.M. 2007. Phenomenological overflow and cognitive access. *The Behavioral and Brain Sciences* 30: 521–522.

Sebastián, M.A. 2014. Not a HOT dream. In *Consciousness inside and out: Phenomenology, neuroscience, and the nature of experience*, 415–432. Dordrecht: Springer.

Sebastian, M.A. Manuscript. Experiential awareness: Do you prefer *It* to *Me*?

Sligte, I.G., H.S. Scholte, and V.A.F. Lamme. 2008. Are there multiple visual short-term memory stores? *PLoS One* 3: 1–9.

Sperling, G. 1960. The information available in brief visual presentation. *Psychological Monographs General and Applied* 74(11): 1–29.

Weisberg, J. 2014. Sweet dreams are made of this? A HOT response to Sebastián. In *Consciousness inside and out: Phenomenology, neuroscience, and the nature of experience*, 433–443. Dordrecht: Springer.